Emerging Space Powers

The New Space Programs of Asia, the Middle East, and South America

Brian Harvey, Henk Smid, and Théo Pirard

Emerging Space Powers

The New Space Programs of Asia, the Middle East, and South America

Mr Brian Harvey FBIS
2 Rathdown Crescent
Terenure
Dublin 6W
Ireland

Mr Henk H. F. Smid
RIBS SC&I/DB&C
Breda
The Netherlands

Mr Théo Pirard
Freelance journalist
Pepinster
Belgium

SPRINGER–PRAXIS BOOKS IN SPACE EXPLORATION
SUBJECT *ADVISORY EDITOR*: John Mason, M.B.E., B.Sc., M.Sc., Ph.D.

ISBN 978-1-4419-0873-5 Springer Berlin Heidelberg New York

Springer is a part of Springer Science + Business Media (*springer.com*)

Library of Congress Control Number: 2009937491

First published as *The Japanese and Indian Space Programmes*, 2000

Cover design: Jim Wilkie
Project copy editor: Christine Cressy
Typesetting: BookEns, Royston, Herts., UK

Printed in Germany on acid-free paper

Contents

Authors' introduction

The early years of space exploration provided the world with a simple narrative: a life-and-death struggle between the Soviet Union and the United States – one that climaxed on the Moon in July 1969. Barely noticed, other countries had already begun to build their space programs. Japan launched its first satellite in February 1970, followed by China (April 1970) and then India (1980). Israel followed in 1988. Earlier (1965), France inaugurated what evolved into Europe's extensive space program.

The opening years of the 21st century provided a more complex narrative for space exploration. China joined the space super-powers of Russia and the United States in achieving manned spaceflight (2003). Not long after, China, India, and Japan launched moon probes (2007–2008). New countries began to invest substantial resources and energy in space development: Brazil developed its own launcher, as well as an Earth resources observation program with China; North Korea attempted to fire a satellite into orbit, while South Korea began its own rocket program; at a time of some political and military tension, Iran put its first satellite into orbit.

In 1999, Praxis published *Two Roads into Space: The Indian and Japanese Space Programs*. Much has happened to the Japanese and Indian space programs since. Rather than simply bring the story of the Japanese and Indian space programs up to date, the authors felt that a more interesting approach for readers would be to look at all the new emerging space powers, examining their different origins, philosophies, paths of development, models, progress, and outcomes. Here, we look at Israel, which developed its space program principally to fulfill intelligence purposes; North Korea, whose path to space has been idiosyncratic and difficult to interpret; and South Korea, a recent but enthusiastic user of space technologies for economic development.

Then there are Iran and Brazil. Narratives of the space endeavors of Iran have been generally partisan. The indigenous launch of the Omid satellite on 2nd February 2009 was an achievement of which the Iranians can be proud. Against all odds, but mostly with the political will to accomplish this, Iran needed only the last 10 years of the "space era" to fulfill its wish of having a satellite of its own in space.

Much can be said about why they wanted it and about how and with whose help they finally made their dream come true.

Since the early space days, Brazil aspired to be a space nation, with the ultimate goal to become independent in the space arena. This book describes the long and difficult way it went and the sacrifices it had to make. At the time of writing, this goal has not been achieved yet, but most Brazilian people are convinced that success is around the corner. Others are of the opinion that profiting of space applications for the benefit of the country and the world could be achieved differently and redistribution of scarce money for space endeavors is at hand. Still, Brazil can be described as an emerging space country.

Chapters 1–6 and 15 were written by Brian Harvey, Chapters 7–11 by Henk Smid, and Chapters 12–14 by Théo Pirard.

Acknowledgments

The authors wish to thank all those who kindly assisted them with the provision of information, advice, and photographs for this book. In particular, they would like to thank:

JAPAN

- Prof. Yaunori Matogawa, Director, Uchinoura Space Center
- Dr Chiaki Mukai, NASDA
- Keiichi Nagamatsu, Director, Industrial Affairs Bureau, Keidanren, Tokyo
- Maki Sato, Japan Satellite Systems Inc., Tokyo
- Dr Nobihiro Tanatsugu, ISAS, Toyko
- Kanako Toshioka, Institute of Space & Astronautical Science (ISAS), Kanagawa
- Ryoko Umetsu, Administration Department, Aerospace Division, Nissan Motor
- Institute of Space & Astronautical Science, National Space Development Agency, Nissan, Rocket System Corporation
- Ryojiro Akiba

We especially wish to acknowledge the assistance of Prof. Makoto Nagatomo, ISAS, and Dr Ryojiro Akiba, ISAS, who provided original material and translations concerning the life of Dr Hideo Itokawa.

INDIA

- Dr Padmanabh Joshi, now of the Nehru Foundation for Development in Ahmedabad
- Mrs Mrinalini Sarabhai
- Prof. U.R. Rao, former director, ISRO
- S.K. Bhan, National Remote Sensing Agency, Hyderabad
- Dr George Joseph, ISRO Space Applications Center, Ahmedabad

- S.M.A.K. Khan, Head, Liquid Propulsion Systems Center, Valiamala, Thiruvananthapuram
- M.Y.S. Prasad, INSAT Master Control Facility, Hassan
- D.P. Rao, National Remote Sensing Agency, Hyderabad
- S. Srinivasan, Vikram Sarabhai Space Center, Thiruvananthapuram
- Indian Space Research Organization, Liquid Propulsion Systems Center
- Mr Krishnamurthy, Indian Space Research Organization (ISRO)
- Staff of ISRO in Ahmedabad

EUROPE

- Rex Hall, London
- NASDA, Paris
- Philip S. Clark, Molniya Space Consultancy, Hastings

Thanks are due for kind permission to reproduce photographs in Chapters 1–3, ISAS, NASDA, and the Japan Aerospace Exploration Agency (JAXA); and in Chapters 4–6, the Indian Space Research Organization (ISRO). The photographs of Kalpana Chawla and Sunita Williams are courtesy of NASA.

IRAN

Much help from official sources in Iran was not expected, nor received. I would like to thank my Iranian friends who supported me in this undertaking, although a lot of them questioned the outcome of it. I would certainly like to thank the young Iranian people I met over the last three years, who believe in a future for their country and encouraged me to write this book, especially Aryan J. Pourbaghery, a young enthusiast Iranian student who, from the first time we met, was convinced that writing the book would show Iran from a different viewpoint and helped me wherever he could. Because of his interpretation of the Farsi language, much became clear to me. Special thanks I reserve for Parviz Tarikhi, a man of stature, who stood by me all the way. His insight into the Iranian space efforts enlightened the path I had to travel to write this book. Thanks are due to those who kindly gave permission to use photographs. Chapters 7 and 8 are dedicated to the people of Iran.

BRAZIL

Thanks are due especially to Dr Gilberto Camara, Director General of the National Institute for Space Research (INPE), and Paulo Moraes Jr, from the Space Directorate of the General-Command for Aerospace Technology (CTA). Thanks are due to those who gave permission to use photographs, referenced books, and the internet. The chapters on Brazil are dedicated to Ria.

ISRAEL

Thanks are due, for her assistance, to Josh Shuman, Shuman & Associates (PR for Spacecom & SatLink Communications).

KOREA (NORTH)

Thanks to Philippe Cosyn, Belgian space chronicler, who visited Pyongyang.

KOREA (SOUTH)

Special thanks to Dr Lee Joo_Jin, President of KARI, to Dr Sungdong Park, President and CEO of SaTReC Initiative, and to Sunae Han, member of the Program Development Division, SaTReC Initiative, for their nice and efficient assistance during the 60th IAC at Daejeon (October 2009).

Last, but not least, we wish to thank Praxis, who made it possible for this book to be published.

Brian Harvey
Henk Smid
Théo Pirard
2009

Note on terminology

The nomenclature of both Indian and Japanese spacecraft presents a number of problems. This book uses the formats likely to be most familiar to existing observers of the two programs. On the Japanese side, following the prevailing custom, the main NASDA rockets have been given Roman lettering and numbering (N-I, N-II, H-I, H-IIA) whereas the ISAS launchers have been given a combination of Greek letters and Arabic numbers (e.g. *Mu 3*, *Mu 5*). Japanese satellites are referred to both by their technical designators (e.g. Engineering Test Satellite VII) and by their Japanese name, given to them once they enter orbit (*Kiku 6*).

Illustrations

Tables

1

Japan: Origins – the legacy of Hideo Itokawa

On 16th April 1944, the low, slender silhouette of a submarine made its way slowly out of the German U-boat base of Lorient in occupied France and headed out into the deep waters of the Atlantic. Nothing unusual, for this scene had taken place hundreds of times over the previous three years. Except that the submarine was not a German U-boat, but the Imperial Japanese Navy submarine *I-Go-29* with Commander Eiichi Iwaya on board. It was heading for Japan with the German Reich's most secret rocket engine designs. Soon after, another Japanese submarine, under the command of Haruo Yoshikawa, left Lorient with more priceless rocket designs.

I-Go-29 reached the Japanese naval base of Singapore on 14th July, where Cdr Eiichi Iwaya disembarked, bringing with him as many blueprints as he could carry and took a plane to Japan. It was as well he did, for *I-Go-29* hit a mine and sank off Taiwan on 26th July. The second submarine was caught by the allied air forces, sunk and Cdr Yoshikawa was drowned.

Japan had become informed of German rocket advances the previous year. Under the Japan–Germany Technical Exchange Agreement, 1943, the two countries had agreed to share technical information. Japan was aware of American plans to bomb Japan with a long-range, high-altitude bomber, the B-29, and asked Germany for help. In response, the German Air Force, the Luftwaffe, had told the resident senior Japanese naval office resident in Berlin, Cdr Eiichi Iwaya, in March 1944 that Germany was developing a rocket-propelled fighter able to climb to 10,000 m in 3.5 min, the Messerschmitt 163, or *Komet*. Germany had high hopes that the *Komet* could reach the marauding American and British bombers and use its extraordinary speed to cause havoc to their formations. Early the following month, on 6th April, Cdr Iwaya saw the Me-163 on test at Augsburg and was astonished by its small size, stubby delta wings and vertical ascent close to the speed of sound. The Germans agreed to hand over the blueprints of the Me-163 and its HWk 509A rocket engine, a manual on how to handle its tricky fuel (called T-stoff) and other information on rocket propellants, hence the submarine mission.

JAPAN'S ROCKET PLANE

Back in Japan, on 20th July 1944, Mitsubishi's Nagoya plant was ordered to go ahead with the manufacture of a rocket frame and engine. The project was called *Shusui*, but given the navy designation of J8M-1 and the army name of Ki-200, the engine being called the Tokuro-2. *Shusui* was to climb to over 12,000 m under rocket power and attack the B-29s with its two 30-mm machine guns: its pilot would then bring it back in a gliding descent to its base and land on a skid. The fuel consisted of 2 tonnes of 80% hydrogen peroxide and 20% hydrazine, methanol and copper potassium cyanide.

As the allies drew closer to Japan, getting the rocket ready was a race against time. The Japanese were not helped by the loss of the two submarines, for they contained more extensive documents and samples that could not have been carried on Cdr Iwaya's plane. Good progress was made with the body of the plane, the first wooden models being built by August 1944. A glider version called *Akigusa* ("autumn grass" in Japanese) was built and flown by Lt Toyohiro Inuzuka at Ibaraki.

The engine proved more problematical. No sooner was the Tokuro-2 tested that November than the factory was hit by the Mikawashima earthquake and then a B-29 raid. The site had to be moved to Natsushima. The engine was not completed until 2nd July 1945. On 7th July 1945, Lt Inuzuka brought the *Shusui* up to 250 m, using 16 sec of fuel for an initial test. However, while gliding back, the wingtip hit an observation tower and the *Shusui* landed hard. Although there was no fire, Lt Inuzuka was badly injured and died early the following morning. Then, the fuelling problems that had dogged the German development of the Me-163 came to afflict the Japanese. In what was to have been the first operational test of *Shusui*, on 15th July 1945, the engine exploded and killed the pilot, Lt Skoda.

The *Shusui* was almost identical to the Me-163. The Japanese army and navy ordered 155 of them and work began on setting up a *Shusui* production base at Atsugi. A *Shusui* corps was formed, the 312 Naval Flying Corps, based at Yokosuka. In the event, only four flight models were made. The third test of *Shusui* was set for 2nd August 1945. It was delayed and by the time it was rescheduled, Japan had surrendered.

INTRODUCING HIDEO ITOKAWA

Although *Shusui* was the high point of Japan's wartime rocket effort, in fact, Japan's use of rockets dated back to the 1920s. The Japanese space program owes its origins to Professor Hideo Itokawa, a teacher at the Institute of Industrial Science at Tokyo University, later the Institute of Space and Aeronautical Science.

Hideo Itokawa was born on 20th July 1912. The name Hi-De-O means, in literal Japanese, "fire coming out of man". Although of diminutive height, the young boy earnestly set about living up to his title. When he was aged four and his Sunday School presented a Christmas play, he volunteered for the part of Moses. With his older brother, in his last year at Azabu Nanzan primary school, he constructed electric canon ignited by alcohol and shot out of glass tubes. He competed with his brother to see whose pellets would fire the furthest, remarkably never injuring himself in the process, to the surprise and relief of his mother. He left primary school two years early to make a precocious entry to secondary school. By the age of 12, he was pondering the problems of electromagnetic propulsion, devoting his daylight hours and apparently some sleepless nights to overcoming the challenge. He persuaded a local blacksmith to make some metal coils for him, though he later commented that it was difficult to get the right materials in Japan in the 1920s.

Entering the newly opened First Tokyo City Middle School, he carried out electromagnetic experiments, prevailing on his physics teacher, Genji Kawashima, to give him the run of the school's science laboratory. Most of the experiments he made there were built from first principles, since, as he said later, the right theoretical books simply were not available. His other passion at school was music and he expressed an interest in becoming a composer. His favorite piece was Schubert's unfinished symphony. He took up the cello and played the symphony as an extra cellist in an orchestra. He suffered from ill-health as a result of an episode during his summer holidays. Collecting a snake to put into a bottle, he suffered a bad snakebite and had to cut open the poisoned finger to wash out the venom – nearly losing his hand in the process, suffering a long fever as a result and being off school for months. He had a second brush with death when he was caught in the middle of the great Kanto earthquake of 1923 and later, when skiing, was afflicted with snow blindness from which he eventually recovered.

At the age of 15, Itokawa was very taken by the flight of American solo aviator Charles Lindbergh across the Atlantic. He entered the school of aeronautical science in the University of Tokyo in 1932, one of the first such university schools in the world. However, Itokawa was far from a mono-dimensional aeronautical engineer and his interests broadened out enormously when in college. He took part in the Tukiji Shogekijo theater and read the works of Shakespeare translated by Tubouchi, most of all liking *Merchant of Venice*, *Hamlet* and *Henry V*. He joined a literary discussion group and learned English in his spare time. As a student during the great depression, he was involved in the turbulent leftist politics of the period and was student class representative. Although disadvantaged by illness and small size, he loved sport and outdoor activities – basketball, boating, swimming and skiing. He graduated in 1935, having written a dissertation on the problem of the sound barrier.

AERONAUTICAL ENGINEER

In the late 1930s, Itokawa applied to join the officer corps of the Imperial Japanese Army, but he was turned down due to his poor health. While contemplating going to aviation graduate school in 1939, he was advised by his professor to take up a post as a designer for the Nakajima Aircraft Company. When the war came, Hideo Itokawa became a designer for the Imperial Japanese Army and was an engineer for the Nakajima Aircraft Company from 1939 to 1945, though he continued his research and teaching work in the university. Although aircraft-makers like Mitsubishi were better known to their American and British opponents, Nakajima was one of imperial Japan's main wartime plane-makers. Nakajima made the *Gekko* and *Saiun* fast reconnaissance planes, the Ki-49, which bombed Darwin, the *Tenyan* torpedo bomber, the outstanding Ki-84 *Hayate* fighter and, toward the very end, the Ki-115 special attack (*kamikaze*) fighter. Itokawa, together with Yasuni Kayama, led the team that developed the Ki-43 *Hayabusa* (Japanese for "falcon"). About 5,900 *Hayabusas* were built (the Americans called them "Oscars") and they saw service in China and the Pacific. Although not as famous as the Mitsubishi Zero, they were a solid design, highly maneuverable and a mainstay of the Japanese Army. One army ace, Satoru Anabuki, scored 39 victories in the *Hayabusa*. At war's end, *Hayabusas* were supplied to three kamikaze units, the 1st, 18th and 19th Shunbutai, when they were used to attack allied ships and ram American B-29 Superfortresses. The series became extinct soon after the war, but, in the late 1990s, collectors salvaged old parts from *Hayabusas* crashed in the Kurile islands and, in the new century, the Ki-43 flew again in Texas. A *Hayabusa* may be found in the American museum in Oshkosh.

In addition to *Shusui*, Japan developed a rocket-propelled kamikaze plane, the *Okha* ("cherry blossom"). The *Okha* was carried to its target by a mother plane. Once it was dropped, the pilot used its rocket engines to speed the plane, with its 2-tonne warhead, into a 800 km/hr dive onto its target. The *Shusui* and *Okha* were the best known of Japan's wartime rockets, but not the only ones. As American air raids began to devastate the mainland, rockets were designed by the Naval Technical Assistance Unit to attack the bombers. The unit's first rocket was a 25-kg surface-to-air rocket. The unit developed, in the Yokosuka dockyard, the *Funryu* anti-ship solid-fuelled missile (*Funryu* 1 and 2) and a more ambitious anti-ship liquid-fuelled guided missile (*Funryu* 3 and 4), flying one to an altitude of 32 km. Kawasaki and Mitsubishi between them developed air-to-surface missiles with a thrust of 250 kg for 75 sec and a range of 8 km. Japan's wartime rockets would have made more impact had they reached mass production. Itokawa himself stayed with conventional aircraft and was not involved in Japan's construction of a rocket fighter. Long before the surrender, Itokawa had been reassigned to the engineering department in the University of Tokyo. He was made associate professor and was now called "Professor Itokawa". He spent most of the rest of the war lecturing students – something he found much less interesting than designing aircraft.

After the war, Japan was not permitted to build either aircraft or rockets, so Itokawa went to work in the medical school in the University of Tokyo, concentrating on neurological problems and the design of scanners to detect brain tumors and epilepsy. In October 1949, he completed his doctoral thesis, one that brought him back to his earlier interest in music. His thesis subject: the acoustic qualities of the flute and violin. At a practical level, he built an improved model of the violin. After completion of the thesis, he developed a meter to measure the how deeply a patient was anesthetized. In December 1953, he boarded the ship *Cleveland* to cross the Pacific and visit medical facilities in the United States, lecturing in the prestigious Chicago University for six months. While in the medical library of the university, he came across text books on space medicine. He was enormously impressed with the level and pace of American rocket development in the mid-1950s.

Hideo Itokawa

In the years that followed the war, the United States and Soviet Union sent re-engineered German A-4 (V-2) rockets high into the atmosphere with small scientific payloads, cameras and even animals. Itokawa hoped that these developments could be matched in Japan. The legal restriction on airplane and rocket development was lifted in 1953 under the San Francisco peace treaty. This delighted Itokawa, who believed that Japan could quickly resume rocketry and wrote a magazine article called "Rockets in Five Years!", but it was sent back, rejected on the basis of being "presumptuous".

Not easily dissuaded, with his colleagues and students, he began to design and build small solid-fuel rockets under the aegis of the Institute of Industrial Science at the University of Tokyo, where he and his colleagues formed what was called the Avionics and Supersonic Aerodynamics Research Group (AVSA). This group took in a mixture of ex-wartime designers, civil engineers, architects and experts in mechanics and physics. Their first studies were theoretical. At the second meeting of the group in April 1954, discussion centered on a feasibility study by Hideo Itokawa on a rocket transport plane able to fly the Pacific in 20 min. Almost certainly unknown to him, his Chinese counterpart, Tsien Hsue-Shen, designed similar rockets at exactly the same time.

What really changed everything (the same was also true worldwide) was the announcement of the International Geophysical Year (IGY) in 1954. Many years earlier, there had been International Polar Years and the IGY was designed as a worthy successor to galvanize the post-war scientific community so as to improve our knowledge of the natural environment of our planet. It exceeded its wildest expectations, achieving its primary purpose of involving 60,000 scientists from 66 nations but had the unintended but profound effect of sparking off the space race.

科学は作る

（2）

ロケット旅客機

20分で太平洋横断

国産機・十五年後の夢ならぬ夢

八万メートルの超高空をゆく

Essay by Itokawa, 1955

Worldwide, governments saw the IGY as a low-cost means of generating interest in science. Japanese scientists Kenichi Maeda and Takeshi Nagata pressed the Ministry of Education for Japan to participate in the year. The Ministry initially refused, but, over the New Year holidays, an official in the Ministry chanced upon an article called "Rocket Transport" in the newspaper *Mainichi.* He persuaded his departmental colleagues to change their minds: AVSA was able to obtain a government grant of ¥3.3m (€29,464) from the ministries of education and international trade, supplemented by Fuji Precision company, to "pursue research".

Unusually for the post-war pioneers, Itokawa chose to go down the road of solid-fuel rockets. The German A-4 and most of the post-war American and Russian rockets were liquid-fuelled. These rockets had two fuel tanks, one containing fuel, and the other an oxidizer which was pumped to great pressure and then fed into a combustion chamber for ignition and burning. Solid-fuel rockets, by contrast, were more akin to the traditional firework, containing a single tank filled with a gray, sludge-like substance that was poured in and later solidified. Solid-fuel rockets were less sophisticated (there was no plumbing involved) and offered potentially greater thrust, but were harder to control and could not be turned off or throttled. Solid fuels were not well known in Japan, but Itokawa and his colleagues had such limited resources that the simpler solid fuels were the only realistic possibility open to them.

FIRST ROCKETS

AVSA volunteered to build a rocket for the IGY and used the title "the Sounding Rocket Research Group". The first such rockets were really tiny. The first was – quite appropriately – called *Pencil*, being just 23 cm long and 1.8 cm in diameter. Using relatively cheap raw materials, Itokawa made 150 firings of *Pencil* and was able to reach conclusions about the best type of fuel, the most suitable configuration of nozzle and the shape of stabilizing fins. They were fired horizontally at first, at Kokubungi, Tokyo, a disused rifle range. On 12th April 1955, Itokawa and his colleagues felt sufficiently confident in their work to give a public demonstration of *Pencil* in a Tokyo suburb. In the end, 29 were launched. Public reaction to the experiments was divided: most people enjoyed the experiments, but some scientists criticized them as silly and meaningless.

The environs of Tokyo were not a safe launch base, so, in March 1955, Itokawa and his university colleagues established a remote beach launch site at Michikawa, in Akita province in the north-west of Honshu island. This was called the Akita range and it faced out west across the Sea of Japan toward Korea. Here, on 6th August 1955, Itokawa launched a *Pencil 300* at the new site. Later that month, a derivative was launched, called *Baby*. This tiny rocket had a diameter of 8 cm and held a full kilogram of propellants – mainly gunpowder. Thirty-six *Baby*s were fired that year and one version reached 6 km. But, as a sign of things to come, there were objections from local fishermen that the noise was disturbing the sea creatures.

Things were very primitive in those days. None of those involved in the project had a car, so rocket equipment was brought in by horse! Pictures show Dr Hideo Itokawa counting down a rocket under a wooden hut as crowds of onlookers gathered on a nearby hillside. The control panel comprised some large electric bulbs. After a few years, Itokawa and his colleagues were able to construct three small concrete buildings and some temporary wooden ones. The rockets blew up often enough and Itokawa tried to laugh away the problems as inevitable in such an esoteric science.

Neither Itokawa nor his colleagues were sure of the best way forward in rocket development. Following the end of the *Pencil* program, Itokawa asked his colleague Ryojiro Akiba to give some attention to the problem of reaching ever higher altitudes. Akiba was able to borrow a mechanical analog computer from an electronics professor and do some of the basic calculations. He suggested multistage rockets and the use of rockoons, the latter idea being invented by Dr James Van Allen of Iowa, the man who made the successful interpretation of Earth's radiation belts.

Pencil

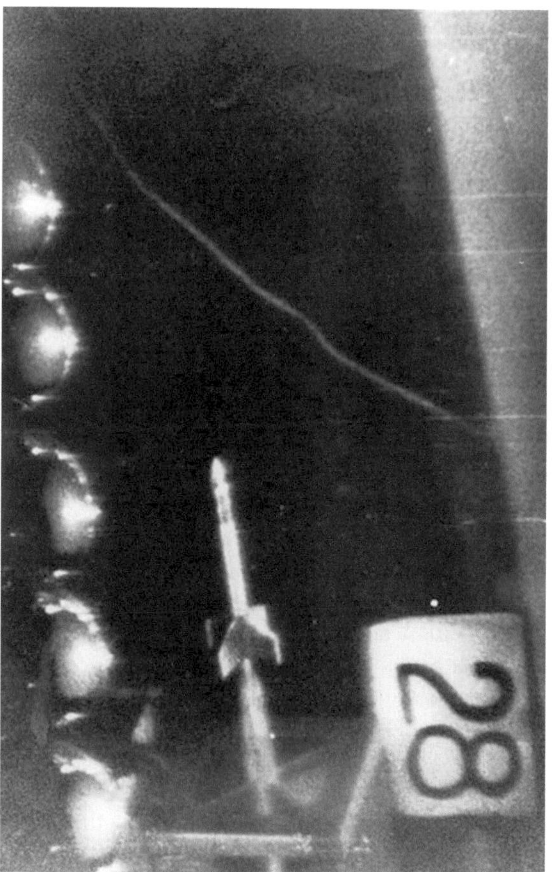

Pencil test

The theory of the rockoon is that a balloon carries a ready-to-go rocket to a high altitude, at which point its engine is lighted to send it into space. The balloon, using passive energy, saves the rocket the energy required for the earliest and most difficult part of its ascent through the atmosphere. The Science Council of Japan approved the rockoon project in spring 1956, though the money in fact came from the Yomiuri newspaper company and the project was to be implemented by the Japanese Rocket Society. Rockoons were successfully launched from 1959 to 1961 from Rokkasho, Aomori, but the project ended at that stage. The rockoon idea was to be resurrected for shuttle tests many years later.

In the event, multistage rockets offered more immediate promise to the altitude problem. Itokawa built a two-stage version of *Baby*, a rocket 150 cm long. In the top part of this *Baby*, Itokawa installed a tiny radio transmitter and, in another version, a small camera and parachute.

SOUNDING ROCKETS

Itokawa and *Pencil*

AVSA was allocated a further ¥17.4m (€155,357) for a series of sounding rockets called the K program, or *Kappa*, to be flown during the IGY. A *Kappa 3* reached an altitude of 18 km in July 1957. AVSA proceeded with the improved *Kappa 4*, but trouble with propellant mixtures led to explosions in September 1957.

The operational version for the IGY was the *Kappa 6*, a two-stage solid-fuel sounding rocket 5.6 m long, 25 cm in diameter and weighing 260 kg, 10 times larger than *Pencil*. At this stage, the university's rockets used letters from the Greek alphabet, *Kappa* being later followed by *Lambda* and then *Mu*. As a sounding rocket, *Kappa* was a radical development, for, until that time, most sounding rockets designed to reach altitude had been liquid-fuelled. The university stuck with solid fuels, using reinforced plastic and aluminum for the nozzles and rocket bodies, respectively. The *Kappa 6* was the first of its kind to have two stages, the upper rocket taking over when the first stage exhausted its propellants. The first *Kappa 6* was launched on 16th June 1958. The second stage shut down early for no apparent reason, but a second launching three days later was successful.

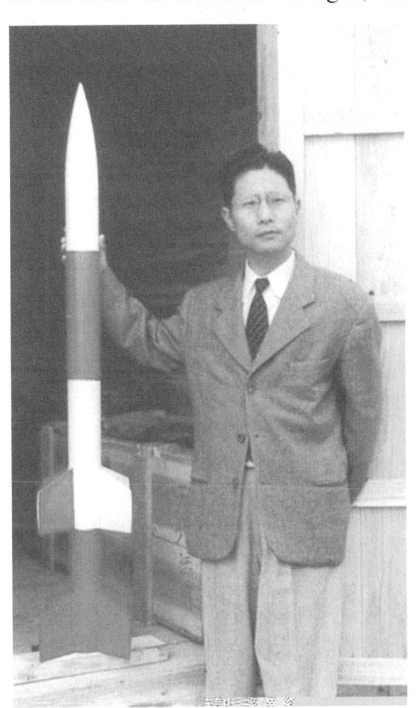
Itokawa and *Baby*

More importantly, the *Kappa 6* could carry 12 kg of scientific instruments. Thirteen *Kappa 6* rockets were launched as Japan's contribution to the IGY, reaching altitudes of up to 60 km – the beginnings of space – where they collected information on the upper atmosphere and cosmic rays. The launchings attracted much public interest and Hideo Itokawa was portrayed in the public press as "Dr Rocket". Itokawa founded the Japanese Rocket Society (JRS) (1956), which became the Japanese affiliate of the International Astronautical Federation (IAF).

Noting their success, the government began to show increasing interest in rocketry. In 1958, the Prime Minister's office established a National Space Activities Council to

Kappa

discuss the best way forward in space research. Later, the government set up the National Space Development Center of the Science & Technology Agency, while the University of Tokyo's space activities were given new form as the Institute of Space and Aeronautical Science (ISAS) in 1959. Its first director was one of Itokawa's colleagues, Professor N. Takagi. ISAS in effect brought together the old Aeronautical Research Institute in the university with the rocket research scientists.

The *Kappa 6* became the basis for a series of new sounding rockets, each more impressive than its predecessor. The *Kappa 8*, weighing 1.5 tonnes, 11 m long, with a payload of 90 kg, could reach an altitude of 200 km, the altitude of a satellite orbit. It was first flown in September 1959 and used steel motor case welding techniques of the type developed by the shipbuilding industry. The *Kappa 9L* was the first three-stage sounding rocket. In April 1961, the month Yuri Gagarin flew around the world in orbit, the *Kappa 9L* soared to 310 km above the Earth. Its successor, the long-lasting *Kappa 9M*, reached the same height, but with four times greater a payload. The last of the series, the *Kappa 10*, launched in late 1965, reached 700 km. *Kappa* rockets were later exported to Yugoslavia and Indonesia.

Japan's next rocket was the *Lambda*. Although originally built as a high-performance sounding rocket, it later became the rocket that first gave Japan access to Earth orbit. The purpose of *Lambda* was to reach altitudes of 3,000 km, well into space. A testing center for solid and liquid-fuel rockets was established by the university at Noshiro.

UCHINOURA LAUNCH SITE

The *Kappa* was able to reach a height of 400 km, but, if it went off course, it could well crash on the other side of the Sea of Japan, causing an international incident. Accordingly, in 1959, Professor Itokawa began to search for a new and more suitable launch site. The criteria were that the perfect site should be on the Pacific coast (so as

Kappa 9M

Beach launch site

to obtain the best eastward trajectories and the stages dropping harmlessly over sea), with clear year-round weather, away from air routes and sea lanes, with few local residents or farms or factories, good communications and away from fishermen and boats. After two years of surveys, the search was narrowed down to seven sites: Monahitohama in Hokkai island, Obuchi beach, Sikashimadai near Tokyo, Kazitorisaki, Toisaki, and in the south in Kagoshima, Uchinoura and Tane. Uchinoura, at the tip of the Japanese island of Kyushu, was not originally a good candidate, because it was remote (31 hr from Toyko by train), rocky (making construction difficult) and the local fishermen objected again, but it scored for weather and, above all, the local government wanted it.

Uchinoura was chosen on 11th April 1961 and construction work started in February 1962. They moved not a moment too soon, for, in May 1962, there was a dramatic accident to a *Kappa* sounding rocket at Akita: the first stage exploded and the second stage, still live, skipped across the sea and headed straight toward a cottage, where it burned out. No one was injured, but everyone got a fright.

Building Uchinoura out of rocky hillsides proved relatively easy compared to the fishing problem. Fishing is a vitally important industry in Japan, and, indeed, the Japanese are reputed to eat more fish per head of population than any other nation on Earth. Local fishermen objected strongly to the noise of ascending rockets, which, they argued, upset the local fish. In any case, range safety officers insisted the area downrange of a launch pad be cleared while a rocket was being fired, lest débris from an explosion fall on fishing boats. The strap-ons for the later H-I rocket, for

Early Mission Control

Constructing Uchinoura

example, crash some 25 km downrange, right in the middle of the fishing zone. The upshot of the stand-off was a bizarre compromise whereby rockets may only be launched in two limited periods of 90 days altogether during the year, during February and September. Whilst a fair compromise on paper, not even the most efficient space industries have ever managed to organize their space launches to fit perfectly into such a schedule; nor, indeed, are the alignments of the Moon and planets always coincidental with the Japanese fishing seasons.

Clearance of the 510-ha site took over a year, but, by the time it was completed for its official opening in December 1963, Japan had a full launch facility, with *Kappa* and *Lambda* pads, control, radar and tracking facilities. The first *Lambda* was launched from Uchinoura in July 1964. This was the *Lambda 3*, a three-stage rocket 19 m long, 73.5 cm in diameter, weighing 7 tonnes. Its first flight lasted 17 min, in the course of which it flew 1,000 km high and impacted 1,090 km downrange in the Pacific. The immediate purpose of the *Lambda* launchings was to enable Japanese participation in the next international year, which was the International Year of the Quiet Sun (IYQS), marking the minimum levels of solar activity. During the solar minimum, Japan sent up 34 *Lambda* and *Kappa* sounding rockets. The third *Lambda* was used to study the radiation of our own Milky Way galaxy and was the first Japanese rocket to be launched by night. Public interest continued to grow.

REACHING EARTH ORBIT

The *Lambda 3* marked the limits of what could be achieved by sounding rockets. By the mid-1960s, it was logical and becoming ever more possible for Japan to proceed to the next step, which was to put a satellite into orbit. In 1960, Hideo Itokawa and Ryojiro Akiba co-authored a paper in which they outlined how a small satellite could be put into orbit by adding a motor to the upper stage of what was in effect a large sounding rocket. In 1962, Itokawa and his colleagues presented a report entitled *Tentative Plan for a Satellite Launcher*. The Science Council of Japan held a symposium on a scientific satellite the following year, one that addressed the following questions:

- Can Japan still make a contribution to science, despite its late start?
- Is a satellite project feasible?
- Can Japan achieve a satellite using indigenous technology, or should it rely on the United States?
- How can a satellite be tracked without a tracking center outside Japan?

Events began to move at a faster pace now. In 1965, the National Space Activities Council gave the go-ahead for a scientific satellite program proposed by ISAS. Professor Itokawa proposed the development of a new rocket, the *Mu*, as an operational satellite launcher and this was authorized in August 1966. Professor D. Mori was appointed project director.

In the meantime, a new version of the *Lambda* was developed, the *Lambda 4S*, under the direction of Professor T. Nomura. Itokawa had the notion that this launcher might be capable to getting a satellite into orbit before the *Mu* was ready, at lower cost and much sooner. The first stage had a thrust of 36,970 kg, augmented by two 13,150-kg thrust solid strap-on boosters. The second stage had 11,800 kg thrust, the third 6,580 kg thrust and the fourth 816 kg. *Lambda* was a small, pencil-shaped launcher, with fins at the bottom and middle. Weighing 9 tonnes, with a thrust of 53.5 tonnes, the *Lambda 4S* was 16.9 m tall. The cost of development was ¥118m (€1.05m). It was as well they concentrated on the *Lambda*, for the *Mu* project was abandoned as too expensive (though a much later rocket was to bear the same name).

The difference with its predecessors was the *Lambda 4S* had a tiny fourth stage comprising a small solid-rocket motor with a tiny capsule and instruments attached. In essence, the launching technique was to fire an unguided sub-orbital rocket but to use a small motor at the peak of the flight to kick a payload into orbit – a technique since used by many of the smaller space nations. In size, the *Lambda* was the smallest, minimalist rocket ever used to get a spacecraft into orbit. Though taller than the later British *Black Arrow*, it was much thinner.

A precursor version of the *Lambda 4S* was tested in ballistic flight in March 1966, but it faltered when the vanes of the third stage failed to control the spin of the rocket. A second test that summer worked, the version being called the *Lambda 3H*. Now the scientists felt they could aim at orbital flight.

Japan's first attempt to orbit a satellite was made on 16th September 1966 but the pyrotechnic devices on the last stage failed to fire. More heart-breaking failures followed. On the second attempt, on 20th December, the de-spin motor failed and the last stage separated prematurely. On the third attempt, on 13th April 1967, the third-stage motor did not fire. At around that time, the United States had made approaches to Japan, suggesting that it might be better off using American rockets instead. Indeed, the State department had expressed alarm at the progress of Japan's solid rockets. Itokawa was adamant that Japan must learn to develop its own technology itself, his comments earning him the displeasure of NASA's administrator, James Webb. The influential newspaper *Asahi Shimbun* suddenly began a campaign against Itokawa, criticizing his leadership to the point that he resigned his post.

For most people, this would have been the ultimate personal disaster. However, Itokawa had always promised to himself that he would move on from rocketry once the *Lambda* project came to fruition. When he left the space program, he set up Japan's first think tank, the Systems Research Institute, in May 1967. He once said, philosophically, that a person should be able to pursue several careers – a change every 10 years was a good thing – and always made it clear that passionate though he might be about rocketry, he intended to pursue several other interests in his lifetime. Hideo Itokawa went from the space program to a project to build huge undersea flying saucer-shaped oil storage in 1 billion-liter tanks. This took up much of his time from 1968 to 1970. The aim was to give Japan a strategic oil reserve (a foresightful project in the light of the subsequent shortages of oil). To do this, he went abroad to study oil storage techniques. Next, he moved on to a project for a nuclear-powered ship, the *Mutsu*. Hideo Itokawa travelled abroad and met the great

Ohsumi launch

Saturn rocket designer, Wernher Von Braun, with whom he shared the same year of birth and played an important role in the development of the early Indian space program (Chapter 4).

Working on without him, Japanese scientists did not meet with success until the orbiting of a satellite on 11th February 1970 by the *Lambda 4S* rocket from the Uchinoura Space Center. The rocket was fired unguided out over the Pacific ocean, the stages dropping off one after the other. The third stage fell off at 85 km, by which time the rocket had reached a speed of 5 km/sec. As the rocket reached the height of its trajectory, the fourth stage solid-rocket motor fired, just enough to give the small satellite the final kick into orbit. Once orbiting, it was named *Ohsumi* after the peninsula where the space center is located. Entering an orbit of 338–5,150 km, inclination 31°, *Ohsumi* weighed 38 kg, including a radio transmitter, battery, thermometer and accelerometer. The signals failed after seven orbits, but Japan had become the fourth nation in space.

But where was Hideo Itokawa? He was far away – driving across the nighttime desert on the Kuwait–Saudi Arabian border in the middle east, planning his oil storage project. His driver heard the good news on the radio and told Itokawa straight away, who later recalled how he "sobbed endlessly with delight" to know that his lifetime's dream had come true.

INTRODUCING THE *MU-4S*

Ohsumi was, to all intents and purposes, a test satellite. It was too small to carry any useful scientific instrumentation. The next step was to put in orbit a real scientific

Tansei

satellite. A similar approach was followed by the Chinese, who launched a demonstration satellite in 1970 (*Dong Fang Hong*) but their first true scientific satellite the following year, *Shi Jian 1*. To launch a scientific satellite with a useful payload required a more powerful launch vehicle, in effect an orbital rather than a sub-orbital vehicle. Development of an orbital launcher was begun in the institute as far back as April 1963. A new launch pad for more powerful rockets was constructed at Uchinoura in 1966, with an assembly room and satellite processing facility. The new rocket, called the *Mu-4S*, used one large set of tail fins for stabilization.

The *Mu-4S* weighed nearly 43 tonnes and its objective was to place in orbit a small scientific satellite that would study solar radiation for a year. Although larger than the *Lambda*, the *Mu* was still one of the smallest in the world, being comparable in size to the French *Diamant* or the American *Scout*, being 23.6 m high, 1.42 m in diameter, and having a thrust of 80 tonnes. Weighing 44 tonnes on the pad, the white and red *Mu-4S* swung outward from the launch tower at an angle. Although

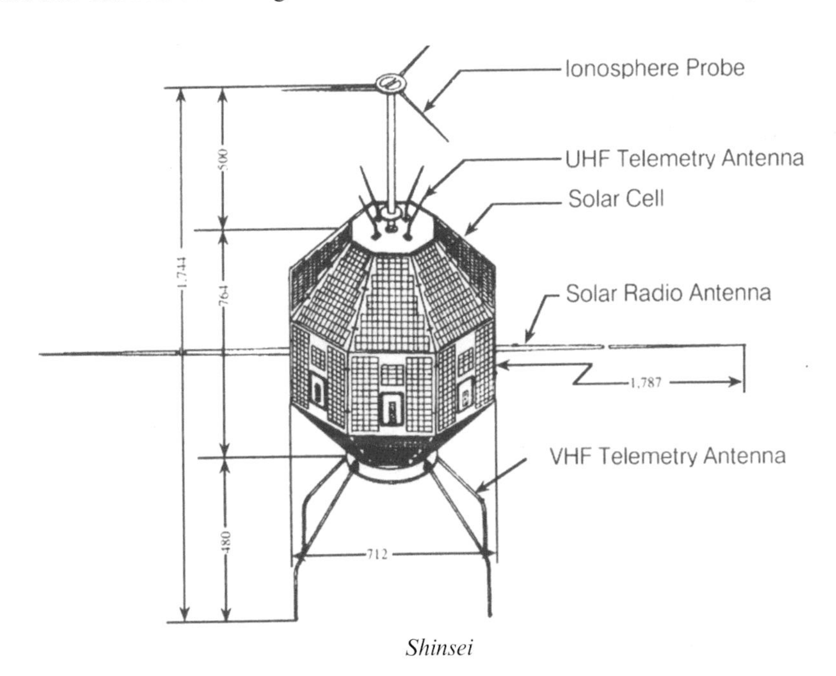

Shinsei

the first *Mu-4S* climbed perfectly into the sky on 25th September 1970, the fourth stage let the designers down and the satellite crashed to destruction downrange in the Pacific Ocean.

Success was achieved the following year. *Tansei* followed *Ohsumi* into orbit on 16th February 1971 and *Shinsei* on 28th September 1971. *Tansei,* weighing 62 kg, entered orbit of 990–1,110 km, inclination 30°. *Tansei* means "light blue" in Japanese, the colors of Tokyo University. *Tansei* was designed to study plasma waves and density, electron particle rays, geomagnetism and electromagnetic waves. It carried a transmitter that operated for a week, though the satellite's orbit should keep it in space for 1,000 years. The main purpose of the launch was to demonstrate the *Mu-4S* rocket. For Itokawa, *Tansei* was the realization of his dream of Japan launching its own small scientific satellites using modest resources.

ITOKAWA POSTSCRIPT

In his retirement, Itokawa wrote books and became a popular philosopher, scientist, economist and educator. He was, reluctantly, prevailed upon to write an autobiography. He found it difficult to do so, describing his character as being that of a boar in the forest – an animal that dashes forward and never looks back. So he wrote what he called a "personal history", not in book form, but as a poem, serialized in *Nikkei Shimbun* newspaper (10th November–6th December 1974).

Itokawa lived on to old age – 86 years – until 21st February 1999, when he died of a brain infection. A year later to the day, his former colleagues gathered to remember him. Most of the world's space programs owe their development to a great father figure – but few can have as a founder as universal a man of his age as Hideo Itokawa.

DISCOVERING A NEW RADIATION BELT

Japan's third satellite was *Shinsei,* meaning "new star" in Japanese. *Shinsei* was 1.2 m high, weighed 65 kg, carried solar panels and was designed to measure solar and cosmic radiation, the ionosphere and solar activity. Entering orbit of 869–1,865 km, inclination 32°, *Shinsei* was expected to remain in orbit for 5,000 years. *Shinsei*'s achievement was to identify a new, small radiation belt around the Earth, adding to those found earlier by the Russian Sputniks and the American *Explorer* spacecraft. The belt was located at low altitude near the equator and emitted a new type of radio wave.

Denpa was Japan's third scientific satellite, the fourth to reach orbit (19th August 1972). The 75-kg *Denpa* entered a highly elliptical orbit of 245–6,291 km, 31.03°, where its instruments analyzed the Earth's ionosphere, geomagnetic field, electrons and plasma. The second and third stages were unable to reach the intended height due to strong head winds and the fourth stage had to be fired earlier and longer than planned. Despite their best efforts, ground control found that the satellite had

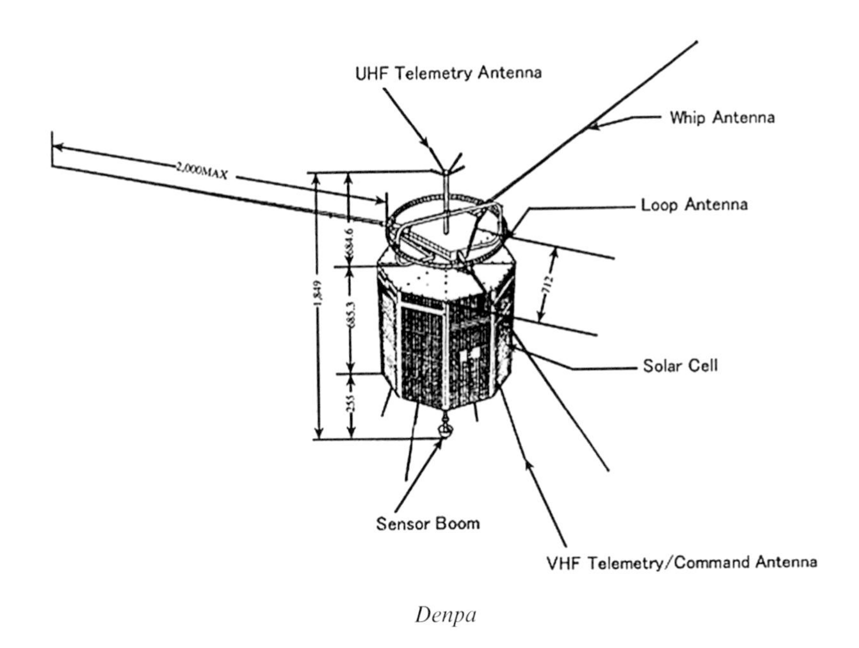

Denpa

entered orbit 40% lower and three times higher than planned. The Sun sensor subsequently failed and there was then a voltage fault in the encoder, so only fragmentary information was returned.

NEW VERSIONS: THE *MU-3C, H*

A problem with the early Japanese rockets, great though their achievements were, was that they were unguided. This meant that everything depended on the rocket being fired at the right angle at take-off and the rocket's fins keeping it strictly to this flight path. Deviations were difficult to correct and the final orbit of the satellite was hard to determine. Accordingly, a new version of the *Mu-4* was developed, one designed to ensure a more accurate insertion of satellites into orbit. The *Mu-3C* had similar dimensions to the *Mu-4S*, but the second stage had thrust vector control with additional control motors; and the third stage was equipped with radio guidance. This achieved the desired results when, on 16th February 1974, the

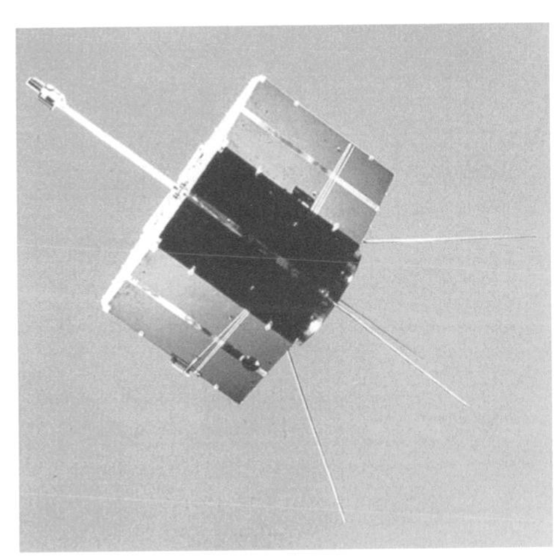

Tansei 2

Mu-3C put *Tansei 2* into an orbit of 284–3,233 km, 31.2°, quite close to the orbit planned. *Tansei 2* was followed by *Taiyo*, which studied solar ultraviolet and X-rays. CORSA, also called *Hakucho*, was a 100-kg satellite launched on *Mu-3C* on 21st February 1979 to study X-rays. It was Japan's first X-ray astronomy satellite. It had 11 X-ray detectors to survey the sky along the galactic plane and to detect gamma ray bursts from neutron stars.

Japan's early work in X-ray astronomy owed much to one of Itokawa's colleagues, Minoru Oda. Younger – he was born in Sapporo in 1923 – he entered Osaka University to study nuclear physics during the war and then worked in the Massachusetts Institute of Technology during the 1950s. When ISAS was founded, he came back to Japan and persuaded his colleagues that X-ray astronomy should be a priority of Japanese space science, devising the instruments for *Hakucho* and later missions (*Tenma*, *Ginga*). He was the most influential leader in Japanese space science until his death in 2001.

Efforts continued to improve on the performance of the *Mu* rocket. For the *Mu-3H*, the first stage was lengthened so as to launch new satellites. Here, *Kyokko* (meaning "aurora") was a 103-kg satellite flown in 1978 to investigate aurora, while *Jikiken* (meaning "magnetosphere") was a 100-kg research satellite carrying equipment to study plasma, charged particles, electric and magnetic fields, launched in September 1978.

By the late 1970s, Japan had realized Itokawa's dream of a country able to put small scientific satellites into orbit using small rockets developed with limited means. Despite setbacks and disappointments, eight small satellites had been put into orbit. A reliable solid rocket booster had been introduced. The first satellites are summarized in Table 1.1.

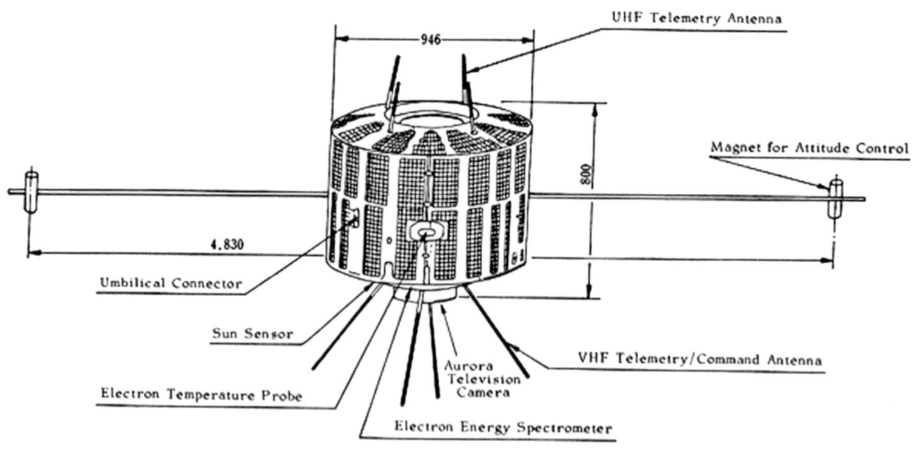

Kyokko

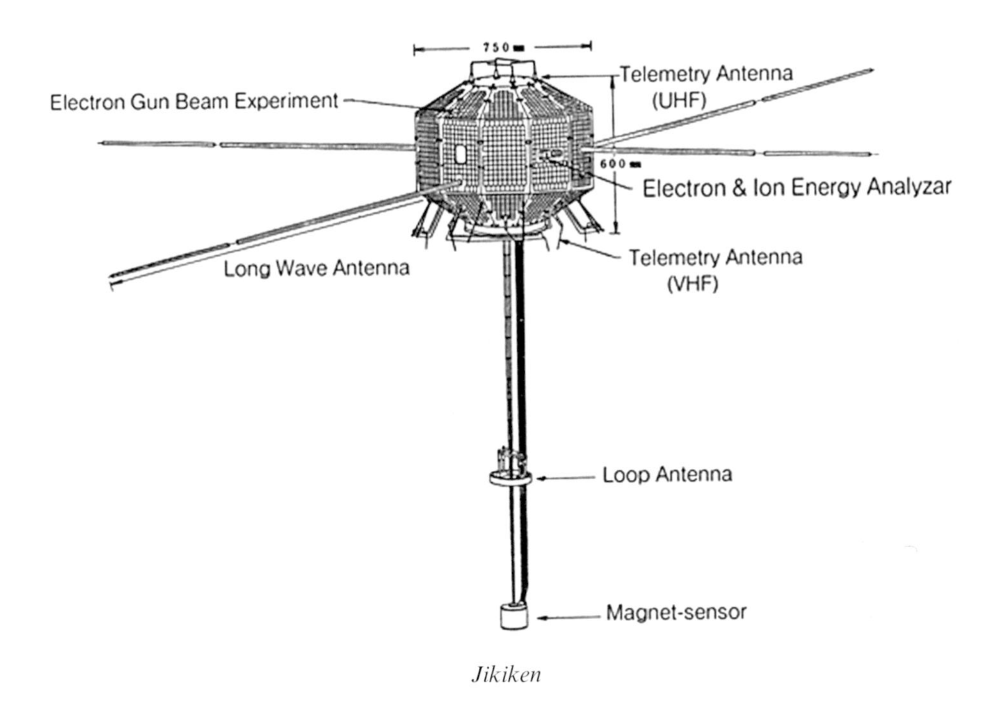

Jikiken

Table 1.1. Japan's early scientific satellites (up to 1980)

11 Feb. 1970	*Ohsumi*	*Lambda-4S*	Uchinoura
16 Feb. 1971	*Tansei*	*Mu-4S*	Uchinoura
28 Sep. 1971	*Shinsei*	*Mu-4S*	Uchinoura
19 Aug. 1972	*Denpa*	*Mu-4S*	Uchinoura
16 Feb. 1974	*Tansei 2*	*Mu-3C*	Uchinoura
24 Feb. 1975	*Taiyo*	*Mu-3C*	Uchinoura
19 Feb. 1977	*Tansei 3*	*Mu-3H*	Uchinoura
4 Feb. 1978	Exos A/*Kyokko*	*Mu-3H*	Uchinoura
16 Sep. 1978	Exos B/*Jikiken*	*Mu-3H*	Uchinoura
21 Feb. 1979	CORSA/*Hakucho*	*Mu-3C*	Uchinoura

FORMATION OF NASDA

Even as Professor Itokawa was developing a small-scale program of scientific space exploration, moves were afoot to develop a more ambitious space program, one that reflected growing Japanese confidence in its economy, manufacturing and science. Up until the late 1960s, spending on space research had gone to a number of different organizations and groups, although the University's successes meant that it attracted the most public attention. The amounts of public money spent had been modest enough – ¥1.369bn (€122m) over the years 1954–1970. The government established the Space Development Council in the Prime Minister's office in May 1960 to present advice and devise policy on space flight, the council evolving in May 1968 into the Space Activities Commission (SAC), whose function was to propose policies, submit proposals to the Prime Minister and bring coherence to work in the field. SAC proposed a 15-year plan covering 75 launchings and the spending of ¥1,479bn (€132m) involving:

- a Japanese space laboratory as part of the American Space Shuttle program, with equipment operated by Japanese astronauts;
- development of a fully hydrogen-powered launcher;
- a deep space program with lunar sample return and probes to Venus, Mars, Jupiter and Saturn;
- participation in a future American space station; and
- study of interstellar probes.

Pressure for a more aggressive space program came from Japanese industry, where there was a parallel set of developments. The federation of the economic organizations, the *Keidanren*, established a space committee in June 1960, leading to the Space Activities Promotion Council being set up (10th June 1968), bringing together 89 companies engaged in space activities. Its role was to present the needs of space-based companies to government, make proposals for the development of space activities and ensure better public understanding of space exploration. Specifically, industrial and commercial interests made an early pitch for the future development of the space industry, the Japan Broadcasting Corporation (semi-governmental) and the Ministry for Posts & Telecommunications arguing that it should urgently address the problem of poor-quality television reception in the Japanese islands. In 1973, business organizations began to campaign for long-term planning to develop a space industry and other forms of advanced technology. The Nikkeiren association proposed a 15-year ¥5.5bn (€49m) plan for a range of applications satellites, a powerful liquid-hydrogen-fuelled rocket and a space laboratory. The industry committee often had the effect of upping the ante on the government-based commission, spurring the government side to think more ambitiously of space program development.

The government responded by the establishment of the National Space Development Agency, NASDA, set up on 1st October 1969 under law #50, 23rd June 1969. NASDA was formed out of the National Space Development Center of the Science & Technology Agency and the Radio Research Laboratory of the Ministry of Posts & Telecommunications. Its brief was to develop artificial satellites, plan space programs, develop and launch rockets, track and control satellites and develop the necessary technologies, facilities and equipment. NASDA was specifically charged with responsibility for the development of launch vehicles; the promotion of technologies for remote sensing; and the promotion of space experiments. The idea was that NASDA would take charge of launch vehicle development, operations and the development of technology and applications satellites. First head of NASDA was a railway engineer, Hideo Shima, developer of

Tanegashima seaside site

the Shinkansen super-fast trains. Meantime, Tokyo University, through the Institute of Space and Aeronautical Science, would continue to retain responsibility for sounding rockets and scientific satellites. Although ISAS could continue to develop its own small solid-fuel rockets, the delineation of responsibilities between the two involved a limit being set on the size of rocket to be launched by ISAS: a diameter of 1.41 m.

The 1969 law institutionalized, rather than resolved, the divided nature of the Japanese space program and this division lasted over 30 years. Japan, in effect, developed two different, parallel space programs – a unique situation in a civilian space program. NASDA and ISAS each had its own fleet of rockets, launch sites, mission controls and tracking systems. Rivalry between them was such that ISAS literature rarely mentioned the work of NASDA and *vice versa* – indeed, it was possible for a novice to read of the work of one without even being aware of the work of the other! Having said this, despite the division of work between ISAS and NASDA, many NASDA scientists came from ISAS. In the event, NASDA always obtained the lion's share of Japanese space spending, generally around 80%, the ISAS proportion eventually falling to less than 8%.

NASDA moved quickly to set up its own launch site and plan the development of its own fleet of powerful, liquid-fuel rockets. The principal objective was to develop a rocket that could deliver a payload to 24-hr orbit – the position, 36,000 km above the Earth, where a satellite takes one day to orbit the Earth and thus appears to hover all the time, the ideal location for communication and weather satellites – comsats and metsats. This is also called a geosynchronous orbit.

Some liquid-fuelled rockets had been tested on Nijima island in Tokyo bay in 1964. This was too close to built-up areas, but a site for sounding rockets had been developed in Takesaki, Tanegashima island, in southern Japan over 1966–1968. Tanegashima had been considered by Hideo Itokawa when he sought a launching base for ISAS in 1959 and was only 100 km from his eventual choice, Uchinoura, with similar advantages and disadvantages. Tanegashima is a picturesque island site, set in rolling hills alongside a rocky–sandy seashore where gentle waves roll in from the Pacific. The initial construction cost was ¥1.466bn (€13m). Like their colleagues in Uchinoura, they were to suffer the same restrictions by the fishermen. Tracking stations were set up at Katsura, on the Boso peninsula near Tokyo and Okinawa.

NASDA'S ROCKET, THE N-I AND ITS FIRST MISSIONS

When NASDA was first formed, the original intention had been for Japan to develop a powerful indigenous rocket capable of putting technology satellites into low Earth orbit and communications satellites into geostationary orbit. First would come the Q rocket to launch a 85-kg satellite to 1,000 km by 1972 and then the N rocket to launch a 100-kg satellite to geostationary orbit by 1974. The majority view in Japanese industry and the research community was that Japan should develop its own capacities and follow a path of indigenization.

Here, broader political factors intervened. In 1965, American President Lyndon Johnson dispatched both Vice-President Hubert Humphrey and NASA Administrator James Webb to extend space cooperation with Japan and Europe. Although the discussions were clouded in the niceties of the term "cooperation", national security and state department memoranda of the period emphasized the need to control Japanese rocketry through the supply of American technology rather than let Japan develop its indigenous solid rockets. Cooperation in communications satellites should be withheld from countries not participating in the American-led INTELSAT consortium. Prime Minister Eisaku Sato, though, heeded Itokawa's advice to stick to indigenous rocketry, but began to change his mind during a summit with President Johnson in November 1967. On the summit agenda was Japan's desire to obtain the return of the Okinawa and Ogaswara islands from American occupation and it appears that a trade-off between the two issues was considered. In January 1968, American ambassador to Japan Alexis Johnson formally proposed the licensing of American rocketry technology to Japan. Negotiations eventually concluded in a summit between Japan and the new American administration of Richard Nixon in November 1969, which ratified the arrangement for the licensing of Japanese rockets in exchange for the return of Okinawa in 1972. In autumn 1970, the indigenous Q and N rockets were duly scrapped. The new licensed rocket would be called the N-I, the license costing about ¥6bn.

The N-I was based on the American Thor booster. The Mitsubishi company was made prime contractor, with the first and second liquid-fuel stages on license from Rockwell and the third solid-fuel stage on license from Thiokol. Design and

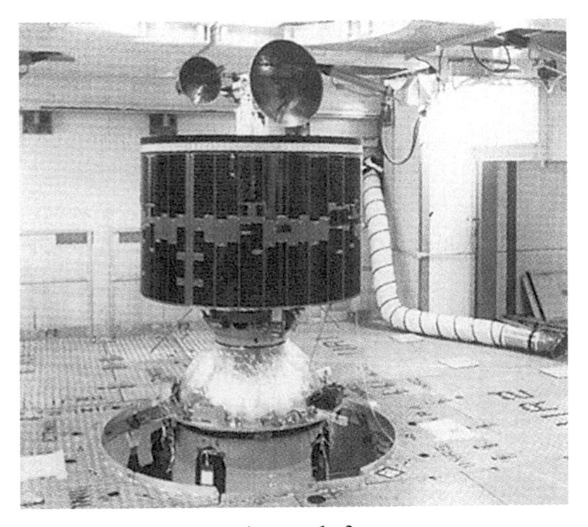

Ayame 1, 2

construction took place in the first half of the 1970s. The N-I was first used to launch the first engineering test satellite (ETS) or *Kiku*, which means "chrysanthemum" in Japanese. The concept of the ETS series was that *Kiku* satellites would test out new technologies that would later be applied to the Japanese space industry, mainly in the field of communications. The small 83-kg ETS-1, on 9th September 1975, was a test of the N-I launcher and its tracking systems. The satellite was 80 cm in diameter and had 26 sides and entered an orbit swinging out to 1,100 km, the first NASDA satellite. Eighteen months later, the N-I achieved its prime objective when drum-shaped ETS-2 was the first Japanese satellite to reach geosynchronous orbit. With *Kiku 2*, Japan became only the third country in the world to reach geostationary orbit. Whatever the political merits of the licensing arrangement, it is doubtful whether Japan could have made as swift progress without the licensed American systems.

The second N-I satellite, *Ume 1* (*ume* means "apricot") was put into orbit by the N-I on 29th February 1976 to monitor radio waves in the ionosphere and use the results to forecast short-wave radio communication conditions. Power failed after one month and the backup 141-kg model was put into 975–1,224 km, 69°, orbit as *Ume 2* on 16th February 1978, where it worked successfully for five years. *Ayame 1* was an experimental, Japanese-built comsat, launched on the N-I on 6th February 1979. There were reports later that it was lost when it collided with the third stage of the launcher. *Ayame 2* was launched on the N-I in February 1980 and was a 260-kg cylindrical experimental comsat, but contact was lost when the apogee motor failed to fire, stranding it in a useless orbit. The N-1 launches are summarized in Table 1.2.

Table 1.2. Launches of the N-I

9 Sep. 1975	ETS I/*Kiku 1*	N-I	Tanegashima
29 Feb. 1976	*Ume 1*	N-I	Tanegashima
23 Feb. 1977	ETS II/*Kiku 2*	N-I	Tanegashima
16 Feb. 1978	*Ume 2*	N-I	Tanegashima
6 Feb. 1979	ECS 1/*Ayame 1*	N-I	Tanegashima
17 Feb. 1980	ECS 2/*Ayame 2*	N-I	Tanegashima

COMMUNICATIONS SATELLITES: *YURI, SAKURA*, JCSAT, NSTAR, *SUPERBIRD*

The Japanese archipelago was so extensive that satellites offered a relatively rapid and convenient means of providing both the main and remoter islands with television, telephone and data relays. An early priority for NASDA was to develop such a system for the island chain, first with experimental satellites and then operational ones of growing complexity and performance. NASDA itself was not permitted to handle commercial operations directly, so a complex set of institutional arrangements were made to have communications contracted out to private companies. Because Japan did not have the domestic skills necessary, NASDA encouraged Japanese companies to buy in expertise from abroad, going to American companies such as Philco Ford and General Electric and using American launchers. Two types of satellite were commissioned: *Yuri* (BS, or Broadcasting Satellite) and *Sakura* (CS, or Communications Satellite) for direct broadcasting and television/telephone, respectively.

With BS 1, *Yuri*, launched on an American Delta in April 1978, Japan became the first country to experiment with direct commercial television broadcast to the home, the overall cost of the experiment being ¥28.8bn (€257m). It was followed by the BS 2 series (*Yuri 2*), built by General Electric and Toshiba. Here, the aim was to provide direct broadcasting to households in the remoter islands that did not have television,

Yuri series

Sakura series

using small dish receivers less than 1 m in diameter. Nothing was taken to chance and, in advance of the tests, the home receivers were subjected to the full range of weather conditions in the Japanese archipelago, from typhoons in the southern islands to the harsh winter snows of the northern island of Hokkaido. Here, direct broadcasting reached out to 420,000 viewers. Six years later, the BS-3, *Yuri 3* set was brought in, this time most of the satellite components being home-made (83%). The BS-4, *Yuri 4* series followed in 1997, completing the roll-out of the direct broadcasting program.

Although *Sakura* carried television, its role was also to build satellite-based telephone systems. The first, an experimental satellite, was launched by an American Delta rocket on 15th December 1977. It was 3.51 m tall, 2.18 m in diameter, weighing 676 kg, with six transponders, the first satellite to use quasi-millimeter waves to transmit signals. This paved the way in 1983 for an operational system providing general area coverage, the *Sakura 2* series, the first to use the high-capacity K-band frequencies (20–30 Ghz), able to handle 4,000 phone calls at a time. *Sakura 3A*, launched on H-I on 19th February 1988, was double the weight, 1,100 kg and provided telecommunications channels for the police.

The early years of Japanese satellite communications were dominated by the *Yuri* and *Sakura* series. Later, they were joined by JCSat and *Superbird*. JCSat was developed by the Tokyo-based Japanese Communications Satellite Company (hence JCSat), a joint venture between Itoh, Mitsui and other large Japanese trading companies formed in April 1985 to bring telephone and television services to business and private users in Japan. JCSat 1 was launched by Ariane 4 on 6th March 1989, the satellite being built by the famous American satellite-maker, Hughes. These were much bigger than the earlier *Yuri* and *Sakura*, the Hughes 393 being 10 m tall, a 3.66-m diameter drum with a 2.4-m antenna able to produce a beam to cover not only the Japanese islands, but audiences further afield in Asia.

Its rival was *Superbird*, owned by the Space Communications Corporation, a company developed by Mitsubishi. The first, *Superbird A*, was launched into 24-hr orbit on Ariane 4 on 5th June 1989, being followed by *Superbird B* in 1992 and *Superbird C* in 1997. By this stage, Japanese communications satellites for the domestic market had become routine, the eleventh JCSat being launched in 2007. Likewise, it served not only the Japanese archipelago, but Korea, Australia, New Zealand, Hawaii, Malaysia and China.

The CS, *Sakura* series evolved into a new generation of satellites called Nstar, with the American Loral and Hughes companies the contractors and Europe's Ariane winning the launch award. These were large satellites, weighing 4 tonnes, with 26 transponders. By now, communication satellite launches had become so routine that, coincidentally, on the same day, 29th August 1995, an Ariane launched N-Star from Kourou while an Atlas 2AS fired aloft JCSat 3 from Cape Canaveral. By early 2000, BS *Yuri*, CS *Sakura*, Nstar, JCSat, *Superbird* and their services between them provided 300 television channels to Japan, as well as a range of related telephone and data relay systems. As a result of the program of state stimulus and private development begun in the 1970s, Japan was introduced to satellite-based television faster than almost any other country.

INTRODUCING THE N-II

The N-I had a number of drawbacks. First, it was able to send only about 130 kg to geostationary orbit. This was fine for experimental purposes, but most operational communications satellites by the late 1970s required a lifting capacity of at least twice that amount. Second, by the time the N-I was developed by the Japanese, it was already well out of date by global standards. The N-I first flew in 1975, but most of its technology (e.g. the guidance system) dated to the 1960s. No sooner was it flying than Japanese space experts were thinking of the need for a more powerful vehicle and this was approved in September 1976. Again, a licensing arrangement was entered into with the United States. Prime contractor was Mitsubishi Heavy Industries.

These licensing arrangements have rarely been discussed much in public, but they appear to be a running sore to at least some Japanese. Whilst the licenses forbade the transfer of the technology to third parties or countries, a normal and reasonable condition, the operation of the agreements also effectively prohibited Japan from offering its American-derived launchers on the world commercial market. Some of the technology associated with the licenses was classified and the Japanese were not allowed to know about some of the components that they were themselves operating. Whenever problems arose in the N-I rocket, American technicians had to be called on to fix faults.

1981 saw the introduction of the N-II launcher. For the Japanese, the N-II was an essential part of their efforts to place larger satellites into 24-hr orbit for the 1980s and reduce reliance on American Deltas. The N-II more than doubled Japan's ability to reach 24-hr orbit, from 130 to 350 kg. Just as the N-I had been based on an earlier American launcher, the Thor, the N-II was based on a more recent version of the same Thor, the Thor Delta. Compared to the N-I, the N-II featured a number of improvements, principally nine solid-fuel strap-ons (rather than three), a longer first stage with 34% more capacity and a new motor for the second stage (the Aerojet AJ10-118F). The strap-ons burned for 38 sec and were dumped 85 sec into the mission. The third stage consisted of a solid-fuel Thiokol TE-M-364-4 motor, with a propellant weight of 1.1 tonnes and thrust for 44 sec. The third stage was supplied directly from the United States rather than built in Japan. The N-II was built in Mitsubishi's factory in Nagoya.

The first launch of the N-II was *Kiku 3*, a 640-kg cylindrical satellite launched on 11th February 1981, carrying a pulse plasma engine. Just before the first N-II went up, five members of NASDA's ruling council went to a Tokyo shinto shrine to pray. The mission objectives were to test the operation of the test satellite, the generation of electrical power and the operation of the N-II launcher. Later, the rocket was used for launching weather and communications satellites to 24-hr orbit, concluding with Japan's first marine observation satellite.

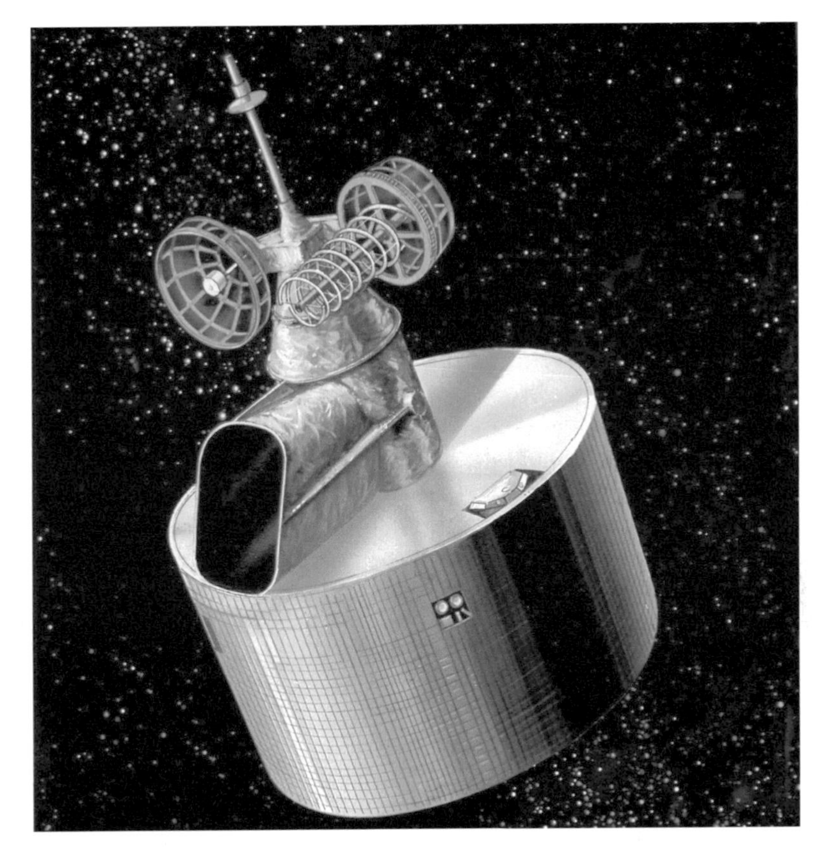

GMS

WATCHING EARTH'S WEATHER

Paralleling the development of communications satellites, an early achievement of NASDA's space program was the orbiting of a system of weather satellites. These were especially important for Japan, being both an island nation and one periodically affected by typhoons.

Japan's first Geostationary Meteorological Satellite (GMS), built by Nippon Electric and Hughes for NASDA, was launched from Cape Canaveral on a Delta 2914 on 14th July 1977 and was positioned at 140°E, just south of Tokyo itself. Called *Himawari*, or "sunflower", the 130-kg cylindrical satellite was 3 m long, 1.9 m in diameter and covered with solar cells at the side. *Himawari* was Japan's contribution to the Global Atmospheric Research Program, or GARP, sponsored by the World Meteorological Organization (Europe contributed Meteosat and the United States two GOES satellites). The satellite itself, and its successors, were based on the American GOES weather satellite model. *Himawari*'s aim was to carry out a global weather watch, collect and distribute weather data and monitor solar particles. *Himawari* was stationed in such a way that it could scan the planet from Hawaii in the east to Pakistan in the west with visible and infrared scanners. Its main instrument was a Visible and Infrared Spin-scan Radiometer, providing high-resolution visible images of the Earth every 30 min. It also carried a Space Environment Monitor to measure ionized gases entering the Earth's atmosphere from the Sun.

A second-generation GMS was launched four years later, this time by the domestic N-II launcher. GMS-2, *Himawari 2*, was 20 kg lighter, due to the slightly lower capacity of the N-II, but it made up for the reduction by the use of lighter materials. GMS-2, like its predecessor, was drum-shaped, with observation camera and visor on top, on which rested the two spacecraft antenna. A Space Environment Monitor recorded solar particles and their effect on Earth's communications. The main instrument was the Visible Infrared Spin Scan Radiometer, which scanned the entire planet: its first picture, relayed in early September 1981, showed the swirling clouds of the Pacific Ocean, Australia showing clearly through and filling the bottom left side of the globe. Placed at 140°E, GMS-2 transmitted pictures eight times a day until November 1983, when its electrical motor began to break down. When GMS-2 finished operations, the Japanese weather agency took GMS-1 out of retirement at 160°E to take over its work until GMS-3 could be launched. GMS-3, *Himawari 3*, built by Hughes for Nippon Electric NEC, followed in August 1984, providing, every 30 min, infrared and visible light views of Japan, China, south-east Asia and Australia.

The first attempt to launch GMS-4, *Himawari 4*, from Tanegashima failed on 8th August 1989, when there was a rare pad abort. The computer detected a potential valve failure in the first-stage motor just after it ignited and halted the countdown just in time. A second attempt a month later on 5th September succeeded. The satellite was placed in geostationary transfer orbit 16 hr after launch and eventually maneuvered into position at 150°E.

The GMS series concluded with GMS-5, *Himawari 5*, in 1995, which replaced GMS-4. It carried a visible infrared spin scan radiometer providing full pictures of the Earth's disc every 25 min in one visible and three infrared bands and was designed to measure sea temperatures and water vapor more precisely than ever before. The instrument had a resolution of 1.25 km (visible light) and 5 km (infrared light), respectively. The N-II launches are summarized in Table 1.3.

Table 1.3. Launches of the N-II

11 Feb. 1981	ETS IV/*Kiku 3*	N-II	Tanegashima
11 Aug. 1981	*Himawari 2*	N-II	Tanegashima
3 Sep. 1982	ETS III/*Kiku 4*	N-II	Tanegashima
4 Feb. 1983	*Sakura 2A*	N-II	Tanegashima
5 Aug. 1983	*Sakura 2B*	N-II	Tanegashima
23 Jan. 1984	BS 2A/*Yuri 2A*	N-II	Tanegashima
3 Aug. 1984	*Himawari 3*	N-II	Tanegashima
12 Feb. 1986	BS 2B/*Yuri 2B*	N-II	Tanegashima
19 Feb. 1987	MOS 1A/*Momo 1A*	N-II	Tanegashima

H-ROCKET: INTRODUCING LIQUID HYDROGEN

The next series of rockets, the H series, attempted to build on the success of the American-licensed N-I and N-II. The purpose of the H series was to double again the payload to 24-hr orbit, this time to 550 kg, to replace American technology with equipment designed and built in Japan, to make Japan independent as a launching country and to offer launches at internationally competitive rates. This time, 80% of the H-I was made in Japan. Approval to begin the project was given in February 1981.

Central to the H rocket concept was the introduction of a hydrogen-powered cryogenic upper stage. Hydrogen-powered upper stages give considerable extra boost for payloads destined for 24-hr orbit or deep space. The technology is called cryogenic because it involves handling the hydrogen fuel at extremely low temperatures (the boiling point for liquid hydrogen is –253°C and –183°C for liquid oxygen). The low temperatures, combined with the explosiveness of hydrogen, posed difficult engineering challenges. The United States had developed the first hydrogen-powered upper stage, the *Centaur*, in the 1960s, though not without difficulty or some spectacular explosions.

For this dangerous enterprise, the two wings of the Japanese space program, ISAS and NASDA, came together. As far back as 1972, Professor M. Nagatomo of ISAS had proposed a 7-tonne cryogenic upper stage for the *Mu* solid-fuel rocket; NASDA, for its part, wished to reduce its dependence on American-licensed rockets and motors. ISAS had carried out model tests of its engine in the late 1970s at the Noshiro test center when the Space Activities Commission asked ISAS to build a larger engine in cooperation with NASDA. The go-ahead for a cryogenic, hydrogen-powered upper stage was given in 1982. The new, restartable motor was called the LE-5, the contract being awarded to Mitsubishi. The 255-kg LE-5 was designed to develop 10.5 tonnes of thrust, with a specific impulse of 447 sec and a combustion pressure of 37 atmospheres. This became the basis of the second stage of the H-I.

In a typical launch profile, six of the H-I's solid rocket strap-ons ignited at lift-off. These burned out at an altitude of 4 km, when the other three strap-ons ignited. All nine were jettisoned 85 sec into the mission to tumble into the Pacific Ocean. The first stage burned out at 4 min 30 sec. Eight seconds later, now soaring above the atmosphere, the second stage would light up. The payload fairing at the top would come off 5 min 14 sec into the mission. The liquid-hydrogen second stage would typically burn for 363 sec, bringing the vehicle to a velocity of 7.8 km/sec. After a 22-sec coast, the solid third stage would light for a burn of about a minute to achieve a geostationary transfer orbit, following which the payload would separate 26 min after take-off. A further minute's burn would circularize the orbit.

The H-I made its first flight on 13th August 1986. A splendid white rocket, with the Japanese flag and *Nippon* written in black on its side, it took into orbit three payloads on its first mission – an experimental geodetic satellite, an amateur radio satellite and a magnetic bearing flywheel experimental system. The experimental geodetic payload, *Ajisei* (meaning "hydrangea"), was a 685-kg passive satellite designed to improve triangulation measurements of the surface of the Earth from an

Ajisei

altitude of 1,500 km. Built by Kawasaki, it entered a circular orbit of 1,479 km. In the shape of a 2.15-m ball, it was covered with 120 glassy reflectors designed to beam back both laser and light beams. Launched with *Ajisei* was *Fuji*, a 26-sided 50-kg amateur radio satellite only 0.5 m across.

ETS V, *Kiku 5*, a 550-kg satellite launched by H-I in August 1987, was a technology development satellite to test the use of C-band and L-band transponders for ships and aircraft and was placed in geosynchronous orbit at 150°E. It was the first to use an indigenous kick motor. In an experiment in Pacific regional cooperation, it was used to transmit educational programs to Fiji and Papua New Guinea. The apogee motor developed by ETS V was later used for the BS-3 series of broadcasting satellites, a good example of how the engineering program paved the way for a commercial application. The H-I launches are summarized in Table 1.4.

Table 1.4. H-I launches

13 Aug. 1986	*Ajisei*	H-I	Tanegashima
	Fuji		
27 Aug. 1987	ETS V/*Kiku 5*	H-I	Tanegashima
19 Feb. 1988	*Sakura 3A*	H-I	Tanegashima
16 Sep. 1988	*Sakura 3B*	H-I	Tanegashima
5 Sep. 1989	*Himawari 4*	H-I	Tanegashima
7 Feb. 1990	MOS 1B/*Momo 1B*	H-I	Tanegashima
	Orizuru		
	Fuji 2		
25 Aug. 1990	BS 3A/*Yuri 3A*	H-I	Tanegashima
25 Aug. 1991	BS 3B/*Yuri 3B*	H-I	Tanegsahima
11 Feb. 1992	JERS/*Fuyo*	H-I	Tanegashima

SOUNDING ROCKETS

Although sounding rockets paved the way for Japan's first orbital missions, they did not become redundant when orbital spaceflight became commonplace. About four are still fired each year. Between the start of the space age and 1996, Japan flew 325 sounding rockets. The *Kappa 9M* series from the 1960s continued in use for many years and 90 were fired altogether. A sounding rocket campaign was conducted with the United States at Wallops Island in 1967.

ISAS continued to use sounding rockets for microgravity experiments. The main sounding rocket developed was the Nissan-built S-520, which first flew in 1980 and replaced the *Kappa M*. It has two stages and can send an experimental payload up as far as 1,000 km.

The rocket weighs 2.285 tonnes, is nearly 9 m long and can reach an altitude of 350 km. It is launched from a platform through the roof of an enclosed pad, reminiscent of the way a telescope sticks through the dome of an observatory. ISAS also uses the MT-135, the smallest sounding rocket, developed by Professor F. Tamaki, used to sample the middle atmosphere for ozone depletion and the S-310, a medium-sized sounding rocket.

The S-310 and the S-520 have been launched abroad – in Antarctica (the Japanese base at Showa) and Andoya, Norway; and launches of the SS-520 have been scheduled for Spitsbergen. On 18th February 1983, a S-520 sounding rocket was used for a Microwave Energy Transmission in Space (METS) experiment, to test the possibility of transmitting solar energy to power the electricity grid on Earth. For the weather service, the MT-135 rocket, made by Nissan, has been launched almost weekly from the Japanese Meteorological Agency's weather station at Ryori on the north-east coast, both for general forecasting and to study specific problems such as the ozone layer. On 4th December 2000, a SS-520-2 sounding rocket was launched from the arctic island of Spitzbergen with 10 experiments to study the ionosphere, electrons and the magnetic field up to 1,000 km, finding escaping oxygen ions and what is called auroral hiss.

S-310 launch *Showa*

THE EARLY JAPANESE SPACE PROGRAM

Like the Soviet Union and the United States, Japan's space program had its roots in wartime Germany and like them, too, it was guided by a father figure, in this case, Hideo Itokawa. The post-war situation necessarily curbed the ambitions of the early space program, but the Japanese proved to be remarkably versatile in building a program around minimalist solid-fuel rockets and small satellites. With the establishment of NASDA, Japan progressed to a full-scale space program and infrastructure with larger, licensed rockets and a vision of developing space applications for telecommunications and forecasting. The fact that Japan was the third country to reach geosynchronous orbit symbolized how far it had come in a short period of time.

2

Japan: Into the solar system

Throughout the 1970s, 1980s and 1990s, ISAS continued its development of Japanese scientific space programs. Although, in terms of budgets and rocket-lifting power, it was the poorer cousin of NASDA, some of its missions were spectacular and attracted more public interest. These headline-grabbing missions are reviewed first, before looking at the range of domestic applications missions closer to Earth.

On 14th April 1981, ISAS changed its name from the Institute of Space and *Aero*nautical Sciences (ISAS) to the Institute of Space and *Astro*nautical Sciences (conveniently also ISAS) and became a national, inter-university research institute, independent of the University of Tokyo and answerable to the Ministry of Education, Science & Culture. It had become too big for just one university and soon set up its own dedicated campus at Sagamihara. The first director of the reformed institute was Professor D. Mori.

NEW *MU-5* VERSIONS: THE *MU-3H* AND *MU-3S*

The program of regular satellite launchings continued with the small, solid-fuel *Mu* launcher. These satellites programs now pursued distinct themes: for example, the Astro series made astronomical observations, the Solar series studied the Sun, Exos the atmosphere and so on.

A new version of the *Mu-3* was introduced, the *Mu-3S*, with first-stage vector control, improved accuracy of orbital insertion and a 50% increase in payload (300 kg). The *Mu-3S* was used to make almost annual satellite launchings in the early 1980s. *Hinotori* was a small scientific satellite launched on 20th February 1981 with an X-ray telescope and spectrograph for solar studies. It produced significant results in X-ray and neutron star astrophysics.

The purpose of Astro-B, or *Tenma* ("flying horse"), was to image celestial X-ray sources, such as nebulae, galaxies and bursts. *Tenma* consisted of a 89.5 by 110.4-cm box weighing 218 kg, with four solar panels fitted at the base providing 150 W of electrical power. *Tenma* transmitted to the ground, both in real time and through stored data, five orbits a day out of its 15 daily passes. *Tenma* was Japan's eighth scientific satellite and second X-ray satellite, carrying five instruments to observe

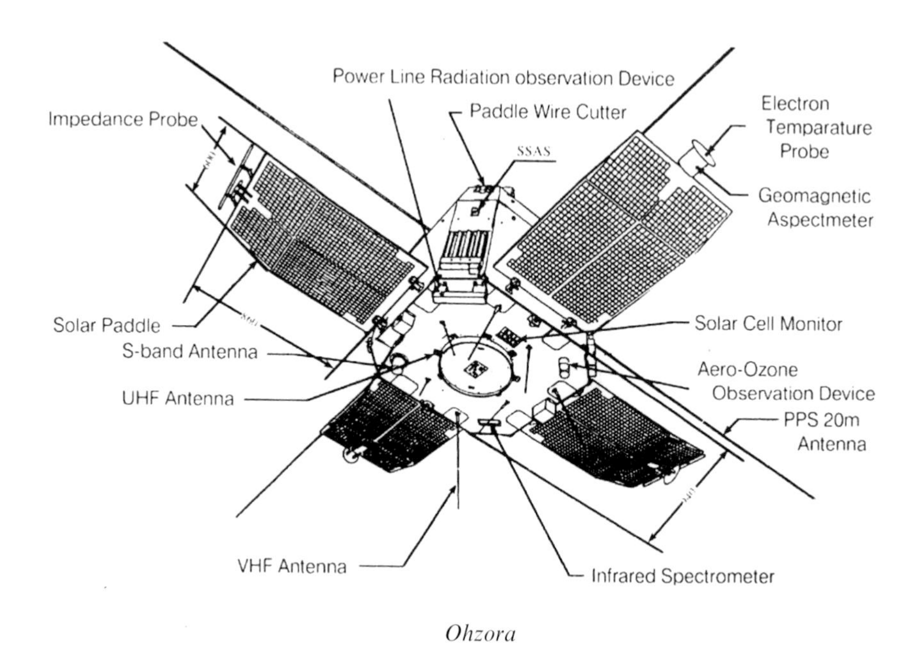

Ohzora

X-rays from stars and galaxies. Launched on 20th February 1983, the instruments were 10 scintillation proportional counters, a transient X-ray source monitor, X-ray focusing collector, radiation belt monitor and gamma burst detector.

Exos-C or *Ohzora* ("big sky") was part of a middle atmosphere program. From an altitude of 300–800 km, it used optical instruments to observe phenomena in the Earth's atmosphere from between 10 and 130 km. One particular area of study was the South Atlantic Geomagnetic Anomaly and an alarm system was fitted to note whenever the satellite overflew the anomaly. The satellite weighed 180 kg and was a cuboid, 1 m each side, with four solar panels.

PROBES TO COMET HALLEY

1985–1986 marked the first year since the start of the space age for a major regular comet to approach the inner solar system. Halley, the most famous of all the comets, named after Astronomer Royal, Edmond Halley, who first characterized cometary orbits, had in fact first been observed as far back as 240 BC. Since then, it had been observed 28 times, every 76 years (the appearance in 164 BC was missed). Several countries prepared space missions to intercept or pass close to comet Halley in the period around its closest approach to the Sun, or perihelion passage, due on 9th February 1986.

The United States, by this time the leader in interplanetary exploration, was, ironically, the only country not to send a probe to comet Halley, though a small *Explorer*-class scientific satellite was dispatched into its distant tail. American spending on deep space missions was very low in the late 1970s and a mission to comet Halley was one of many casualties of the glacial domestic financial atmosphere of the period. The Soviet Union mounted a spectacular double mission, sending probes *Vega 1* and *2* to Venus (where they dropped probes and balloons) before altering course to intercept the comet. Most dramatic of all, the European Space Agency fired the *Giotto* probe right into the head of the comet.

For Japan, this was the opportunity to organize its first deep space mission. Two missions were organized – a pathfinder and a main probe. Mission director was Professor K. Hirao of the Planetary Research Division of the ISAS Tokyo laboratory, while science director was the X-ray astronomy pioneer, Minoru Oda. The spacecraft used lightweight carbon fiber-reinforced plastic and were assembled at the ISAS Sagamihara space facility. These would be the first deep space probes to use solid-fuelled rockets. Because comet Halley had not timed its visit to the inner solar system to suit the southern Japanese fishermen, delicate negotiations took place for a breach in the normal regulations so as to permit out-of-season launches to chase the comet.

The pathfinder, MS-T5, was to fly eight months in advance of the main probe, Planet A. MS-T5 weighed in at 141 kg, had 2,000 solar cells for electrical power and carried three experiments to detect plasma wave instability, measure the solar wind and analyze the structures of the interplanetary magnetic field. Planet A was to close to within 10,000 km of the comet. It was small, able to carry only 10 kg of instrumentation, drum-shaped, 70 cm in height and 1.4 m diameter, weighing 135 kg, with cells around the side, charge coupled device cameras, scientific instruments (principally charged particle collectors) and an 80-cm-diameter high gain antenna on top. The camera was designed to take pictures to a resolution of 30 km during the encounter. Planet A had low-thrust gas jets that gave it the ability to carry out limited course corrections – 10 kg of hydrazine propellant for six 3-N thrusters for trajectory correction, attitude and to settle the spin of the spacecraft. Two thousand solar cells provided between 67 and 104 W of electrical power, depending on the distance from the Sun.

A special version of the *Mu-3* launcher was devised, the *Mu-3SII*, with two strap-on boosters to give additional thrust. The *Mu-3SII* weighed 61 tonnes, 12 tonnes more than the standard model. The SII model was a substantial improvement on the S version, being heavier (61 tonnes compared to 48.7 tonnes), longer (27.8 m compared to 23.8 m), with a doubled payload (up from 300 to 770 kg). The SII benefited from the first stage of the "S" but had improved second and third stages with movable nozzles and an optional kick fourth stage. Although introduced for one mission, it became the basis of a series of later scientific flights.

The probes were sent directly into solar orbit, without the use of an Earth parking orbit. MS-T5 was launched from Uchinoura on 8th January 1985, followed by Planet A on 19th August 1985 at the start of a 20-day window. Three days into its mission, MS-T5 made the first of a series of mid-course corrections with its small pulse jets. MS-T5, now renamed *Sakigake* or "pioneer", was able to concentrate on the area downstream of the Sun and study the solar wind, magnetic fields and plasma waves, approaching to within 7m km of the comet on 12th March 1986.

Uchinoura launch over sea

Planet A, renamed *Suisei*, made its closest approach 200,000 km from the head of the comet, on 8th March 1986, just a month after the perihelion passage. *Suisei*'s instruments noted clearly the moment when the little spacecraft crossed the bow shock generated by the comet against the solar wind. The camera was able to focus on the comet's nucleus and its hydrogen cloud for a month following the interception and found the rotation period for the comet. The two probes were tracked by a new 64-m deep space antenna constructed at Usuda, in the radio-quiet Nagano valley, 170 km north-west of Tokyo, completed on 31st October 1984. After the mission was over, Usuda tracked the American *Voyager 2* flyby of the planet Neptune in 1989.

Seven years later, *Sakigake*'s orbit intercepted that of Earth, the small spacecraft passing 90,000 km over the Indian Ocean. Its motor was used again to reset the spacecraft's orbit in such a way that it could better study the Earth's magnetic field and the solar wind. The Japanese probes were probably the least publicized of the earthly armada that flew to Halley in 1986. They did attract considerable public interest in Japan itself and the Tokyo Broadcasting Service made 16 television programs on the probes.

MU-3SII SCIENTIFIC MISSIONS

The *Mu-3SII* went on to become the most used rocket of the *Mu* series. Exos-D, renamed *Akebono* or "dawn", but sometimes also known as ASCA (Advanced Satellite for Cosmology & Astrophysics), entered a highly elliptical orbit out to 10,460 km at 75° on 21st February 1989 and was designed to study particle acceleration in the southern lights, the *Aurorae Australis*.

Sakigake *Suisei*

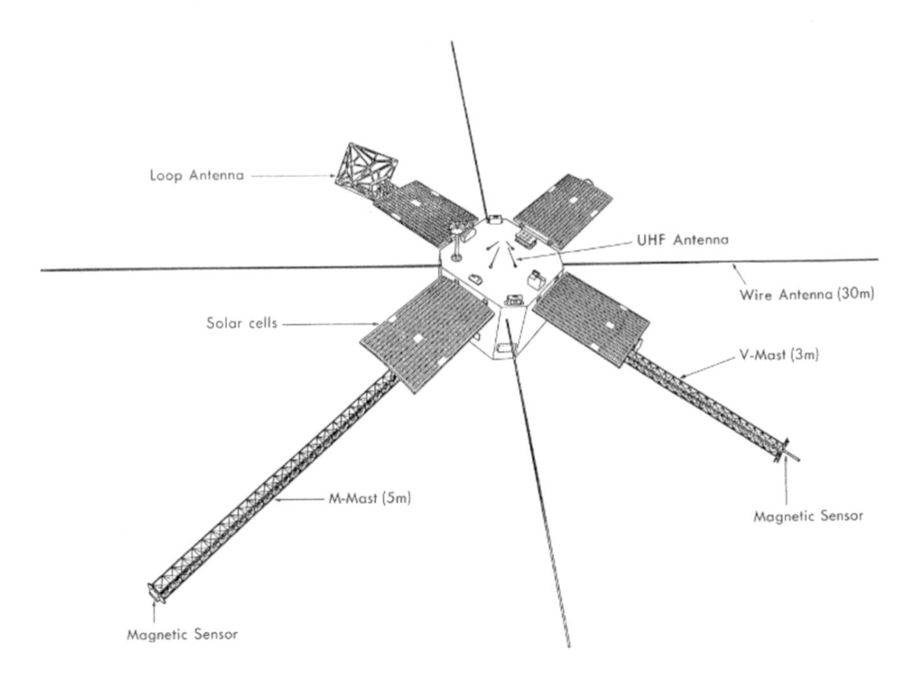

Akebono

Astro-C (430 kg) was developed in collaboration with British scientists at the University of Leicester and the Rutherford Appleton Laboratory. Astro-C, known as *Ginga*, was launched in February 1987 and carried what was then the largest satellite-borne X-ray detector, weighing over 100 kg. Its aim was, over five years, to

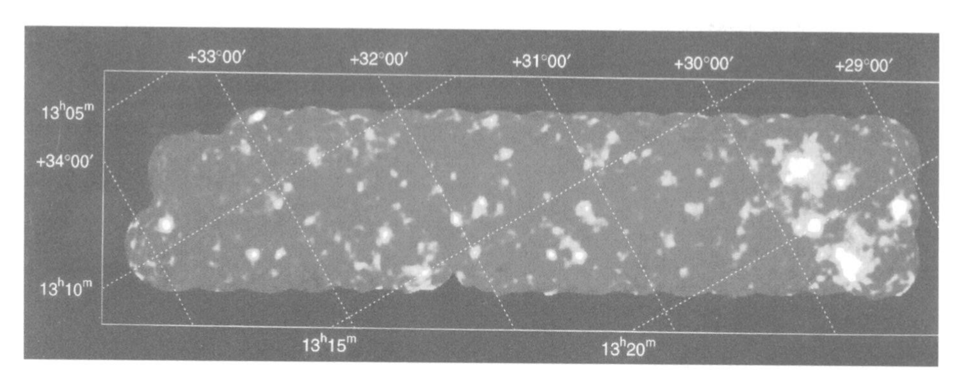

Akebono sky survey

obtain important new information on X-rays released from the vicinity of neutron stars and black holes.

Astro D, known as *Asuka*, was a 420-kg X-ray satellite launched on 20th February 1993 carrying four telescopes, the successor to *Ginga*, which had completed its mission at the end of 1991. *Asuka* entered orbit of 538–647 km, 96 min. It was a 4.7-m tall, 1.3-m diameter box with 2.8-m panels to provide electrical power for five to six years. The four X-ray telescopes were 10 times more powerful than those carried on *Ginga* and could pick up light from objects 10bn light years distant – rays that were normally absorbed by the Earth's atmosphere and could not be observed on the ground. Its X-ray camera was designed to provide clearer, sharper images than any previous X-ray camera. *Asuka* carried an American telescope called BBXRT, Broad Band X-ray Telescope, which had flown in space earlier on the American Astro-1 Spacelab mission and had provided excellent results on active galaxies, hot interstellar gases, supernovae, black holes and quasars. It transmitted images of the structure of the bright supernova SB1006 and of dark matter lying between galaxies. *Asuka* was the fourth of the program of X-ray satellites dating back to *Hakucho*. Between 1993 and 1995, *Asuka* found 119 distant active galaxies and faint X-ray sources, compiling a map of the north galactic pole. *Asuka* crashed back into the atmosphere over New Guinea in March 2001.

In addition to the *Mu-3SII*-launched satellites of this period, the Japanese–American *Geotail* was launched by American Delta on 24th July 1992, to study the structure and dynamics of the Earth's magnetic tail. *Geotail* used lunar swingbys to steer itself repeatedly into the tail of Earth's magnetosphere. *Geotail* was a drum-shaped spacecraft with two 6-m masts and two 100-m antennae, designed to measure the magnetic field, electric field, plasma, energetic particles and plasma waves.

SOLAR STUDIES: *YOHKOH* AND *HINODE*

The *Mu-3SII* marked the start of the "Solar" series of scientific satellites. Japan launched the Solar A *Yohkoh* ("sunlight") probe on 30th August 1991, designed to study solar activity on a three-year assignment, entering an orbit of 519–787 km, 31°.

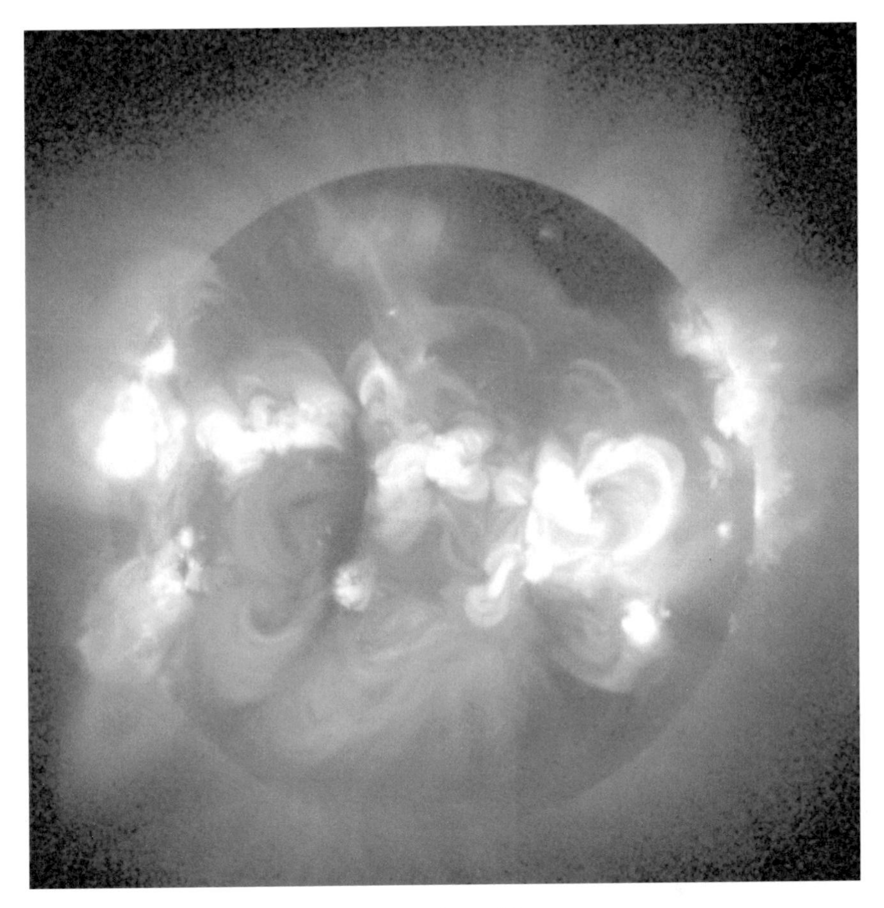

Yohkoh view of Sun

The 390-kg satellite had four instruments to study the high-energy phenomena of solar flares: two telescopes (NASA built the soft X-ray telescope and the Japanese the hard X-ray telescope) and two spectrometers (the British Bragg crystal spectrometer and the Japanese wide-band spectrometer). The Bragg instrument was built by the Mullard Space Science Laboratory and Rutherford Appleton Laboratory. *Yohkoh* was targeted to make observations during the solar maximum but proved equally valuable during the solar minimum of 1996 and the subsequent resurgence of activity. Pictures of the Sun during the maximum period showed a bright yellow, red and black surface gushing hydrogen – by contrast, only a glowing rim a few years later.

So successful was Solar A, *Yohkoh*, that Solar B was agreed as a mission in 1998, though it used a more powerful and later launcher, the *Mu-5*. The 900-kg Solar B was launched from Uchinoura on 23rd September 2006 and was called *Hinode* ("sunrise") once it entered orbit (a small amateur satellite, *Hitsat*, was carried piggyback). That November, it began a three-year solar survey from its 630-km-high orbit, using its three telescopes (optical, X-ray, ultraviolet) developed with Britain and the United States. The solar telescope was designed to follow the Sun's magnetic field, the X-ray telescope the corona and the ultraviolet the velocity fields, temperature and density of the corona, transition region and plasma of the Sun.

Hinode was a spectacular success, with a much better performance than *Yokhoh*, sending back extraordinary yellow and orange images of the Sun, its surface and wispy waves. In the 1940s, the Swedish scientist, Hannes Alfven, had put forward a theory that magnetic waves caused winds to blow across the face of the Sun and spread outward at speeds of nearly 3m km/hr – but he could never prove they existed. They were called Alfven waves. *Hinode* proved that he was right and interpretation of the *Hinode* data suggested that the magnetic waves also, through acceleration of energy, transformed the temperature of the Sun, 6,000°C, into the 1,000,000°C of the solar corona.

Next, *Hinode* found X-ray jets spewing hot gas out of coronal holes in the Sun at enormous speed hundreds of times a day. These had been spotted by *Yokhoh*, which had found one or two X-ray jets, but they were so infrequent as to be considered only a curiosity. The difference was that the *Hinode* X-ray telescope worked much faster and could pick out jets up to 240 times a day. They came from all over the Sun –

Hinode

coronal holes, sunspots and at random – and were so many that they constituted a major area of solar activity. Scientists soon speculated that there was a close connection: were the X-ray jets driving the solar wind?

THIRD TO REACH THE MOON: MUSES A

The success of *Sakigake* and *Suisei* four years later gave Japan the confidence to consider further missions outside Earth orbit. In 1987, ISAS formally sought approval from the Space Activities Commission for its first Moon probe to start a wide-ranging program for lunar, planetary and cometary exploration. The project was called Muses A and the project manager was Professor Kuninori Uesugi. It was ready within three years.

Japan's first Moon probe was launched from Uchinoura on 24th January 1990. It was the first spacecraft launched to the Moon by any country since the Soviet Union's *Luna 24* had landed in the Sea of Crisis in August 1976 to bring rock samples back to Earth. Muses A made Japan the third country to launch a Moon rocket. A new 20-m antenna was built at Uchinoura to track Muses A. The ¥4.3bn (€38m) Muses A project comprised two spacecraft:

- mother craft, 193 kg, 1.4 m in diameter and 79 cm high, called *Hiten*. Its weight included 42 kg of hydrazine fuel using 12 thrusters. Solar cells provided 100 W of power;
- lunar orbiting subsatellite, 11 kg in weight, 40 cm in diameter, 27 cm high, called *Hagoromo*.

The mother craft was called *Hiten* after one of the subjects of Buddha, whose duty was to play music in heaven. It was a drum-shaped object with two antennae underneath and the little lunar orbiter placed on top. The only scientific instrument carried was a micro-meteoroid detector made in Munich, Germany. The mission was more an engineering than a scientific one.

The first attempt to launch Muses A on 23rd January reached 18 sec before launch, when technical problems halted the countdown. However, it got away the following day, and the *Mu-3SII* put the spacecraft into a 240–400-km parking orbit around the Earth. After two circuits, the solid rocket motor fired to place the spacecraft in a highly elliptical Earth orbit, looping out to 16,000 km from the Moon two months later. In fact, the impulse given to *Hiten* was 50 m less than planned, which meant that *Hiten* had to take an extra orbit to reach the Moon. Worse, during the translunar coast, the transmitter on the *Hagoromo* subsatellite failed, which meant that no signals would be returned from lunar orbit.

On 18th March, 14,900 km behind the Moon and 54 days after launch, the mother craft released the orbiter, which fired its own motor so as to enter lunar orbit. *Hagoromo* entered lunar orbit of 7,400–20,000 km. The Kiso tracking observatory at the University of Tokyo was able, with its fine Schmidt camera, to spot the engine burning and confirm that *Hagoromo* had indeed entered lunar orbit.

Hiten meantime continued in its slow, lazy, Earth–Moon curving orbit. By October 1990, the distance from Earth stretched to 1.34m km. On 19th March 1991, its return trajectory took it 120 km back into the Earth's upper atmosphere to perform what was in effect the first ever aerobraking maneuver at high velocity. Japanese mission controllers decided, in consultation with the American Jet Propulsion Laboratory in California, to take advantage of the fact that *Hiten* still had residual fuel so as to devise an extended mission. The aerobraking adjusted the orbit in such as way as to swing *Hiten* back out to the Moon for eventual lunar capture on 15th February 1992. *Hiten* entered a lunar orbit of 422–49,200 km, one amended by a plane change three months later. The orbit was not a stable one and the small spacecraft began to spiral inward. Rather than let *Hiten* crash at will, ground controllers used the very last fuel to guide *Hiten* to its final impact. *Hiten* eventually hit the Moon near the crater Furnelius on 10th April 1993, the following year, at 38°S latitude, 5°E longitude. Japan had become (1993) the

Hiten and *Hagoromo*

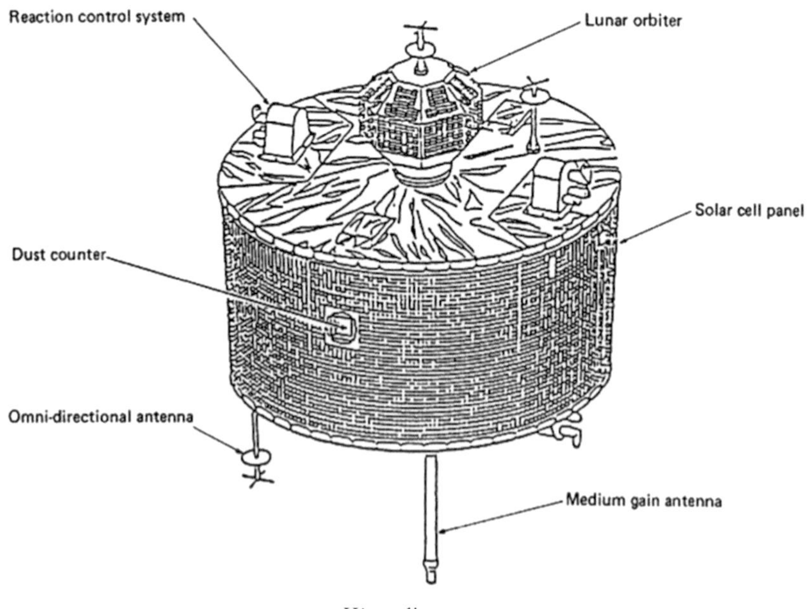

Hiten diagram

third nation to hit the Moon after the Soviet Union (1959) and the United States (1962); and the third to enter lunar orbit after the Soviet Union and the United States (both 1966).

EXPRESS: FROM PACIFIC SEACOAST TO THE JUNGLES OF AFRICA

The last of the *Mu-3SII* launches was *Express*. The story of the *Express* capsule is one of the strangest in the Japanese space program, involving five other countries: Germany, Russia, Australia, Britain and Ghana. The German company, OKB, in conjunction with the German space agency, DARA, had the idea of flying some microgravity experiments and returning them to Earth. The completion of the design studies coincided with a period of retrenchment in German public spending, not least in the space industry. The price tag became unaffordable.

In an effort to cut and spread costs, DARA sought out international partners. It persuaded the Khrunichev company in Russia to let them have a re-entry capsule used for its now defunct military Fractional Orbital Bombardment System (FOBS). This was a project, tested in the 1960s, to send small warheads around the Earth on a three-quarter orbit to descend on the United States from Mexico. The 405-kg warhead, civilianized as a space cabin, was called *Express*. The *Express* cabin had sufficient volume to carry up to 165 kg of payload and was fitted with six experiments, three from Japan and three from Germany. The next international partner was Japan, which offered a free launch on the *Mu-3SII* in exchange for full access to the research results. Japan supplied an additional module to attach the *Express* to the top of the *Mu-3SII*, but, in doing so, the total payload weight had now risen to 762 kg, the extreme limit of the performance of the launcher. Where to land the cabin posed a further challenge, but one that introduced the next international partner, the Australians, who had long been searching for ways to bring the Woomera Test Facility back into business. It was therefore agreed to recover the satellite near the Woomera launching base in the Australian desert some 5.5 days after lift-off. The mission would be tracked by stations in Germany, Santiago (Chile), Bermuda and Woomera.

Express rode on top of the last *Mu-3SII* rocket to be fired from Uchinoura on 15th February 1995, blazing into a nighttime sky. An attitude control problem with the second stage 130 sec into the mission meant that control was lost for 20 sec and fuel was depleted as the rocket tried to compensate. The problem was later attributed to the aerodynamic effects of an unusually heavy payload during ascent. This meant that the planned orbit of 210–398 km was not achieved. The resulting orbit was so low – possibly as low as 110 km – that it was considered that the cabin must burn up. United States Space Command never detected the launch nor entered it in its logs. However, the downlink signals confirmed that the Russian computer had stabilized the cabin and the mission had begun. Uchinoura established radio contact with the satellite at 3.30 pm, Santiago at 4.01 pm, and Uchinoura again at 4.51 pm.

Attempting to salvage something from the mission, German ground control, in Oberpfaffenhofen, Munich, asked the tracking center in Santiago, Chile, to

Express

command the satellite to re-enter, hoping that they might at least recover the capsule rather than wait helplessly for it to be destroyed. When *Express* made its next pass over the center at 5.30 pm, Santiago received no signals and had no idea whether the re-entry command had been received.

So the story ended. It was presumed that *Express* had burned up somewhere over the Pacific after about three orbits. The failure of the *Mu* was blamed on the heavy payload – twice that of the *Mu*'s previous heaviest assignment. Extra propellant had been loaded to all the *Mu* stages, computer simulations indicating that the profile would work. The computer must have been wrong.

Several months later in Britain, Geoffrey Perry, the science teacher known for his role in listening to signals from early Soviet space probes, was alerted to a report in the *Ghanaian Times* of 3rd February 1995 by-lined by Gariba Ibrahim in Tamale, which reported that a strange object with Russian markings on an orange parachute had descended from the skies at Kotorigu in west Mamprusi, near the border with Togo. The newspaper recorded the fact that the deputy commissioner of police, Patrick Agboda, had visited the area and noted that the bushes surrounding the fallen object were burned. Kotorigu is very rural, lacking electricity, water or telephone, and the road can only be used by four-wheel-drives. The people living in the traditional African villages in the area must have been startled by the sonic boom that preceded the red-hot object descending under its parachute. The local chief, it was later learned, told people to keep away from the object from space. He ordered his brother to guard it while he went to the district chief, Mr Gumah of Walewale.

A photograph later appeared in the *Ghanaian Chronicle* on 20–23rd February, bylined by Abdulla Kassim in Tamale, which showed the strange object hanging from a tree and local people posing behind the spread-out parachute. It was clear to Perry that it looked very like a FOBS re-entry capsule and, indeed, the article quoted the commander of armed forces in the region, Group Captain Aryetey's conviction that it was Russian. Perry noted that the ground track of *Express* brought it over west Africa four hours after launch. Perry contacted the Ghanaian authorities a number of times in the course of the year to ascertain whether the strange object was indeed the *Express* capsule, but fruitlessly. Having made no progress, he then sent an article on the missing capsule to the November 1995 monthly news bulletin of the Western Australian Astronautical Society with his speculation as to the true fate of *Express*.

In fact, more had gone on than he had realized. District Chief Gumah had organized transport to take the *Express* away. Ten strong men had been required to put the capsule onto a truck. Hundreds came to look at the object in Walewale, where it had become something of a local attraction and was it was handed over to the army. Two weeks later, the Ghanaian air force took the object to the town of Tamale, 50 km south, where it was stored in a huge aircraft hangar. This hangar had been built by the Russians in the 1960s during a period of intense Soviet–Ghana cooperation and the base commander, who had been there at the time, at once recognized the cyrillic script on the parachute (lettering on the cabin itself had burned off).

Meanwhile, Perry's article in the news bulletin of the Western Australian Astronautical Society had been read by some members of the recovery team in Woomera, who must have been wondering what had happened to their seriously overdue cabin. They sent the article to DARA in Germany, who phoned the German embassy in Accra. After a few quick enquiries, it was ascertained that the strange object was almost certainly the *Express*. In January 1996, officials from the German space agency DARA arrived in Ghana, identified the *Express* and asked for their satellite back. They found that the capsule had been moved to Tamale, where it was still lying, untouched and unopened, in a corner of its hangar. It was a little dented, not from its brief sojourn in space, but from its truck ride along rutted African roads. Until the Germans arrived, no one had known quite what to do with it. The military authorities had set up a scientific commission to determine whether it was a radioactive warhead (they drew a negative). They had called the Russian embassy in Accra, who quickly said it was not their satellite. The articles in the Ghanaian press about the strange object were well read in the small European community in Ghana, though the German members seem to have missed it and when the cabin had been declared lost, no one had thought to alert Germany's wide-flung outposts throughout the world.

Luftwaffe squadron 62 duly sent a Transall plane out to Ghana, which brought back the *Express* to Germany in a wooden container. Subsequent examination of the cabin found that *Express* had entered the atmosphere at 5.50 pm on that January day, 80 km above the Earth. Barreling nose first into the atmosphere, the cabin survived the intense heat. Sensing the onrush of air 6 km high, the barometric system commanded the parachutes to open. The experimental packages were in perfect order and the experiment to test ceramic materials on the nose cone was declared a complete, if belated, success! Thus ended the *Mu-3SII* series.

MUSES B: INTRODUCING THE NEW *MU-5* LAUNCHER

Muses A, the first Moon flight, was the first of three Japanese spacecraft devoted to unconventional missions. The next, Muses B, was an experiment in very long baseline inferometry and marked the first launch of the *Mu-5* launcher, successor to the *Mu-3*, the aim being to double the lifting power. Approval for the new launcher had been given in 1989. Built by Nissan, *Mu-5* weighed 135 tonnes at lift-off, making it the largest ever *solid-only* propellant launcher flown. *Mu-5* was 2.5 m in diameter and could lift 1,800 kg into Earth orbit. Originally, the division of responsibilities between ISAS and NASDA had limited ISAS to launchers of 1.4 m diameter or less, but this restriction was at last waived.

The first stage, the M-14, used a mixed fuel of 68% ammonium perchlorate, 20% aluminum and 12% PBHT inside high-strength maraging steel. Efficiencies were achieved by using a collapsible nozzle on the first stage, extendible nozzles on the third and fourth stages and side jets. Another improvement was the use of fiber optical gyros to sense vehicle attitude. Development of the rocket was slow and, by the time it was launched, had fallen several years behind schedule, the main cause being the difficulty in developing retractable nozzles. *Mu-5* required the upgrading of the *Mu* pad at Uchinoura. Although much larger, the *Mu* still followed the same launch procedure as the older series, being tilted at an angle on a crane beside its launch tower. The cost of a launch was ¥13.20bn (€118m).

Shooting aloft on a pillar of bright yellow flames against a calm sea and clouds, the *Mu-5* placed 830-kg Muses B into an elliptical 573–21,402-km 6.5-hr orbit on 12th February 1997. Once in orbit, it was renamed *Haruka*, meaning "faraway".

Haruka's basic shape was conventional enough – a box with solar panels. What made it unusual was the 8-m-wide wire-mesh golden-colored antenna, which, on the 17th day of the mission, unfurled like a giant petal, topped by a reflector that peeped through the thin structure. *Haruka* worked on the principle that its radio telescope in space could be combined with another one to Earth to make a very long baseline, thus obtaining a wide measuring base and improving the resolution. It was designed to have a resolution of 90 micro arc seconds, or a hundred times better than the Hubble Space Telescope (though in the radio, rather than the visual medium).

The ¥8,664m (€77.3m) *Haruka* mission was developed by ISAS in cooperation with colleagues in the United States, Canada, Australia and Europe. *Haruka* was tracked by stations in five locations – Usuda, Greenbank, Goldstone, Madrid and Tindbinbilla, whence it sent data at 128 megabytes a second. It was designed to record active radio sources in the universe, in galaxies, quasars, pulsars and masers.

Mu-5 lift-off

One of its first observations was of quasar 1156+295. Whereas ground pictures showed a reddish-yellow blob, *Haruka* images were much more precise, showing clearly a jet of material being ejected from the core.

The *Mu-5* enabled a much larger size of astrophysical satellite to be orbited compared to the earlier Astro series, though the first such mission turned out to be the only *Mu-5* failure. Astro E was an X-ray astronomy mission, with two instruments for soft X-ray observations and a large counter array for hard X-ray observations, designed between them to ensure both a wide range and variety of sources and high sensitivity. Astro E weighed 1,650 kg and was due to enter a circular orbit at 550 km. Launched on 10th February 2000, the thrust began to fail

Astro E

only 5 sec into the mission and sparks began to come out of the nozzle area at 25 sec, indicating that the heat-resistant graphite had burned through. Although there was no explosion, Astro E failed to make orbit and probably crashed into the Pacific Ocean. The failure came at a really bad time, for NASDA's H-II rocket had crashed only four months earlier, meaning that the Japanese space program was now effectively grounded.

Astro E was eventually replaced on 19th July 2005 when its replacement was successfully launched and named *Suzaki* once in orbit, called after a bird god. *Suzaki* was a 1,700-kg, 6.5 m tall observatory carrying an X-ray spectrometer, X-ray imaging spectrometer, hard X-ray detector and five X-ray telescopes. The *Mu-5* put *Suzaki* into an initial 247–560-km, 31.4° orbit, later circularized at 550 km.

The series continued on 21st February 2006 with the 950-kg Astro F infrared astronomy satellite. In 1983, the European IRAS infrared observatory had revolutionalized astronomy by detecting the infrared radiation normally absorbed by the atmosphere. Astro F was Japan's first infrared observatory and its main purpose was to follow the way in which drifting hydrogen gas and dark clouds formed into star nurseries. Astro F used mechanical cooling techniques designed to keep the 67-cm aperture telescope functioning for much longer than had been the case with previous infrared telescopes, being cooled at the extraordinary temperature of –267°C, just 6°C above absolute zero. On arriving in its 304–733-km, 98.2° orbit, the 952-kg Astro F was given the name of *Akari*, the Japanese word for "light", and it deployed a small, 3-kg nanosatellite, *Cute 1.7*, built by students at Tokyo Institute of Technology. On the way up, the launcher deployed a 15-m solar sail, but it did not

Akari

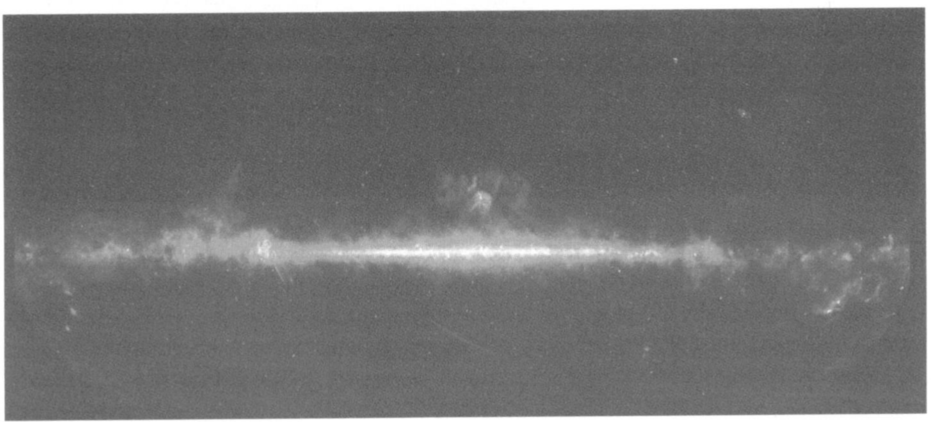

Akari all-sky map

open properly. *Akari* later raised its orbit to an operational altitude of 695–710 km and soon sent back spectacular images of newly formed stars in the Milky Way and in distant nebulae. Astro G will follow in 2012.

The liquid helium coolant on *Akari* lasted until August 2007, ending far infrared observations, by which time it had surveyed 94% of the sky. It was able to continue to use a near infrared camera, which made significant findings. The most remarkable images were of the ageing star, Belegeuse, 200 light years from Earth, which was crossing the star-forming region of Orion's belt at 17 km/sec, spewing out gas as it did so and creating its own bow shock in the interstellar medium. *Akari* then observed eight supernova remnants in the Large Magellanic Cloud, 160,000 light years distant, finding warm dust grains in the region left behind from the huge supernova explosions, but none of the expected cold dust.

Nozomi

NOZOMI TO MARS

The Planet A mission to comet Halley was, as the named suggested, the first Japanese planetary-type mission. The Planet B mission was defined as Japan's first venture to a terrestrial-type planet and was originally a plan to launch a small payload to Venus, the nearest planet and the easiest to reach. The objective changed when it became clear that Mars, although further away, was scientifically more promising.

Planet B became Japan's first spacecraft to Mars and was launched on 4th July 1998. Soon after its launch, it was renamed *Nozomi*, or "hope". The objective was to orbit Mars at an altitude of 300–47,500 km, the low point later adjusted to 130–150 km, and transmit data for one Martian year. Fourteen instruments, weighing 33 kg, were designed to send back information on the magnetic field, the structure of the atmosphere (turbulence, motion and seasonal variations) and on energetic particles, with a particular view to answering the question as to why Mars lost its water. The scientific instruments came not just from Japan, but also from France, the European Space Agency, the Swedish Institute of Space Physics, the Canadian Space Agency, NASA and Munich Technical University. NASA had a neutral gas mass spectrometer to measure the chemical composition of the Martian atmosphere. A Japanese/French camera was carried. Several passes were to be made of Mars's tiny moons, Phobos and Deimos. In focusing on the Martian atmosphere, *Nozomi* planned to complement rather than compete with the American and European probes of that period.

The project cost ¥18.6bn (€166m). Although the start of the mission attracted little attention abroad, there was much more excitement at home, where 270,000

Nozomi view of Earth and Moon

people had their names inscribed in miniature on the spacecraft. Its weight was 540 kg, 258 kg without fuel, making it one of the smallest spacecraft to travel to Mars. *Nozomi* was a hexagonal box shape measuring 1.6 by 1.6 by 0.6 m, with a bell-shaped motor underneath and a dish antenna on top. *Nozomi* had a 500-kg thrust engine to carry out a complex set of maneuvers, first in Earth orbit, then in trans-Mars coast and then to subsequently enter orbit around the red planet.

Twenty-three minutes after lift-off, the spacecraft entered a swinging orbit of 703–489,382 km. Its solar panel sprung open. Thin 25-m-long radio wire antennae were deployed. Also sticking out was a 5-m magnetometer. At a far point out in the trajectory, 300,000 km from its home planet, its cameras were turned on to capture an unusual view: the Earth–Moon system in the same frame. A thin crescent of Earth filled the bottom left of the picture, while in the far right could be made out the crescent of Earth's Moon. On 24th September, as it came near to the Moon, it used lunar gravity to swing the spacecraft into an even more extreme orbit, one as far out as 1.7m km. Cameras clicked to return spectacular pictures of the far side of the Moon, picking out the Moscow Sea and the pitch-black floor of crater Tsiolkovsky.

This was in preparation for a maneuver on 20th December to nudge it out of the Earth–Moon system and on its way to Mars. *Nozomi* flew around the Moon at a distance of 2,809 km on 18th December and then, firing its little engine, passed 1,003 km from Earth two days later, but the burn at perigee proved insufficient to kick the spacecraft onto a trans-Mars path. It later transpired that the valve supplying the oxidizer had failed to open sufficiently, wasting fuel and not producing sufficient thrust either. As a result, a second burn had to be made. This failed to rectify the situation and it turned out that the fuel supply had been severely depleted at this stage.

Despite this, ground controllers calculated that they could still find a way of reaching Mars orbit, but it would take many years. The burn had placed *Nozomi* on a slow solar orbit – one that would involve three orbits of the Sun, a swing by Earth in December 2002 and June 2003, sufficient to eventually reach Mars some time in either December 2003 or January 2004, four years late. Ground controllers were still confident that the spacecraft's equipment would remain in sufficiently good condition to carry out its delayed mission in full.

Here, nature intervened and a severe solar flare on 21st April 2002 badly damaged the electrical systems on *Nozomi*. The computer command and control system survived, but were working on much reduced power. Worst, the flare shorted the

Nozomi passes Mars

heating system, causing the hydrazine fuel to freeze, and this was ultimately to prove the fatal blow. The hydrazine would thaw out in the heat of the inner solar system, but it was more than likely to freeze up again at Martian distances.

Nozomi swung by Earth at 30,000 km on 21st December 2002 and 19th June 2003, before heading to its decisive encounter with Mars. *Nozomi* approached Mars at a distance of 1,000 km on 9th December 2003. Controllers still hoped they might be able to fire its motor for a Mars orbit insertion. The plan was to test fire the engine on 9th December and carry out a Mars Orbit Insertion maneuver on the 15th at a distance of 894 km from the planet. When the time came, though, *Nozomi* did not respond to commands and was presumed lost. It seems likely that communications had actually broken down some time before, but ground controllers clung to the hope that they could somehow retrieve something from the mission. It was not the only victim of what the space community called the "galactic ghoul" that lurked near Mars, snaring unwary spacecraft, for Britain's *Beagle 2* explorer was lost only 10 days later. An investigation into the mission concluded in May 2004 that the final failure was not due to the solar storm, but a technical one, the precise nature of which it was not possible to ascertain.

The third in the Planet series, Planet C, is intended to be a mission to Venus and is called the Venus Climate Orbiter. A feature of the mission is the deployment of a second, smaller (300-kg) spacecraft called *Ikarus*, a 20-m solar sail, the first demonstration of a solar sail on an interplanetary mission. As Planet C heads inward toward Venus, *Ikarus* will head outward to the trojan asteroids and Jupiter.

Japan also joined the Bepi Colombo mission, developed by the European Space Agency and named after an Italian space engineer, with JAXA providing a third of the funding for the mission. This involves a Mercury Planetary Orbiter (MPO) carrying piggyback a smaller Japanese orbiter, the Mercury Magnetospheric Orbiter (MMO), carrying five instruments of which one would be contributed by Austria. Its purpose is to explore the planet's magnetosphere and how it reacts with the Sun. Launcher will be the *Ariane 5* rocket from Kourou, Guyana, in 2014, with arrival in 2019. The MMO will explore the planet's magnetosphere, exosphere and their interaction with the solar wind.

RENDEZVOUS WITH AN ASTEROID: *HAYABUSA*

Despite the disappointing outcome, *Nozomi* attracted plaudits for trying to achieve a lot with a small spacecraft through the use of imaginative trajectories. Even more ambitious was the next interplanetary probe, which also encountered its fair share of difficulties. This was the third in the Muses series, Muses C, named *Hayabusa*, the same as Hideo Itokawa's most famous aircraft. And, by delicious irony, it travelled to a 535-m across asteroid, 1989ML, named *Hideo Itokawa*. This was a typical S-type (stony) chondrite asteroid, so it was hoped that the sample would tell us much about the class as a whole, the smallest visited by a spacecraft up to then. Never likely to make things easy for themselves, the Japanese chose to introduce electric xenon propulsion for the mission and to try to recover the asteroid samples back on Earth afterwards.

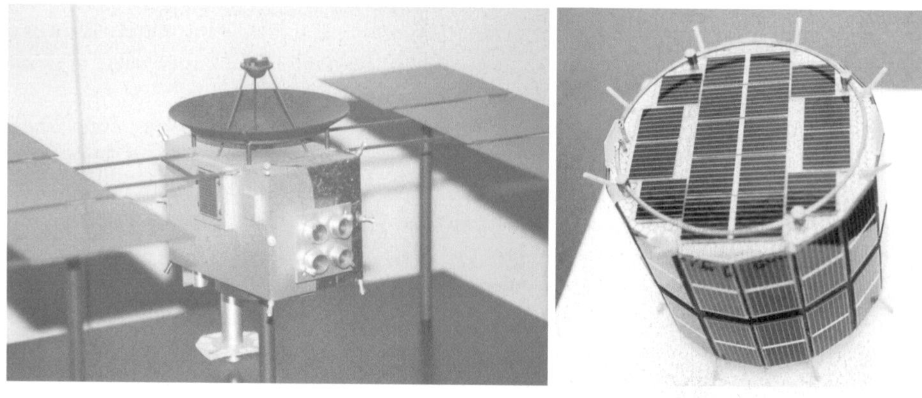

Haybusa *Minerva*

Minerva landing

Muses C, *Hayabusa*, was launched 9th May 2003 from Kagoshoma. The *Mu-5* sent the 510-kg, €150m *Hayabusa* on a huge loop around the solar system that would swing out to the asteroids, firing its ion propulsion system almost continuously. Mission Control was Sagamihara, near Yokohama. It soon began to use its ion engines to propel its way outward into the solar system.

Like *Nozomi*, *Hayabusa* followed a complex series of maneuvers to reach its target. *Hayabusa* flew past Earth at a distance of 3,700 km in May 2004, Earth's gravity giving it a push outwards and testing its cameras as it did so. By the end of

Hayabusa approaching Itokawa

August 2005, it was 35,000 km from asteroid Itokawa and closing. On 12th September, the chemical thrusters were fired to stop it in its track, slowing it down to 20 km from the asteroid, where it began to take its first pictures, finding that the potato-shaped object was slowly rotating. *Hayabusa* entered a 18–20-km orbit around Itokawa, with two months to prepare for the crucial landing operation. Asteroids such as Itokawa have such a low gravity that one cannot land there in the normal sense: investigating them required a hovering descent to the surface, assisted by the prior deployment of markers and a small 1-kg nanoprobe lander called *Minerva*, the size of a child's toy.

Hayabusa made its first approach on 4th November 2005, when it closed in to 3,500 m. Things began to go wrong now, for one of the reaction control wheels failed, which meant that the altimeter could not be pointed. In order to avoid a collision, the spacecraft was recalled at 700 m and backed away again. A second approach was made on 12th November. This time, at 500 m, the first of three 10-cm markers was released to descend gently to the surface and provide a reference point for the sampling operation, but it missed the asteroid and drifted away into space. Nevertheless, the 4-cm *Minerva* lander was released, but was deployed at too high an altitude, 55 m above the asteroid, missed and was lost. *Hayabusa* then retreated to 5 km, much further out.

On the third approach, on 19th November, the second marker was released, this time reaching the surface. This was a sufficient guide for *Hayabusa* to make its own descent to the surface, touching down on Itokawa at 20.10, bouncing back and settling down again (21.30) and then taking off again (21.58). No sampling operation seems to have been carried out, though. All this time, *Hayabusa* was 280m km from

Earth. Not only did the sheer distance present control problems, but *Hayabusa* was in line of sight with Usuda Mission Control in Japan only 8 hr a day.

The fourth and final descent was made on 25th November. This time, *Hayabusa* closed in from its parking orbit at 12 cm/sec, cutting its speed to 6 cm/sec when it was 40 m out. The laser range finder was turned on at 30 m and soon after 7.00 am, *Hayabusa* reached the surface, touching down in an area called the Muses Sea. Once contact was reported, a small charge was detonated to blast débris into a receiver cone, followed by a command to ascend. Things went wrong again at this stage, for one of the thrusters malfunctioned and two of the three remaining reaction control wheels began to operate erratically. It was by no means certain whether samples had been correctly collected.

No one is still really sure what happened. It is possible that the spacecraft span out of control when lifting off the asteroid. We know that there was a major anomaly, contact was lost and the spacecraft went into safe mode. Not until 29th November did the probe acknowledge a signal through its low-gain antenna. There was a burst of transmission, weak and interrupted at times, at 8 bytes/sec, on 1st December, and with bad news: there had been a fuel leak, attitude control trouble, a loss of electrical power, a battery discharge and a dangerous loss of temperature. On 2nd December, the command was sent to re-start the chemical engine, which flared into life briefly and then turned off. The high gain antenna, though, was drifting out of alignment with Earth and, the following day, an emergency command was sent to use the electric engine to restore the alignment. This worked and data levels were restored to 256 bytes/sec on the 5th, sufficient for ground control to get the probe to re-play the record of the landing. The computer memory did not show that samples had been collected; it is possible that they had, but the memory was wiped by the anomaly on the 27th. On the 6th, Mission Control was able to calculate that *Hayabusa* was 550 km from Itokawa, 290m km from Earth and moving back to Earth at a leisurely 5 km/hr. Over the next few weeks, mission controllers gingerly tried to get all the spacecraft systems working once again. According to the project manger, the probe was still alive, but communications were seriously damaged: "It is almost a miracle that it functions at all."

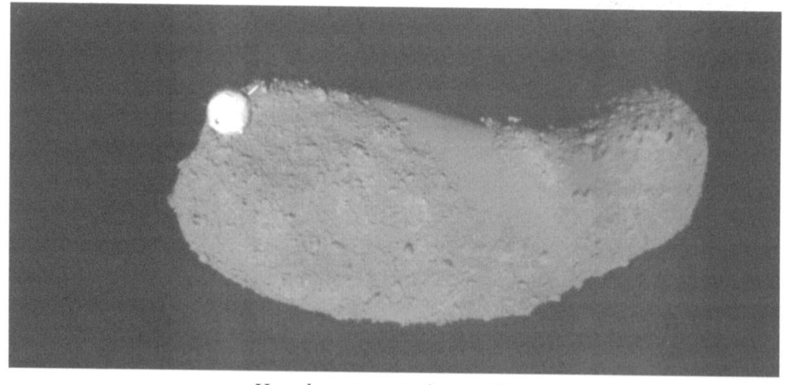

Hayabusa target descending

The next few months for Mission Control were very difficult. Controllers managed to get *Hayabusa* to a safe distance away from the asteroid, but the probe was tumbling slowly. Getting it back to Earth from 290m km away was challenging enough in the best circumstances, but the spacecraft was barely under control. Controllers figured out various ways of conserving fuel, recovering stability and fuel-efficient trajectories that might get *Hayabusa* back to Earth. In December, celestial mechanics were such that *Hayabusa* would drift out of sight from the ground for a three-month period. Would they ever even recover the spacecraft? The odds seemed to be against.

Much to their surprise and delight, mission controllers in Usuda received fresh signals from *Hayabusa* in March, although some 330m km distant. Not only that, but the spacecraft unloaded 1,500 high-definition images of the asteroid. Power levels were low, but the ion engines still appeared to be in working order. Cruise back to Earth began on 25th April, even though only one attitude control reaction wheel was available, rather than three.

In May, controllers managed to purge the errant fuel leak, meaning that the remaining fuel could be used without loss. They worked out a trajectory that would return *Hayabusa* to Woomera, Australia, in June 2010, barreling in through the atmosphere at 8 km/sec for a desert landing. By this time, Itokawa had become one of the best known asteroids in the solar system, scientists being surprised by the porous nature of its surface, some 40% like a giant pumice stone. What took them aback most was the absence of craters: instead, the asteroid was covered in boulders. Itokawa had a mixture of smooth and rough terrain, covered in gravel rather than

Hayabusa lands on Itokawa

powder. The asteroid looked as if it was the product of collisions and it was called a "rubble-pile asteroid". Its density was 1.9 g/cm^3, the surface uniform but lighter and darker according to the level of space weathering. The outcomes of the mission were put up on a mission science website to facilitate global access. Analysts were able to construct an evolutionary history.

Without even waiting for *Hayabusa* to return to Earth, JAXA began to make plans for *Hayabusa 2* to visit a C-type asteroid and, in a project with Europe, a visit to a D-type ("dormant comet") asteroid, this mission being called Macro Polo after the Italian merchant who sailed to Japan in the 13th century.

BACK TO THE MOON: *KAGUYA*

The adventures of *Hayabusa* occurred only a year before Japan returned to the Moon, this time with a truly scientific mission to follow the pathfinders of the 1990s, *Hiten* and *Hagoromo*. Following these early missions, ISAS proposed a two-stage set of scientific missions: the first, called Lunar A, would be an orbiter carrying penetrators, while the second, called Selene, would carry a lander. The ISAS move was timely, for it coincided with a revived scientific interest in the Moon, the discovery of water in the late 1990s by the American probe *Lunar Prospector* and the first European mission there, SMART. The mission was costed at ¥10.5bn (€93.75m).

Lunar A with penetrators

The original plan for Lunar A was to drop two 13-kg penetrator probes into the lunar surface. Shaped liked torpedoes, 15 cm in diameter and 83 cm long, each had a heat flow probe and a seismometer to measure moonquakes to solve the mystery of the Moon's core. Heat flow experiments had been carried by the Apollo missions, but the Japanese ones were five times more sensitive. The heat flow experiment was designed to detect whether any heat was still emanating from the Moon's core, how much and the size of the core. The Apollo seismometers detected three types of moonquakes – shallow ones from the impact of meteorites, those sparked off by Earth's tidal forces and others coming from deep in the Moon's core – so Lunar A would add to this knowledge. The plan for Lunar A was to deploy one penetrator 45 km north of the *Apollo 14* landing site of Fra Mauro and the second on the lunar far side in the middle of the relatively flat crater. Each penetrator would burrow up to 3 m into the Moon at a speed of 285 m/sec at a rate of 10,000 G. Half of the penetrator was motor and half was instrumented, with seismometer, battery and transmitter. After deploying the penetrators, the mother craft would act as a data relay and provide television on its lunar imaging camera. Every 15 days, it would pass over the penetrators and unload their data to Usuda Deep Space Center.

Although the idea of a penetrator sounded simple enough, designing equipment to survive a 300-m/sec impact that caused up to 10,000 G was no mean task (1 G = the force of Earth's normal gravity). By the early 1990s, Japanese scientists had tested several models of penetrators, firing models from a nitrogen gun into sand boxes designed to resemble the lunar surface. Separating the penetrator from the orbiter proved to be problematical: on 17th September 1995, an S-520 sounding rocket was launched to test the separation method for the orbiter and the penetrators, but the mission went awry and telemetry was lost after 25 sec. Tests also suggested the danger of a wire breaking between the seismometer and the data relay system and that the design had insufficient protection against battery leaks after the violent impact. Serious question marks were raised during design reviews in 1997 and the project put on hold while these were sorted out. The outcome of a stormy design review was to cut the number of penetrators from three to two, substituting the third with extra battery power. The second mission, Selene, also experienced many design problems. Selene was SELenological and ENgineering Explorer, a ¥21bn (€188m) project led by Nippon Electric to orbit a large, 4.8 m tall, 2.8-tonne spacecraft with 14 instruments and a 350-kg lander. The original mission plan envisaged the placing of Selene into 100-km lunar orbit, the deployment of a 40-kg relay and gravity field satellite and the sending down of a lander shaped like a multisided box, 2.2 m across and 1.1 m high. With four landing legs, it would come down from orbit in a 28-min powered descent like the American lunar lander: on arrival, it would make a television survey of the lunar terrain for two hours and then transmit further information for two months more.

Eventually, the Lunar A mission with its troublesome penetrators was canceled in February 2007. The mission had been over 10 years in preparation and had suffered from repeated cost overruns in the course of engineering tests and re-designs, but better that than a mission failure. It was decided to bring the delayed Selene 1 mission forward without a lander, but with some of Lunar A's payloads instead, with Selene 2 to follow in 2015. One could say that the two projects were merged: after such prolonged delays, it was surprising that the new Selene mission was made ready in the following six months. At this stage, the new mission was quoted at a cost of ¥755bn (€400m), but it is not known whether this included the accumulated but now written-off costs of Lunar A. A feature of the project was that news on the project was quite difficult to obtain. The Japanese tended to be tight-lipped, launch dates frustratingly vague and no real sense being given of the state of preparations. Then, suddenly, a press announcement was made that the rocket was on the pad, ready to go. Appointed as mission director was Yoshisada Takizawa.

Selene 1 was launched on 14th September 2007 on the 13th H-IIA from Tanegashima, which took off into a clear sky with a stiff easterly breeze. Selene was separated 45 min later and once its solar panels deployed soon thereafter, it was given the name *Kaguya*. On board was a disk with messages recorded from 412,627 well-wishers. The name *Kaguya* was the favorite in a public consultation, named after the Japanese fairy moon princess who descended to Earth and was found in a bamboo plant, where she was reared by a couple called Okina and Ouna. *Kaguya* was a rectangular box shape with a motor at one end, two small subsatellites at another, three long aerials, a long magnetometer, communications antenna and solar panel. *Kaguya* carried these instruments:

- X-ray spectrometer (composition of lunar soil);
- gamma ray spectrometer (detection of key elements in lunar soil);
- multiband imager (nine-channel imager, visible to near infrared);
- spectral profiler (near infrared);
- terrain camera (10 m detail);
- lunar radar sounder (under-surface radar sounder);
- laser altimeter (topographical mapping);
- lunar magnetometer (to measure lunar magnetic field);
- charged particle spectrometer (cosmic and alpha rays in lunar orbit);
- plasma energy angle and composition experiment (electrons and low-energy ions);
- radio science detector (lunar ionosphere);
- upper atmosphere and plasma imager (measurement of Earth's atmosphere from Moon);
- subsatellites to measure and map the Moon's gravity field;
- high-definition television.

Like previous Japanese deep space probes, *Kaguya* followed a complex trajectory to reach its destination. The H-IIA placed it first in an elongated Earth orbit, 281–232,805 km, 29.9°. After two major maneuvers, *Kaguya* reached the Moon two weeks later, arriving on a historic day, 4th October 2007, exactly 50 years to the day since the launch of the first Sputnik. At 6.20 am that morning, *Kaguya* fired its motor to enter a highly elliptical orbit of the Moon of 101–11,741 km, 16 hr 42 min, 95°. *Kaguya* was the first of a three-horse Asian Moon race to reach its target. China followed with *Chang e*, its first moon probe, a month later, with India's *Chandrayan*

Kaguya in lunar orbit

following a year later. Over the next month, *Kaguya* gradually reduced the height of its orbit to the intended operational altitude of 100 km.

Kaguya then deployed its two 50-kg subsatellites, duly named after the characters in the fairy story. The first, *Okina*, was ejected into a 115–2,399-km orbit on 9th October and the second, *Ouna*, on 12th October into a 127–795-km orbit. By the end of October, *Kaguya* had reached its operating orbit of 80–123 km. First images were now published in the media. Superior Japanese cinematography was now evident, for the internet images could be followed both as high-definition still images and as an 8-min movie, with a commentary on the swathes of the Moon under observation (the south-western edge of the Ocean of Storms). Later, JAXA, the Japanese Space Agency, published spectacular images of a full Earth rising over the lunar surface, the best images of their kind since Russia's Zond 7 went round the Moon in 1969. Further images followed in November from the far side south pole and the Mare Orientale. These were calibration tests before the scientific program began in earnest. The camera's 10-m resolution provided a fabulous level of detail of the lunar surface: it was even able to image the scorch marks of the Apollo 15 lunar module left from where it came in to land near Mount Hadley in 1971.

The first scientific results were published in February the following year. The radar sounder and altimeter had bounced waves into the Mare Imbrium, from which

Kaguya Earthrise

scientists were able to calculate that it comprised a mixture of lava, volcanic ash and blankets of ejected material. By spring 2008, *Kaguya* had taken over 300,000 low and high-resolution pictures of the Moon. The camera was so sensitive that it was able to map the dim polar regions of the Moon, which, because of their oblique angles, never saw sunshine, but were painted by Earthshine, such as crater Shackleton, which had been singled out as a potential lunar base. The altimeter had made measurements with an accuracy of 5 m.

A year after arriving in lunar orbit, *Kaguya* had compiled the most detailed maps ever of the north and south poles. Multicolor interpretative maps illustrated the different chemical elements of craters and slices of the terrain passing below. Strong traces of uranium were found. A study was made of lunar craters with a central peak, the main craters of interest being Tsiolkovsky, Pythagoras, Jackson, Daedalus, Antoniadi and Gassendi. From this, scientists were able to estimate the age of these craters to be between 2.49 and 2.56bn years. They believed that most volcanism ended on the moon 2.4bn years ago, but there was a final surge on the near side 1.2bn years ago that filled up the lunar seas. The radar sounder was able to indicate the structure of the lunar crust up to 5 km down, suggesting that a magma ocean may have lain beneath. The plasma analyzer mapped the way solar wind ions settled near the lunar surface and found alkali ions for the first time. Circling the moon on the far side and free from Earth's radio noise, the analyzer picked up radio emissions from Jupiter and Saturn.

The other big learning point from the *Kaguya* mission concerned the mascons. These are "mass concentrations" of heavier and lighter rock below the lunar surface that push and pull spacecraft out of their trajectories. First discovered by Russia's Luna 10, they were still not well understood during the Apollo landings and may have caused Apollo 11 to descend off course, whence Neil Armstrong had to take control and fly the lunar module to a safer landing spot. The last lunar gravity map was compiled by Konopliv in 2001. Here, *Okina* and *Ouna* measured the lunar gravity field. Konopliv will have to start all over again, for *Ouna* and *Okina* between them revised the map. The Sea of Serenity, 880 km across, is a strong positive field, while the 550-km-wide Apollo basin is a negative field. Mascon patterns are quite different on the far side compared to the near side. Analysis suggests this is something to do with the age of formation of these parts of the Moon, the uplift of the mantle and cooling time of the lunar rock.

Kaguya completed its mission on 11th June. Over the spring, its orbit was lowered to 50 km, then to 30 km and then to 10 km, so as to better measure the magnetic field. A final 140-sec burn brought it on a gentle 10° descent, crashing near crater Gill at 80.4°E, 65.5°S. Although the mission was long in the making, patience was rewarded by its extraordinary success and scientific haul.

With Selene 1 (*Kaguya*) safely in lunar orbit, Japanese engineers and scientists could speak more confidently of Selene 2 in around 2015. The purpose of Selene 2 was to drop off a communications relay orbiter, make a soft landing and deploy a small rover. This would be followed much later by Selene X, designed for a multitude of possible missions, ranging from a seismic network and astronomical observatory to sample return and the landing of demonstration structures to pave the way for a manned base. Table 2.1 summarizes Japan's later scientific missions.

Table 2.1. Japan's later scientific missions (after 1980)

20 Feb. 1981	*Hinotori*	*Mu-3S*	Uchinoura
20 Feb. 1983	Astro B/*Tenma*	*Mu-3S*	Uchinoura
14 Feb. 1984	Exo C/*Ohzora*	*Mu-3S*	Uchinoura
8 Jan. 1985	MS-T5/*Sakigake*	*Mu-3S*	Uchinoura
19 Aug. 1985	Planet A/*Suisei*	*Mu-3S*	Uchinoura
5 Feb. 1987	Astro C/*Ginja*	*Mu-3S*	Uchinoura
21 Feb. 1989	Exos D/*Akebono*	*Mu-2SII*	Uchinoura
24 Jan. 1990	Muses A: *Hiten* *Hagoromo*	*Mu-3SII*	Uchinoura
30 Aug. 1991	Solar A/*Yohkoh*	*Mu-3SII*	Uchinoura
20 Feb. 1993	Astro D/*Asuka*	*Mu-3SII*	Uchinoura
12 Feb. 1997	Muses B/*Haruka*	*Mu-5*	Uchinoura
4 Jul. 1998	Planet B/*Nozomi*	*Mu-5*	Uchinoura
9 May 2003	Muses C/*Hayabusa*	*Mu-5*	Uchinoura
19 Jul. 2005	Astro E/*Suzaki*	*Mu-5*	Uchinoura
21 Feb. 2006	Astro F/*Akari* Cute 1.7	*Mu-5*	Uchinoura
21 Sep. 2006	Solar B/*Hinode* Hitsat	*Mu-5*	Uchinoura
14 Sep. 2007	Selene/*Kaguya*	H-IIA	Tanegashima

H-II ROCKET: "MOST ADVANCED OF ITS KIND"

In the 1970s and 1980s, using American-based rockets, the N-I, N-II and H-I, the Japanese laid the groundwork for a modern space program. The 1990s saw the introduction by NASDA of a powerful domestic-built rocket, the H-II, and significant advances in technology, engineering and applications.

The H-II was, in effect, NASDA's declaration of launching independence, for the H-II used almost entirely indigenously produced technology, unlike the bought-in designs of the N-I, N-II and H-I. Approval was given for the project in August 1986. The H-II could thus legitimately be offered on the world launcher market without breaching licensing arrangements. To do this, a complex structure was built up in which NASDA funded the H-II's development, but the rocket was built and marketed by the newly created Rocket Systems Corporation for a fee, the company then selling it abroad if possible, or, in the domestic market, selling it back to NASDA for use on a mission-by-mission basis. Rocket Systems Corporation included many staff formerly with NASDA.

It was also madly expensive – one of the consequences of the Japanese drive for quality assurance and elaborate testing. The engineers did not want the H-II to be the first Japanese rocket to explode and it was not, but they were later criticized for their over-indulgence in using expensive materials and multiple backup systems. The Japanese rocketeers also suffered from the smaller size of the Japanese market, which meant that economies of scale available in the United States were not possible in Japan, so it was unlikely to reach such a production run as to make it profitable.

H-II in test

The H-II had an unusual design feature: both the first and second stages were fuelled by liquid hydrogen. To improve reliability, the same engine was used on both stages, the LE-7, though called the LE-5 on the upper stage, a powerful engine of 1,100 kN thrust. The LE-5 could be turned off and on – a considerable advantage in getting payloads into the right orbit – and could be tilted for pitch and yaw. But the big difference was the use of a liquid-hydrogen first stage with the new LE-7 high-thrust, high-temperature high-performance engine able to generate nearly 100 tonnes of thrust, in design since 1987.

The building of the cryogenic fuel tanks posed special engineering challenges. The tanks were built by Mitsubishi with the assistance of the Sciaky Corporation of Chicago and used the same alloy as that employed for the American Space Shuttle, aluminum 2219. Temperatures were kept down by foam–resin insulation. Additional thrust was provided by two large solid rocket boosters, like those of the American Titan IV and the Space Shuttle. The nozzles on the solid rocket boosters could be swiveled 5°. The H-II was imposing and the authoritative American journal *Aviation Week & Space Technology* described it as "the most advanced expendable launch vehicle sitting on the pad today in terms of its integration of modern materials, electronics, computers and propulsion". When the Americans came to construct

H-II rockets in preparation

their Delta 3 in the mid-1990s, they used fuel tanks made in Japan: technology transfer between Japan and the United States was now flowing the other way. With the H-II, Japan was able to send 10 tonnes into low Earth orbit or 4 tonnes to geostationary transfer orbit. The latter figure was better than China's Long March 2E (3.4 tonnes), the American Delta 3 or Atlas 2AS (3.8 tonnes), but inferior to the European Ariane 4 or 5 (4.5 and 6.8 tonnes, respectively) or Russia's Proton (4.8 tonnes).

For all its innovation, the H-II had an unhappy development history. The engine proved problematical, a series of fires, explosions, welding problems and mechanical failures putting the program two years behind schedule. In 1989, a bad mixture of liquid oxygen and hydrogen caused an engine fire that destroyed the test stand for the new LE-7 engine. A fire then broke out during one of the early test firings of the engine in July 1990, melting piping and pumps. In April 1991, four test firings of the first-stage engine failed. On the last one, the engine shut down 14 sec into a 350-sec firing, having failed to reach adequate combustion pressure. On 9th August 1991, an engine valve burst, causing an explosion and the death of a technician. The main problem appeared to be the turbopumps, whose job was to supply high-pressure propellants to the combustion chamber. In 1992, the LE-7 failed 5 sec into a major engineering test and caught fire because of a fuel leak. Then, in June 1992, during a 10-sec firing of the LE-7, fire broke out 5 sec after ignition, when a weld broke and this permitted liquid hydrogen to escape. Although the fire was extinguished in a minute, it was so intense that the engine and turbopump broke away from its mounting to fall 23 m into the water cooling pond! The following year, the LE-7 aborted a test firing on the stand 132 sec into a 350-sec run, not operating smoothly until the summer.

Typical of Japanese thoroughness was the use of a sounding rocket to test elements of the H-II design. The TR-1 series was a sounding rocket designed to pave the way for the H-II rocket. The TR-1 was a quarter-scale model of the H-II, though it had to be equipped with large rocket fins that were unnecessary for the H-II itself. These rockets were 14.3 m long, 1.1 m in diameter and weighed 11.9 tonnes. The purpose of the launchings was to obtain important data on rocket stresses, pressures, heating and exhaust plumes, as well as to test the mechanisms for the release of the strap-on boosters. The first TR-1 test was made on 6th September 1988. A second mission, using dummy solid rocket booster strap-ons to simulate the real ones, was launched 27th January 1989 and recovered from the sea several hours later. The program concluded with a third launch on 20th August 1990. These rehearsals provided valuable testing information – but probably pushed costs up.

SHOOTING STAR

The H-II eventually went for its first real test on 4th February 1994. As if to lay to rest the ghost of its unhappy history behind, H-II rose smoothly from the ultramodern Yoshinuba complex of the Tanegashima launch center on an ambitious space experiment. The twin solid rocket boosters dropped off 1 min 34 sec into the

H-II dawn lift-off

mission at an altitude of 37 km. The first-stage engines burned out at 6 min when the H-II was 227 km high and 630 km downrange, by which stage the vehicle was nearing Christmas Island. The second stage burned for 5 min. Executive director Tomihumi Godai was still in the control room when a congratulatory phone call came through from Prime Minister Mohiro Hosokawa. The H-II placed in orbit a 2.4-tonne prosaically termed "vehicle evaluation payload", more poetically *Myojo*. Its role was to monitor the rocket's performance, acceleration and vibration and circle the Earth in a high-altitude trajectory for four days. *Myojo* was a box-shaped object carrying 1.5 tonnes of water, which it pressurized, depressurized and repressurized in a series of tests. Also carried was a spaceplane experiment, OREX (see *Spaceplanes*, Chapter 3).

For Japan, the successful launch was a great moment. "First large launch vehicle using only domestic technologies," trumpeted the media. The H-II rivaled the best of what the Americans, the Russians and the Europeans could offer and it had all been developed at home this time.

The second H-II launching was a disappointment. The payload was the sixth engineering technology satellite, ETS VI or *Kiku 6*, which was intended to test lasers (the first such Japanese tests), advanced telecommunications systems and new propulsion methods, paving the way for data relay satellites and operational ion engines. The ion engine was rated at 23 mN and was designed to work for 6,500 hr. On the first launch attempt, the rocket engines failed to fire. When the H-II did get off the pad on 28th August 1994, the ¥52.75bn (€471m) 1.6-tonne satellite was left in an initial 236–36,338-km orbit instead of a geosynchronous one. The apogee burn had generated only 10% of the thrust needed. Despite the setback, *Kiku 6* was able to test its ion engine, new nickel hydrogen batteries and electro-thermal-hydrazine thrust. Intersatellite communications systems were tested out. The ion engine reached the designed level of thrust (23 mN). The remaining thrust was used to raise the perigee to 8,560 km by the end of November. Because the orbit passed through the Van Allen radiation belts, the solar cells and electrical equipment degraded quite rapidly. Making a virtue of necessity, the opportunity was used to relay information on electrons and protons in the radiation belts. The mission ended in January 1996.

H-II BRINGS IN ERA OF ILL-LUCK AND UNCERTAINTY

The *Kiku 6* failure began a run of ill-luck that dogged the Japanese space program for the rest of the 1990s. Two years later, the HYFLEX mini-shuttle was lost at sea after a suborbital mission (though on a different launcher). The next year saw the loss of the ADEOS 1 satellite and, in 1998, the placing of another, COMETS, in a virtually useless orbit. These four losses were valued at €471m, €42m, €943m and €353m, respectively (total: €1.809bn (¥202.6bn)). In 1997, the Science and Technology Agency publicly criticized the use of large, expensive satellites where a single fault could jeopardize a multi-billion yen project, arguing instead for smaller, cheaper, less ambitious projects in which a single failure would prove less costly. In a program in which quality control was emphasized throughout, this was an alarming series of events and the government established a high-level panel to examine the problems. Formed of eight Japanese and eight foreigners, the panel was headed up by Jacques-Louis Lions, president of the French Academy of Sciences, and Tokyo University professor, Jiro Kondo.

The panel reported in 1999 and it was unable to find any common thread in these failures (a conclusion common to similar investigations of a run of space disasters in other countries at the same time). The panel had many positive comments to make on NASDA's programs, such as the performance of the H-II and the advanced technology used on Japanese communications and meteorological satellites. This had been achieved with limited resources. The panel noted that although NASDA's annual spending had risen from ¥50bn in the 1970s to ¥100bn in the 1980s and ¥200bn in the 1990s, it had declined as a proportion of Japan's gross domestic product from 0.04 to 0.035%. Staffing had not increased since the 1980s, even though the responsibilities and commitments of the agency had grown. NASDA's budget was less than half that of the Europe's and less than a tenth of that of NASA. Much had been achieved with little.

Critically, the panel found poor lines of management communication, a lack of clear responsibilities for project management, the need for more technical resources in some projects and the need for improved systems of quality assurance. It criticized

NASDA for taking on too many projects, for failing to concentrate its resources and for being the junior partner in too many international projects. The program was failing to generate commercial benefits at home: although Japan had led many of the new communications technologies, Japanese companies still preferred to buy in foreign communications satellites.

All this came at a bad time. The financial crisis which hit the Far East in the mid-1990s took its toll on the Japanese space program. In 1997, the Space Activities Commission agreed, following pressure from government, to cut space spending by 14%, through schedule slippages, the importation of foreign technology and the redesign of several programs, like the H-II and the planned spaceplane.

To cap the tail of woe, another H-II failed, veering off course on 15th November 1999, and had to be destroyed – Japan's first launch failure since 1970. MTS was a project of the Ministry of Transport and was the first major satellite project conducted by a body other than ISAS or NASDA. Built by Loral, MTS stood for Multifunctional Transport Satellite. It was a box-shaped 4 by 4-m, 1,600-kg satellite designed to provide weather observation and air traffic control, navigation and communications from 24-hr orbit. MTS was to replace the last of the GMS series, *Himawari 5*. MTS had four infrared sensors (one with a 1-km resolution) and one in the visible channel to provide weather data on cloud height and distribution, wind conditions and temperatures. The ground-breaking aspect of MTS was that it was intended to anticipate the continued rapid growth in air travel over the north Pacific region by ensuring a much higher level of air traffic control safety through improved communications and navigation. Its 1999 launch was postponed by a series of problems in the H-II rocket, such as a leak detector failure and electronic malfunctions.

When it did eventually take off, the first stage cut out at 3 min 59 sec when the LE-7 engine shut down. Staging went ahead anyway at 5 min 30 sec, half a minute earlier than scheduled. The rocket began to veer off course, telemetry was lost at 7 min 35 sec and the destruct command was reluctantly sent in at 7 min 41 sec. Wreckage came down 380 km north-west of the Ogasawa islands, where the Japanese dispatched a research vessel, the *Karei*, with a submersible, the *Kaiho*, which retrieved remains from 3,000 m down.

Table 2.2. H-II launches

4 Feb. 1994	OREX/*Ryusei*	H-II	Tanegashima
	EP/*Myojo*		
28 Aug. 1994	ETS VI/*Kiku 6*	H-II	Tanegashima
18 Mar. 1995	SFU	H-II	Tanegashima
	Himawari 5		
17 Aug. 1996	ADEOS/*Midori 1*	H-II	Tanegashima
	Fuji 3		
28 Nov. 1997	ETS VII/*Kiku 7*	H-II	Tanegashima
	TRMM		
21 Feb. 1998	COMETS/*Kakehashi*	H-II	Tanegashima

H-IIA first-stage engine

AUGMENTED: H-IIA

The H-II, although a great technological achievement, was much more expensive than ever anticipated. When the H-II first flew, the Japanese were confident that satellite companies would pay the extra money involved in exchange for its reliability. In the end, the reliability was not achieved either. Japan entered the H-II in a number of international commercial competitions for satellite launches in the early to mid-1990s, only to be repeatedly beaten by Europe's Ariane and the Chinese Long March. The financial pressure became even more acute when Russia's Proton entered the world launcher market, with early success. New American launchers arrived. The Japanese had to accept that with the H-II, their attempt to compete head to head with the other countries had not come off. Table 2.2 summarizes the H-II launches.

It was decided to phase out the H-II in the late 1990s and replace it with a more reliable, modernized, simpler, economical version, the H-IIA, "A" standing for "augmented". Each H-II launch had cost around €188m, twice that of an equivalent launch by an Atlas or Ariane and it was NASDA's objective to bring the cost of the H-IIA launch down to between €94m and €141m (¥10.5bn to ¥15.7bn), though a more ambitious figure of €80m (¥8.9bn) was later set. Japan had the ambition that the H-IIA gain up to 17% of the world launcher market by 2003. There was high confidence that the H-IIA would succeed and a first production run of 23 rockets was ordered.

To achieve these competitive economies, NASDA decided, reluctantly, to use imported technologies, with American-based solid rocket boosters, able to offer 10% more power. The redesign involved a new engine, the LE-7A, a very much simplified version of the H-II's LE-7, with new cryogenic tanks, redesigned plumbing, pre-burner chamber and turbopump systems, able to offer 13% more power. Other economies were made: the weight of the second-stage tank was reduced, copper tubing replaced stainless steel tubing, the amount of engine welding was reduced and production time for the engines was halved. New solid-rocket motors were selected, supplied by the American Thiokol corporation, the company that built the boosters for the Space Shuttle. Typical of the economies made by the H-IIA, the solid rockets were shorter and fatter than those of the H-II, comprising a single segment rather than four, all of which had previously to be stacked separately – an expensive and time-consuming process. There were lighter electronics, a reduced number of engine parts and fewer welds, potentially a point of weakness. There was a customized nose fairing for different payloads so as to improve performance. The interstage was 200 kg lighter, being made of carbon composites and with 140 parts, rather than 1,200 on the H-II. Overall, the H-IIA had 20% fewer parts than the H-II, but that was still 200,000 parts. Five versions of the H-IIA were originally announced, all with different combinations of solid-fuel rockets on the side, so as to achieve maximum flexibility for the rush of expected customers.

The development program for the new rocket was set to cost ¥77bn (€696m). The first new solid motors for the H-IIA were successfully tested in 1998. Development of the new LE-7A first-stage engine proved to be problematical – principally the wiring systems and nozzle vibration. The third test failed due to a hairline crack, while another crack caused the fifth test to be called off 3 sec into a 350-sec firing. There was a further premature shutdown in a June 1999 test. Eventually, long-duration engine tests were successfully completed in September 1999, putting the program back on course. Tests of the new second-stage engine were more encouraging. The LE-5B was designed to achieve 14,000 kg of thrust compared to 12,500 kg for the LE-5, but to be lighter at the same time. Thirteen successful firings were carried out by Mitsubishi at its Akita, northern Japan test stand in 1996, including continuous firings of more than 300 sec. In order to prepare for its introduction on the H-IIA, it was decided to give the LE-5B a live test on the H-II on the MTS launch in late 1999, leading to the launch being renamed "the half H-IIA".

The completion of the H-IIA came not a moment too soon, for 1999 saw the loss of the last H-II with MTS and, the following year, the failure of the hitherto unblemished *Mu-5* with Astro E. But when the H-IIA was rolled out, it was an impressive sight and, hopefully, a worthy rival for the Ariane, Proton and Delta.

Japan returned to space with the H-IIA on 29th August 2001, the 285-tonne rocket standing 53 m tall on its Tanegshima pad. The brown and silver rocket rose on the twin pillars of fire of the solid rocket boosters, soared over the Pacific and placed in orbit a laser test payload. The second flight, also a test flight, took place on 4th February 2002, when the H-IIA put three test payloads into orbit, including *Tsubasa*, meaning "wings", to test the durability of commercially viable semi-

Testing the LE-7A

USERS

conductors. One payload, an atmospheric re-entry test, failed to deploy, but this was not a launcher problem.

These two tests cleared the way for operational missions. The first took place on 10th September 2002, going flawlessly and placing in orbit the USERS free-flier and the DRTS relay satellite. The USERS (Unmanned Space Experiment Recovery System) free-flier was a ¥31.698bn (€283m) project to orbit an 800-kg box-and-wings service module carrying technological experiments and recover a 900-kg re-entry module. The purpose was to grow crystals for six months in a Superconductor Gradient Heating Furnace. USERS entered orbit of 440–455 km, 30.4°. After a 255-day mission, the results from the furnaces were transferred to the 1.6-m-diameter cone-shaped re-entry module. This then propelled itself away and, on 25th May 2003, fired a solid-fuel re-entry rocket. A storm delayed recovery and it was not picked up by the recovery ship until 30th May. This was a further advance for Japan, the first time it had recovered a satellite from orbit. The main module continued in orbit on a three-year mission.

The second operational mission was on 14th December, when the H-IIA orbited three payloads: ADEOS 2, the Whale Ecology Observation Satellite (WEOS) and Fedsat, for Australia. This was the first Australian satellite since 1967, when an American Redstrone rocket had put a small Australian satellite into orbit from Woomera. As for WEOS, this was a ¥42m (€377,000) project devised by Dr Takeshi Ori of the NEC Corporation and Dr Tomonao Hayashi of the Chiba Institute in 1995, the idea being to use satellites to monitor the migration of the blue whale, the largest and most endangered species of whale. The whale would be harpooned and attached to the harpoon would be a 10-kg float that would have a satellite transmitter sending reports beamed up to a 50-kg WEOS. Electric power was to come from a system developed by Seiko watches that would recharge the battery in the float from the whale's swimming motion. The satellite was to pick up data from a number of blue whales, telling of their movement, the water temperature, the periods of time spent by the whale on and under water, sending the data down in a daily communication pass.

H-IIA LOSS: BACK TO THE DRAWING BOARD

These successes gave confidence that the changes since the H-II had been successful. In a major setback, the H-IIA was lost on its sixth mission, on 29th November 2003. The H-IIA was attempting to put into orbit a pair of optical and radar observation satellites, the fifth mission having put into orbit the first pair. The problem began at 62 sec, when one of the two large solid rocket boosters generated excess temperature and then lost thrust. This, of itself, was not fatal, but the flames burnt through the electric cabling that sent the command to separate the solid rocket from the main booster. At 105 sec, the other large solid-fuel rocket separated, on schedule, but the faulty booster was still attached, now providing 10,000 kg-worth of asymmetrical thrust. Downward-facing cameras had been fitted on the upper stage, so all this could be clearly seen. The rocket veered off course and it would never have sufficient thrust to reach orbit. At 11 min, the destruct command was sent, with the rogue rocket then at an altitude of 422 km. It turned out that the nozzle of the solid rocket booster had burned through. Wreckage from the H-IIA came down in the Philippines and a robot was sent down to retrieve the payloads before someone else did.

This was a big disappointment. With the H-IIA, the Japanese had thought they had put behind them the problems of design and quality control that had plagued the H-II, but this was not the case. The loss of the two surveillance satellites was a setback, for the existing two required companions to form an effective constellation. As was the habit with the Japanese, the setback did not come on its own, coinciding with the loss of *Nozomi*.

The H-IIA was grounded for well over a year. Over the following 14 months, 80 changes were made. The Japanese introduced a two-stage re-qualification campaign: a first launch (flight 7), followed by an intentionally demanding two launches 25 days apart (flights 8 and 9), with flight 9 carrying an exceptionally heavy payload. After a gap of over a year, the H-IIA returned to flight with MTS-1R in February 2005, renamed *Himawari 6* once in orbit.

The two-launches-in-25-days campaign put a considerable level of stress on Tanegashsima, for the intention was to improve on the three-month launch campaign of the H-II. Flight 8 was launched on 24th January 2006, a calm, cloudy day, with its Earth resources satellites, ALOS (*Daichi*), separating 16 min 6 sec later. Flight 9 was MTS 2, which, at 4.6 tonnes, was at the extreme end of the capability of the H-IIA. MTS (Multifunctional Transport Satellite) was dual use, being designed for weather forecasting (it carried a five-channel imager) and air traffic control, for power needing a 30-m-long solar panel. The payload arrived in a huge wide-load convoy during the night and was stacked on top of the H-IIA in the hangar, being blessed for good luck by a shinto priest. The heaviness of the payload meant that four strap-ons must be used. Before the launch, all the workers and managers gathered together for speeches to focus their attention on the forthcoming task. A large press assembled all along the seashore, with bright red hard hats, in case the worst were to happen. They need not have worried. At T–270 sec, the count went automatic and at T–30 sec, onto internal power. Like its predecessor, flight 9 disappeared into cloud, but cameras on the rocket itself recorded the ascent, the blue Earth receding in the distance and the strap-ons tumbling away. The fairings came off at 3 min 53 sec and second-stage ignition at 7 min. At 28 min, MTS 2 was safely in orbit and there were cheers in the control room, though the overwhelming sensation was one of relief. MTS 2 was the 98th Japanese satellite and the 6,381st in the world.

The re-qualification of the H-IIA marked a major milestone in the Japanese launcher program. Later, it was joined by the H-IIB, whose main objective was to launch the HTV cargo craft to the International Space Station, but whose extra power will make possible the launch of two satellites at a time. The H-IIB clusters four solid-fuel rocket engines at the bottom, rather than the normal two of the H-IIA. The first stage is fatter, 5.2 m in diameter, compared to 4 m on the H-IIA, and is 1 m longer. As a result, its lifting potential is much higher, enabling it to bring 16 tonnes to low Earth orbit and the total amount of fuel carried is 70% more. Development costs were kept down by using existing H-IIA technologies and design. Table 2.3 illustrates the progress, over the years, of Japanese launchers, with Table 2.4 describing the most recent in the series, the H-IIA.

Table 2.3. Japanese launcher evolution

a. Solid-fuel rockets (ISAS)

	Stages	Length (m)	Launch weight (kg)	Payload (Earth orbit, kg)
Lambda 4S	4	16.52	40,000	40
Mu-4S	4	23.6	43,600	75
Mu-3C	3	20.2	43,600	100
Mu-3H	3	23.8	48,700	200
Mu-3S	3	23.8	48,700	300
Mu-3SII	3	27.8	61,000	770
Mu-5	3	24	140,200	1,800

b. Liquid-fuel rockets (NASDA)

	Stages	Length (m)	Launch weight (kg)	Payload (24-hr orbit, kg)
N-I	2	32.57	90,260	130
N-II	2	35.35	135,000	350
H-I	2	40.3	139,900	550
H-II	2	49	266,000	4,000
H-IIA	2	53	285,000	4,000
H-IIB	2	56	551,000	8,000
J-1	3	24.6	91,174	1,152 (Earth orbit)

Table 2.4. H-IIA launches

29 Aug. 2001	Laser test payload	H-IIA	Tanegashima
4 Feb. 2002	*Tsubasa*	H-IIA	Tanegashima
10 Sep. 2002	DRTS/*Kodama* USERS	H-IIA	Tanegashima
14 Dec. 2002	ADEOS 2/*Midori 2*	H-IIA	Tanegashima
28 Mar. 2003	Optical 1, Radar 1	H-IIA	Tanegashima
26 Feb. 2005	MTS 1R/*Himarawi 6*	H-IIA	Tanegashima
24 Jan. 2006	ALOS/*Daichi*	H-IIA	Tanegashima
18 Feb. 2006	MTS 2	H-IIA	Tanegashima
11 Sep. 2006	Optical 2	H-IIA	Tanegashima
18 Dec. 2006	ETS VIII/*Kiku 8*	H-IIA	Tanegashima
24 Feb. 2007	Optical 3 Radar 2	H-IIA	Tanegashima
14 Sep. 2007	Selene/*Kaguya*	H-IIA	Tanegashima
23 Feb. 2008	WINDS/*Kizuna*	H-IIA	Tanegashima

EARTH AND MARINE OBSERVATIONS: *MOMO*

Japan's first satellites were for scientific (ISAS), communications and engineering purposes (NASDA). During the 1960s, an awareness grew of the potential of space observation platforms to watch the seas and land masses for remote sensing. Satellites could be used to make maps, spot pollution, find fish, assess crops and study water resources. Application of imaging data from satellites could be used in thousands of ways to assist agricultural, marine and economic development. The United States had launched the world's first remote sensing satellite, *Landsat*, in 1972.

Responding to these developments, in July 1975, Japan established a Remote Sensing Technology Center, RESTEC, to lead the country's endeavors in Earth resources work. Three years later, Japan built an Earth Observation Center at Hatoyama, Saitama, to receive *Landsat* data and started to take these data from 1979. In 1978, the Space Activities Commission took the decision that Japan should develop its own land and sea observation satellites and move away from dependence on American data. Preliminary designs were carried out in 1979–1980 and the final design was settled in 1981.

The first Japanese remote sensing satellite was the Marine Observation Satellite, MOS, also called MOS 1A, built by NEC. Weighing 750 kg, it was designed to carry four instruments to study the surface of the Earth's oceans. Once in orbit, it was renamed *Momo*, or "peach blossom". It was launched into 903–917-km orbit, 99.1°, period 103 min on 19th February 1987 on the last N-II rocket.

MOS 1A was a box-shaped satellite with one solar panel. It carried a 70-kg multi-spectral electronic self-scanning radiometer using a charge coupled device, a 54-kg microwave scanning radiometer to observe temperature and water vapor and a 25-kg

Momo

visible and infrared thermal radiometer. The scanning radiometer could compile a colored sea map that can indicate pollution in fish-rich zones. *Momo* was able to measure atmospheric water vapor, ice floes, plankton, ocean currents and sea temperature. A gas jet system was devised to maintain station-keeping on orbit. A data collection system was installed in order to collect information from automatic monitoring systems and relay them back to ground control. As it crossed the oceans, its multi-spectrum radiometer covered a swath of 100 km, its microwave scanning radiometer 317 km and the visible and infrared radiometer 1,500 km. This orbit was one which made 14 revolutions a day, repeating the ground track every 17 days – perfect for observing changes below.

Tracking was done by Tsukuba Ground Center, with support from other tracking stations at Katsura, Masuda and Okinawa. Under cooperative agreements, MOS data were received by Thailand (Bangkok), the National Institute of Polar Research in Showa Base, Alice Springs in Australia and the Canadian Center for Remote Sensing (Prince Albert and Gatineau).

The satellite worked for two-and-a-half years, until June 1989. It sent back outstanding pictures of tropical storms. Using false color imaging, the swirl of a typhoon could be made out in dark-blue colors, surrounded by red sea, the red color being due to the warm temperatures from which typhoons sucked up their moisture. Right in the center of the typhoon could be seen a small, 4 km-wide red spot – the eye of the hurricane. Other pictures from *Momo* showed the snow on the top of mountains, floating ice around the northern Japanese islands, a volcanic plume streaming from Mount Fuji. Photographs of the Antarctic ice sheet were relayed to the Japanese observation party at Showa base there.

Three years later, the second marine observation satellite, MOS 1B or *Momo 1B*, was launched by H-I from Tanegashima on 7th February 1990, rising on a pillar of smoke into Pacific clouds as waves lapped the island launch pad. The satellite was actually the backup model for MOS 1A. Also launched into virtually identical orbit with MOS were *Orizuru* (meaning "beginning"), a deployable boom and umbrella test, and *Fuji 2*, an amateur radio satellite. The French ground station at Kourou on the South American coast picked up the new 780-kg *Momo* on its first pass and, not long afterward, so did Japanese stations at Katsura, Masuda and Okinawa. Within an hour, the 5.2-m-long solar panel had unfurled and the satellite began a 60-day period of checkout. MOS 1B carried a dish-shaped microwave scanning radiometer, an X-band antenna and a tube-shaped multi-spectrum electronic self-scanning radiometer. Its instruments were able to identify red tide (jellyfish infestations), the distribution of snow and ice and volcanic ash. Detailed color maps of sea temperatures were compiled for the seas around the Japanese islands, blue for cold currents and red for warm sea. MOS 1B continued to operate until 1996, when it was closed down after a battery failure.

The 50-kg *Orizuru* was an unusual test. The concept was to test out, in miniature, the deployment of free-flying microgravity platforms from orbital stations. These would park some distance from the station, carry out experiments in zero gravity, unperturbed by Space Station operations, and then return. A special purpose of the test was to verify whether umbrella devices could be used, combined with atmospheric drag, to maneuver back to an orbital station. In the course of a week, *Orizuru* deployed its boom 34 times and its umbrella 52 times in what was apparently a successful test.

JERS *FUYO*: INTRODUCTION OF SPACE-BORNE RADAR

The next step after MOS was a land observation Earth resources satellite. This was JERS, or the Japanese Earth Resources Satellite, the first radar-based Earth-resources satellite. Ninety-six percent of the satellite was built domestically.

Weighing 1.4 tonnes, it was equipped with a 12 by 2.5-m Synthetic Aperture Radar (SAR) and multi-spectral imaging radiometers. Because of the technical complexities involved, the design period was unusually long – 12 years. It was launched successfully on 11th February 1992 from Tanegashima, the H-I heading due south toward a 568-km orbit, repeating its ground path every 44 days. Separation took place over Argentina 50 min after take-off and the satellite was duly renamed *Fuyo*.

Problems arose when the SAR failed to deploy because one of six pins holding the antenna in place stuck. It seems that extreme cold may have contributed to the problem, because when the satellite was pointed at the Sun several weeks later, the pin suddenly popped open. Even when it did work, radar images were spoilt by stripes appearing on the pictures – a problem that was eventually overcome by programming them out during processing. When this was done, ground controllers were able to get razor-sharp images of Mount Fuji volcano and the Japanese islands. Later, its instruments were used for land surveys, studies of fisheries and agriculture, natural resources work and disaster prevention. *Fuyo* was able to track forest fires in Mongolia and the crustal movement of the Earth near Iwate volcano.

The main receiving center for *Fuyo* information was the Earth Observation Center at Hatoyama, the center being responsible for processing, distribution and building an archive. The center had three 11.5-m dishes to receive signals. *Fuyo* relayed stored data both to Hatoyama and the University of Alaska in Fairbanks and to a further 10 Earth stations equipped to receive real-time data – Kumamoto in Japan, Bangkok, Showa Base in Antarctica, Fucino in Italy, Kiruna in Sweden, Maspolamos in Spain, Tromso in Norway, Gatineau and Prince Albert in Canada and Beijing in China. By the late 1990s, the center had built up considerable expertise through its handling of data from Japan's MOS 1 and 1B, JERS and ADEOS, Europe's ERS, the United States' Landsat 5 and France's SPOT.

Fuyo was shut down in 1998, having far exceeded its design lifetime. By this stage, ground stations had received 90,000 photographs and 140,000 radar images. The batteries started to malfunction in October and then two of the three gyros switched themselves off. It was unable to acquire the Sun and, starved of electrical power, the electricity system failed. It was later formally switched off and it burned up in December 2001.

ADEOS/*MIDORI*: ATMOSPHERE OBSERVER

ADEOS (Advanced Earth Orbiting Observation Satellite) was the third Japanese remote sensing satellite, following *Momo* and *Fuyo*. ADEOS was a 3,560-kg gold-foil-covered platform designed to note global changes in the Earth's surface, atmosphere and oceans, in particular, the ozone layer, the tropical rainforest, carbon dioxide at the poles and the greenhouse effect. ADEOS carried eight sensors, of which five were developed by Japan, two by NASA and one by the French space agency, CNES. It was an expensive satellite, costing ¥112bn (€1bn), but sizeable, measuring 8 m tall and 4 m wide, and one of the heaviest launched by Japan. A small 50-kg amateur radio satellite developed by Nippon Electric and the Japan

Midori

Amateur Radio League was launched piggyback on the mission by the H-II on 17th August 1996.

Renamed *Midori* ("green observer"), it circled the Earth at nearly 800 km, encountered a routine range of teething troubles but was declared operational in November 1996. Data began to flood into the NASDA Earth Observation Center in Hatoyama. The first set of pictures was superb and presented in a variety of formats (true color and false color). They identified concentrations of chlorophyll, tropical storms off Japan (typhoons Violet and Tom), the El Nino current stretching across the Pacific and the southern ozone hole. The Improved Limb Atmospheric Spectrometer noted the rise and fall of ozone concentrations and another meter recorded the distribution of greenhouse gases. In early 1997, *Midori* tracked the spread of spilt oil from a tanker off Japan, the pictures being put up on the internet.

Important data back came from the AVNIR, or Advanced Visible and Near Infrared Radiometer. Although designed for soil, vegetation, pollution and energy studies, the 8-m resolution instrument sent back outstandingly clear pictures of urban areas such as Hiroshima. Early pictures from the eight-band ocean color and temperature scanner provided images of plankton in the sea off the Japanese coast. ADEOS also carried a TOMS, one of the environmentally most important instruments in Earth observation. TOMS, or Total Ozone Mapping Spectrometer, was invented by NASA in the 1970s and, when flown on *Nimbus 7* in 1978, it found the ozone hole over the Antarctic. A second TOMS was flown on a Russian *Meteor 3* satellite and ADEOS was its third assignment. Within a month, ADEOS had sent back new and worrying images of both the Antarctic and Arctic ozone holes.

Contact with the spacecraft suddenly ended on 30th June 1997. Solar power from its wide 30-m wings was quickly lost, the batteries drained in four hours and it was declared abandoned the following month. Space débris was blamed at first, but a subsequent investigation found a more mundane, more human and more likely cause: a weld had broken at the base of the solar panel.

Midori was replaced by ADEOS 2, which entered orbit of 803–820 km, 98.7°, 101 min on 14th December 2002. Following tradition, it acquired the same name, *Midori 2*. This was a similarly large satellite, 3.7 tonnes in weight, with a solar panel able to deliver 6,200 W of power. Continuing the work begun by its predecessor, it also had a specific role to measure water vapor, sea temperatures and global warming. *Midori 2* carried an advanced microwave scanning radiometer with a wide swath (1,600 km), an optical sensor working in 36 spectral bands able to look sideways as well as down and a limb atmospheric spectrometer that can search for pollution (aerosols and ozone) in the atmosphere. As well as these Japanese instruments, overseas countries contributed equipment: NASA a sensor to measure sea winds and direction, the French space agency CNES an instrument called Polder to measure how the atmosphere reflects solar radiation. Sadly, *Midori 2* suffered a catastrophic failure on 25th October 2003. The satellite lost transmission following a malfunction in the single solar wing, possibly due to intense solar electromagnetic storms.

Midori 2 will eventually be replaced by Gosat (Greenhouse Gas Observing Satellite), a 1.65-tonne satellite designed to measure carbon gases, aerosols and methane in the atmosphere. Hitherto, greenhouse gases were measured at only a limited number of locations, but Gosat was designed to take measurements at up to 56,000 points uniformly across its 98° polar orbit every three days. In the early days of space exploration, names had been chosen by mission managers. Now, in a spirit of electronic democracy, people were invited to suggest names through an internet site and *Ibuki* ("breath") was selected.

Gosat

ALOS: DAY AND NIGHT, CLOUD-FREE

JERS (*Fuyo*) was replaced in January 2006 by ALOS, or Advanced Land Observation Satellite, renamed *Daichi* when it reached orbit, the Japanese word for "mother Earth". This ¥47bn (€350m) project was the largest Earth resources project devised in Japan, the satellite weighing 4,200 kg, one of Japan's heaviest and requiring 7 kW of electrical power, for which a single 22-m-long panel was used. It was intended to produce $\frac{1}{25,000}$ scale maps, observe the Earth, survey natural resources, and also have a role in mapping and disaster management, revisiting each site in the Japanese islands every two days. The spacecraft carried a stereo mapping remote sensing instrument using three telescopes: a 2.5-m resolution mapping camera (PRISM), a 240-kg visible and infrared radiometer with 10 m resolution (AVENIR) and a 475-kg phased array synthetic aperture radar (PALSAR), for day and night, cloud-free observations. From its 700-km-high Sun-synchronous orbit, the data relay system was designed to send back highly compressed information at 1.36 gigabytes a second.

At first, it was reported that *Daichi* images were too blurry for the government's Geographical Survey Institute. Later pictures were of good quality and showed the Japanese islands, ice in the Sea of Okhotsk, fresh volcanic ash on the slopes of an Indonesian volcano and enabled a new profile to be made of Japan's own Mount Fuji. In July 2007, *Daichi* imaged the subsidence and cave-in on the ocean floor resulting from an offshore Earthquake 60 km off Nigata. *Daichi* will be replaced in 2011. From 2008, *Daichi* was used to monitor and protect 10 United Nations Educational, Social and Cultural Organization (UNESCO) world heritage sites, such as Angkor Wat in Cambodia, the Sichuan giant panda sanctuaries in China, the ancient Maya city of Mexico and Manchi Picchu in Peru.

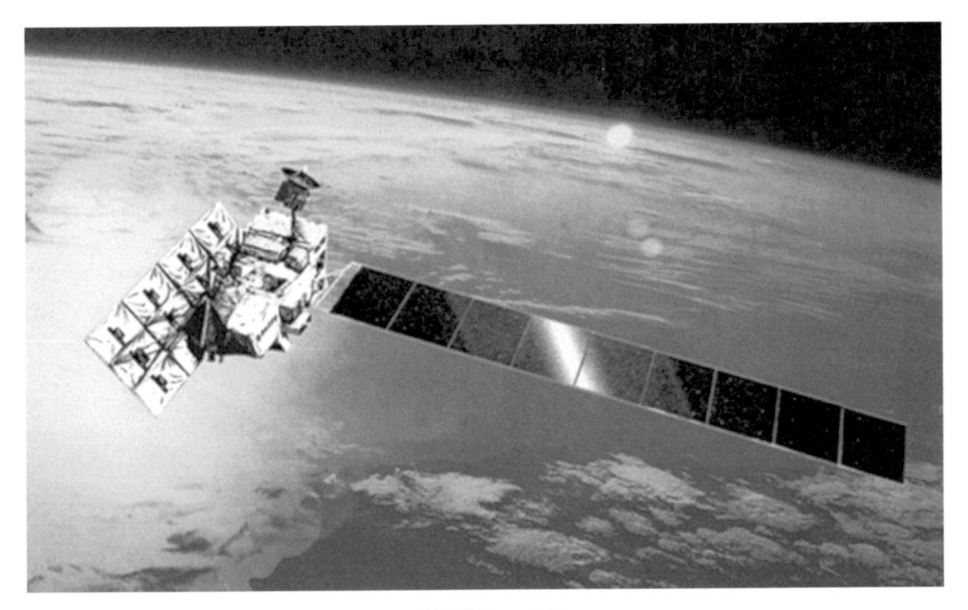

ALOS in orbit

ALOS profile Mount Fuji

TROPICAL RAINFALL

JERS *Fuyo* and ALOS *Daichi* concentrated on land observations, *Momo* on marine observations and *Midori* on the atmosphere. Now, Japan developed a more specialized observation satellite to respond to its particular geographic location with tropical sea masses to the south of the Japanese archipelago. When, on the morning of 28th November 1997, the sixth H-II blasted off from Tanegashima, one of its most important payloads was TRMM, released 14 min into the mission. So important was TRMM that it was the first to be launched outside the limited 90-day launch window imposed by local fishermen. Agreement to do so followed prolonged negotiations with the five fishermen's unions involved. The Science & Technology Agency already paid ¥400m (€3.57m) a year to the fishermen for harbor works and compensation for the 90-day window from Tanegashima and the extension of the window cost a further ¥300m (€2.67m) compensation. TRMM entered orbit of about 350 km, circular at 35°.

The 3.6-tonne TRMM or Tropical Rainfall Measuring Mission was a joint NASA/NASDA mission, Japan supplying the all-important rain radar. NASA paid €283m and NASDA €188m for TRMM. The purpose of the mission was to provide the first comprehensive picture of tropical rainfall, which comprised three-quarters of the world's rain and atmospheric energy. TRMM had an unusual shape of boxes,

TRMM

trusses and rectangular containers. ¥52bn (€471m) TRMM carried a microwave imager to peer through clouds, rain radar, high-resolution infrared scanner, lightning meter and a sensor to measure the Earth's energy. The rain radar could make a three-dimensional picture of rain over a swath width of 150 km with a resolution of 4 km. This was a giant instrument, the shape of a bee's honeycomb, twice the height of a person and three times the width. An early target was the El Nino current, which had made such an impact on the western Americas. A long-term function of the mission was to estimate climate change.

TRMM immediately proved its value in tracking typhoons and measuring their cloud height – information vital to an assessment of the danger they posed. The first images from TRMM were stunning – showing the band of tropical rain around the entire southern hemisphere, swirling cyclones (with heavy rainfall marked by false red colors) and three-dimensional cloud profiles showing where torrents of rain were falling down. Within 18 months, TRMM had improved global rainfall measurements by 25%. Scientists used TRMM to develop computer models of cloud and rainfall so as to make forecasting more accurate. Whereas most weather satellites saw a tropical storm as a white ball of cloud, TRMM's instruments could identify the eye wall, see how it was developing and the direction in which it was traveling.

The satellite was designed to operate for 18 months. The satellite was supposed to be sent into a controlled de-orbit in 2003, but, because it was still working perfectly and providing such useful data, NASA controllers re-boosted the orbit to buy extra time. By July 2004, TRMM's altitude had fallen back again and NASA announced that it was time to carry out a controlled de-orbit maneuver. There was a howl of protests from meteorologists, who argued that they should keep the satellite aloft and take their chances on an uncontrolled re-entry. Why kill off a perfectly good satellite when, for a small risk, you could get another two years of quality data? TRMM remained in orbit sufficiently longer for the Japanese space agency to publish an internet-based quasi-real time (4-hr delay) global precipitation map using data from TRMM and NASA's later AQUA satellite. It was updated hourly and could be directly accessed by any internet user in Asia or elsewhere to locate typhoons or any other form of heavy rain. Table 2.5 summarizes Japan's Earth observation satellites.

Table 2.5. Japanese Earth observation satellites

19 Feb. 1987	MOS 1A/*Momo 1A*	N-II	Tanegashima
7 Feb. 1990	MOS 1B/*Momo 1B*	H-I	Tanegashima
	Orizuru		
	Fuji 2		
11 Feb. 1992	JERS/*Fuyo*	H-I	Tanegashima
17 Aug. 1996	ADEOS/*Midori 1*	H-II	Tanegashima
	Fuji 3		
28 Nov. 1997	TRMM	H-II	Tanegashima
14 Dec. 2002	ADEOS 2/*Midori 2*	H-IIA	Tanegashima
24 Jan. 2006	ALOS/*Daichi*	H-IIA	Tanegashima

ENGINEERING SATELLITES

As well as observation satellites like TRMM, Japan developed engineering and test satellites in the ETS and other series. TRMM was launched as a companion to Engineering Test Satellite 7, ETS VII or *Kiku 7*, costing ¥33bn (€245m). This was a radically different mission – one designed to explore Earth orbiting rendezvous techniques that Japan would need to learn whenever it supplied the International Space Station. Seven maneuvers were planned.

ETS comprised two satellites – *Hikoboshi* (meaning Altair, the active satellite, weighing 2.5 tonnes) and *Orihime* (meaning Vega, the target, 400 kg), named after lovers in a Japanese fairytale. *Hikoboshi* was box-shaped with twin solar arrays and 220 kg of fuel; *Orihime* was much smaller, with one array. The program called for *Hikoboshi* to follow a number of rendezvous and docking profiles using ground control, radar and the global positioning system. After rendezvous, *Hikoboshi* was to test its remote manipulator arm for typical operations that would be carried out on the Japanese module of the International Space Station.

The mission got off to a bad start, for it was hit by a gyroscope failure that prevented it from achieving proper orientation and stability was lost for a worrying six hours. Then, attempts to connect ETS to the American tracking and data relay satellite system broke down. Despite this, Mission Control in Tsukuba went ahead and masterminded the first rendezvous and docking tests, which began on 7th July 1998 – by coincidence, the festival of Vega – when the two spacecraft separated to a distance of 2 m for 30 min. Using sensors and lasers to detect one another's position, the smaller satellite inched toward the mother craft, which then gripped it with three pincers 550 km above Earth.

On 6th August, *Hikoboshi* again separated from *Orihime* to a distance of 2 m and re-docked a few minutes later. Later that day, they separated to 525 m in preparation for more ambitious experiments. At this stage, attitude control on *Hikoboshi* failed and *Hikoboshi* retreated to a distance of 5 km while trouble-shooting took place. Two attempts to re-dock failed on the 8th and 9th, the satellites coming to within 110 m of one another. The satellites lost one another's position and seemed not to receive their instructions from Tsukuba. Again, on 13th August, the satellites came to within 145 m, when attitude controls failed. After two further unsuccessful attempts when thrusters gave problems and attitude controls failed, *Hikoboshi* eventually pulled off a successful re-docking on 27th August, the re-docked combination then being in an orbit of 542–544 km, 34.97°, 95.51 min. In doing so, *Hikoboshi* used up most of its maneuvering fuel. All this was televised as the Earth rolled below.

The docking was hailed by the Japanese and Western press as the first ever unmanned docking (which it was not, for the USSR had done this in 1967) and as an important step in testing out procedures for sending Japanese spacecraft to future space stations (a justifiable claim). Although the maneuvers were full of difficulties, such as thruster anomalies and attitude failures, the purpose of the mission had been to trouble-shoot such problems. When they did occur, collision avoidance procedures were followed successfully and ground controllers were able to park

ETS VII rendezvous

the chaser while new procedures were worked out and fresh software loaded up to the spacecraft. The thrusters and attitude control systems of NASDA's planned relay satellites for the International Space Station were radically overhauled as a result. The engineers responsible were showered with rewards from the Society of Mechanical Engineers, the Society of Control Engineers and the Robotics Society of Japan.

This was far from the end of the *Hikoboshi/Orihime* experiment. The robot arm and hand on board, miniatures of those to be used on the International Space Station, were used to inject gas into a simulated tank, deploy and disassemble a truss structure, fasten bolts, connect electrical leads and capture floating objects. Communications were relayed through NASA's system of tracking and data relay satellites, far out in 24-hr orbit. In spring 1999, German Space Agency (DLR) engineers used virtual reality computers to make further tests of the arm and hand. In September 1999, *Hikoboshi* and *Orihime* again separated, though only to a 200-mm distance, and *Hikoboshi* was instructed to use its hand to recapture its companion without ground assistance. Despite the very small quantity of fuel remaining, and despite some propulsion anomalies, a further set of release-and-recapture tests were carried out on 27th October 1999. The two years of ETS VII experiments were vital in laying the groundwork for Japan's participation in the International Space Station.

ETS VIII: A GIANT, HOVERING INSECT

The next ETS VIII, or *Kiku 8* (December), was the eighth in a series of Engineering Test Satellites (ETS), a huge 6.5-tonne satellite, designed to test the development of

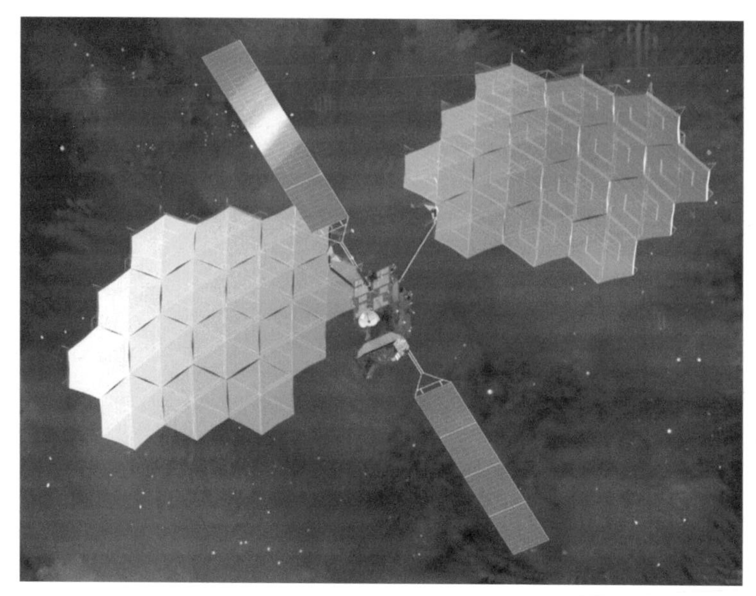

ETS VIII

high-speed, CD-quality satellite-based mobile telephone communications from 24-hr orbit and ultimately lead to the elimination of ground masts. Built by Toshiba and Mitsubishi, the ¥32bn (€292m) ETS VIII was intended to ensure that Japan remained at the cutting edge of new telecommunications technologies. ETS VIII had large deployable reflectors, 19 by 17 m, twice the size of the Muses B experiment, each the size of a tennis court, 40 m across, making it look like a giant hovering insect. Deploying the antenna was, of itself, a difficult and delicate undertaking involving the development of carbon fiber-reinforced struts and computer-controlled small electric motors to push the reflectors into shape, all tested out in zero-gravity flights in Airbus aircraft. Because of these difficulties, two tests were run in advance called LDREX (Large Deployable Reflector), the second succeeding on the European Ariane 5 launch of 13th October 2006.

ETS VIII was eventually launched on 18th December 2006, using the most powerful version of the H-IIA with four solid rocket boosters. The launch, the 11th for the H-IIA, had been delayed for two days due to coastal fog but, in the end, went off perfectly. Separation was confirmed at 27 min 35 sec and the Santiago tracking station picked up signals at 55 min, indicating solar wing deployment. All went well for a year, but, in January 2008, two of its four ion thrusters failed. The mission was able to continue functioning on two other ion engines and its chemical engines.

WINGED BIRD: COMETS/*KAKEHASHI*

ETS VIII was not the only satellite to test advanced systems of communication in orbit. An earlier attempt was made by COMETS – or COMmunications and broadcasting Engineering Test Satellite – renamed *Kakehashi* ("bridge") when it reached orbit on 21st February 1998. Constructed by NEC, COMETS was intended to use transponders to test communications with mobile phone and other satellites, demonstrate tracking and data relay systems, test high-definition television and the use of four xenon propulsion motors, each of 25 mN thrust. Like a winged bird, it had enormous solar panels, 30 m long. The cost was ¥45.2bn (€403m) and it was the most advanced communications technology satellite built in Japan.

The mission got off to a bad start when the H-II second stage shut down early, after only 44 sec into a 192-sec burn, making it impossible for the satellite to reach its intended destination of geosynchronous orbit. COMETS was left in a 247–1,883-km orbit instead of the 36,000 km planned. The NASDA investigation panel later found that a tiny hole had burned through the LE-5A's motor nozzle casing, igniting wires, which caused the engine to shut down. Fortunately, of the 3,900-kg weight of COMETS, 1,900 kg comprised station-keeping fuel. In March, a 90-sec burn lifted the perigee from 250 to 390 km, the first in a series of seven maneuvers to raise the orbit to 500–17,700 km and thus enable about 60% of the original mission to be carried out. COMETS did manage to relay signals for the ETS VII automatic docking operations, but the mission as a whole could not be saved and was abandoned in mid-1999, two years ahead of schedule. Its orbit will last at least 1,000 years.

COMETS ground test

OICETS

BEAMS ACROSS SPACE: *KIRARI* AND *KIZUNA*

The successor to COMETS was the much smaller, 550-kg OICETS, or Optical Inter Orbit Communications Engineering Test Satellite. After many delays, it was eventually launched by a Russian Dnepr rocket on 24th August 2005 into a 608–611-km orbit, 97.8°, 96.6 min and acquired the name *Kirari*. Accompanying OICETS was INDEX, the Innovative Technology Demonstration Experimental Satellite, which was named *Rimei*.

Kirari was box-shaped, with two solar panels 9.36 m long and on the spacecraft box is a laser, shaped like a camera with a zoom lens intended to transmit large volumes of data up to 45,000 km through open space in narrow beams and s-band links with an accuracy of 0.0003°. It was designed to communicate with the €1bn European Space Agency satellite ARTEMIS (Advanced Relay and TEchnology MIssion Satellite), built with data relay systems to link to satellites in lower orbits, navigation transponders, lasers for inter-satellite communication and ion engines for station-keeping.

ARTEMIS was launched on the Ariane 5 rocket on 12th July 2001. In a disappointment, the Ariane's third stage shut down early due to combustion instabilities, leaving ARTEMIS in an orbit of 592–17,528 km, far below the 36,000-km circular orbit intended (also stranded was a commercial Japanese communications satellite, *Bsat 2b*).

Against the odds and using up almost all the fuel for its ion thrusters, European

OICETS laser

space controllers managed to deliver ARTEMIS to its intended orbit in January 2003, a year and a half later. The first successful test run of the laser communications system took place on 9th December between ARTEMIS and *Kirari*.

The next technology demonstration satellite was WINDS (Wideband Inter Networking Demonstration Satellite), launched by H-IIA into spring skies on 23rd February 2008. There was a stiff breeze from the north-east, but all went smoothly and the spacecraft separated 28 min 3 sec later, to be renamed *Kizuna*. As it came over the tracking station in Santiago, Chile, a picture was beamed back to show that the solar panels had deployed properly. By 31st March, its 20-N thruster had brought *Kizuna* to its geosynchronous orbit at 143°E. *Kizuna* was part of the *i-space* part of the *e-Japan* information technology program of the Japanese government. The idea was to demonstrate high-speed internet access, with *Kizuna* providing internet connections at 6 MB/sec to 45-cm vehicle mobile dishes and at 1.2 GB/sec to larger 5-m antennae. *Kizuna* was expected to be especially useful in the areas of telemedicine and disaster operations. Table 2.6 summarizes Japan's technology demonstrators.

Table 2.6. Japanese technology demonstrators

9 Sep. 1975	ETS I (*Kiku 1*)	N-I	Tanegashima
23 Feb. 1977	ETS II/*Kiku 2*	N-I	Tanegashima
11 Feb. 1981	ETS IV/*Kiku 3*	N-II	Tanegashima
3 Sep. 1982	ETS III/*Kiku 4*	N-II	Tanegashima
27 Aug. 1987	ETS V/*Kiku 5*	H-I	Tanegashima
28 Aug. 1994	ETS VI/*Kiku 6*	H-II	Tanegashima
28 Nov. 1997	ETS VII/*Kiku 7*	H-II	Tanegashima
21 Feb. 1998	COMETS/*Kakehashi*	H-II	Tanegashima
24 Aug. 2005	OICETS/*Kirari*	Dnepr	Baikonour
18 Dec. 2006	ETS VIII/*Kiku 8*	H-IIA	Tanegashima
23 Feb. 2008	WINDS/*Kizuna*	H-IIA	Tanegashima

Kizuna

SPY SATELLITES: THREAT ACROSS THE SEA OF JAPAN

The early space programs of the big powers had a strong military orientation – such as the *Corona* program in the United States and *Zenit* in the Soviet Union. By contrast, the Japanese program was entirely civilian, reflecting the post-war and constitutional settlement that restricted Japanese defense spending and prevented the country from building up armed services (though a limited "self-defense force" was permitted).

This changed, following the strange events of 31st August 1998. Across the Sea of Japan, North Korea converted a military rocket, attached a small solid rocket motor and launched it from Musadan-ri in north Hamgyong on its east coast. North Korea hailed what it claimed was its first satellite to be put in orbit. The small sputnik, with a weight estimated at 50 kg, was called *Kwangmyongsong* or "bright star", carrying a transmitter broadcasting "the immortal revolutionary hymn of General Kim Il Song" and the "Song of General Kim Jong Il". North Korea called the long, thin launcher the *Taepo Dong 1* and announced the orbit to be 219–6,978 km, 41°, 165 min. However, the launch was not tracked, neither at the time nor subsequently by United States Space Command, nor by the Russian space tracking system. No one outside Korea ever saw the satellite or heard the immortal hymns relayed to Earth for their benefit. A year later, the North Koreans still insisted that it had broadcast for nine days and would orbit the Earth for another year. The American intelligence community appears to have been caught napping and originally ridiculed the North Korean claims to have launched a satellite. More mature consideration led to the conclusion that a serious satellite attempt had been made, but that it had probably not succeeded.

For the Japanese, the launch of *Kwangmyongsong* had sinister implications. Japan responded initially by vocal diplomatic protests to North Korea. The parliament, the

Diet, was convened. The Japanese noted that the ascent path of the *Taepo Dong* curved over the northern Japanese islands, bringing with it the danger of rocket stages or débris falling to Earth (parts did indeed fall near Sanriku in northern Japan). More to the point, Japan was now unmistakably within missile range of North Korea. In its report on the incident at the end of October 1998, the Japanese Defense Agency argued that North Korea's intentions were to test military missiles rather than put a satellite into orbit. For the Japanese, though, the *Kwangmyongsong 1* episode emphasized their dependence on American intelligence. The United States had a huge electronic signal intelligence ("sigint" in the trade) eavesdropping facility at Misawa on the north-east tip of Honshu and up to 20 advanced spy satellites in orbit at any time. Whilst this system could undoubtedly keep a close watch on the North Koreans, the real question was the preparedness of the Americans to pass on information and warnings in a timely manner, for it transpired that the flames of the ascending launcher had been spotted by an American reconnaissance satellite, but the information had not been relayed to Tokyo. This enraged the Japanese and prompted them to develop their own system. Later, it emerged that the Americans had been asked to share images of nuclear power stations in North Korea, but had declined.

Mitsubishi Electric Company (Melco) proposed a reconnaissance satellite program. Several plans were considered – a multi-purpose Earth observation program with military capabilities; the conversion of the proposed Advanced Land Observation Satellites (ALOS) to part-military use; and, the option chosen, a dedicated military reconnaissance program to monitor North Korean rocket developments that might in the future threaten Japan. Within months, the government of Prime Minister Keizo Obuchi had drafted plans for a program for up to four 85° polar-orbiting spacecraft, 850 kg each, two to be radar-based and able to see through cloud and darkness, and two to be optically based (resolution 1 m), circling the Earth at 500 km. The radar satellite had the advantage of being able to see through clouds and at night, when its companion optical satellite would not be able to take images. The double set would also give both morning and afternoon coverage, valuable to check if, for example, military vehicles had moved. The spacecraft were expected to carry heat-sensing detectors to spot the flame of a rising rocket. There was pressure on the government to buy in the spy satellites directly from the United States to save time, but the administration's view was that the technology should be developed domestically and built on the experience of the ALOS spacecraft. In 1999, Melco was duly selected to carry out preliminary work on the spy satellite's optical sensor, synthetic aperture radar system and maneuvering systems. The Ministry of Transport was put in charge of the program, rather than NASDA. Later that year, it was decided that, in order to save time, Japan would buy in a number of key parts of the reconnaissance satellite project from the United States – principally data recording and interpretation systems and the mechanism for controlling the optical sensors – and a cooperation agreement was signed between the two countries. It was an expensive project, with costs of ¥253.8bn and €2bn being quoted.

Preparations to launch the military satellites were especially secretive. Components arrived in Tanegashima in unmarked crates, riot police ringed the launch site (they were not needed in the event) and coast guard ships patrolled off shore. Even their name appeared to be secret, the nearest to an official name being IGS (Intelligence Gathering Satellite). The first two military reconnaissance satellites were put into orbit on H-IIA from Tanegashima on 28th March 2003 and given the simple designators of Optical 1 and Radar 1. Japan said very little about the launch and would not even give the orbital parameters (amateurs tracking the satellites gave them at 500 km and NASA inadvertently revealed the orbits as 485–510 km). A companion pair was lost with the failure of the sixth H-IIA on 29th November 2003. The second optical satellite was launched on its own on 11th September 2006, the 10th launch of the H-IIA. The next double launching was on 24th February 2007, when the H-IIA put in orbit Optical 3 and Radar 2, both going into Sun-synchronous orbits of 490 km.

The Japanese provided minimal details, following the well established American practice, contrary to international law, of not disclosing the orbits of military satellites. Amateur observers made it their business to follow the satellites and the two were tracked over Honshu within a matter of days of their launch, tracking north to south in close formation, passing the same point every four days. Some even managed to take digital pictures of the satellites from the ground. In summer 2003, pictures of the two satellites appeared in an unexpected place – military magazines – the *Gunki Kenkyu* (August 2003 edition) and *Sekai no kansan* (September 2003). The military magazines gave plenty of further details, such as the operating altitudes (400–600 km), the orbital path (crossing the Sea of Japan each mid-morning), dimensions (box-shaped, 1.7 by 1.6 by 1.3 m), design legacy (the USERS satellite), power supply (solar panels), sensors (the same as ALOS) and weight (between 1.5 and 2.5 tonnes). The optical satellite had solar panels 16.5 m wide, able to generate 3 kW power, while the more power-demanding radar satellite had a 22.5-m solar panel requiring 4 kW, the system able to achieve a resolution of 3 m, they said.

These details were also reproduced on a website run by a science writer, Shinya Matsura (*www.spaceserver.com*). The website mischievously carried directions as to how to make origami paper models of the two satellites. Strangely, the website with the extracts from the military magazines was quickly taken over by squatters and the author disappeared, at least electronically speaking. Table 2.7 summarizes the reconnaissance satellite program.

Table 2.7. Japanese reconnaissance satellites

28 Mar. 2003	Optical 1, Radar 1	H-IIA	Tanegashima
29 Nov. 2003	[Optical 2, Radar 2]	H-IIA	Tanegashima [failure]
11 Sep. 2006	Optical 2	H-IIA	Tanegashima
24 Feb. 2007	Optical 3 Radar 2	H-IIA	Tanegashima

CONCLUSIONS: SCIENCE AND APPLICATIONS

Following the launch of its first small satellites in the 1970s, Japan developed an extensive program of space science and applications. Hideo Itokawa's ISAS continued the tradition of small scientific probes with astrophysical satellites (Astro), solar observatories (Solar), lunar (*Hiten*, *Hagoromo*) and planetary missions (*Suisei*, *Sagigake*, *Nozomi*), all using relatively small solid-fuel rockets. The formation of NASDA paved the way for an expansion of the Japanese space program into the field of large applications satellites for marine observations (*Momo*), Earth observations (*Fuyo*, *Daichi*), the monitoring of the atmosphere (*Midori*, TRMM) and the development of communications (*Kiku*, *Kakehashi*, OICETS). These much larger satellites required Japan to develop its own fleet of medium-sized rockets. Although doing so proved to be troublesome, with many setbacks, Japan moved purposefully from licensed rocket technology (N-I) to its indigenously built launchers, the H-II and the H-IIA. Nothing symbolized Japan's advances as much as the arrival, on the day of the 50th anniversary of the first Sputnik, of Japan's first spaceship to orbit the Moon, *Kaguya*, named after a fairy. To have imagined in 1957 that a Japanese spacecraft would have reached the Moon a half-century later would indeed have been considered a fairytale.

3

Japan: *Kibo* and the Space Station

The various unmanned space ventures undertaken by Japan in the 1970s and 1980s, spectacular and adventurous though they were, lacked a human presence. Public interest in spaceflight is always higher when men and women are directly involved, with human drama and astronauts putting their lives at risk in often dangerous adventures. In the 1990s, Japanese astronauts began to fly in space on the Shuttle and Japan became an important contributor to the most ambitious international scientific project of all time, the International Space Station.

JAPAN'S FIRST ASTRONAUT

Japan was a long-standing economic and scientific partner of the United States, reflecting the close ties established in the post-war years. Ultimately, Japan always wanted to have astronauts in orbit performing experiments of interest to Japanese science, although it recognized that manned Japanese space shots were a distant prospect. Encouragingly, the American Space Shuttle offered the prospects of a Japanese person reaching Earth orbit in the late 1980s.

The Space Shuttle had a large payload bay. This was intended to carry satellite payloads into orbit, but could also house a small laboratory, where scientists could work for a week at a time, then the limit of a Shuttle mission. This was called *Spacelab*. The United States were financially very constrained by the high costs of building the Shuttle, so they invited their international partners to share in the cost of building a Shuttle laboratory, the *quid pro quo* being that their astronauts could fly on board the Shuttle.

The *Spacelab* was a mini-space station module built in Europe and designed to fit snugly into the payload bay of the Shuttle. It was actually a combination of modules that could be configured a number of different ways, depending on the mission required, the largest being a pressurized module with ready-to-fit equipment racks, containers and platforms, adapted according to the type of mission required. Connected by a tunnel to the main cabin of the Space Shuttle, the idea was that astronauts would float down the tunnel to conduct experiments in a shirt-sleeve environment. Between 1983 and 1998, NASA flew *Spacelab* 22 times, kitted out for different countries (Europe, Germany, Japan) or for different sets of scientific experiments (e.g. microgravity, materials science, biology). Japan was to be involved in four *Spacelab* missions. One was a dedicated Japanese space mission, *Spacelab J*, and Japanese experiments flew on three other *Spacelab*s, including the first one.

Japanese equipment flew on the first *Spacelab* mission (though not a Japanese astronaut) and again the following year. In December 1982, the first main *Spacelab* crew and its support astronauts – Byron Lichtenberg, Michael Lampton, Ulf Merbold, Wubbo Ockels, Owen Garriott and Robert Parker – traveled to Japan to test equipment in the NASDA vacuum chamber outside Tokyo. The Space Experiments with Particle Accelerators were designed to help scientists better understand the relationship between the magnetosphere and the upper atmosphere by injecting charged particles into plasma.

Japan then negotiated a deal with the United States for a *Spacelab* mission to be devoted entirely to Japan. Japan would kit out the lab and set up experiments that would be operated by Japanese astronauts. The mission was called *Spacelab J* (J for Japan, comparable to *Spacelab D*, dedicated to Germany (D for Deutschland)). The mission was first set for February 1988 on the Space Shuttle *Challenger*, with the title of mission STS-81G.

This required the selection of Japan's first group of astronauts and a call for candidates duly went out. Three Japanese astronauts were selected from 533 applicants: Dr M. Mamouri Mohri, Dr Chiaki Mukai (both NASDA) and Dr Takao Doi. Dr Mamouri Mohri was a chemistry scientist, born in 1948, who graduated from Hokkaido University and subsequently studied in Australia. Dr Chiaki Mukai was born on 6th May 1952, went to a girls' high school and graduated from Keio University School of Medicine in 1977, acquiring a doctor's license with a specialization in cardiovascular surgery. Takao Doi, born on 18th September 1954 in Minamitama-gun, Tokyo, was a engineer with primary, masters' and doctoral degrees from the University of Tokyo. In the end, it was 44-year-old Mohri who was selected for *Spacelab J*, which was renamed *Fuwatto*, a Japanese word for "floating free" or "weightless". For the first mission, 34 experiments were selected – 22 in the area of materials processing and 12 in life sciences (all but three were Japanese). NASDA flew a number of materials processing experiments on sounding rockets to pave the way for melting and crystal growth studies to be carried out on the laboratory. The first sounding rocket, called the TT-500A, was fired on 14th September 1980 from Tanegashima. The TT-500A climbed 320 km into the Pacific sky, providing 7 min microgravity as the payload described a giant arc 500 km downrange. Tracked by the Ogaswara station, parachutes deployed at 6 km, with splashdown 16 min after launch. Disappointingly, the second sounding rocket launch on 15th January 1981, carrying a metallic compounds experiment called Ni-TiC whisker, was lost when the beacons failed on splashdown.

The *Spacelab J* mission was delayed indefinitely when the Space Shuttle *Challenger* exploded over Cape Canaveral on 28th January 1986. The Soviet Union offered to fly the intended experiments on board its new *Mir* space station launched a month later, but the Japanese decided to keep to their

Chiaki Mukai

arrangements with the United States, even though it meant a long delay. The Shuttle program got back on track again in late 1988 and the first parts of *Spacelab J* arrived at Kennedy Space Center just over two years later.

INSTEAD, A MISSION TO *MIR*

In the event, Japan's first astronaut flew on a Russian, not an American, spaceship. In 1989, the Tokyo Broadcasting System (TBS) decided to celebrate its 40th year of broadcasting by sending one of its own journalists into space. TBS was the largest private broadcaster in Japan, with 26 stations, 1,600 staff and 8,000 affiliate staff and 16 overseas bureaux. The station had a long-standing interest in spaceflight: TBS had been the second Western company permitted to film at Baikonour Cosmodrome and had covered the flight of the Soviet Space Shuttle, *Buran*.

TBS approached the Soviet space agency, then called Glavcosmos, to strike a deal whereby TBS would pay ¥1.26bn (€11.3m) to fly a journalist to the *Mir* space station for a week (much cheaper than NASDA paid NASA to fly Japanese astronauts on the Shuttle, at ¥3bn each). In the event, TBS sub-sponsored its own costs, inviting advertising from Sony (whose letters were displayed prominently on the rocket at take-off), Minolta, American Express Japan and various healthcare, chemical and insurance companies.

In a competition to find an appropriate reporter, 145 men and 18 women applied, ranging in age from 23 to 55. The applicants included newscasters, announcers, correspondents, TV directors, field staff and even members of the sales and accounts departments. The first round of medical tests reduced the field to 21 people. Four Soviet doctors then arrived to assist in the next stage of selection, which reduced the group to seven. They were sent to centrifuge training. All failed: the Soviet doctors were adamant that they would not lower their standards just for this mission. One had a stomach condition that produced excessive gastric juices when under stress; another had too fine blood vessels. A particular problem for the Japanese was that 80% of the population have a deviated nasal septum (a bend in the inside channel of the top of their nose): whilst the Japanese did not regard this as a problem, conventional Caucasian science was that this imperfection invited infection.

So, a second round of recruitment was organized, this time attracting 55 male and nine female applicants. They were exercised on bikes, made to lie down with their heads at a lower angle than their body, decompressed and whirled around in centrifuges. Following the tests, the Soviet doctors again expressed their unhappiness. The Japanese doctors then went back to the first recruitment round and selected a different group of seven finalists: Moscow correspondent Toshio Koike, Toyohiro Akiyama, Ryoko Kikuchi, Atsuyoshi Murakama; and from affiliated TBS companies, Nobuhiro Yamamouri, Kouichi Okada and Naoki Goto. Once again, this group was sent to the doctors and the centrifuges. This time, Soviet doctors found two who met their standards: 48-year-old editor Toyohiro Akiyama and 26-year-old camerawoman Ryoko Kikuchi. Their selection was duly announced in September 1989. Koike, Yamamouri and Goto were also allowed to train, provided that they had their tonsils taken out. They all went to Moscow on 1st October 1990. Murakama and Okada were dismissed.

Akiyama, born in 1942, was a sociology graduate who had previously been a Washington bureau chief, a reporter in Vietnam and had worked in the Japanese service of the BBC in London. Now, he was a senior man within the company and editor-in-chief of the foreign news division. He had been commentator on a number of Shuttle missions, including the *Challenger* disaster. Getting to the final selection was a major challenge for him, for he had to kick the two typical journalistic vices of drinking and smoking, in his case four packets of cigarettes a day. Kikuchi was much the fitter of the two, her recreational choices being mountain climbing, cycling, skiing, basketball, swimming and kabuki theater. Born in 1964 in Zama, Kanagawa, she had studied Chinese at the Tokyo University for Foreign Studies. She had less overseas journalistic experience, limited to China and the Seoul Olympics in Korea. The mission of the journalist was to make two 10-min television broadcasts each day, several 20-min radio broadcasts and contact amateur radio hams. In the course of the broadcast, film would be taken of the behavior of six Japanese tree frogs in orbit. Six television cameras were delivered in advance by the unmanned Progress M-5 cargo ship.

The timing of the mission had never been entirely clear, 1992 being suggested as the likely date. At the time of the selection of Akiyama and Kikuchi, the *Spacelab J* mission had been set for June 1991. The Soviet Union brought the journalist mission forward to December 1990, in what was considered by observers as a cunning move to ensure that the first Japanese in space rode a Soviet, rather than an American, rocket. The setting of the date did not – publicly, at least – upset NASDA, who congratulated TBS on its achievement, but instead it enraged Soviet journalists, who protested that a foreign journalist would fly in space before one of their own.

On 2nd November 1990, the final team was chosen, Akiyama emerging as the winner. Even had the choice been otherwise, Kikuchi would have been robbed of her flight in any case, for she fell victim to appendicitis three weeks later, just days before the scheduled mission. Toyohiro Akiyama was duly launched on 2nd December 1990 on Soyuz TM-11 with Viktor Afanasayev and Musa Manarov. A hundred Japanese media technicians attended the launch, setting up their cameras close to the rocket and its flame trench, giving spectacular views of ignition. Emblazoned in Soviet and Japanese flags, Soyuz TM-11 soared into a clear sky with live coverage on Japanese television. Akiyama carried with him a small Japanese mascot – a doll dressed in a bright kimono. Going on air as he entered orbit, the journalist radioed to Earth, echoing the famous first words from the orbit of Yuri Gagarin: "This is Akiyama! The Earth is blue!" Akiyama was the 237th person in space.

But things did not go quite as planned. Soon after getting into orbit, Akiyama had a bout of space sickness – a problem that afflicts half of all space travelers. Akiyama was ridiculed in the probably jealous Western press as a chain-smoking, whiskey-swilling idiot. This was unfair, for Akiyama shot some of the best film ever taken of life on board *Mir*, conveyed a sense of what many ordinary people must feel in orbit and his medical parameters were comparable to his professional cosmonaut companions. TBS attracted record viewing figures during the week-long mission. He came back with stunning color film that made a top-class video.

In the course of his eight-day mission, Akiyama made live broadcasts from the Space Station, much of it timed for peak viewing time. He filmed activities inside the station and pointed his cameras outside the station to pick out landmarks below, like Mount Fuji. He noted how small was Japan, compared to the large land masses of Africa, the Americas and Siberia. He returned to Earth on the Soyuz TM-10 spacecraft with Soviet cosmonauts Gennadiy Manakov and Gennadiy Strekhalov, who were coming back after four months in orbit. The retro engine fired at 5.13 am on 10th December and the cabin reached the ground at 6.08 am near Arkalyk, Kazakhstan. Emerging from the capsule, Akiyama announced that he was hungry and joked that his needs could be met by some beer and cigarettes. He had been in orbit for 7 days 21 hr and returned to his broadcasting career in Tokyo two weeks later.

This was not the end of the relationship between Japan and *Mir*. In 1996, under an agreement between NASDA and the Russian Space Agency, RKA, two pieces of equipment were later installed on the *Mir* space complex – one outside to test the effects of space radiation on cell cultures (radiation monitoring) and the other on the inside to examine how tiny organisms reproduced in weightlessness (mircroflora). Japan paid ¥1bn (€943,000) for the experiments.

FUWATTO'S SUCCESS

The way was now clear for the *Spacelab J* mission, now named *Fuwatto*. The Japanese had now waited a long time for this mission and, for America's partners, the program had proved an expensive and frustrating one. The numbers of *Spacelab* missions were restricted because the Shuttle never achieved the frequent launch rates originally projected. The number of visiting, non-American astronauts was limited to one or two per mission. Even the studiously reticent NASDA hinted its disapproval of the unequal arrangement when it noted that *Spacelab J*, paid for by Japan, carrying entirely Japanese equipment, was operated by four Americans and "only one Japanese payload specialist will be allowed to board the shuttle". The Americans later responded to their partners' unease by making many of the later American *Spacelab*s quite international by nature and flying a number of visiting scientists.

Forty-four-year-old Mamouri Mohri was selected for the first *Spacelab J* mission, which duly took place on 12th September 1992 on the second flight of the Space Shuttle *Endeavour*, the mission designation being STS-47. Commanded by Hoot Gibson, Mamouri's companions were pilot Curt Brown, mission specialists Jan

Japan's early astronauts

Davis, Jay Apt, Mae Jemison (the first black woman in space and who later appeared in *Star Trek*) and payload specialist Mark Lee. *Endeavour* climbed into a 280-km orbit of 57° – one that passed over Tokyo. They returned to Cape Canaveral on 20th September after their seven-day mission had been extended by one day to permit additional experiments. Mission commander Hoot Gibson later described it as a trouble-free mission. *Spacelab J* carried 30 kg of souvenirs – a flag of the rising sun, commemorative rubber stamps, the NASDA flag, and "*Fuwatto '92*" emblems.

The life sciences experiments concentrated on investigations of human health in space, but a number of other animals were also carried: two carp fish, 36 chicken embryos, 7,600 fruit flies, 1,800 hornets, fungi, plant seeds, frogs and frogs' eggs. Other experiments used a furnace to test gas evaporation in low gravity, low-level acceleration and protein crystal growth. Jemison relayed video pictures of frogs hatching tadpoles under weightlessness. Mohri gave a microgravity lesson to Japanese school children. For Gibson and Mohri, there was a television relay to the Japanese prime minister and the head of NASDA. In order to get maximum benefit from the laboratory, two 12-hr shifts were worked with two astronaut teams, the red and the blue teams.

Mohri was in space for 7 days 22 hrs and was the 282nd person to go into orbit. As soon as *Endeavour* landed, the life sciences experiments were removed – such as fruit flies, plant seeds and eggs (some hatched out later). There was a panic when the laboratory's electricity failed after touchdown and the carp were left without fresh oxygen for 27 mins. But they survived. All the *Spacelab* racks were shipped back to Japan a month later and post-flight analysis of the mission began. Later, Mamouri Mohri became director of the Miraikan science museum in Tokyo and a frequent broadcaster on spaceflight.

The success of the *Fuwatto* mission was widely hailed in Japan. Although no further dedicated *Spacelab J* mission was in prospect, there was the possibility of more seats being available for NASA on future *Spacelab* and Shuttle missions and, later, on the International Space Station. Accordingly, a second call for astronauts was issued. In the second round of selection, held in 1991, 372 people applied to be astronauts. The requirements of candidates was that they should be Japanese, less than 35 years old, speak English, be a graduate in natural sciences and have three years' research experience, be between 149 and 193 cm in height and give a 10-year commitment. Applicants were advised that they could expect to fly in four of five years on a Shuttle mission before an operational assignment to the Space Station would follow. Of the 372 applicants, 331 were men and 41 were women. Japan's fourth astronaut was selected – Koichi Wakata, a 28-year-old Japanese Airlines engineer with a degree in applied mechanics.

In a supplementary recruitment, Soichi Noguchi was selected in a one-off round of astronaut selection in May 1996, the third round. Born in 1965, Soichi Noguchi was an engineer, with degrees from Tokyo University and experience working for Ishakawajima-Harima Heavy Industries. He was sent straight after his selection to NASA Johnson Space Center in Texas to train as a mission specialist. Later, in 1998, he became the first Japanese astronaut to train in the Yuri Gagarin Space Center in Star Town, Moscow. In summer 2005, Noguchi participated in the return-to-flight of the Shuttle, the first after *Columbia* was lost.

Noguchi flying T-38s

INTERNATIONAL MICROGRAVITY LABORATORY 1, 2: NEWTS, FISH, CELLS

Japan did indeed get further opportunities apart from *Fuwatto*. The International Microgravity Laboratory (IML-1) *Spacelab* flew on the Shuttle *Discovery* in January 1992. Japan provided two of the 42 experiments – one for crystal growth for the organic superconductor and the other for the investigation of biological effects of cosmic radiation in a spaceflight environment. Japan got its second Shuttle seat on the second International Microgravity Laboratory (IML-2), also called Shuttle mission STS-65 (Space Transportation System 65). The honor went to Chiaki Mukai, who flew for 14 days 17 hr on the first of the Shuttles to fly, *Columbia*. This mission, launched on 8th July 1994, flew 80 experiments, of which 12 came from Japan. The laboratory carried newts, fish and cell cultures; on the materials side, a large isothermal furnace and free-flow electrophoresis unit. In the course of the two-week mission, the newts laid eggs and the medaka freshwater fish spawned in orbit.

Chiaki Mukai later flew again on STS-95 (Shuttle *Discovery*) on 29th October 1998. This was popularly known as the "John Glenn flight" because of the presence on board of America's first astronaut to enter orbit, John Glenn, in 1962. This time, Glenn flew partly as a test of how older people could survive a spaceflight, partly (though this was not admitted) to regenerate public interest in spaceflight. Chiaki Mukai's role was to operate five Japanese payloads on the flight in the *Spacehab* module – in effect, a mini-*Spacelab*. These involved a study of neural signals in toadfish so as to better understand space sickness, the role of gravity in the growth of

Chiaki Mukai in Shuttle

cucumbers, the use of melatonin to enable astronauts to sleep better and the testing of blood samples.

SPACE FLIER UNIT

The concept of a free flier dated to the 1980s: the idea is to launch a space platform on one spaceship, which, after a period in orbit, is retrieved many years later by another. Ideally, a free flier is re-useable and may be kitted out for a range of different missions. Approval for a Japanese free flier was given in 1986. Japan set up the Institute for Unmanned Space Experiments with Free Fliers – a joint project between the Ministry of International Trade and Industry and 13 leading industrial and technological companies.

The 4-tonne Space Flier Unit (SFU) was launched by H-II from Tanegashima on 18th March 1995 and planned for later recovery by the American Space Shuttle. The flier took the form of a platform 4.46 m wide and 2.8 m high, with six payload bays on top. Electrical power was provided by two 24-m solar panels. During its period in orbit, thrusters maintained attitude control and adjusted its orbit, while an S-band antenna relayed data down to Earth control. This was not the first time that a space platform had been put into orbit for subsequent Shuttle recovery, for this had been done with the Long Duration Exposure Facility in the 1980s. Japan's intention was to re-launch the SFU for further microgravity experiments five times. Also launched with the SFU was *Himawari 5* metsat.

The ¥60bn (€535m) SFU carried experiments in astronomical observations, the laying of eggs by water lizards and the growing of diamond crystals in weightlessness. The unit carried four experiments to test the possibility of transmitting solar energy from space to power electricity systems on Earth. Following ideas proposed by Peter Glaser for solar power systems, a group of enthusiasts began to explore how such ideas could best be explored in Japan – an important point granted Japan's dependence on foreign raw energy sources. These were tested out first on sounding rockets and then the SFU. During its mission in

Space Flier Unit

Earth orbit, the space flier was able to survey 7% of the sky using an infrared telescope, providing useful information on zodiacal light, interstellar dust and background cosmic radiation.

The Space Flier Unit was retrieved by *Endeavour* (STS-72), launched on 11th January 1996. The Shuttle entered orbit almost 20,000 km behind the flier but it closed in rapidly. Two days later, mission commander Brian Duffy steered the Shuttle in for the final approach while mission specialist Leroy Chiao used a laser to call out precise distance data. Astronaut Koichi Wakata was aboard, the third Japanese to ride the Space Shuttle (the 341st person to go into orbit). During the final approach, he waited in the back of the flight deck, ready with the remote manipulator arm to capture the free flier.

Then problems began. In advance of retrieval, ground controllers at Sagamihara Operations Center near Tokyo commanded the 10-m solar arrays on the flier to retract so that the unit could fit the Shuttle payload bay. After many further commands, they still refused. Accordingly, a drastic backup plan had to be put in place: explosive bolts were fired to blast the panels away as the Shuttle and flier, in formation, passed the west coast of Africa. An orbit later, the Shuttle closed in for the final maneuver. Wakata extended the arm to the now panel-less flier, grappled it with the remote arm lever and pulled it into the payload bay, where it was safely berthed for return to Earth. Without panels, the batteries of the flier would last only an hour. When Wakata grappled the flier with the Shuttle's robot arm, only three of the four latches engaged, which would make retrieval impossible. Wakata persevered, the fourth latch engaged and he brought it on board with just five minutes' battery remaining. They never found out the cause of the problems: probably equipment had been damaged by the hot-and-cold cycles of orbit.

Wakata put his experience with the arm to further use the following day to release a NASA free flier. Unlike the Japanese one, this free flier, called OAST or Office of Aeronautics and Space Technology, was for a short-term deployment and was recovered by Wakata after two days.

Koichi Wakata and his crew went on a tour around Japan on their return to visit schools, factories and civic associations. American astronaut Dan Barry recalled many years later how he was taken aback by the level of enthusiasm with which they were greeted, with Wakata treated like a rock star by teenage girls screaming his name and rushing forward with flowers. It had originally been intended to fly the SFU up to five times, but it never flew again.

PREPARING FOR THE INTERNATIONAL SPACE STATION

Next up was 43-year-old Dr Takao Doi, who flew on mission STS-87 on 10th November 1997 on the Space Shuttle *Columbia*. He was the fifth Japanese in space, the fourth on the Shuttle, the first Japanese to walk in space and the 367th person to enter orbit. His companions were Kevin Kregel, commander, pilot Steven Lindsey, mission specialists Winston Scott and Kalpana Chawla (Indian-born) and payload specialist Leonid Kadenyuk of the Ukraine. STS-87 was a dedicated microgravity mission, but the highlight of the mission was the spacewalk by a Japanese astronaut to test assembly techniques for the forthcoming International Space Station. The spacewalk was modified because of a problem experienced by *Columbia* in retrieving the free-flying *Spartan* satellite and this had to be dealt with first. Winston Scott and Takao Doi donned spacesuits on 24th November, the sixth day of the mission, clambered down the payload bay and stationed themselves to grab the *Spartan* as the Shuttle maneuvered close. Television cameras showed them waiting for the right moment as the blue Earth rolled by. They were able, with the help of the Shuttle's remote manipulator system, to lock the *Spartan* down. This key task accomplished, they proceeded to test out a crane of the type to be used to move equipment – some of it awkward and heavy – around the International Space Station. The astronauts returned to the cabin after 7 hr 43 min, one of the longest spacewalks performed from the Shuttle.

In order to gain more space station assembly experience, a second spacewalk was added to the mission on the 13th December. Scott and Doi went out for five hours to again practice moving objects around the payload bay. In the course of the spacewalk, one of the astronauts in the cabin used a joystick to maneuver a beach ball-sized subsatellite called Aercam Sprint with two television eyes, designed to hover around and inspect difficult-to-reach parts of a space station. The Shuttle returned to Earth after over 15 days with a trawl of microgravity research.

Takao Doi on EVA

JAPAN AND THE INTERNATIONAL SPACE STATION

In 1985, Japan was invited to take part in the project for the American space station *Freedom* announced by President Ronald Reagan the previous year. Japan was asked by NASA to contribute a laboratory for experiments in scientific, microgravity and materials processing research. The components would be launched by the American Shuttle in its payload bay. The original cost of this program of participation was then estimated at ¥310bn (€2.76bn). On 14th March 1989, a memorandum of understanding (MOU) was signed between Japan and the United States through their respective space agencies, for the "design, development, operation and utilization" of the space station *Freedom*. Similar MOUs were signed at around the same time between the United States on the one hand and Europe and

Canada on the other. Japan quickly began preliminary design of its Japanese Experiment Module (JEM). JEM effectively comprised three parts, each requiring a separate Shuttle mission to launch: a Logistics Module (LM), Pressurized Module (PM) and Exposed Section (ES). Over time, the pressurized module acquired a name, *Kibo*, to accompany the European module (*Columbus*), the American modules (*Destiny*, *Harmony*) and the Russian ones (*Zvezda*, *Zarya*).

Despite the many ups and downs of its design and development, the building of the JEM remained steadily on course throughout the period from approval in 1989 to arrival at the Tsukuba space center in 1997. For Japan, its own module offered the opportunity for permanent participation in manned space flight (albeit as a partner rather than a leader) and a platform where research could be carried out into manufacturing technologies in weightlessness and vacuum. It was intended that of the station's permanent crew of six, one should always be Japanese – a permanent presence. Meantime, Japan proceeded methodologically with the preparation of scientific experiments for its module. The first call for scientific experiments was made in 1992 and led to the selection of the first 49 Japanese experiments. In 1997, the first payloads were selected for the Exposed Facility – a pulsed laser beam to detect space débris smaller than 10 cm, an X-ray detector, an instrument to study trace gases in the atmosphere and a high-capacity space communications system, all to be operated by the Space Environment Utilization Research Center in Tsukuba.

In advance of the launch of JEM, sounding rockets were used to test some of the experiments to be done on board the space station, the first in September 1991. On 20th August 1992, the TR-1A2 sounding rocket was fired to 227 km to carry out preliminary microgravity experiments of the type that would later be flown on the JEM. On 25th September 1996, NASDA launched the third TR-1A test rocket from Tanegashima to test six microgravity experiments. The sounding rocket's engine burned for 64 sec, reached an altitude of 262 km, the payload splashing down in the Pacific Ocean 14 min after lift-off 320 km downrange. The launching offered 361 sec of microgravity, permitting the study of the growth of coloidal crystals, nucleate boiling, the combustion of air–fuel mixtures, the melting of germanium and welding. A fourth JEM precursor sounding rocket was flown on 25th August 1995.

By 1995, construction of an engineering module had been completed and, in the following year, the JEM passed a week-long critical design review. The review was attended by astronaut Koichi Wakata, who had just returned from orbit. The reviewers closed 344 review item discrepancies and sent 14 onward for further evaluation. The review process gave the go-ahead for manufacturing. Nine Japanese companies had a major involvement with the International Space Station, led by Mitsubishi (Heavy Industries and Electric) and followed by Toshiba, Ishakawajima, Nissan, NEC, Kawasaki, Hitachi and NTT.

In Tsukuba Space Center, a Space Station Operations Facility (SSOF) was completed in 1996 to support the work of the JEM (each country was responsible for the control of its own module). The SSOF comprised a 24-hr operation control room, astronaut training center, space experiment laboratory, space station test facility and zero-gravity test center. The astronaut training center comprised a 10-m-deep swimming tank, isolation chamber and high-altitude chamber.

Underwater training, Tsukuba

Although Japan carried out its preparatory work with exemplary efficiency, the same could not be said for the project leader. *Freedom* went through an endless series of redesigns between 1984 and 1992, eventually exhausting all the money allocated for its construction, although not a single component was ever built. The many redesigns caused nothing but frustration to the United States' international partners, whose designs remained sound despite the many American changes.

Newly elected President Clinton called a halt to the chaos in 1993 and ordered a final, cost-conscious redesign. This, too, came in over budget. At around the same time, the Russian *Mir-2* space station was becoming delayed by Russia's accelerating financial crisis and the only way to save both was by an effective merger of the two programs – a decision reached remarkably quickly by both countries. A new Japanese–American international agreement was initialed to take account of the new arrangements and this was signed by Japan in November 1998, just before the launch of the first module of the station, the Russian *Zarya*. What had been the American space station *Freedom* was renamed, more neutrally, the International Space Station (ISS).

THE ELEMENTS

Although the pressurized module was the largest Japanese contribution to the station, first due up was the logistics module, a 3.9-m-tall, 5-tonne cylinder placed like a hat atop the main pressurized section. Originally, this was intended as a full-sized research module, but this had to be downsized when the station took a higher latitude orbit as a result of Russian participation. The logistics module became a storage facility, with eight equipment racks instead, two being for experiments, called *Saigo* and *Ryutai*. *Saibo* was a biological experiment for animal cell cultures and to grow plants both in zero gravity and simulated gravity. *Ryutai* was a physics experiment for fluids, crystals and chemical solutions.

The pressurized module, or PM, was the center of Japanese participation in the ISS, built by Mitsubishi Heavy Industries. In appearances not unlike *Spacelab J*, the 15.2-tonne module is 11.2 m long, carries 23 experiment and storage racks, requires between 5 and 25 kW of power and can be used by two to four astronauts at a time. It is actually the largest individual pressurized module space on the station.

Astronauts can move in an out of the PM and the LM. Between them, they offer space of 325m^3, almost doubling the total volume of working space available on the station at that point.

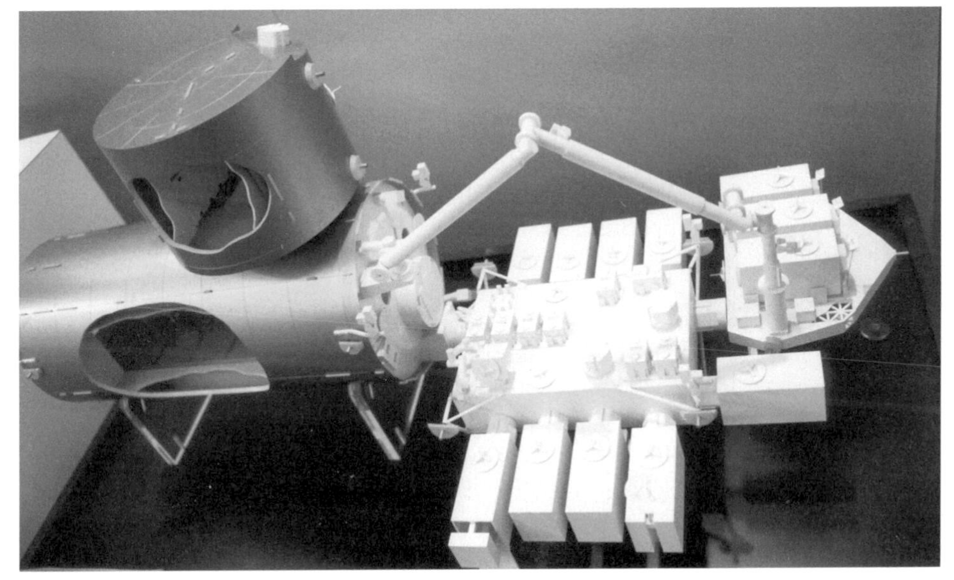

Kibo

One of the first objects to be designed was the JEM Remote Manipulator System (JEM RMS), a robot arm that would be used to move large and small payloads around on the outside of the facility. The JEM RMS comprised a 10-m-long arm able to handle 7 tonnes and a 1.5-m fine arm planned for lighter objects up to 700 kg, the main arm being launched with the pressurized module and the fine arm later. A fine arm prototype was tested out successfully on the Shuttle mission STS-85 in August 1997: astronauts Jan Davis and Steve Robinson put the system through its paces over four days of tests, learning to move objects as it would do on the station.

The Exposed Facility or EF weighs 3.8 tonnes and is attached below the airlock of the pressurized module. Using the remote manipulator arm from the inside of the module, Japanese astronauts will be able to conduct experiments outside, whilst still in a shirt-sleeve environment in the comfort of the module. The platform, which measures 5 by 5.2 by 4 m, can take payloads of all different shapes and sizes and is suited for up to four plug-in modules. The main experiments to be conducted on the outdoor porch will be astronomical and materials processing.

Kibo's activities will focus on basic science, with a particular interest in space biology. It is planned to bring up an aquarium with zebrafish, so chosen because they have a transparent embryo, which makes it easy to follow their stages of development. Among the experiments anticipated in the pressurized module will be the development of semiconductors and alloys, the creation of new materials and medicines, the performance of liquids in weightlessness and life sciences.

SUPPLYING *KIBO*

Kibo would require continuous re-supply once in orbit. Similarly, Europe's *Columbus* module required a re-supply system, called the Automated Transfer Vehicle, the first of which was called *Jules Verne* and arrived at the ISS in spring 2008. Here, Japan originally planned to build its own space shuttle, HOPE (see below). This proved to be too challenging and Japan eventually settled on a more conventional, utilitarian design called the HOPE Transfer Vehicle, HTV. HTV was shaped like a cylindrical can, 10 m long and 4.4 m in diameter, weighing 16.5 tonnes when fully loaded. HTV comprised a propulsion module, an unpressurized middle section with an exposed scientific pallet (1.5 tonnes) and a pressurized section (4.5 tonnes) for astronauts to enter (in an alternate configuration, the pressurized section can incorporate the exposed section). It has two hatches, one 1.2 m in diameter for pressurized cargo and another 2.7 m in diameter for unpressurized cargo. The cost of the HTV program was estimated in 1998 at ¥68bn (€500m), with each production module costing ¥14bn.

The new H-IIB, developed as the carrier of the HTV, is expected to place the HTV in a 350–400-km orbit, 20,000 km behind the International Space Station for a three-day chase.

As it catches up with the ISS, HTV will use the global positioning system to close to 20 km from the complex. At 500 m, radar lasers will be used to bring the HTV in for the final phase, using techniques developed by *Kiku 7*. Coming up to the station

HTV entering orbit

from below, it will halt at 10 m in what is called a capture box, where it will be grabbed by the manipulator arm. HTV will bring up to 7 tonnes of food, clothes, water, batteries, consumables and scientific equipment up to the *Kibo* module. The HTV will spend a month docked to the station, where it will be unloaded and then filled with rubbish before being undocked and burned up in the atmosphere. One flight to the station is envisaged each year through 2016. After two weeks docked to the station, it will be separated for a destructive re-entry. The first HTV was unveiled in spring 2008.

KEEPING IN CONTACT: DATA RELAYS

The final piece in the *Kibo* jigsaw is a data relay system, or Data Relay & Tracking System (DRTS). Early manned spaceflight suffered from problems of limited communications – astronauts could communicate downward with the ground only when they passed, generally quite quickly, over ground or sea tracking facilities. Early in the Shuttle program, NASA introduced the tracking and data relay system, whereby Shuttles communicated *outward* to 24-hr communications satellites, which relayed voice and data back to Mission Control in Houston. Three such satellites were sufficient for global coverage, making it possible for the Shuttle to communicate with the ground throughout each flight. The Soviet Union introduced a similar system for *Mir*, called *Luch*, although, due to financial shortages, it operated only sporadically and China later introduced its own system, the *Tian Ling*.

The Japanese system envisaged two DRTS, Data Relay and Tracking Satellites, called E and W, respectively, because of their positioning at 90°E east and 160°E

DRTS

west respectively, enabling *Kibo* to send 50 MB/sec down and take 3 MB/sec up through their large 5-m dish antennae. The first DRTS, later renamed *Kodama*, was launched on the second operational return-to-flight mission of the H-IIA on 10th September 2002. *Kodama* encountered early difficulty when an apogee motor shut down 2 min early during the third maneuver to lift it to 24-hr orbit, temporarily stranding it at 33,900-km altitude. A fourth and final 39-min burn was successful, leaving it on the station to await the arrival of *Kibo*.

ASTRONAUTS FOR *KIBO*

By 1998, a mockup of the ISS, including the Japanese JEM, had been assembled in the Mockup and Integration Laboratory in Houston, Texas. Visitors could walk into the spacious module, admire the wide range of experimental equipment and storage racks installed on the walls and note the large airlock designed to facilitate access to the exposed facility. They could examine the robotic arm on the outside of the module and the drum-shaped logistic module on top. The module was painted with a red image of the Japanese flag, accompanied by the letters "NASDA" in English and Japanese. As for the real pressurized module, it was completed in the Mitsubishi factory in Nagoya in 2001, brought to Tsukuba for quality inspection and then shipped to Cape Canaveral.

Japan's full astronaut corps

So much for the hardware, but who would fly to *Kibo*? A fourth group of astronauts was selected in 1999. Those chosen were Satoshi Furukawa, born in Yokohama in April 1964, a surgeon in the Department of Medicine hospital attached to the University of Tokyo; Akihiko Hoshide, born in December 1968 in Setagaya-ku in Tokyo, an engineer in the Space Utilization Promotion Department at NASDA; and Naoko Sumino, the only woman of the group. She was born in December 1970 at Matsuda and was an aerospace engineer specializing in centrifuges at NASDA's Tsukuba Space Center. Later, the group was joined by Naoko Yamazaki. Strangely, her interest was sparked off when, as a teenager, she had seen on television the Shuttle *Challenger* explode and, inspired by the courage of the astronauts, this sparked off her interest in space exploration.

They were sent on a training program that consisted of 18 months' instruction at the Tsukuba Space Center, followed by intensified study of ISS systems in the United States. First, they received induction, lecture-based instruction in such areas as engineering and space science. This was followed by specialized training, which covered such areas as the use of spacesuits, spacewalking (done through underwater experience), zero gravity and survival training. Because they would be on board the station for three to six months at a time, candidates were required to pass a long-duration aptitude test. They also went to the Institute for Biological and Medical Problems in Moscow. In anticipation of an emergency return to Earth and landing far off course, Japanese astronauts Akihiko Hoshide and Satoshi Furukawa went to

Sochi on the Black Sea for summertime splashdown and water recovery exercises and later for more rigorous winter survival training in wintertime Siberia. This group was joined by two more astronauts in February 2009, airline pilot Takuya Onishi and air force pilot Kimiya Yui, bringing the astronaut squad to 10.

ARRIVING AT THE SPACE STATION

The Japanese modules had a long wait at Cape Canaveral – five years altogether – as the Space Shuttle was prepared to return-to-flight after the *Columbia* disaster in 2003. The long delays invariably pushed up the cost of Japan's participation in the ISS: whereas the original estimate was ¥310bn, this now climbed to over ¥1 trillion.

The lengthy period of waiting for Japan finally came to an end during the spring night of 11th March 2008. At 2.28 am that morning, the Shuttle *Endeavour* soared into the night sky over Cape Canaveral. Its trajectory took it along the east coast and observers in Massachusetts a few minutes later spotted *Endeavour* as a star tracking over to the east, even noticing the burns of its thruster jets as it maneuvered into orbit. On the Shuttle with six American colleagues on the STS-123 mission was Takao Doi on his second mission. He had been an astronaut since 1985 and had lived in Houston since 1995. In the payload bay was the first of three components for the International Space Station, the Logistics Module (LM). Shortly after arriving at the station, the Japanese module was lifted out of the payload bay and, guided by Takao Doi, temporarily installed on the American *Harmony* module of the station.

Inside the ISS

Takao Doi was the first to enter the new module on 15th March and turned the equipment on two days later. Doi returned to Earth on the *Endeavour* in a night-time landing at Cape Canaveral on 27th March.

The second mission was the Pressurized Module (PM), flying on STS-124, *Discovery*, this time with Akihiko Hoshide, launched on the evening of 31st May 2008. Nineteen minutes later, it had crossed the Atlantic and could be seen in the skies of north-western Europe, its bright orange tank trailing alongside, following the Space Station, which had crossed over only 10 min beforehand. Two days later, *Discovery* had docked with the ISS. The following day, Hoshide manipulated the

HTV arriving

station's robotic arm to gingerly lift *Kibo* out of the Shuttle cargo bay and positioned it on the *Harmony* pressurized module, close to the European *Columbus* laboratory, which had arrived earlier that year. It was then locked in place with 16 motorized bolts. On the fifth day of the mission, after checking the seals for pressure, Akihiko Hoshide became the first person to enter *Kibo*, followed by American astronaut Karen Nyberg, both wearing goggles in case of floating débris that might hurt their eyes. He hung a small curtain outside the hatch, typical for a Japanese home. The rest of the Shuttle crew followed, delighted with the huge space of the empty module. They began to turn on the computer, power and cooling systems. Mission control for *Kibo* was formally transferred to Tsukuba. The module did not remain empty for long, for they quickly began to load it with equipment and experiment racks. Fire extinguishers were installed, cables connected and emergency breathing devices attached.

The next day, Shuttle astronauts Mike Fossum and Ron Garan made the outside of *Kibo* ready to receive the logistics module, which, the following day, was transferred to its place on top of *Kibo*, looking like a hatbox atop the cylinder. The spacewalking astronauts deployed the débris shield around *Kibo*, installed cameras and uncovered the shutters on its windows. Looking through the two portholes at the far end of *Kibo*, Hoshide then tested out its 10-m-long robotic arm. Between the logistics module and *Kibo* itself, Japan now had the largest new laboratory area on the Space Station. For the astronauts, the uncluttered space on the new *Kibo* was a joy, for they could turn weightless somersaults there without fear of colliding with sensitive equipment, which was still stowed in the racks. There was a congratulatory phone call from Prime Minister Fukuda.

Kibo was soon operational. The first experiments were in fluids and convection research, with the next phase focusing on crystallization and biological experiments. In a significant step forward for Japanese manned spaceflight, Koichi Wakata was brought up to the station on the Shuttle *Discovery* in March 2009, spending four months there. When it was time for him to come home in July, the Shuttle *Endeavour* had just arrived there with the porch, or Exposed Facility, which it successfully put in place. For the future, Japan was promised places on three future missions on the Space Station: Soichi Noguchi and Satoshi Furukawa, as well as a mission for Japan's second woman astronaut, Naoko Yamazaki. Table 3.1 summarizes manned Japanese space missions.

Table 3.1. Japanese manned space missions

2 Dec. 1990	Toyohiro Akiyama	Soyuz TM-11/*Mir* space station
12 Sep. 1992	Mamouri Mohri	STS-47, *Endeavour* (*Spacelab J/Fuwatto*)
8 Jul. 1994	Chiaki Mukai	STS-65, *Columbia* (IML-2)
11 Jan. 1996	Koichi Wakata	STS-72 *Endeavour* (Space Flier Unit)
10 Nov. 1997	Takao Doi	STS-87 *Columbia* (Spartan 201)
29 Oct. 1998	Chiaka Mukai	STS-95 *Discovery* (John Glenn mission)
11 Mar. 2008	Takao Doi	STS-123 *Endeavour* (Logistics Module)
31 May 2008	Akhiko Hoshide	STS-124 *Discovery* (Pressurized module *Kibo*)
16 Mar. 2009	Koichi Wakata	STS-119 *Discovery* (Wakata on long-stay mission)

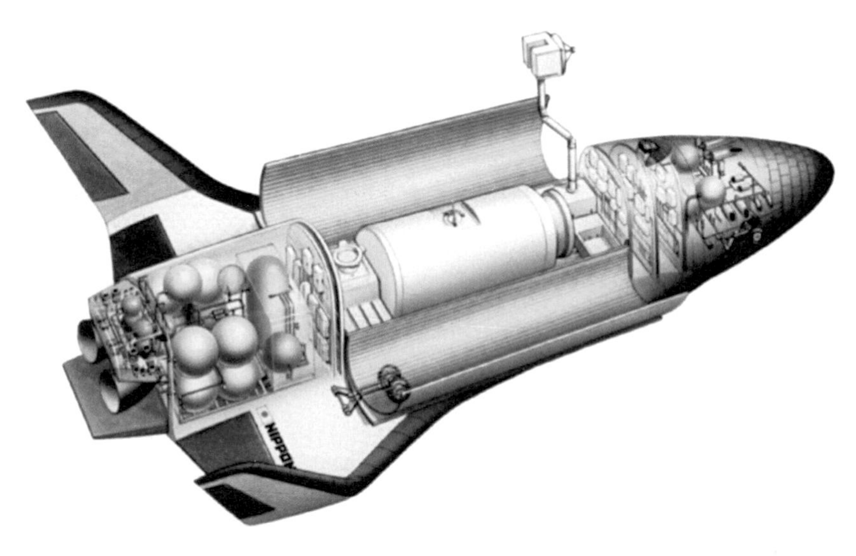

HOPE plan

JAPANESE SPACEPLANES: ORIGIN

Japan originally intended to supply *Kibo* with its own spaceplane, first an automated shuttle and then manned. Spaceplanes date to the beginning of the space age. In 1960, the United States Air Force developed, though never flew, a manned spaceplane called the *Dyna-Soar* (short for dynamic soaring). The Soviet Union developed a plethora of spaceplane projects at the same time, like *Spiral*, but none flew into orbit with a human crew. Both countries eventually developed large space shuttles, the space transportation system or shuttle (1981) and *Buran* (1988), respectively. In the 1980s, the European Space Agency flirted with a small manned spaceplane called *Hermes* to take off on the *Ariane* rocket from French Guyana and return to a runway near Toulouse in France. The promoters of a Japanese spaceplane argued that it would be an economical and convenient way to resupply the Japanese module at the International Space Station; bring back research results; supply other parts of the Space Station; and launch small satellites into space. A spaceplane held out the prospect of radically falling costs for access to space.

NASDA first researched the idea of spaceplanes in the early 1980s, looking at a range of different designs, such as ramjets and scramjets. In 1987, the Space Activities Commission proposed the development of a manned space shuttle, preceded by a series of unmanned tests. Wind tunnel tests were carried out in 1988 in the National Aerospace Laboratory. Research was begun on the types of materials that could withstand re-entry at temperatures of 1,700°C – such as carbon fiber-reinforced polyamide, carbon-reinforced carbon composites, ceramic tiles and titanium. A model was even sent to the Central Aerohydrodynamics Laboratory in Russia for testing.

NASDA defined an unmanned spaceplane, HOPE, 11.5 m long, to ride the forthcoming H-II rocket into orbit, in scale not unlike the manned *Hermes* project. HOPE was closely linked to the arrival of the H-II – indeed, HOPE stood for H-II Orbiting Plane. In fact, the term "HOPE" became a general one used to describe Japan's spaceplane project, sometimes confusingly, covering a range of projects associated with Japanese spaceplane development. At its core was a small spaceplane that would rendezvous with the space station, dock with the help of its manipulator system and return to a runway landing. Later versions would be manned. The Japanese developed the concept in four phases:

- OREX (Orbital Reentry Experiment), testing materials to survive re-entry;
- HIMES (Highly Manoeuvrable Experimental Space vehicle), a 500-kg sub-orbital model designed to be launched from balloon at a height of 20 km to reach Mach 4 and an altitude of 80 km, providing transonic data;
- ALFLEX (Automatic Landing Flight Experiment), a lifting body tested in the Australian desert, providing subsonic data and testing landing systems;
- HYFLEX spaceplane, to provide supersonic data.

DEVELOPMENT AND TESTS

Japan made a considerable investment in its spaceplane project, starting with precursor tests in the 1980s. There was an initial setback, for the first HIMES failed on 21st September 1988 when its balloon collapsed at 18 km before the mini-shuttle was even fired. The gondola separated and the spacecraft crashed into the Pacific. This was a 2-m model, weighing 185 kg, with a solid rocket motor for a Mach 4 flight and gliding test. Four years later, a one-seventh model was successfully tested in February 1992. Launched by a balloon off Uchinoura, it reached an altitude of 70 km before impacting 400 km downrange. By way of historical footnote, the rockoon concept, first advocated as far back as 1956, had now been vindicated.

The next test was OREX, launched on the first ever H-II on 4th February 1994. OREX was released 14 min into the mission, a flattish flying-saucer-shaped dome, 3.4 m in diameter and 1.46 m high, weighing 865 kg, designed to test the carbon and ceramic tile materials that one day might be used on a Japanese spaceplane. OREX, renamed *Ryusei* or "shooting star" once in space, made a single orbit around the Earth. Following a five-minute re-entry burn, thrusters adjusted the angle of re-entry, where the surface temperature glowed white hot to a temperature of 1,570°C, hotter than re-entry on the Space Shuttle. The shooting star blazed through the atmosphere over the Pacific, came through radio blackout and splashed down 460 km south of Christmas Island, though the payload was not recovered.

ALFLEX, 9 m long, 760 kg in weight and a third the size of HOPE, was a lifting body with a squat body, dumpy nose and bent-up wings. NASDA was anxious to learn about the landing characteristics of spaceplanes and, with the cooperation of the Australian authorities, built tracking facilities in the almost deserted Woomera Missile Base in South Australia. Woomera, 500 km north of Adelaide, had been the

Ryusei

center for Britain's rocket efforts in the 1960s: all that was left was the airport and a minor Australian–American joint defense facility. The population had dwindled. Now, NASDA converted a hangar to accommodate ALFLEX and its KV-107A mother helicopter, built a flight control center and installed new navigation systems. Its test program cost ¥422.4m.

The first ground tests of ALFLEX were made in April 1996, tended tests in May and the first glide test on the morning of 6th July. Towed in the air at a speed of 130 km/hr, ALFLEX was released at an altitude of 1,500 m some 2.5 km from the airfield. It dived at 30°, reached a speed of over 200 km/hr and then flattened out for the final stages of landing. Satellite and microwave systems guided the body in to a perfect, computerized desert runway landing.

a.

b.

c.

a. ALFLEX in the desert, b. ALFLEX drop, and c. ALFLEX descent

HYFLEX made only one flight. It was chosen as the payload for a new rocket designed to replace the *Mu* series, called the J-1, but on an up-and-down sub-orbital flight, not an orbital mission. HYFLEX was, in effect, a one-seventh scale model of the HOPE spaceplane, a snub-nosed spaceplane with stubby tail wings made out of aluminum and ceramic tiles. The mission took place on 12th February 1996. The long, thin rocket, decorated with red stripes and the word "Nippon" running down the side, took off from the old N rocket Osaki pad at Tanegashima. The J-1 sent the 1,048-kg, 4-m-long lifting body through the full profile of Mach 14 during a 7.6-min mission over 100 km high. Released at 3.9 km/sec at an altitude of 110 km, HYFLEX came through sub-orbital re-entry, glowing red-hot at a 47° angle. After passing through re-entry, HYFLEX dived nose-down at 30°, making a series of banking maneuvers, until the parachute opened 300 km north-east of Chichinjima, near the Bonin Islands. All went well until splashdown, when the flotation harness broke free of the spaceplane, which then sank in 4,000 m of water, too deep to be salvaged, 200 km north-east of the Bonin Islands. Of the 14 key areas of data required, NASDA obtained information on 12 by radio-telemetry, but missed out on two because of the sinking.

J-1

REVIEWED AND REVISED

The overall development of the HOPE concept proved to be expensive. Between 1988 and 1998, ¥43bn (€383m) was spent on these precursor programs and the subsequent designs. In the late 1990s, the space program went through a number of reviews in the wake of Japan's financial problems. Trying to save something from the all the previous investment, HOPE-X was devised as an intermediate program "to establish the basic technologies for each flight phase of HOPE, using nearly the same size and configuration vehicle as HOPE ... the final stage of developing HOPE", as the coded bureaucratic language explained. The HOPE-X program became a small-scale HOPE precursor, a 12-tonne automated space shuttle able to deliver 3 tonnes to the International Space Station. HOPE-X was 15 m long, with a 9.7-m wing span, a height of 5 m and with twin tails, to be placed vertically on the H-IIA. The idea was that after its mission to the International Space Station, it would head nose-first into re-entry, protected by its carbon and ceramic tiles, and make an automatic touchdown, much like the Russian *Buran* Shuttle.

Tests of a HOPE-X model began in Woomera on 14th July 2002, but went wrong when incorrectly fitted wiring caused a spurious signal to separate the payload. The

HOPE-X

next test was moved to Kiribati, otherwise known as Christmas Island, and took place on 18th October 2002. The site was significant, because it had now been chosen as the place of return for the Japanese spaceplane. Domestic landing sites in the Japanese islands (Magejima Island, Uchinoura and Kamaishi in Iwate) had been rejected because they required overflights of other countries on the descent (China and Korea, respectively) and were in areas of already crowded airspace. By contrast, Kiribati, thousands of miles downrange in the Pacific, offered a good climate and already hosted a Japanese tracking station. Kiribati had a long uninhabited peninsula with an abandoned 1,800-m-long runway built by the British to fly in instruments for the nuclear tests carried out there from 1956 to 1963. NASDA agreed to take responsibility for the development of a new runway, buildings, roads, water and electricity. Japan made an agreement with the Kiribati government, taking a lease on the landing field until 2019.

In its first flight from the field, HOPE-X used a jet engine to rise to 600 m in much the same way as the Russian jet-powered *Buran* tests. Then, in a 9-min 30-sec flight, it used the global positioning system to glide back to the runway automatically but bumpily. The flight took place at 5.50 am, deliberately early before later-rising seabirds would be up and in the air later in the morning to pose a bird strike danger. Several tests were planned with ever greater altitude, speed and approach angle (up to 25°). The next test was on 5th November, when it flew back from 3,000 m in the course of a 20-min flight and came in with a landing angle of 15°. What became the final test in the series took place on 2nd July 2003 from Esrange, Kiruna, northern Sweden, under a balloon, Kiruna being a long-standing site for the launch of balloons and sounding rockets. The HOPE-X model was lifted to the intended height of 18 km, the plan being to drop it for a Mach 0.8 dive and a soft parachute landing five minutes later. Only one of the three parachutes opened and the model was badly damaged when it reached the ground. The nose was dented and the left wing broken.

This appears to have marked the end of the project and little information was published about it since. The Japanese have been equally coy about what, if anything, would replace the J-1 launcher that sent HYFLEX on its mission. Like many space projects, neither appears to have been officially closed down. These were not the only Japanese spaceplane projects of the period. The National Aerospace Laboratory made a number of paper studies of an aerospaceplane in the 1990s and as many as 330 staff were assigned to the project at one stage. The idea was to build a Concorde-shaped aerospaceplane, 94 m long, weighing 350 tonnes, which would take off and land like a conventional jet but, once in space, fly a mission similar to the Space Shuttle. The body of the aerospaceplane would contain a giant fuel tank of slush hydrogen, flying on a mixture of rocket and scramjets.

The only concrete outcome of all this effort was the modest but utile HTV (HOPE Transfer Vehicle) to re-supply *Kibo* on the ISS, but its connection to the spaceplane program is more semantic than real, for its relies on traditional technologies. Space development is littered with failed spaceplane projects among all the main space powers and the Japanese were no more successful than the others in breaking out of this unforgiving field of research. The seagulls on Kiribati are likely to be left undisturbed.

HOPE at the ISS

HOW THE JAPANESE SPACE PROGRAM IS ORGANIZED

We conclude by looking at Japanese space organization, facilities and budgets. The key organizational feature of the Japanese space program from the beginning was the division of responsibility between ISAS and NASDA, the former concentrating on small scientific missions, the latter on technology development and receiving by far the largest share of the budget. Running two space programs within one was arguably wasteful, since NASDA and ISAS each had separate launch bases, launchers, designs, personnel and facilities. The idea of merging the two was the central theme of the Japanese space debate in the 1990s and, after much agonizing, was eventually accomplished on 1st October 2003 as JAXA, the Japanese space agency, its first head being NASDA's Shuichiro Yamanouchi, aged 69.

JAXA has seven directorates (transportation, applications, human flight, aerospace, aviation, lunar and planetary, ISAS) and 21 departments, from finance to public affairs to space education.

The space agency is not the only player in the Japanese space industry, for five other government departments and agencies receive budgets for space development (the Ministry of International Trade and Industry (MITI), the Ministry of Posts & Telecommunications (MPT), the Ministry of Education (originally the parent body of ISAS), the Japan Meteorological Agency and the Telecommunications Advancement Organization of Japan). About 9,500 people are directly employed in space activities in Japan, tens of thousands indirectly through the contracting companies building hardware, many being household names, like Mitsubishi. The Japanese space program operates under the office of the Ministry of Education, Culture, Sports, Science and Technology, while policy is determined by the Space Activities Commission (SAC), which sets the broad outline of space developments. The Space Activities Commission was set up by the Minister for Science & Technology in 1968 and built on the previous work of the National Space Activities Council, set up in 1960. The Commission normally comprises four members or Commissioners. The Commission issued, in 1978, the first 15-year plan, *Outline of Japan's Space Development Policy*, one that is regularly updated, is the working document for the program and outlines key aims such as the promotion of science and technology, meeting social need, the reduction of launcher costs, the promotion of international cooperation, the development of industry, the preservation of the space environment and the balanced use of manned and unmanned systems.

A further important player is the Council for Science & Technology Policy. Generally, new space programs must have the approval of both the Commission and the Council. A third influential policy body is the Space Activities Promotion Council, set up in 1968, which is an industry body, comprising 96 companies involved in the space business.

MAIN FACILITIES

JAXA has 18 field facilities as well as offices abroad in Paris, Bangkok, Houston, Washington, DC and Cape Canaveral. The most important facilities are for launching, tracking, control and testing. Despite the merger in 2003, JAXA continued to operate the two original launch centers and the program's facilities remained largely unchanged. The main facilities were the two launch centers (Tanegashima and Uchinoura), the Tsukuba space center and the tracking network.

TANEGASHIMA RANGE: LAUNCH SITE BY THE OCEAN

Tanegashima is Japan's main launch base and home of its only operational space rocket, the H-IIA, and its variant, the H-IIB. Tanegashima is on an island below the end of the Japanese peninsular and when Portuguese navigators first reached Japan, they made their landfall in Tanegashima. Around a nearby lake may be found a power plant, range control center, telemetry station and checkout facility. The Tanegashima launch site consists of three main pad areas. These are:

- Takesaki, to the south-east, which is used for sounding rockets, solid rocket booster tests and range control;
- Osaki complex, originally built for the Q, N, N-I, H-I and which may be used for future light rockets; and
- Yoshinuba complex, now used for the H-IIA, a double pad able to handle two larger rockets at a time.

The launch pad called Osaki comprised a pad, high mobile service tower and cold propellant storage facilities. The rocket was prepared in a mobile service tower, before being rolled back 100 m before launch. For the final stages of the countdown, the rocket was serviced by two umbilical towers. The pad then lay idle until it was adapted and simplified to take the J-1 solid rocket booster for its only flight with HYFLEX in 1996. In preparation for a launch, rockets and payloads are tested in the adjacent Solid Motor Test Building, the Spin Test Building, clean room and Third Stage & Spacecraft Assembly Building.

Less than a kilometer away, but closer to the headland, is the modern Yoshinuba launch complex, set up to handle the H-II and then the H-IIA. It is one of the world's most impressive launch sites and dwarfs the old Osaki site. Construction of this 150,000-m^2 site began in the mid 1980s and was completed in 1992. Yoshinuba has its own vehicle assembly building, mobile launcher, pad service tower, control center and cryogenic storage services. Rockets are brought the short distance to the pad on the mobile launcher, which rolls down an apron, with parallel roads on either side and a grass centerline with the large letters of the space agency carved into the greenery. The vehicle assembly building is a smaller version of the famous Apollo vehicle assembly building at Cape Canaveral and there is just enough space to stack two H-IIAs side by side. Spacecraft and rockets are put together vertically on a mobile tower in the vehicle assembly building, which has a number of bays. The tower is 22 m high, weighs 800 tonnes and, once everything is ready, makes the 500-m rail journey down to the pad. The crawler has 56 wheels on 14 axles, travels at 2 km/hr and takes 25 min to reach the pad, which is not far away. Here, the rocket meets a pad service tower, a fixed section with two rotating parts. This is 67 m tall, with 12 floors, and uses a 20-tonne crane. A second pad, on a slip road parallel and close to the original pad, was completed in 1999.

Tanegashima includes solid propellant storage facilities and a static firing test center that enables the first stage to be held down on the pad for test firings of its main engines. The test stand for the LE-7 was fitted out in the course of 1989 and has a big cement deflector trench, a water coolant system and, nearby, liquid hydrogen

Tanegashima control

and liquid oxygen storage tanks and a high-pressure gas storage facility for helium and nitrogen.

Launches at Tanegashima are controlled from an underground blockhouse 170 m away, which one enters through a hexagonal, tent-shaped building some 500 m from the pad and controllers take a lift downward. Because the H-IIA carries 275 tonnes of fuel, any explosion is likely to be a large one, so the center is 10 floors underground, with a 1.2-m-thick ceiling, but is bright and airy. The air-conditioning system can provide air for 100 controllers for four hours (there is an evacuation tunnel in any case).

Fishing restrictions historically proved to be a significant restraint on the operation of the Tanegashima launch site. These became increasingly problematical as the program expanded and greater launching frequency was required. New agreements were reached between the government and the fishermen's union to extend the launch window from 90 to 180 days, while still preserving the prime fishing period of March to mid-June.

UCHINOURA LAUNCH CENTER

Uchinoura, also called Kagoshima after the administrative region where it is located, was Japan's first launch facility, used to fire the solid rocket boosters of the ISAS scientific program. It is built on plateaus and in valleys on a hilly mountain site in

Uchinoura launch pad

southern Japan. Inland lie the control, tracking and payload integration centers. The rocket pads are close to the coast: the Lambda center, used to launch the first scientific satellites and then the later Mu center, comprising an assembly building, control tower and launch stand. Satellite preparation buildings and administrative complexes complete the launch site. Optical trackers are located 2 km away on the Miyabaru plateau.

When the *Mu-5* rocket launched *Hinode* in September 2006, it was announced that this would be the last *Mu* mission. Originally, it had been intended that the *Mu* be replaced by the J-1 rocket, but this made only one flight, a sub-orbital one, and its base was set for Tanegashima in any case. *Hinode* may turn out to be the last orbital launch from Uchinoura, which may, over time, be only a sounding rocket base. Sounding rockets continue to be fired from Uchinoura. They are erected in their tower and then tilted toward the sea at the angle at which they should be fired, normally an angle of 80° out over the sea. But, as the launch site for Japan's first satellite ever, its place in the history books is undisputed.

The importance of the sounding rocket missions should not be underestimated. In recent years, they have been used to test solar sails that could be used to enable spacecraft to travel long distances across the solar system using the solar wind. The first solar sail, a mere 300 micrometer thick, was tested by balloon in August 2003.

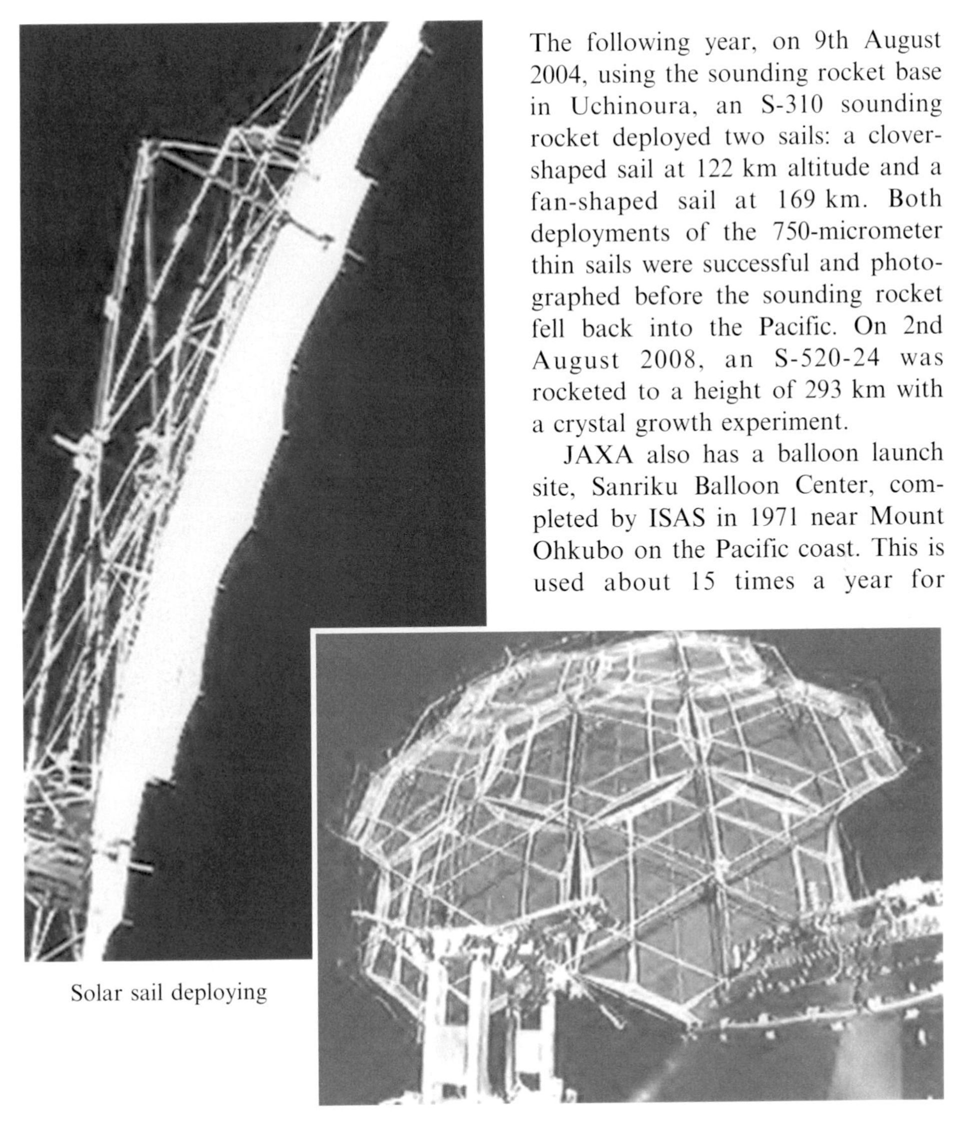

Solar sail deploying

Solar sail deployed

The following year, on 9th August 2004, using the sounding rocket base in Uchinoura, an S-310 sounding rocket deployed two sails: a clover-shaped sail at 122 km altitude and a fan-shaped sail at 169 km. Both deployments of the 750-micrometer thin sails were successful and photographed before the sounding rocket fell back into the Pacific. On 2nd August 2008, an S-520-24 was rocketed to a height of 293 km with a crystal growth experiment.

JAXA also has a balloon launch site, Sanriku Balloon Center, completed by ISAS in 1971 near Mount Ohkubo on the Pacific coast. This is used about 15 times a year for balloons for weather and astronomical observations and several hundred have been launched from there. Some balloons have flown as high as 46 km and others have reached China.

TSUKUBA AND SAGAMIHARA SPACE CENTERS

Tsukuba is to the Japanese what Mission Control Houston is to the Americans, Star Town is to the Russians and Oberpfaffenhofen to the Europeans. Tsukuba is a 53-ha science city 60 km north east of Tokyo, opened by NASDA in June 1972, a combined tracking, control and testing center, equipped with simulation chambers, structural test building, weightless environment test building, acoustic chamber, solar simulation chamber, a magnetic test facility and a 25-m-tall thermal vacuum test chamber. There is a spacecraft clean room integration and test building for payloads on the H-II. The acoustic test facility is 17 m tall, 10.5 m in width and depth, and has a volume of 1,600 m^3; its purpose is to anticipate the full wavebands of vibration experienced in the course of a space mission, the sounds being fed in by an electro-pneumatic sound converter modulating pressurized air. The vibration test facility is able to shake satellites and modules up to 8 tonnes in weight in three dimensions (roll, pitch, yaw) to anticipate the stresses of launch.

Balloon launch

Tsukuba Mission Control had a staff of 60 working in 8-hr shifts, using a mixture of traditional consoles and flat-panel displays. Tsukuba became a manned mission control center in March 2008, when the pressurized module *Kibo* arrived at the International Space Station with astronaut Takao Doi. People listening in the exchanges between the station and the ground became used to hearing new words: KCC (*Kibo* Control Center), J-Com (Japanese Experiment Module Communicator) and J-Flight (JAXA flight Director). Other personnel filled key positions for electrical systems, robotics, payloads and spacewalking. A payload room was built beside the Mission Control room so that experiment leaders could be on hand as experiments were conducted on board. There is a separate control room for the HTV.

Japan's other space center is Sagamihara, the original home of ISAS. It is 40 km west of Tokyo and back-dropped by mountains. The center comprises a mixture of

Sagamihara

high and medium-rise complexes, not unlike a modern university campus, hosting facilities for astrophysics, plasma, planetary science, basic space science, systems engineering, transportation, propulsion, spacecraft engineering, advanced systems, utilization and applications. Sagamihara includes facilities for the development of advanced systems and a data center to process, store, analyze and distribute space science data through a range of databases and supercomputers.

TRACKING FACILITIES

Tsukuba hosts the main Japanese ground station, but tracking stations are located in Katsura (south of Tokyo), Okinawa, Masuda (near Tanegashima), Uchinoura and downrange in Chichi-Jime island (Ogasawara). Weather and Earth resources data from Japanese satellites are received separately by the Japanese Meteorological Center, set up in 1977 under the Japan Meteorological Agency, in turn responsible to the Ministry of Transport. The center operates a ground station at Hatoyama, 35 km north-west of Tokyo, where it has two 18-m dishes, assisted by distant stations in Ishigaki Island and Crib Point in Australia. Once signals are picked up, they are computer-processed and distributed to their wide range of users.

For deep-space missions, there is a large tracking dish at Usuda, 1,450 m high up in the mountains in the middle of the island of Honshu. The core of Usuda Deep Space Center is a 64-m dish originally used to track *Sakigake* and *Suisei* to comet Halley, but deployed for all the subsequent deep-space missions and even for the American *Voyager* mission to the outer planets.

Usuda dish

ROCKET TEST CENTERS

As noted, some rocket testing is carried out at the launch centers, but most space programs require dedicated rocket engine test facilities, like Marshall Space Flight Center (Alabama, United States) or Sergeev Posad (Russia). Here, there are two facilities. First, ISAS built the Noshiro Test Center in 1961 near Asani beach in the north-west of the main island of Japan, Honshu, away from population centers. It is used for testing solid and liquid-fuel rocket motors, including cryogenic engines, and has vertical and horizontal test stands and fuel tanks. Second, NASDA built the Kakuda Propulsion Center, finished in July 1980. Here, the LE-5 and LE-7 series of engines were perfected. The center has a high-altitude test firing stand to simulate firings in the space environment and a tank to simulate vacuum, heating and solar radiation.

KEY COMPANIES

As is the case in Russia, Europe and the United States, the production of spacecraft, rockets and equipment is contracted out to large industrial, defense and technology contractors. Many of the Japanese companies are household names in the West because of the wide range of their products, such as cars and electronics.

The principal industrial companies involved in the Japanese space program are Ishakawajima Heavy Industries (IHI) (upper stages and small engines), Mitsubishi (liquid-fuelled rockets), Nissan (solid-fuelled rockets) and Rocket System Corporation (launch services, e.g. the H-IIA). Some of these companies have large subsidiaries and so are involved in a very full range of rocket and spacecraft activities – for example, Mitsubishi has a large electrical division (Mitsubishi Electric Company, also called Melco).

The two main companies building rockets are Mitsubishi and Nissan. Mitsubishi had a long involvement with the N series and the H-I and was prime contractor for the H-II, building both the rocket bodies and the engines, and even has its own test stands and facilities (e.g. Tashiro Test Field, Nagoya). Nissan made the *Mu-5* satellite launcher and many of Japan's other solid-fuel rockets and strap-on boosters and also has its own test facilities – a research and development center at Kawagoe and a test facility at Taketoyo, where solid rocket engines may be static tested. For satellites, the main Japanese industrial companies are Mitsubishi Electric (Melco), Toshiba, Kawasaki and Nippon Electric Company (NEC). Many of these companies are large global trading companies with substantial in-house research and development experience. For example, Nissan has a large 480,000-m^2 plant in Tomioka, with 800 staff devoted to the design, development, production and testing of space equipment.

The main companies involved in the provision of satellite-based television services are Space Communications Corporation of Japan, Japan Satellite Systems, the Broadcasting Satellite System Corporation and NHK (Nippon Hoso Kyokai) and these have, in turn, a long history of inter-relationship, merger and rivalry.

JAPANESE SPACE BUDGET AND AMBITIONS

Space spending by Japan got off to a slow start, only about ¥2.940bn (€26.2m) being committed a year in the mid-1960s. Spending began to rise in the mid-1970s, associated with the formation of NASDA and the substantial costs of developing the N rocket. Substantial development costs were associated with the H rocket, spaceplanes, *Kibo* and participation in the International Space Station Program. Space budgets then plateaued in the late 1990s with Japan's financial crisis, leading to cancellations (HOPE), delays (lunar programs), economies (replacement of H-II by H-IIA) and some programs consigned to an uncertain future (J-1 launcher). At one stage, fanciful designs had been circulated of a Japanese colony on the Moon (*Lunar city 2050*). Instead, Japan had to abandon its ideas of even building a spaceplane. The most striking single feature of the program is how little was spent on the space science part of the program, granted what it achieved. Only a few years before the merger, ISAS received only 7.3% of the space budget, but not only was able to run an impressive deep-space program, but managed its own launchers, launch sites and facilities, demonstrating a remarkable level of efficiency.

The Japanese space budget has improved little in the early 2000s, at a time of much increased space spending by its Asian rivals, China and India. The budget rose hardly at all, from ¥275bn in 2000 to ¥277.8bn in 2006 (€2,013m). Still, it is the third space-faring nation by spending, after the United States by far the largest, and Europe. This figure does not include the military reconnaissance satellites.

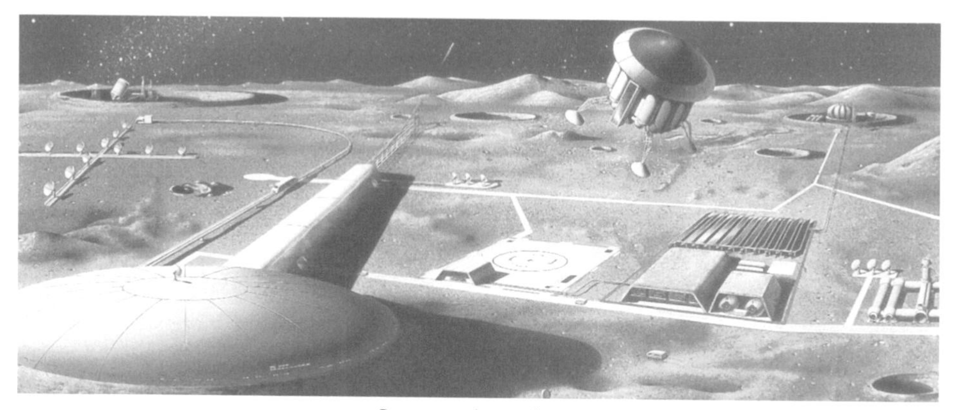

Japanese lunar base

The future Japanese space program works according to an aspirational long-term plan, *Vision 2025*. Signs that retrenchment could not last forever came on 22nd October 2003, when the science and technology minister decided to set up a commission to review Japanese space aims for the following 30 years, including long-term participation on the International Space Station and manned flight. The decision was almost certainly prompted by the success the previous week of China in orbiting its first astronaut, Yang Liwei, marking out China as the leading Asian space-faring nation and was given an added edge by the later decision of India to launch its own astronauts by 2014. China's achievement raised a real question as to Japan's hitherto leading role in Asian space exploration and where it saw its future priorities.

CONCLUSIONS

During its rapid expansion in the 1980s, it was natural that Japan should consider an involvement in manned flight and that it should do so in collaboration with its traditional partner, the United States. Japan recruited an astronaut squad in the 1980s and its members subsequently flew on the Japanese space laboratory, *Fuwatto*, and on other shuttle missions, including the retrieval of its Space Flier Unit. It was logical for Japan to become involved in the International Space Station, again through the United States as intermediary. With its previous experience, Japan was in a good position to plan and build a substantial module, *Kibo*, with a logistics module, exposed facility, one of the largest components on the station, accompanied by a re-supply system (HTV). As a result, Japan was set to achieve a permanent manned presence in space, albeit one reliant on other countries to deliver its astronauts to the station. Japan's own frustrated attempts to build a spaceplane had shown that its reach was beyond its grasp. Despite that, a permanent presence on the station, the largest international scientific endeavor ever mounted, is no mean achievement.

In the competition to launch an Earth satellite, Japan had beaten China by just a few months in 1970. Now, Japan had been overtaken by China and faced a strong challenge from India, to where our story moves next.

4

India: The vision of Vikram Sarabhai

The ancient texts of Sanskrit – the *Vaimanika Shastra*, *Rmanyane*, *Rig Veda* and *Mahbharatha* – tell of how, many years ago, the whole sky blazed when a rocket ship called the vimana ascended into the heavens. The best translation of the Sanskrit tells us that the vimana "gave forth a fierce glow, the whole sky was ablaze, it made a roaring like thunderclouds" and took off.

Interpreting the fragments of ancient Sanskrit is a dangerous exercise. These texts have received as many interpretations as there are interpreters. They range from those who see them alternately as a rocket user manual, an Indian equivalent of the Bible's Book of Revelations, to an early new-age fantasy, to clear proof of the landing (and subsequent return) of an alien civilization from another galaxy. Either way, they are the earliest known written account of anything approximating to a rocket.

India was a great and ancient civilization, with medical texts dating to 800 BC. Indian astronomy reached its peak in the 5th century under the great astronomer and mathematician, Aryabhata, who figured out that the Earth revolved around the Sun 1,000 years before Europeans did. Waves of invasion and colonization led to the decline of science in India. Ironically, the last weapon used by Indians against the British colonizers was the rocket. Hyder Ali and his son, Tippu, used small bamboo gunpowder rockets against the British army in the battles in Srirangapatna in the 1790s: they could travel up to 3 km but were not very accurate. They were probably derived from China, where the rocket had been invented as far back as the 13th century. Tippu's rockets were brought back to England for testing (they can be found today in Woolwich Museum). Using these Indian designs, Britain then deployed the first ever "rocket troops" against Napoleon during the Peninsular War.

Not until independence in 1947 was India able to make a fresh start. The concept of harnessing science to build the modern India was in many ways the achievement of modern India's first Prime Minister, Jawaharlal Nehru, who declared:

> "Science alone can solve the problems of hunger and poverty, insanitation and illiteracy, of superstition and deadening custom and traditions, of vast resources running to waste, of a rich country inhabited by starving people."

FATHER OF INDIAN ASTRONAUTICS, DR VIKRAM SARABHAI

Although Vikram Sarabhai's legacy is principally seen as the Indian space program, he was one of the builders of modern India. He was an institution-builder, creating no less than 42 institutes and agencies in fields as diverse as science, industry, management and education. India's independence was important to him: his family was in the textile milling business and his home town, Ahmedabad, was known as "the Manchester of India". But, with independence, the industry could move from just production to doing its own research and development. He appealed to Indian scientists abroad to return home and build the new nation.

Vikram Sarabahi was born into a wealthy Indian industrial family on 12th August 1919, his father Ambalal being a mill owner. Vikram went to Montessori school and then had English private tutors. He grew up in a highly political atmosphere, being part of the élite that led India to independence and subsequently formed its first governments, personally meeting many leaders of the Congress Party, such as Jawaharlal Nehru. When Vikram was three, Mahatma Gandhi began his campaign against the caste system: among the mill owners, only Ambalal Sarabhai supported the campaign and even bankrolled it, to the disgust of his fellow industrialists. Vikram's mother, Sarladevi, was a social activist, feminist and promoter of Montessori schools in India and his sister was a political activist. His aunt, Ansuyaken, was leader of the national labor movement. Mahatma Gandhi frequently visited – indeed, his *ashram* was in Ahmedabad.

Vikram Sarabhai graduated from Gujarat College in 1937 and went to St John's College, Cambridge, whence he graduated in physics and mathematics in 1939. Returning to India at the outbreak of World War II, he went to study in the Indian Institute of Sciences under Chandra Shekara Venkata Raman (CV Raman) (1888–1970) and, in 1942, published his first scientific paper, "Time Distribution in Cosmic Rays", in the *Proceedings of the Indian Academy of Sciences*, vol. 15. The family used to holiday in the Himalayas: Vikram Sarabahi saw the advantages that high altitudes (4,237 m) provided for the interception of cosmic rays, so next time, he brought with him cosmic ray counters, which were transported up there on a pony. During a visit to his mountain station in Gulmarg, near Alpathari lake, in 1943 to carry out high-altitude experiments, he conceived the idea of a dedicated laboratory for cosmic and atmospheric physics, even though he was aged only 23 at the time. Trying to get a broad range of data across India, he also installed cosmic ray counters in

Vikram Sarabhai

Bangalore in the set of meteorological instruments at the base of the RAF's 225 squadron.

After the war, Vikram Sarabhai returned to Britain for two years to complete his doctorate in the Cavendish laboratory, entitled *Cosmic Ray Investigations in Tropical Latitudes*, which he was awarded on 24th May 1947. The post-war period saw several countries establish cosmic ray institutes – Ireland, Norway, Sweden and Belgium – and he looked at comparable institutes during his stay in Europe. On his return to India in 1947, Vikram Sarabahi duly established the Physical Research Laboratory (PRL). It started modestly enough, in his own home, moving to a temporary office, with the foundation stone laid by C.V. Raman on 15th February 1952 and the formal opening by Prime Minister Jawaharlal Nehru on 10th April 1954. The PRL developed outstations in Kodaikanal, Thiruvananthapuram and Gulmarg. Vikram Sarabhai was a hands-on researcher, insisting on personally making his own experimental equipment. He brought his students with him to the observing station in Kashmir. He was a director of the PRL from 1947 to 1971, supervising Ph.D. students over the entire period. He himself was author of no fewer than 85 scientific papers. From 1947 to 1966, in between all his other activities, he managed the family textile business and its network of companies.

Vikram Sarabhai as a young man

SPUTNIK AND THE IGY

Vikram Sarabhai's entry into the space arena came with the International Geophysical Year (IGY), then set for 1957–1958, which had been an important stimulus in Japan. In his capacity as director of the PRL, Vikram Sarabhai drafted the proposal for India's participation in the IGY and the Indian government sent him to Brussels, Belgium, from 30th June to 3rd July 1953, to present the proposal to the international community.

The launch of Sputnik during the IGY completely changed the public and governmental attitude toward space research: Sarabhai immediately wrote to the government to suggest that India, too, should now consider building an Earth satellite. Sputnik was tracked by the Utar Pradesh state observatory and, several years later, the Physical Research Laboratory in Ahmedabad built equipment to receive satellite telemetry.

But what path should India follow? Although the two space powers had now

Vikram Sarabhai with Nehru

embarked on a race to the Moon, Vikram Sarabhai felt that this was not an appropriate path for India to follow. Vikram Sarabhai met with people such as Arthur C. Clarke, who came to live in Sri Lanka and who had a vision of space applications. Japanese space pioneer Hideo Itokawa visited him soon after Sputnik and he also informed his thinking. "My priorities are different . . .," said Sarabhai in presenting his proposal to the Indian government in 1961, suggesting a small, but highly focused space program. Here, Sarabhai articulated his vision of a space program orientated toward development, using television for agriculture, health and family planning and for telephone, by-passing microwave systems. The government agreed and in August 1962, set up the Indian National Committee for Space Research (INCOSPAR), formally under the Atomic Energy Authority, with himself as first chairperson. In reality, the space program was run by Sarabhai from the Physical Research Laboratory in Ahmedabad. INCOSPAR's task was to advise the government on space research, promote international collaboration and participate in international activities.

The program would begin with sounding rockets – but where could they find a launching site? Ever since 1841, when the first magnetic map of the sub-continent had been made, it was known that India's magnetic equator, on its south-west coast, was full of electrical anomalies. Two sites there were considered near the west coast town of Thiruvananthapuram: one was at the seaside, a place called Vellana

Vikram Sarabhai 1960s

Thuruthi, but there was hesitation because the name meant, when translated, "white elephant". They settled instead on Thumba, 16 km away, a site owned by Christian missionaries. The scientists explained to the local bishop that the ultimate purpose of the space program was to benefit the poor of India and the bishop explained it so to the congregation in the church. The congregation gave them permission. Conditions were primitive: whereas Russian rockets went down to the pad on railways and American ones on road crawlers, Indian rocket parts were transported around by bicycle. Snakes chased rats in the undergrowth round about.

FIRST ROCKET LAUNCH, 1963

With the help of the American and French space agencies, NASA and CNES, the launch of India's first sounding rocket, a home-built version of the French Centaure,

Early rocket on a bike

India's first sounding rocket

took place on 21st November 1963, launching 25 min after sunset so that the engineers could follow it into the night sky, where it released a sodium vapor cloud at its high point of 200 km. Nationally, the impact of India's first sounding rocket achievement was lost, for it was eclipsed by the shocking news of the assassination of President Kennedy the following day.

Sarabhai proposed that the center be turned into an international launch center and its formal title became the International Equatorial Sounding Rocket Facility. This was agreed by the United Nations in January 1964, but the formal dedication did not take place until February 1968, being carried out by Prime Minister Indira Gandhi. Partners in the venture were NASA, CNES and the USSR Hydro-meteorological Service, emblematic of India's desire to be even-handed in its choice of partners abroad and not be part of one of the world's superpowers or the other. Under the agreement, NASA was responsible for the telemetry and the ground units. CNES supplied Centaure rockets and radar. Russia provided computers, ground testing and a Mil helicopter. Meteorological rockets were typically fired to 70 km, with two-stage scientific rockets to 180 km, many carrying barium releases.

Sixty-five sounding rockets were fired from Thumba over 1963–1968: British Skua Petrel (for the Met Office), the French Centaure, the American Nike and the Soviet M-100. Firings continued into the 1970s: at one stage, the M-100 was fired once a week to collect data in

Early Indian sounding rocket

the 55–80-km altitude range and 20 Skuas were fired for the Met Office in London. The French Centaure 2 was modified for tropical conditions and a propellant tank built at Thumba for its fuel, later enabling India to become self-sufficient in solid rocket fuel. It began flying in 1969 and was able to reach 400 km. The first entirely domestically built sounding rocket, the *Rohini 75*, was launched from there in November 1967. It was just 75 mm in diameter, the *Rohini 75*. This was the first in a family of Indian sounding rockets, from the 2.3-m-tall *Rohini 125*, the 4-m *Menaka* to the 10-m-tall *Rohini 560B*. From 1969, Indian-made propellants were used. In the course of their evolution, the payload of Indian sounding rockets grew from 16 to 90 kg and the altitude reached rose from 90 to 360 km. Sounding rockets, although they might look simple, had up to 300 components, all of which had to be made to a high level of reliability and precision – a vital first step toward making rockets for orbit. The diameter of the domestic sounding rockets grew from 75 mm eventually to 1,000 mm, with up to four stages.

Construction on site brought India advances in welding and the use of materials such as fiberglass and graphite and, by 1972, 3,000 people were working there. In other advances, the country began to receive American weather satellite information (TIROS) (1964) and more than 100 high-altitude scientific balloons were launched from Hyderabad with scientific payloads of 250 kg up to 36 km on eight-hour flights, many of them cosmic ray missions.

SPACE PROGRAM FOR EDUCATION

One of the most important experiments initiated by Sarabhai was SITE (Satellite Instructional Television Experiment). Originally, Sarabhai had wanted to test the idea of village educational television through an Indian-launched satellite, but the opportunity to do so arose earlier through an American satellite, the Applications Technology Satellite-F (later called ATS-6). Eighty percent of Indians lived in villages then: Sarabhai wanted to prioritize the villages because they were the least developed parts of the country and posed the greatest challenges. In "Television for Development", a talk given to the international development conference in Delhi on 14–17th November 1969, Vikram Sarabhai said:

> "We should consciously reach the most difficult and least developed areas of the country and, because they are in this state, we should reach them in a hurry."

Although Delhi had television from the 1950s, its reach was only 30 km and there was still no national service, so all-India TV proposed a national system, starting with ground masts in the country's 17 largest cities and then spreading outwards. Vikram Sarabhai took the opposite approach, proposing instead that television be first relayed to the villages, not from expensive masts, but instead from geosynchronous satellites. In cooperation with NASA, space scientists undertook a cost–benefit study of the various possible combinations of ground and space-based relays. A core question was the cost of receivers. Here, Sarabhai argued that receivers could be made out of cheap, local materials, the dishes themselves out of nothing more complicated than chicken wire. The feasibility study found that a satellite-based system was actually cheaper, Rs577m, while a terrestrial system would be Rs1,851m.

Vikram Sarabhai leaving for meeting

SITE was the key phase of a multi-stage idea. The idea was to do a pilot experiment, then SITE, then a foreign-built communications satellite (INSAT 1), then a domestic-built comsat (INSAT 2) and then a domestically launched comsat. Leader of SITE was B.S. Rao, a physics and electronics expert who went to France and the United States to study direct broadcasting.

This discussion led to the signing on 18th September 1969 of the *Television for Development* agreement between NASA and the Atomic Energy Commission (the state body with overall responsibility for the space program).

The arrangement was that the NASA satellite ATS-F would be moved to 20°E to begin an experimental program of direct broadcasting to the villages of India, with a special focus on the most backward areas: Orissa, Bihar, Utar Pradesh and Rajasthan. ATS programming would cover farming, occupational training, health, hygiene, and family planning. Vikram Sarabhai was very aware of the power of television:

> "TV can provide a qualitative improvement in the richness of rural life and thereby reduce the overwhelming attraction of migration to cities and metropolitan areas through education and entertainment of a high standard."

While awaiting the launch of ATS-F, a pilot was done using conventional television with farmers close enough the main transmitter in Delhi. An agricultural instruction program was transmitted, called *Krishni Darshan*, with 20 min a week instruction on poultry, fruit and vegetables and caring for animals, accompanied by question-and-answer sessions, with the instruction animated by two farm characters, Chandhariji and Panditji. To make the process properly scientific, the results were compared between participating farmers and a control group not exposed to television instruction. The outcome: those who saw the programs improved their use of fertilizer, weed control and had higher wheat yields. In 1970, French television made a film in Ahmedabad about the role of space applications in development. Sarabhai spoke of how satellite television would enable the villages of rural India to progress economically, so that village living should be as modern as city living, imagining a day in which such things as bank transfers could be done electronically by satellite. In *INSAT – A National Satellite for Television and Development*, Vikram Sarabahai outlined, with his colleagues, how, in time, India would design, build and launch its own geosynchronous satellites.

India would still require Earth stations to transmit to ATS-F. Here, Vikram Sarabhai brought together a team of engineers at Ahmedabad: some had experience of microwave stations, but essentially they had to start from scratch, searching the existing literature for clues and ideas. The United Nations Development Program (UNDP) and the International Telecommunications Union provided expertise and equipment. Transmitters and receivers were provided by Nippon Electric in Japan.

Originally, the Indian minister for communications decided that the first Earth station would be built by the American company RCA. When he found out that the contract was ready for signature, Sarabhai was furious, believing this to be well within the capacity of Indian engineers. A "nervous, angry, red and trembling" Sarabahi contacted the Prime Minister, who overruled the minister.

Ahmedabad Earth station, with a 14-m dish, duly opened in August 1967, being called the Experimental Satellite Communication Earth Station (ESCES) and had been built in 87 days. As soon as it was working, it made contact with the American Applications Technology Satellite 2 (ATS-2) communication satellite. Sarabhai turned it into a training school, to which other countries sent their experts for instruction. Sarabahi insisted that the station be built by Indian engineers rather than a foreign construction company.

Indian engineers next built the first operational Earth station in India, at Arvi,

Earth station Ahmedabad

Vikram Sarabhai turning first sod

Early satellite television

near Poona. This had a dish more than twice the size (29 m), designed for overseas satellite communication via the international INTELSAT system. Consideration was given to buying in a system from abroad, but even though it took longer, it was decided to build all the equipment at home. A third station was then opened in New Delhi.

SPACE PROGRAM FOR REMOTE SENSING: THE "HIGH ROAD"

Even though Vikram Sarabhai was keen to develop an indigenous space applications program, he knew that it would be a long time before India could do so. In the meantime, India must rely on other countries. Vikram Sarabahi was very taken by the quality of the American Gemini pictures taken by astronauts from Earth orbit. In "Remote Sensing in the Service of Developing Countries", an address given to the Indian Geophysical Union on 29th December 1970, he outlined the possibilities of space-based oceanography, geology, hydrology, cartography for India. His idea was to send 12 top students abroad, train them up in remote sensing and, on their return, have them teach 40–50 people, who would then constitute the core of an Indian national remote sensing center. Already, India was testing out the use of French sensors in remote sensing from a Russian-built Mil helicopter with a NASA scientist, 300 m over coconut plantations in Kerala, to detect root plant disease from its infrared signature.

Vikram Sarabhai was the first chairperson of the United Nations conference on the peaceful use of outer space, where he spoke in Vienna, Austria, on 14th August 1968. This conference was the one that, more than any other, marked the use of space research for Earth applications. On his return, Sarabhai at once created a unit in the Physical Research Laboratory in Ahmedabad as the starting point for Earth resources work by India.

Vikram Sarabhai at time of UN speech

Here, he urged that developing countries use space exploration as a means of escaping dependence and poverty and, instead, leapfrog ahead. He extended atom scientist Homi Bhabha's idea of "the technology high road". You have to move ahead of the developed countries, he said: merely trying to keep up all the time means you will always use obsolete methods and lose before you even start. Sarabhai was an advocate of indigenization: developing countries should learn from advanced countries, match them and then develop the next stage of that technology themselves. The key elements of indigenization were, he said, to selectively follow foreign example, but to retain management, research and the building of equipment at home, persuade Indian students to study abroad but then return and emphasize high standards of domestic quality control and zero-defects. In 1963, he was appointed chairperson of the Electronics Committee, dedicated to the indigenization of India's electronics. In writing the preface to the Indian space plan for 1970–80, he declared:

> "We cannot have 20th century space research with 19th century industry or antiquated systems of management and organization .... We have to rise from an in-built culture within which a major departure from an existing well-proven system and anything which is innovative in character is regarded with suspicion."

His most famous, enduring articulation of Indian space objectives, around concrete economic and social purposes, was given at the formal dedication of Thumba to the United Nations on 2nd February 1968, when he said:

> "There are some who question the relevance of space activities in a developing nation. To us, there is no ambiguity of purpose. We do not have the fantasy of competing with the economically advanced nations in the exploration of the Moon or the planets or manned space flight. But we are convinced that if we are to play a meaningful role nationally and in the community of nations, we must be second to none in the application of advanced technologies to the real

> problems of society which we find in our country. The application of sophisticated technologies and methods of analysis of our problems is not to be confused with embarking on grandiose schemes whose primary impact is for show rather than for progress measured in hard economic and social terms."

The records show that Dr Itokawa from Japan was there, for he had tried to bring his country's space program into orbit at around the same time. Indeed, this was the beginning of a connection between the early space programs of India and Japan. At an international astronautical conference, he had met Vikram Sarabhai and they became personal friends. Sarabhai persuaded Indira Gandhi that the Indian government should invite Itokawa to be advisor to the fledgling Indian space program for three years, then at a crucial stage of development. Itokawa regularly traveled with Sarabhai between Delhi, Bombay, Ahmedabad, Madras and Thiruvananthapuram, supervising the setting up of the institutes of the Indian space program. He found Sarabhai a hard man to keep up with and could not match his 21-hr working day. It was Hideo Itokawa who proposed to Sarabhai the name Rohini, one of the Hindu gods, for India's satellites. At the end of his life, Hideo Itokawa wrote of this period in his 1994 book, *Third Road – India, Japan and Entropy*.

SUDDEN END

Vikram Sarabhai was a man of phenomenal personal energy. He rose routinely at 5.30 am every morning and often didn't get to bed until 1.30 am. His family urged him to take time off, to relax and Prime Minister Indira Gandhi even ordered him to, with equally little result. His widow, Mrinalini, remembers him as someone who "was always burning the candle at both ends", in a hurry, constantly imagining, always thinking ahead to new projects. On 30th December 1971, he attended a conference in Kovalum, Thiruvananthapuram, that ended at midnight. He was due to return to Bombay early the following morning – a hectic but quite typical schedule for him. He died suddenly during the night, aged only 52, apparently suffering little pain.

His early death caused widespread shock and grieved his colleagues and the country. Later, his stunned colleagues spoke of him much as Soviet space scientists recalled the organizer of the Soviet breakthrough into space – Sergei Korolev. They spoke of his child-like enthusiasm for space travel, his vision of space applications in the service of Indian economic development, his ability to organize, lead, manage and inspire. At a personal level, they remembered his consideration, his humanity, his warmth, patience, his trust of people to do the job assigned to them. They admired his interest in art, music, literature and politics, which matched his passion for science, technology and social reform. He always urged scientists to be involved in their countries and implored them to put science to work in the service of practical needs. He himself endlessly proposed projects for communication, agriculture, family planning and education. He was showered with awards during his lifetime and

Vikram Sarabhai's last picture

many more posthumously. He was admired for his professional and personal qualities: informal in manner, never pretentious, self-assured but not confrontational, a man who left no enemies: "No one had a bad word about him," everyone said. Many of his students became leaders of the Indian space program. The close connection between rocketry and state building was evident when his student, Abdul J.P. Kalam, was elected 11th president of India for a five-year term on 25th July 2002, though to describe Kalam as only a rocketeer would be unfair, for he was also a poet, spiritualist, writer, educator and philosopher, a man after Sarabhai's broad interests, too.

According to Dr Padmanabh Joshi, who studied Vikram Sarabhai's role as an institution-builder, the role of a founder of an organization is critical, for the founder sets the design, structure, values and the culture of everything that follows. The full story of Vikram Sarabhai has still not yet been told, for his extensive papers have not yet come into the public domain. They have probably gone to the archive of the Atomic Energy Commission, which is a far from open institution. Even the current President of India was unable to make any progress in finding or releasing them.

Crater Bassell, on the Moon at 21°E, 24.7°N in the Sea of Serenity, was re-named crater Vikram Sarabhai. His widow, Mrinalini Swaminadhan, whom he married in 1942, founded India's leading classical dance academy, Darpana, which continued to thrive.

ATS: VILLAGE TELEVISION

Sadly, Vikram Sarabhai did not live long enough to see the *Television for Development* agreement come to fruition. The satellite which made it possible, ATS-6, was launched in May 1974. It had an unusually large 9.1-m antenna for direct broadcasting and was used for pioneering such diverse activities as teleconferencing and broadcasting medical advice. NASA now moved ATS over India for the period August 1975 to July 1976, the programming being uplinked from India's first ground station in Ahmedabad. One thousand and two hundred hours of ATS programs, made by the Space Applications Center in Ahmedabad, were relayed to 2,400 villages and were seen by an estimated 5m people. The signals were relayed up from Ahmedabad and New Delhi to ATS and then retransmitted for four hours every day. Programs were put together in a tiny studio made available by a friendly education officer in Bombay city council.

Transmissions went out for three hours a day, of which half an hour was science education for children. The programs covered such subjects as agriculture, health, hygiene, family planning, rural development and literacy. The 3-m-diameter ground sets were made out of inexpensive equipment like chicken wire and signal converters. Television sets were established in the designated areas, often in schools, with equipment ferried around the country by jeeps and trucks. According to Vikram Sarabhai, SITE was intended to be designed to be operated by "villagers with no

SITE school

SITE jeep crossing rural India

SITE village television

previous experience of tuning knobs, with fluctuating electricity supplies, able to withstand the onslaught of rodents, insects and mould growth and finally be repaired with locally available components".

The highest uptake of ATS programming was from women and illiterate viewers. Villagers and their children learned how to operate and fix television sets. Contrary to expectations that people would only watch entertainment programs, the highest uptake was for "hard" instructional educational programs. The benefits flowed in many different ways. Sarabhai had written beforehand: "Scientists and engineers need to get cow dung on their feet ... while social scientists need to understand the spirit of technology." SITE was evaluated by a hundred social scientists, who went to 27 selected villages before, during and after the experiment, with resident anthropologists in nine villages for 19 months. Villagers, including their children, learned how to fix faults. In time, SITE was called "the largest sociological experiment in history".

In a follow-up experiment called STEP, or Satellite Telecommunications Experimental Project, tests of ground communications were made from 1977 to 1999 with Europe's satellite *Symphonie*. This concentrated on the testing of truck and jeep-based sending and receiving stations, the development of multi-lingual educational satellite television, the use of satellite communications in response to natural disasters and for broadcasting other matters of national importance (viewers noted that cricket matches fell within this definition).

The success of SITE gave Indian space scientists the confidence to develop wider applications programs. The country lost each year Rs50bn (€1bn) to pests, Rs15bn (€319m) to droughts and Rs7.7bn (€163bn) to floods – but satellites could give

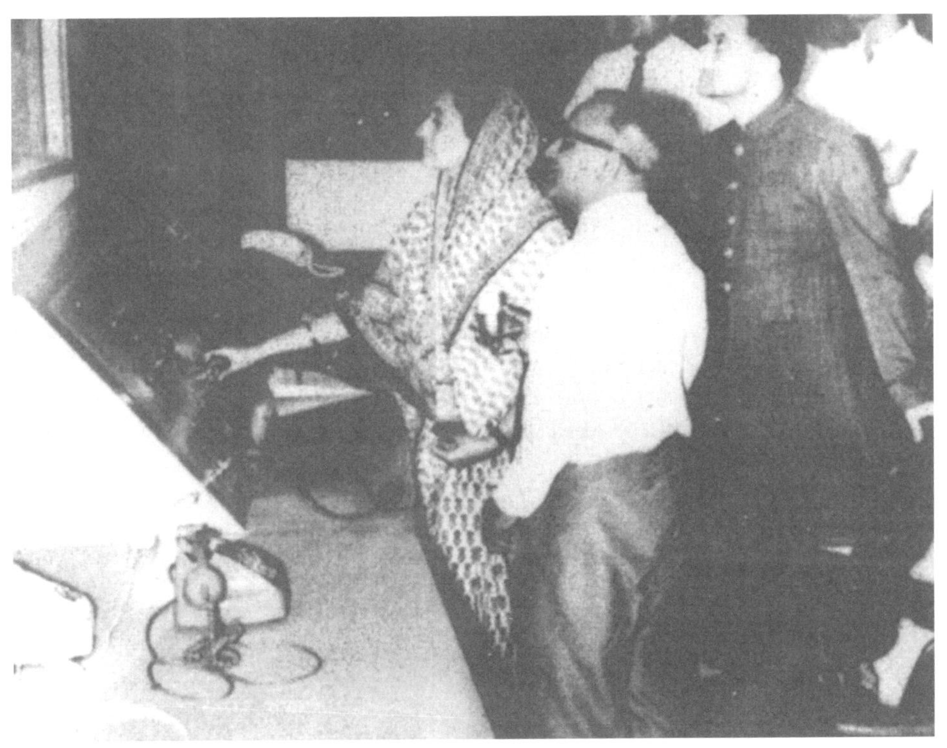

Indira Gandhi at Thumba

warning of all these dangers. One of the world's worst storms took place in November 1970, when a 9-m-high sea surge headed into the Bay of Bengal at over 224 km/hr: over half a million people were drowned, mainly in Bangladesh – partly because there were no satellite warnings then. Seventy-five percent of India's population lived off agriculture and 60% lived in rural villages – and satellites offered a quicker means of reaching and warning them than any other method. And, although this was poorly appreciated in the West, baseline Indian economic data were poor: maps in Asia were much less accurate than in Europe and few had been updated since the war. Again, satellites could rectify these gaps.

In 1970, scientists used a helicopter to detect coconut blight. In 1972, the United States had orbited *Landsat*, the first of seven pioneering satellites devoted to such Earth resources studies. In 1974, as part of the ARISE project (Agricultural Resource Inventory Survey Experiment), aircraft were used to survey agricultural fields in Andhra Pradesh and Punjab. Not long afterwards, India began to use *Landsat* data on a systematic basis. In 1978, India began to build a station in Hyderabad in Andhra Pradesh designed to receive data directly from American *Landsat* Earth resources satellites. This became the base of the Indian National Remote Sensing Agency, which analyzed foreign and domestic remote sensing data, led by Prof. P.R. Pisharoty, later called "the father of remote sensing in India".

THE IDEA OF AN INDIAN EARTH SATELLITE

Following the launch of the first sounding rockets, Vikram Sarabahai concentrated on laying the paths of the future development of the Indian space program. The aims of the program: satellites for development, education and remote sensing, coupled with India learning how to develop these skills itself (indigenization). Sarabhai held a brainstorming session on the idea of an Indian Earth satellite in the old school house in Thumba in 1967. By 1970, the engineers had settled on a 30-kg satellite. Although it was originally intended that Indian satellites would carry television, telephone and remote sensing equipment, such equipment was heavy, so the early small Indian satellites would instead carry useful small scientific instruments of the type developed by the Physical Research Laboratory. India had no experience of operating space experiments until, in 1961, it made arrangements to receive ionospheric data from beacons on DS satellites in the Soviet Union's Cosmos program. It was also clear that although it would take India some time to develop an indigenous launcher, it might be possible to develop an Indian satellite sooner, to be launched by one of India's space partners at Thumba (France, the Soviet Union, Britain or the United States).

An important institutional element was put in place at this time. INCOSPAR had operated under the aegis of the Atomic Energy Authority, chaired by Vikram Sarabhai after the sudden death in a plane crash on Mont Blanc on 24th January 1966 of its founder, Dr Homi Bhabba, aged only 57. The space program had now outgrown its parent, so it was formally established as the Indian Space Research Organisation (ISRO), on 15th August 1969, the operating body, and later the Space

Commission in June 1972, the main policy-making body, set up under the Department of Space (DOS) under the office of the Prime Minister.

PREPARATIONS FOR FIRST SATELLITE, *ARYABHATA*

Following the death of Vikram Sarabhai, the space program was led by Satish Dhawan (1920–2002), appointed chairman of ISRO by Prime Minister Indira Gandhi in 1972. Satish Dhawan had been professor of aeronautical engineering at the Indian Institute of Science in Bangalore, where he had become its youngest director at the age of 42. He was a passionate teacher and only agreed to the assignment on the condition that he could continue his educational work.

Indian scientists and engineers were familiar with the Soviet space program through Russian participation in Thumba and through the Cosmos program, so it was natural to ask the USSR for assistance with the launch of an Indian satellite. Negotiations were opened by Professor U.R. Rao, leading to an agreement between India and the Soviet Union, signed on 10th May 1972. This provided for the launch by the USSR of an Indian satellite and for Soviet use of Indian ports by tracking ships and vessels launching sounding rockets.

Professor Rao was given, with his colleagues, 36 months to conceive, design, build and test the satellite. About 200 were involved in the satellite team. "We worked night and day in asbestos-roofed sheds with practically no infrastructure," he recalled later. There were more bullock carts on the road than cars. The X-ray experiment he developed himself in the Physical Research Laboratory in Ahmedabad. For the first satellite, Russia made the solar panels, batteries, spin systems and tape recorders. The satellite was 1.2 m tall, 1.6 m in diameter, with 26 flat faces. It was orientated by means of magnetometers that sensed the Earth's magnetic field and stabilized by nitrogen gas fed from bottles pressurized at 200 atm. Several models were manufactured in India and tested in the Bhabha Atomic Research Center. Ground stations to track the satellite were built in Sriharikota and Poona, but the satellite also had the benefit of the new Russian tracking station at Bear's Lake. The satellite was designed to be picked up at a distance of 2,500 km.

Satish Dhawan

India's first satellite, named

Professor U.R. Rao

Aryabhata, 360 kg, was launched by the Soviet Union on 19th April 1975 from Kapustin Yar cosmodrome near the river Volga into an orbit of 563–619 km and dedicated to the study of stellar X-rays, neutron and gamma radiation from solar flares and particles and radiation fluxes in the Earth's ionosphere. *Aryabhata* was named after the 5th-century Indian astronomer and mathematician. The Soviet Union used its *Cosmos 3M* rocket from the family of launchers that had been putting scientific and military satellites into orbit since 1964.

Aryabhata worked until April 1980. One of its 14 regulators went

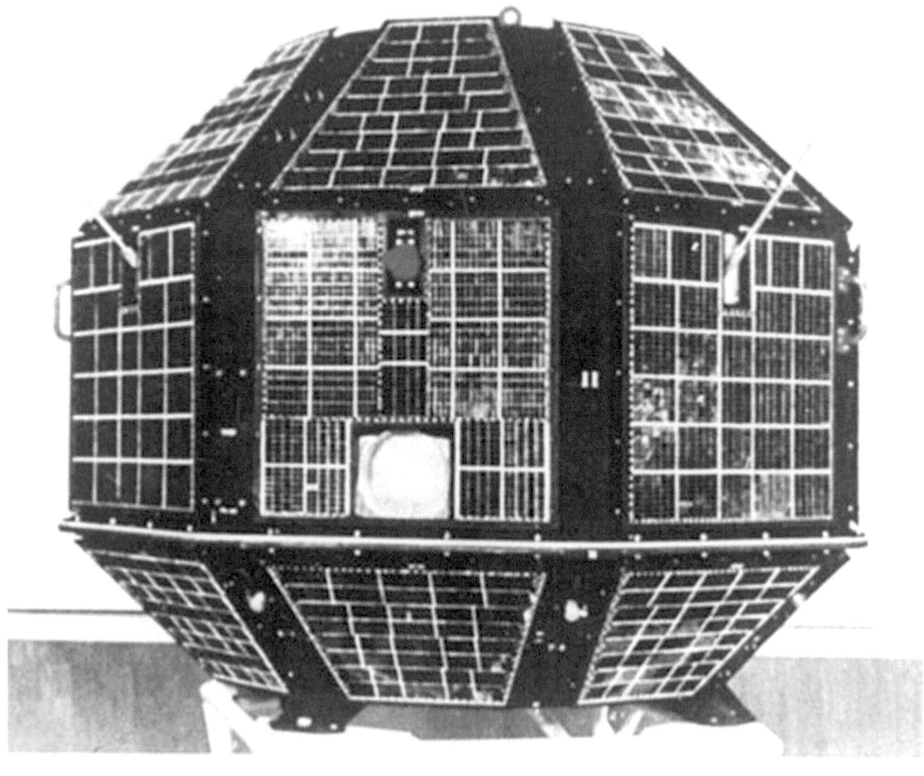

Aryabhata

out of operation on the 41st orbit (as a precaution, a standby transformer was installed on the next satellite). This led to reports that the mission failed at this point and had to be abandoned. Soviet media reports long insisted that the mission was going perfectly and returning vast quantities of data and this is confirmed by Indian accounts. The cost of the mission was Rs40m (€800,000). *Aryabhata* eventually burned up in Earth's atmosphere on 10th February 1992, on its 92,875th orbit. A ground station at Sriharikota picked up the last 30 sec of transmission before it fell silent forever.

BHASKHARA

India's second satellite, *Bhaskhara*, launched on a *Cosmos 3M* rocket on 7th June 1979, also from Kapustin Yar, was, in reality, the backup model of *Aryabhata*. *Bhaskhara* was named after a 7th-century Indian astronomer. It was a 26-sided polyhedron, 1.19 m tall, 1.55 m diameter, weighing 442 kg, with a surface area of 6.5 m^2. Energy was supplied by the 26 solar panels that covered the outside of the spacecraft, these cells feeding nickel cadmium storage batteries. The spacecraft was

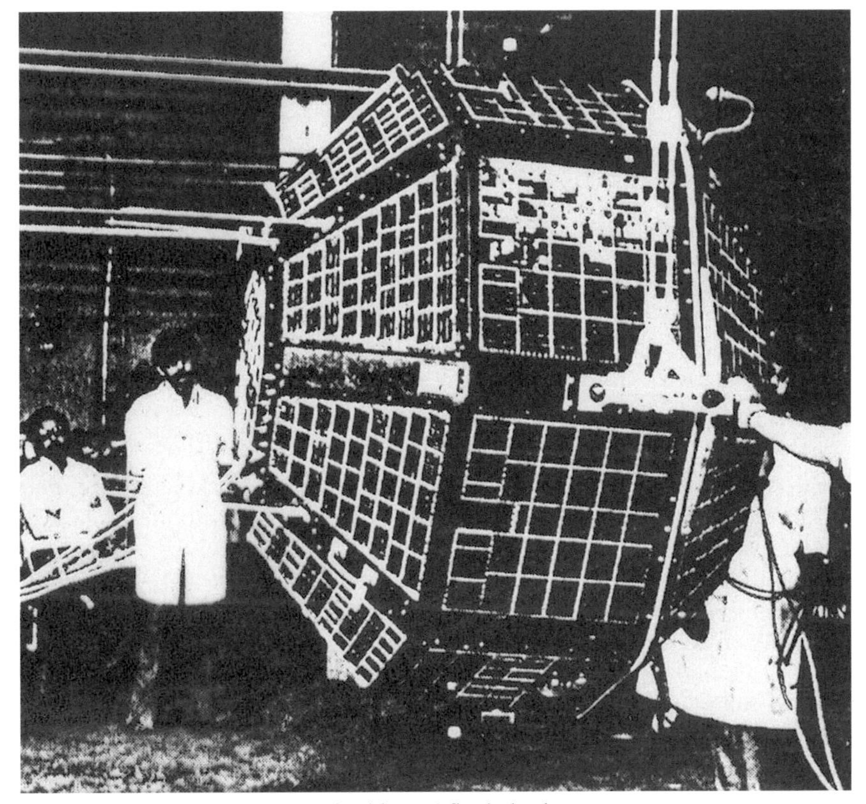

Bhaskhara 1 final checks

Bhaskhara 1 finishing touches

controlled by an infrared horizon sensor, a sun sensor and triaxial magnetometers. The USSR supplied 3,500 solar cells and the nitrogen pressurization system. Cost of the project was Rs65m (€1.38m).

The satellite had four experiments. *Bhaskhara* carried two low-resolution television cameras, designed to show to best advantage changes in water and vegetation. Each picture was to show an area of 341 km^2 with a resolution of 1 km, providing black-and-white pictures from which false color could be derived using digital processing. *Bhaskhara* carried three radiometers to study the oceans around India, obtaining information on sea surface temperature, ocean wind velocity and moisture content. The third experiment was a data collection and relay package, designed to pick up data from eight meteorological stations and retransmit them to a central receiver station in Ahmedabad. The fourth experiment was a pinhole camera to identify X-rays.

Bhaskhara entered orbit of 519–541 km, 50.67°, 95.16 min. Although the radiometers worked properly, the television cameras were not turned on for nearly a year because of gas being trapped in the camera system. They were used only during passes over India and were switched off otherwise in order to save power. About 10 pictures were received every day. The cameras provided information on snow melting in the Himalayas, river flooding in northern India, desertification in Rajasthan, rainfall off the coast of India and mineral resources in Gujarat. Data were downloaded in 10–15-min passes over Sriharikota, the transmission rate being 91,000 bits/sec. Transmissions ceased on 1st August 1981.

In order to suit Indian tracking systems, the Indian satellites flew a south-eastern trajectory out of Kapustin Yar, the only satellites to do so: the Russians normally avoided this route, which overflew China, with the practical danger of losing fallen spacecraft to the Chinese and the diplomatic danger of upsetting them.

Bhaskhara 2

Bhaskhara 2 was launched by the Soviet Union in June 1981 into an orbit of 525 km, 50.7°. Its cameras provided information on agriculture, weather and vegetation. Its images were used in preparing land use maps in west Bengal and the microwave sensor contributed to knowledge of sea winds, water vapor and rainfall rates. It seems to have been much the most successful of the first three launches and returned the most data. *Bhaskhara 2* burned up on 30th November 1991.

The *Aryabhata* and *Bhaskhara* satellites were constructed by Hindustan Aeronautics Ltd, HAL, with the solar cells, batteries and thermal paints being supplied by the Soviet Union. The launches were provided free, on the condition that the USSR had access to the data collected. Table 4.1 summarizes the early Indian satellites.

Table 4.1. India's first satellites, launched by the Soviet Union

19 Apr. 1975	*Aryabhata*	Cosmos 3M	Kapustin Yar
7 Jun. 1979	*Bhashkara*	Cosmos 3M	Kapustin Yar
20 Nov. 1981	*Bhashkara 2*	Cosmos 3M	Kapustin Yar

AN INDIGENOUS INDIAN ROCKET

Ultimately, India did not wish to depend for long on other countries, however well meaning, to launch their satellites (the USSR) or develop satellite programs (the United States). Even as the *Aryabhata* launch was planned and the ATS experiment took place, efforts were underway to develop a home-built rocket launcher. There was a rapid expansion in the space industry in the mid-1970s. The numbers employed rose to 10,000, many of whom were professional scientists persuaded to return from abroad. By this stage, 30 universities were involved in space research projects stimulated by ISRO.

The 1970 feasibility study had defined an objective of placing a small, 40-kg satellite into an orbit of 300–900 km. For a launcher, the Indians felt that the best approach was to build a rocket based on the smallest and simplest launching system known to them, the proven American small satellite launcher, the Scout. The Scout used simpler solid fuels – an advantage the Japanese had also appreciated (the Russian Cosmos rocket used liquid fuels). In effect, India hoped to build a rocket comparable to the American Scout, which had been placing small payloads in orbit since the early 1960s (although superior American engineering techniques gave the Scout a payload of 180 kg).

The Indian version of the Scout was called the Satellite Launch Vehicle (SLV). Appointed as project director was Abdul Kalam, later President. Years afterward, he recalled how basic things were: "In those days, we didn't have computer aided design, push button phones, computers, internet, e-mails or even a colour television monitor" and the space engineers had to rough it, traveling from one facility to another in trains, third-class, unreserved. Long-distance phone calls had to be booked in advance and they often waited until midnight for them to come through.

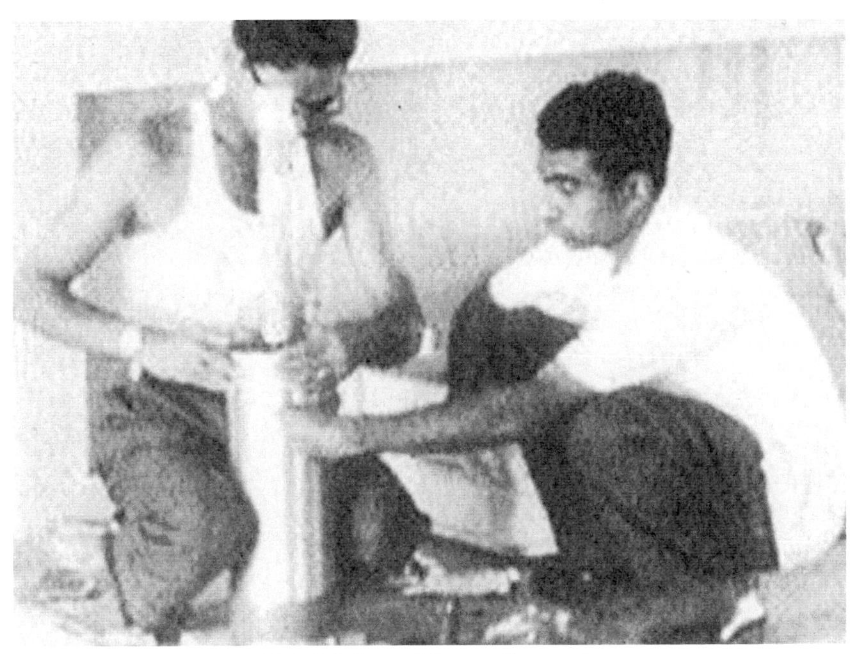

Abdul Kalam with sounding rocket

Precious equipment parts were ferried around on cycle rickshaws. The development of the SLV began in 1973. The SLV had 10,000 components, of which 85% were developed in India itself. The SLV was a four-stage rocket, and was 22.7 m high and 1 m in diameter, making it unusually slim. The SLV was, by world standards, tiny in size, half the size of the Japanese *Lambda*. Still, it was a huge advance on the biggest Indian sounding rocket of the time, the *Rohini 560*, which was only 560 mm in diameter. Forty-six public and private enterprises were involved in the venture. The SLV had 44 main systems, all of which had to work for it to put a satellite into orbit. Parts failed during tests and had to be redesigned and rebuilt.

The United States were willing to help and Indian scientists visited Wallops Island, Virginia, the launch base of the Scout, to see the famous launcher in operation. Although the SLV looked like the Scout and was inspired by the Scout, the eventual SLV design moved on quite along way from the Scout. Assistance was also provided by Germany, which made available the Porz Wahn wind tunnel in Cologne, which could simulate various altitudes and pressures, with speeds up to Mach 8.

The SLV had a take-off weight of 17 tonnes, of which 13 tonnes were fuel. The first stage was 10 m long, with one 63,500-kg thrust motor and the solid fuels packed in three segments. The second stage was 6.4 m long, with a single motor, and the third stage only 2.34 m long. Finally, the 1.5-m-long fourth stage had a motor of just over 2 tonnes' thrust. Fuel was made at the launch site of Sriharikota, comprising 70% ammonium perchlorate oxidizer and 30% aluminum. Years later, the Americans chose similar fuels for the solid rocket boosters of the Space Shuttle.

SLV in the factory

Preparing the first SLV was hazardous. Six people were injured with burns during a second-stage static test when nitric acid got into the oxidizer tank. Later, the umbilical did not free at launch, so a mechanic climbed up the tower beside the live rocket and removed it by hand – a reckless thing to do that would not be permitted by health and safety authorities nowadays.

PLANNING THE FIRST HOME-LAUNCHED SATELLITE

The first domestically launched satellites were designed as technology demonstrators. They were called *Rohini*, a suggestion by Hideo Itokawa. The first series comprised 35-kg multi-sided capsules with a charge coupled device (CCD) camera with 1-km resolution for Earth observation. A key objective was to test the Indian-made solar cells.

The establishment of a domestic rocket and satellite program required the building of a rocket launching base. Hitherto, launches had been made from Thumba and another site, Belasore, was later to be used for sounding rockets. For the national rocket program, a new site was chosen – Sriharikota Island on the south-east coast.

Sriharikota is 180 km^2, a barrier island sandwiched between the Bay of Bengal and Pulicat Lake. The island is 44 km long, 7.8 km broad at its widest and has two streams in the middle, the Peddavagu and the Chinnavagu. The island is a sand

SLV Launch Control

ridge, at no stage more than 10 m above sea level. The site is not unlike Cape Canaveral, which is also a sand ridge with waterways. Annual rainfall is 1,200 mm, moderate from July to September but very heavy from October to December. Like Cape Canaveral, Sriharikota abounds with wildlife, with 400 species belonging to 110 families, some unique to the island. A herbarium was established to preserve the flora and fauna there in 1989 with scientists from Sri Venkateswara University.

Equipment was brought in by rail and by air through Madras (now Chennai). Conditions were very basic and engineers personally escorted their equipment in trains all across the continent to Sriharikota. One arriving plane crashed and burned in Madras: thankfully, all the crew and passengers survived, but last out was one of the scientists hugging the receiver for the first satellite. It was a hectic, busy period, with long days and late nights. Equipment parts were tested out live on sounding rockets.

FIRST LAUNCHES: INDIA – A SPACEFARING NATION

Originally, India had planned its own first satellite for 1974, but there were long delays in the supply of key components and ISRO engineers had to be seconded to the companies concerned to speed things up. They also missed the deadline of 1978, the 30th anniversary of Indian independence.

The first attempt to launch was made on 10th August 1979, the payload being the

SLV on pad

first 36.7-kg *Rohini* satellite. The computer held the launch at T–2 min, telling the ground controllers that there was insufficient fuel in the second stage. Ground controllers checked their records and knew that they had installed sufficient fuel, so they overrode the computer and commanded the launch to go ahead. When it did, the second stage duly failed. Even though the third and fourth stages fired, they did not have enough thrust to make good the earlier failure, so the rocket crashed into the Bay of Bengal 317 sec after launch, 500 km downrange. The dejected rocketeers were heading back home on the bus when Abdul Kalam overtook them and stopped them. He had found traces of a nitric acid leak from a valve on the launch pad. Live black-and-white surveillance had not detected this, but the incriminating leak showed up when Abdul Kalam examined color film later. The computer had been right after all. Space engineers formally declared the mission to be "70% successful", though the press was less forgiving, countering with "a flop is a flop". Moral of the tale 1: install a color monitor. Moral of the tale 2: listen to what the computer tells you! The valve was modified.

The first successful *Rohini* was eventually launched on 18th July 1980. Although it was a thundery day inland, conditions out to sea were calm. The rocket rose straight into clouds. Engineers squeezed against each other in their hostel a safe distance away to watch. R.B.K. Menon recalls:

> "All of us were looking toward the seaside, not knowing what to expect. We became impatient. We felt as if our watches had suddenly slowed down. But we held on. At six o'clock sharp, we could see the lightning streak piercing the sky from amidst greenery on the western front of the sea. That was the rocket striding upward. We felt that was all. We were about to disperse. Then came within moments the thundering sound which now rings in my ears."

This time, the second stage behaved perfectly, bringing the ascending rocket to 92 km. The third stage brought it on an ever flatter trajectory to 127 km. After a long coast, the fourth stage fired for the orbit insertion at 300 km some 8 min after lift-off. *Rohini* began transmission at once, immediately reporting spin, direction and speed. It was a 40-kg 0.5-m spheroid, orbited at 325–950 km, 97 min, 44.75°, higher than the planned orbit of 276–472 km, and it lasted until 24th July 1981. The solar cells, all indigenously made for the first time, managed to develop 3 W of electrical power and the satellite spun at the rate of 165 rpm. The satellite was tracked by ISRO stations in Sriharikota, Nicobar, Thima and Ahmedabad. The satellite's television camera was operated to give three complete sets of India. Signals continued until 1st August 1981. India had become the seventh space nation.

The second *Rohini*, also a 40-kg spheroid, reached orbit of 187–418 km, 46.27°, 90.49 min on the SLV from Sriharikota on 31st May 1981. It was hoped that its imaging camera would provide quality Earth resources pictures. The second launch was covered live on Indian television and radio. This time, the guidance unit was made in India and the satellite carried new solid-state components for a remote sensing camera. Everything went well at first but a slow spin began to develop at 6.28 sec, causing the launcher to move off course. The second stage tried to correct the error, but, in doing so, lost 6% of the intended velocity. The third stage was able

Rohini on SLV

to dampen the pitch and yaw errors but the vehicle was still spinning and at 3% less velocity than intended. The second Rohini entered orbit, but the low point was 183 km, not enough to sustain operations for long. Ground controllers decided to use the onboard equipment as much as they could in the time available to them, such as the camera. After nine days and 131 orbits, contact with the satellite was lost on 9th June while it was over Fiji. Later analysis found that one of the four fins on the first stage had not responded to guidance commands, causing the spin that restricted the mission.

The second satellite was celebrated by a full-scale model on the Republic Day parade. It was brought to the Science Museum in London, where the "Festival of India – From Bullock Cart to the Space Age" was opened jointly by Indira Gandhi and Margaret Thatcher.

The third satellite was much the most productive. *Rohini 3* entered orbit on 17th April 1983 from Sriharikota. There was a modified fourth stage, made of kevlar-reinforced fiber which led to a weight saving that enabled 50 kg more fuel to be carried. As a result, the perigee could be much higher, with a longer orbital period, 99 min. An hour and a half later, it was tracked for 11 min on its first pass over India. It flew on a red-and-white rocket, with the Indian national flag on its side, watched by Prime Minister Indira Gandhi. Afterwards, she congratulated the 800 personnel of the launch team and told them they had done a "great job". At a subsequent press conference, she justified Indian space spending on the basis that it would solve agricultural problems. Like a child's education, she said, the results

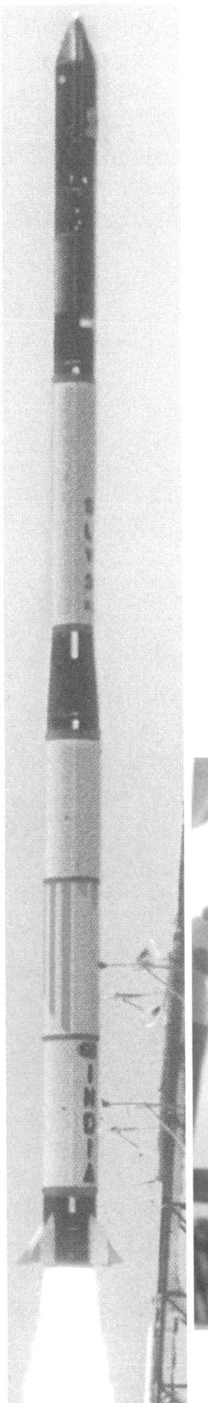

come later. The satellite had a camera capable of picking out water, vegetation, clouds and snow. Weighing 41.5 kg, its two cameras sent back over 5,000 images before the satellite was turned off 18 months later. It was a smart camera, because it was programmed not to take images when there was too much cloud. Table 4.2 summarizes the early Indian satellite launches.

Table 4.2. India's first satellite launches

18 Jul. 1980	*Rohini*	SLV	Sriharikota
31 May 1981	*Rohini* 2	SLV	Sriharikota
17 Apr. 1983	*Rohini* 3	SLV	Sriharikota

CONCLUSIONS

Like Japan, India's space program owed much to the vision of its founder, Vikram Sarabhai, and two men who knew and helped one another. India's space program started from equally primitive conditions, a deficit made good by Sarabhai's particular leadership, focus and adaptability. India brought to a conclusion a decade that saw the arrival of the three Asian

SLV about to be lifted

SLV take-off

space powers: Japan, China (both 1970) and now India (1980). Japan and India started very similarly, with sounding rockets and then small, solid-fueled satellite launchers. Now, India faced the challenge Japan had met earlier, that of broadening its base, developing its facilities and designing more powerful launchers. Eight years after its first satellite, India was the proud host of the 39th world congress of the International Astronautical Federation in Bangalore, where over 600 technical papers were presented at an event attended by 900 delegates from countries all over the world.

5

India: Space technology and the villages

India had become a spacefaring nation. By the early 1980s, the country had launched three satellites of its own and had three Sputniks orbited by the USSR. The experimental phase was over: now was the time to put space technology to the operational service of Indian economic and rural development, the purpose for which the enterprise had originally been intended. Two main systems were developed: the Indian Remote Sensing Satellite System (IRS) and the Indian Satellite System (INSAT).

INTRODUCING IRS

In Earth observations, the next stage of development was the IRS or Indian Remote Sensing Satellite. IRS was part of the Indian Remote Sensing Satellite System, itself, in turn, part of the National Natural Resource Management System.

IRS was designed to weigh up to a tonne, cross the same stretch of Earth every 21 days and carry out systematic surveys of the Earth's surface. IRS was built to carry three linear imaging self-scanning high-resolution sensors, one 70 mm, two 35 mm, for the use of agriculturalists, geologists and hydrologists. This was a considerable advance on *Bhaskhara*, which had 1-km resolution in two visible wavelengths. Typically, IRS orbited at 904 km, crossing the equator at 10.25 am each day, transmitting data at 20 MB per second to the ground station in Shadnagar. For orientation, IRS had 16 small thrusters with 80 kg of hydrazine fuel.

These satellites were still far too heavy for the SLV, so the first IRS, IRS-1A, was launched by the Soviet Union in March 1988 on a Vostok rocket and operated successfully for three years. The 940-kg satellite was put into a 1,112-km polar orbit of 103.2 min. Fifty Indian scientists, accompanied by the press, travelled to Baikonour to supervise the launch. The launch pad was obscured by thick fog and falling snow, so they presumed there could be no take-off. They did not realize that Russian launches paid little attention to weather conditions. The visitors were taken aback to hear the sound of the take-off, for the launch had gone ahead anyway, even though conditions were so impenetrable that nothing could be seen. India is believed

IRS-1 design

to have paid about €2m for the launch and the satellite cost R650m (€13.8m) to build.

On its third anniversary, a postal stamp was issued in its honor. By then, IRS-1A had made 50 complete maps of India and taken part in a number of remote sensing experiments. The all-India maps provided disturbing information that the national level of forest had decreased from 14 to 11%. On the positive side, water maps had led to an increase in the success rate of bore-drilling from 45 to 90%. A full national inventory of wasteland had been made on a 1:250,000 scale – one which suggested that at least half could be reclaimed. Crop yields had been estimated on the basis of its data.

By its fifth anniversary in 1993, it had completed a salt map of the country on the same 1:250,000 scale and it had returned a total 400,000 images in 25,470 orbits, of which 3,700 were passes to map India. IRS-1A operated for six years, twice its design life. It was eventually retired in 1995 and put in reserve.

The second, IRS-1B, weighing 908 kg, was put into orbit by a Vostok rocket from Baikonour cosmodrome into 857–918-km polar orbit on 29th August 1991. This time, the price charged by the now cost-conscious Russians had risen to Rs629m (€13.39m). IRS-1B had a bank of linear imaging scanning cameras and sent back its first pictures the next day. It was specifically designed to focus on forecasting the crop yields of tea and coffee. Three years later, IRS-1B had covered India 50 times on cycles of 22-day passes (synchronized with IRS-1A, this gave 11-day revisits). IRS-1B made the first transmissions to the Earth Observation Satellite company in Norman, Oklahoma.

IRS-1B lasted three years. The tracking stations used for IRS-1A and B were Shadnagar, near Bangalore, Lucknow, Sriharikota, Nicobar and Thiruvananthapuram, with a mobile station in Mauritius and in Russia, Bear's Lake. Data were stored in high-density tapes. When sorted, the images were made available in standard, stereo, coded vegetation and urban mapping formats and mosaics as required.

IRS-1A

SECOND-GENERATION IRS

The third, IRS-1C, 1,250 kg, was launched from Baikonour on 28th December 1995 using a Molniya M rocket, using an unusual 901-sec profile to put the satellite in a 816–818-km polar orbit. Solar panels sprung out 93 sec after orbital injection and the spacecraft swiveled to acquire Sun lock so as to start generating electrical current at once. Four years later, it had completed its primary mission and still had sufficient fuel to go on.

IRS-1C was an advance on 1A and 1B, offering improved resolution, stereo viewing and frequent site revisits, making it the most advanced remote sensing satellite in the world at the time. IRS-1C and its later companion 1D were the second generation, heavier, at 1,250 kg. Resolution was improved to 400 m and transmission was designed to be sent to worldwide destinations, not just India, but also Hartebeeshoeck, South Africa, Weilheim, Germany and Poker Flat, Alaska. They flew somewhat lower than 1A and 1B, and had a 24-day revisit period. Each 101.35-min orbit shifted westward by 2,820 km at a time. IRS-1C carried a tape recorder so that data could be stored for later transmission.

IRS-1A view of Mumbai

IRS marked the introduction of the Linear Imaging Self-Scanning Sensor, or LISS, which operated in the visible and infrared wavebands using French-made charge-couple device scanners, with a panchromatic camera able to provide 6-m resolution from 600 km, then thought to be the highest available in the world. The LISS operated in four bands and a wide-field sensor with a swath of 810 km and a resolution of 188 m, with filters for Earth observations in the green, red, near infrared, short-wave infrared bands, which covered, respectively, vegetation, plants, landforms and crops, while the panchromatic camera was ideal for geological imaging and mapping urban areas. Analysts use a technique called "false color imaging", with an internationally standardized index to make subtle distinctions between, say, healthy crops or plants (red or pink) and diseased ones (green). Early IRS pictures clearly showed flooded areas, different kinds of grasses, water holes for animals, different types of soils, vegetation types, soil erosion, forest fires, sediments off coasts, damage done by mining and human settlements. The outcomes were fed

LISS camera system

into the beginnings in India of the Geographic Information System (GIS), a modern tool of geographers and planners to integrate data into computerized modeling systems able to show up a variety of information by category or combination of categories. The much improved resolution enabled the Digital Mapping Center in Hyderabad to begin work on a project of mapping the country with a new 1:250,000 topographic map.

Some of the detail available on the satellite images was extraordinary: the camera had sufficient resolution to pick out passenger airplanes parked at airports. In 1993, India gave a franchise to EOSAT, an American company based in Lanham, Maryland, to process data covering countries other than India and to sell images to the world market (it already held the rights on American *Landsat* data). From June 1994, EOSAT began to receive the IRS data at its station in Norman, Oklahoma. Another indication of the quality of IRS data was that in 1998, the Japanese National Space Development Agency, NASDA, a world leader in the area, applied to receive IRS data at its own Remote Sensing Technology Center. The first IRS missions are summarized in Table 5.1.

Table 5.1. IRS missions

17 Mar. 1988	IRS-1A	Vostok	Baikonour
29 Aug. 1991	IRS-1B	Vostok	Baikonour
28 Dec. 1995	IRS-1C	Molniya	Baikonour

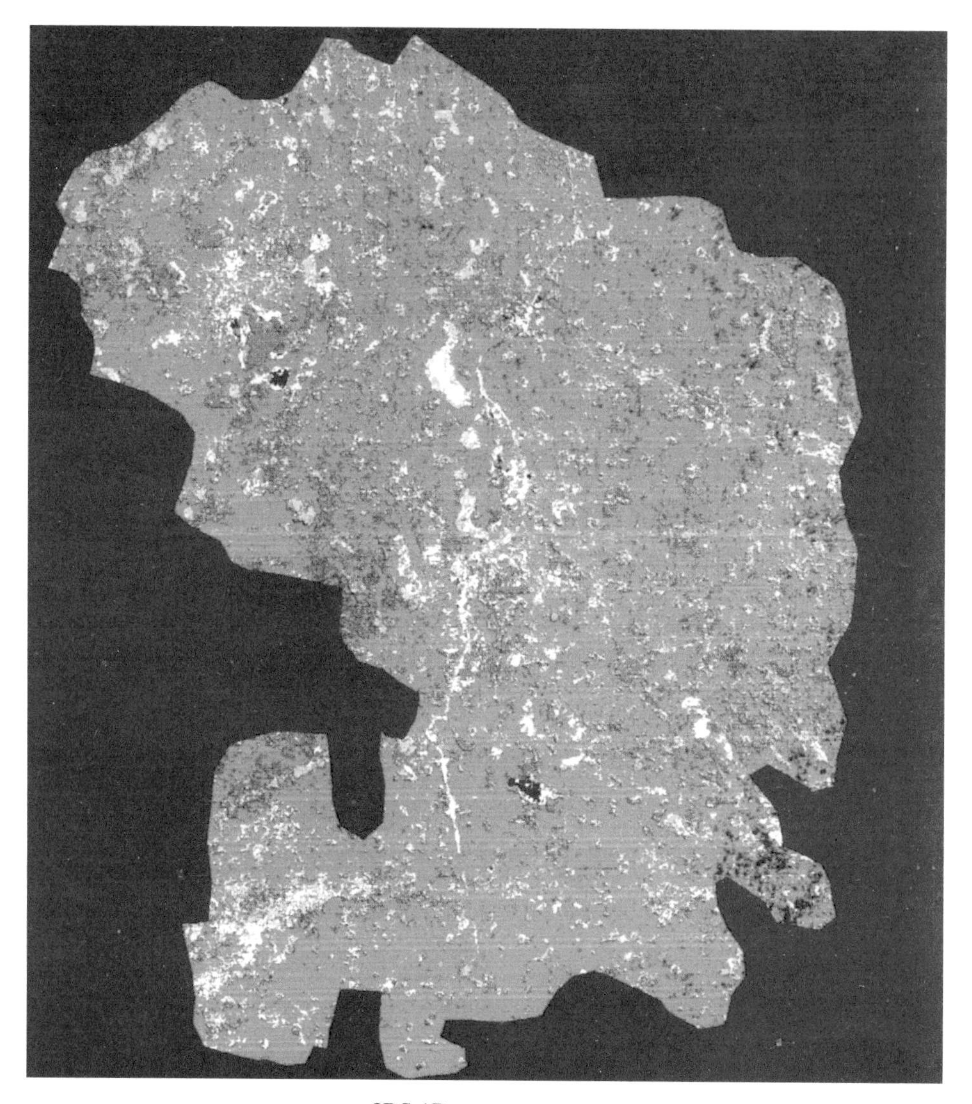

IRS-1B resources map

INDIAN REMOTE SENSING: A BALANCE SHEET

The use of remote sensing by satellites in India was part of the national system for managing natural resources in India. This is called the National Natural Resources Management System (NNRMS), which has the brief of compiling satellite and ground data for purposes of combating drought, mapping wasteland, estimating crop yield, water resources management, flood prevention, mineral development, land-use mapping, forestry management and ocean development (including coasts and fisheries). The Department of Space has overall responsibility for the NNRMS. Local development plans for sustainable development have been drawn up for 45% of the country using IRS data.

Among the benefits of remote sensing from satellites, the following have been cited:

- satellite maps have tracked sewage entering rivers (e.g. the Yamuna, near Delhi) and have planned the location of treatment plants in 38 towns;
- studies of forest cover have mapped the space available to tigers and the boundaries of the reservations;
- satellite pictures have showed the devastating effects of deforestation and mining;
- chemical, mining and engineering companies were asked to use satellite pictures before locating new sites, in order to prevent pollution of ground or surface water;
- ocean maps have indicated where fish are likely to be found, doubling catches;
- the volume of water in ecological wetlands has been measured;
- satellite photos have identified poor embankments on rivers, where breaches are most likely in the event of future flooding;
- soil maps have been compiled showing the respective levels of salt and lime;
- remote sensing satellites have been important in mineral development. IRS-1A located deposits of diamonds and rubies near Najranagar, zinc in Wantimata and copper in Karnataka. Other IRS satellites helped to find base metal, tin, iron ore, bauxite and oil;
- glaciers and the volume of snow melt coming from the Himalayas have been measured;
- new islands formed from sediment were identified off the coast of Tamil Nadu;
- satellite data have been used for integrated water planning, which involves a mixture of dams, water drilling, tree planting and water table maintenance;
- hot spots of urban air pollution have been identified (e.g. Bangalore).

The availability of water is one of the great problems of India, as it is in most developing countries. Under the National Drinking Water Mission, set up by the government in 1987, there is a commitment to ensuring every village has access to up to 40 liters per head of clean, drinking water a day. The Department of Space was charged with making space-based maps to identify the presence of underground water. Accordingly, water maps of 447 districts on a 1:250,000 scale were compiled. These are colored maps, with blue for areas of good ground-water potential, green for areas of moderate potential, depending on the geology, red for no potential and yellow for doubtful. At the other extreme, IRS satellites have tracked flooding, enabling warnings to be provided and danger areas to be predicted.

The biennial national forest survey of India is now satellite-based. India has a target of 33% forestry, but the level was still only 12% in the 1990s and falling. Encouragingly, the 1993–1995 survey, for example, showed a marginal increase in

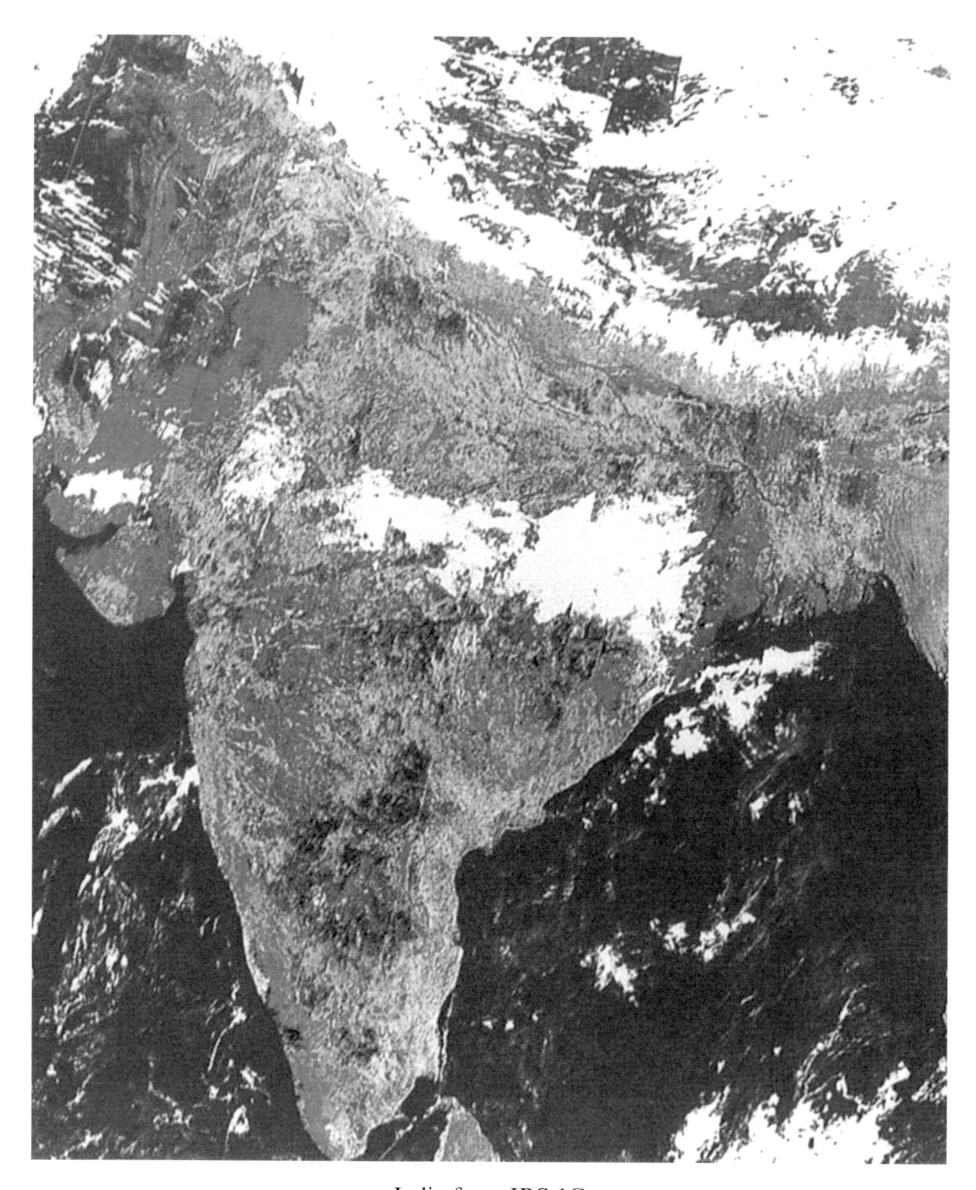

India from IRS-1C

the total forest cover in the country. Related to this is wasteland mapping. Here, satellites carried out a national mapping exercise for the National Wasteland Development Board at 1:250,000 scale, enabling many areas to be reclaimed for food or forestry. One of the early achievements of the IRS system was the compiling of 1:250,000 maps on agricultural use, able to distinguish cropping patterns, grazing land, waste areas and waterlogged parts.

The 60-km-long ring road for Bangalore was planned using satellite data. This

Earth resources for urban planning

approach had the advantage of being more up-to-date than conventional maps (which had not including new housing in the path of the proposed road), enabled the road to follow the geological terrain, was faster than ground-based surveying and realigned the road on poor-quality land.

Satellite imaging was used in an experiment to promote the silk industry in India. Despite great demand for Indian silk, production is limited to a small number of areas. IRS-1A was used to estimate crop yields in four districts in Bangalore, Mysore, Manya and Kolar, to identify new sites for mulberry growth and to forecast silk cocoon production.

In an experiment off the three maritime states of Gujarat, Maharashtra and Andhra Pradesh, IRS data of sea conditions were used to predict the location of fish. Generally, fish may be found in areas where temperatures change and not in areas of uniform temperature. Local fishermen headed for the identified areas, with dramatic results in increased catches. By 1997, the IRS system was used systematically for

Resourcesat view of Himalayas

assessing likely national crop yields for seven main crops and was used to test for others (e.g. chilli).

In an unusual exercise to protect wildlife, the IRS system was used to map elephant trails in north Bengal so as to better assess the suitability of their habitats. In Junagadh, satellite imagery found an unexplored archaeological site that subsequently yielded pottery, bones and stone walls.

IRS data have been used to warn of locusts, especially the desert locust, which thrives in Rajasthan, Haryana and Gujarat. A particular form of stratus cloud is a known carrier of wheat rust spores, so tracking these clouds by satellite can give 20–25 days' warning of the arrival of wheat rust, thereby enabling precautions to be taken. Interpretation of IRS data was able to identify conditions favorable to the spread of brown plant hopper, a nasty pest that can ruin rice production. Similarly, areas infested by cotton fly could be identified on satellite photos (they show up as dark red on a lighter red background) – the only remedy being crop rotation to break the lifecycle of the pest. Satellite data have been used to check on pests in tomatoes, coconuts and oranges.

In the course of 1994–1995, the parliamentary standing committee on science, technology, environment and forests made a detailed review of the work of the Department of Space in remote sensing applications and satellite-based early warning. In a lengthy report examining all aspects of the program, the parliamentary committee concluded that the program was "well thought-out and implemented on a steady long-term basis so as to derive optimal advance [for India] from this high-tech area". It was a unique program in the world and commended the efforts of the Department of Space.

At the 1999 United Nations Conference on the Exploration and Peaceful Uses of Outer Space (UNISPACE for short), the head of the Indian delegation and director of ISRO, Dr K. Kasturirangan, told delegates that space technology had an important role in finding sustainable paths to development, promoting social equity and enabling all citizens to reach a minimum quality of life. Prof. U.R. Rao, former director of ISRO, subsequently pointed out that 25% of the world did not have adequate drinking water and that 40% of its people were illiterate. Half the world had never made a phone call. Eighty-four percent of mobile phones, 91% of faxes and 97% of internet access was in the developed world. By 2050, 9bn of the world's 11bn population will be in developing countries. "Space technology must contribute to their health and food," argued Prof. Rao. "Where space technology has been put at the service of sustainable economic development, there has been great success. Satellite television education has liberated women and challenged the old divisions of labour." Ninety-five percent of deaths from disaster happen in developing countries: satellites offer these countries the tools of disaster management and the possibility of leap-frogging older technologies to bring health and literacy. At the conference, India formally represented the G77 – the group of 77 poorest developing nations pressing the richer countries to support satellite-based systems for Earth resources monitoring, disaster warning and telecommunications.

The old and the new in space applications came together with the development of the new discipline of space-based archaeology. In Ahmedabad in Gujarat, analysts noted the way in which images from Earth resources satellites could pick up subterranean structures normally invisible above ground. From that, they were able to reconstruct what Gujarat looked like thousands of years ago: rivers followed quite different courses and towns beside them were located in places where no one suspected they had ever existed.

INSAT: INDIA'S COMMUNICATIONS AND WEATHER SYSTEM

The IRS system was one leg of the Indian applications program: the other was INSAT. The INSAT (Indian National Satellite System) program was planned in 1976 as a Rs19bn (€405m) project, a joint venture between the Department of Space, the Indian Meteorological Department, the Department of Telecommunications and All-India Radio. The aim was to provide a communications and weather satellite system for India. A head office for the system was set up in Bangalore. Weather warning was especially important for India. Although meteorological recording

INSAT 1

stations had been installed since 1849, it only recorded weather after it happened: there was no warning system and India suffered an average of two severe cyclones a year for which there was little warning. Approval of the INSAT program was given by the government in 1977 and slots in 24-hr orbit registered two years later.

Weather warning satellites were most effective in a 24-hr synchronous geostationary orbit hovering over the sub-continent – something far beyond India's capabilities. Accordingly, foreign launchers must be used. Some time was spent in these negotiations. An agreement was signed in July 1978 between the Space Commission and NASA whereby the United States would launch the first Indian National Satellite System satellite, INSAT 1, on the Space Shuttle in 1981. Nor did India have the know-how to build such a system itself, so the first round of INSAT satellites, called the INSAT 1 series, was also sent out to tender abroad. The first two satellites would be built by Ford Aerospace and Space Systems Loral and was the first agreement whereby NASA would launch a commercial satellite on the Shuttle for another country. Two would be build and flown, INSAT 1A and 1B.

With the INSAT series, India attempted to combine the two great advantages of geosynchronous orbit: communications (as demonstrated by ATS-6) and a high-altitude platform from which to observe the Earth. Most other countries have used

24-hr satellites for one purpose or the other, but not both at the same time. India was the first country to combine these two functions – an approach that it reckoned was 40% cheaper in the long run than using separate satellites.

At that time, both television and telephone services in India were poorly developed. Even by the late 1970s, television reached less than 30% of the people. Although television had started in India in 1959, there had been little investment in the system and a second station had opened in Bombay only in 1972. Those programs that were not imported were of poor quality and in black and white, contrasting unfavorably with India's robust film industry. Television sets were in short supply – a function of restrictions on imports and the lack of domestic production. The ground station system was undeveloped and India had to build 30 Earth stations in quick order, achieved through importing American and Japanese equipment at a cost of Rs2.664bn (€56.7m). The aim of the INSAT series was ambitious: to more than double the proportion of Indians receiving television to 75% of the population.

India set up ground stations to handle INSAT 1 in Hassan (two 14-m dishes), New Delhi and Madras and a number of truck and jeep-mounted terminals for use during natural disasters like floods. Weather pictures were to be sent down to the Earth station in Delhi for passing on to the weather ministry for further relaying to secondary stations.

INSAT PRECURSOR: APPLE

Fortuitously, an opportunity arose to test out some of the principles of the INSAT system before the first Shuttle mission was due to fly. Europe was then in the early phases of testing its new commercial launcher, the Ariane.

The first four flights of Ariane were development flights. The European Space Agency (ESA) offered free space on these missions, on the understanding that, being test flights, there was a risk of failure. India already had good relationships with the ESA, having signed its first agreement with Europe in 1971, followed by further major agreements in 1977 and 1978. India responded promptly to the ESA opportunity, developing an experimental communications satellite on a tight time schedule. The satellite was called the Ariane Passenger Payload Experiment (APPLE). Design began in 1978, fabrication in early 1979 and the flight model was completed in 1980. The cost of the project was Rs150m (€22.9m), the spacecraft being built by Hindustan Aeronautics. APPLE had to be brought from its factory to the electromagnetic test center and it was important that it be transported there by a specialized non-metallic vehicle. No such vehicle existed in India, they thought, until someone quickly realized that India was actually full of them: ox-pulled wooden carts. The picture of a white ox pulling the shiny satellite through a field told so much of the practicality and frugality of the Indian space program. Building APPLE was challenging, the engineers having only 36 months available between design and shipment of the flight model. Five models were built to squeeze out design errors: structural, thermal, engineering, prototype and flight models.

APPLE

The purpose of APPLE was to test out means of stabilizing satellites in 24-hr orbit, C-band communications transponders, the kick motor, solar panel and batteries. It was a major opportunity to develop the experience necessary for controlling satellites in 24-hr orbit, and later in constructing a domestic communications satellite. Already, Indian space planners were thinking ahead to a domestic-built comsat, the INSAT 2.

APPLE was a 630-kg cylindrical structure, 1.2 m tall and 1.2 m in diameter, with two 1.2-m^2 solar panels and two transponders. The fourth stage of the SLV rocket was used as an apogee kick motor to get APPLE into its final orbit. This was a delicate moment, one that had often gone wrong in launchings, and this was the first time that India had attempted to use an apogee boost rocket. This time, a solid-fuel rocket was used and because solid rocket boosters fire only once, the maneuver must go right and there were no second chances. The motor weighed 320 kg, of which 272 kg was solid-fuel propellant designed to burn for 33 sec and give a thrust of 2.2 tonnes. APPLE was launched by Ariane V3 (V = *vol*, or "flight" in French) on 19th June 1981, accompanying the main payload, the *Meteosat 2* weather satellite.

Getting APPLE into its final correct orbit was a torment. The apogee boost motor fell short and could only achieve a 22-hr orbit of 31,000–35,800 km and the rest was to be achieved by the 16 thrusters on board, using only 42 kg of hydrazine fuel. Then, one of the two solar panels failed to deploy. Worse, it obscured the Sun,

The APPLE team

preventing the satellite from locking on into the right attitude. One crisis meeting followed another. After eight thruster firings, the last of 85 sec, APPLE eventually reached its intended station, 102°E.

Although they never freed the jammed solar panel, APPLE relayed television programs and educational teleconferences. Emergency communications for cyclones were tested out. Newspaper facsimile pictures of *The Hindi* were test-transmitted. Republic Day celebrations were broadcast nationwide. There were tests of satellite-based banking transactions and the monitoring of railway wagons.

The APPLE experiment lasted just over two years. APPLE involved the testing out of small portable terminals as little as 1 m across in states such as Orissa and Gujarat. The jammed solar panel prevented the operation of the radiators on one side of the spacecraft, causing the risk of overheating, so ground controllers took the risky decision to pitch the satellite regularly, even though this meant losing lock on Earth. APPLE operated until 19th September 1983, when it had to be abandoned.

FIRST INSAT 1: A SYSTEM ESTABLISHED

The Shuttle suffered many delays before it entered service, so INSAT 1A was lofted into orbit instead by a Delta 3910 on 10th April 1982 from Cape Canaveral. Due to the lack of a suitable orbital slot, it had to be parked over Indonesia, which was far from ideal. INSAT 1 was built with 12 television transponders, two TV direct broadcasting antennae and, for weather forecasting, a Very High Resolution Radiometer to image the Earth every 30 min. The radiometer had a resolution of 2.5 m in the visible band and 10 km in infrared. The TV transponders could reach up to 100,000 small Earth terminals, or, instead of transmitting television, the system could handle up to 8,000 telephone calls at a time or be used for radio. INSAT 1A was a box-shaped body weighing 1,152 kg (fuelled, about 550 kg unfuelled) with a single solar panel with 12,942 solar cells at one end, counterbalanced by a boom at the other – the first such arrangement at the time. INSAT had a 445-N booster rocket to reach apogee and six 22-N thrusters for station-keeping. INSAT was intended to make a big difference, for, in 1982, India had only 12 television transmitters.

Initially, the main antenna failed to fully deploy, threatening the broadcasting aspect of the Rs5.7bn (€122m) mission, but these problems were overcome by bathing the antenna in sunlight and by the brief firing of a thruster. Like APPLE, the mission was full of further difficulties and drama. First, the C-band antenna failed to deploy and blocked four of the 12 thrusters. Then, the boom failed to deploy, which meant that thrusters would have to be used to balance the solar panel, reducing the lifetime of the satellite from seven years to two. The undeployed boom also obscured radiators on one side of the spacecraft, leading to overheating and reducing broadcasting to 5 hr a day. On 31st August, the satellite accidentally locked onto the Moon instead of the Earth, turned away from Earth and control was lost. The confused satellite apparently tried to re-orientate itself, but depleted all its hydrazine fuel in an hour. The satellite drafted off station and was lost, being taken out of commission on 4th September after being operational for only four months. The timing was especially unfortunate, for it coincided with the opening of the Asian games, hosted by India, and the country had to buy in communications satellite lines from the United States and Soviet Union to make good the loss.

Weather warning was the other function of INSAT, being designed to provide half-hourly pictures of cyclones and sea conditions, warn of floods and disasters and collect and transmit data from remote observation stations. The first pictures were received in early May 1982 with a nighttime image and then a daytime one. The satellite worked well initially and produced good pictures. Its replacement, INSAT 1B, was launched from the payload bay of the Space Shuttle *Challenger* on the eighth Shuttle mission, STS-8, on 31st August 1983. The astronauts spun the blue and gold INSAT out of the Shuttle's payload bay at 40 rpm. The astronaut responsible, Guion Bluford (also the first black person in space), reported that "INSAT was deployed on time with no anomalies and the satellite looked good". Forty-five minutes later, the payload assist module fired to send INSAT into a geostationary transfer orbit of 296–38,173 km. Following several firings by the satellite's own liquid-propelled engine, INSAT was stabilized at 74°E (though there were difficulties with extending the solar array and this was not accomplished for over a month). INSAT 1B extended television coverage to 70% of India's population. In 1991, experiments were made in videoconferencing to train adult education officials. ISRO is reported to have paid NASA Rs1.485bn (€31.6m) for the launch. No one was apparently more relieved than Prime Minister Indira Gandhi, who was anxious to enlist television in her campaign to be re-elected the following year. 1B became an early star of the program and soldiered on until 1993, when it was finally retired, lasting almost 10 years compared to the few months of 1A.

INSAT 1 had originally been intended as a two-satellite system. Because of much higher than expected demand on satellite television, two more were ordered. The first was INSAT 1C. Originally to be launched by an Indian astronaut on the Shuttle, the mission was delayed when *Challenger* exploded and the Shuttle was grounded. INSAT 1C satellite was reallocated to Europe's Ariane rocket and sent into orbit on Ariane V24 on 21st July 1988, being subsequently stationed at 93.5°E. Disappointingly, not long afterwards, the satellite lost a quarter of its power due to the failure of a diode. Although the weather observation platform was kept fully going, only half the telecommunications systems could be used. In November 1989, the satellite lost attitude, Earthlock and had to be abandoned. India filled the gap by leasing capacity from the Soviet Intersputnik system.

Although it had been planned to fly INSAT 1D on the Shuttle in 1990, the new post-*Challenger*-disaster Shuttle program moved away from commercial satellite launches. INSAT 1D was transferred to a conventional Delta rocket instead. In a bizarre accident on the launch pad on 19th June 1989, a crane hook tangled with cables and crashed into the satellite, destroying the C-band antenna and damaging the rest of the satellite so badly that there were fears it would be written off. The manufacturers, Ford, deemed the satellite repairable and took it away for a year's repairs, which cost as much as Rs443m (€9.4m). In a doubly bizarre turn of events, Ford brought it to its workshop in Palo Alto, California, just in time for the big earthquake there, where it suffered a further Rs.5m (€140,000) damage. INSAT 1D was eventually launched on Delta 4925 on 12th June 1990. The four INSATs had used, between them, three different launchers (Delta, Ariane and Shuttle). INSAT 1D was much the most successful of the four and was still operational five years later, having sent back 24,500 weather images. Television had been relayed to 2,000 community receiving sets.

INSAT 2: MADE AT HOME

Preliminary studies of a successor system were commissioned in 1979, even before the APPLE satellite, and were completed 1981, with an in-service date set of 1989, though, in the event, this slipped until 1992. The technical specifications required increased capacity, improved reliability and additional meteorological equipment. Their weight was 50% greater than the INSAT 1 series (1,906 kg at launch, 911 kg once all its propellant had been depleted, compared to 550 kg) with the number of C-band channels up from 12 to 18. A lot of work went into addressing the earlier problems with solar panels. Instead of a staged, partial and then full deployment, a new system was devised to open the panel or panels in one go, when facing the Sun. INSAT 2A was set to operate at 74°E and 2B at 93.5°E. 2A and 2B were joint communications and weather satellites, while 2C and 2D were lighter (747 kg) and communications only. The Rs20bn (€425m) INSAT 2 program was approved in

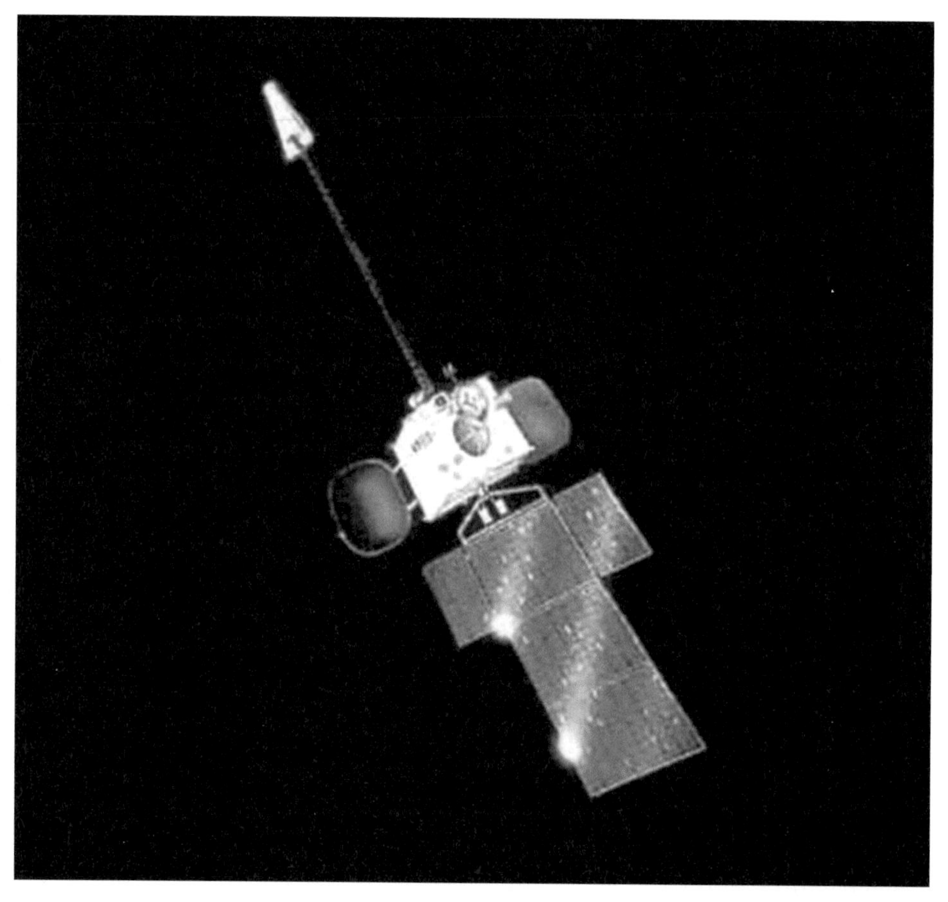

INSAT 2

April 1985, the principal aim being to build the satellites within India itself, although foreign launchers would still be required. Ultimately, the intention was that later versions would be fired by Indian geosynchronous launchers.

Here, though 2C and 2D broke new ground by having the first higher-frequency transmitters in the Ku band so as to suit electronic news gathering and higher-density traffic. Second, transmission could go to small rooftop antennae. Third, wide-beam antennae were used, extending transmissions beyond India to south-west Asia. Fourth, the satellites were used for mobile communications. This required a higher power level: two solar panels (span 23 m) rather than one and power of 1,200–1,600 W rather than 1,000 W. Lifetime was extended to 10 years. They also carried a beacon for an international maritime distress system. The station was to be maintained by 16 attitude thrusters. The Very High Resolution Radiometer had improved definition compared to INSAT 1 and was able to send a full Earth scan every 33 min or examine more select areas for cyclones every 7 min.

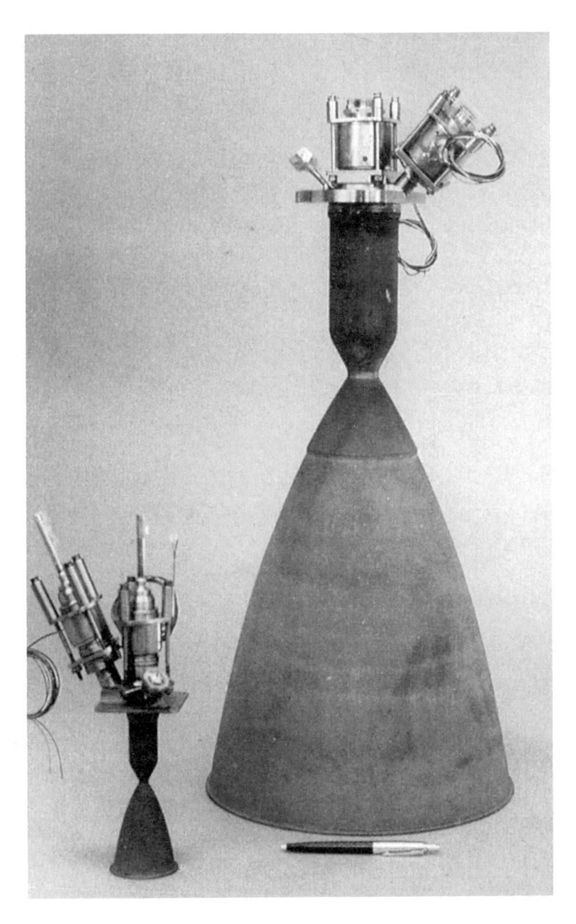
Liquid apogee motor

Europe's Ariane was chosen as the launcher for the series. This time, the INSAT satellite was built in India itself, with some limited assistance from abroad. For example, British Aerospace supplied titanium-made helium pressurization tanks. The electronic modules and high-speed, high-density integrated circuits were designed by Racal-Redac, based in Reading. The final stage was Indian-built, developed by the Liquid Propulsion Center in India. Called the Liquid Apogee Module, or LAN, its function was to provide sufficient thrust in one burn or more to get the communication satellite from an orbit of 200–36,000 km to a circular one of 36,000 km. It had a thrust of 440 N, a specific impulse of 310 sec and used mixed oxides of nitrogen and monomethyl hydrazine.

Launched by Ariane 4 on its 51st flight, V51, on 10th July 1992, INSAT 2A entered a geostationary transfer orbit

10 min later. The master control facility at Hassan took over control and maneuvered it into position for the critical firing to raise its perigee and place it in geostationary orbit. The next day, the liquid apogee motor fired twice, once for 3,900 sec and then for 806 sec. Three days later, the 15.5-m^2 1,200-W solar array was deployed, followed by the Sun tracker, the antennae and the boom. On 23rd July, Hassan commanded the orbit to be trimmed and for the satellite to drift toward its parking slot at 74°E, which it reached on 29th July. By this stage, the transponders and the weather-imaging system had been switched on.

INSAT 2A began operations in August 1992 and the Minister for Science & Technology, Mr Kumaramangalam, was able to tell the Indian parliament, the Lok Sabha, that India could be justifiably proud of having developed such a state-of-the-art communications satellite by itself. With INSAT 2A, India was able to operate a series of telephone trunk routes across the sub-continent, provide television for 65% of the Indian landmass and 80% of the population and signals for 127 radio stations. In September, the weather imager on INSAT 2A sent back a picture of the Earth's globe: to the north were the dusty browns of Arabia; to the south, the swirling clouds of the southern ocean; and in the middle, India, with three large cloud areas off the coast. On the communications side, annual rental revenues quickly reached in the order of Rs4bn (€85m) a year.

INSAT 2B was launched by Ariane 4 from Kourou, French Guyana, on 23rd July 1993. On 6th August, it beamed its first weather images back to Earth and it was declared operational shortly thereafter. It carried the search-and-rescue transponders under the international COSPAS/SARSAT system. This is a global network whereby satellites can pick up a distress beacon, generally from ships at sea, and quickly relay the signal to the nearest ground station. Because it uses a global network of polar-orbiting satellites, most distress calls can be notified to a ground station and then to a national rescue service anywhere in the world within half an hour and is much more reliable than the traditional ship-to-shore system. India installed its first COSPAS/SARSAT ground station, called a local user terminal, in Bangalore in 1989 and a second in Lucknow the following year. India installed a 2.4-m dish, which not only served distress calls adjacent to India, but acted as a relay service for Bangladesh, Indonesia, Kenya, Malaysia, Maldives, Singapore, Somalia, Sri Lanka, Tanzania and Thailand as well. The most publicized rescues under the system have been those of round-the-world yachts, but the unpublicized work of the system goes on all the time, with alarms triggered by cargo ships either sinking or in danger of doing so. In 1998, a beacon was set off when a trekking expedition in the Himalayas got into trouble and had to be rescued.

INSAT 2C, with 23 transponders, was launched by Ariane on 7th December 1995. It introduced Ku-band technology to India and used such a wide footprint that the satellite could be used as far afield as Australia, China and central Asia. By then, India had 70 transponders in space. INSAT 2D was launched by Ariane from Kourou on 3rd June 1997 but, due to problems with electrical supply, it suddenly lost its ability to lock on the Earth: communications to 83 terminals in the north and north-east of the country (Jammu, Kashmir and Uttar Pradesh) went down. Even the national stock exchange in Bombay was knocked out. The disrupted stations

INSAT 2C

were transferred to the old INSAT 1B; the national stock exchange declared a holiday while the problems were sorted out. Ground Control in Hassan managed to restore three, later seven transponders on INSAT 2D and even reconnect the stock exchange, but this provided only temporary relief and the satellite was abandoned in

October, a demoralizing outcome. India then bought the Arabsat 1C satellite from an Arab consortium. Arabsat was moved to INSAT's orbital position, where its 26 transponders could be brought into play by Hassan master control (it was posted the name of INSAT 2DT). This was a temporary measure and the decision was then taken to bring forward the launch of the first INSAT 3.

INSAT 2E was launched by Ariane on 3rd April 1999 on its 117th mission, V117. ISRO scientists in Kourou cheered twice – at separation 21 min after lift-off and some hours later, when the first signal was acquired. Their celebrations were short-lived, for the circularizing of the orbit to 36,000 km went astray when the liquid apogee motor cut out 16 min into a 75-min burn. Fortunately, with three further firings commanded from Hassan, it eventually arrived on station and the first weather pictures were received in Hassan Master Control Facility in mid-April. INSAT 2E was positioned at 83°E, over the Indian Ocean, and declared operational a month later. INSAT 2E weighed 2.55 tonnes, had a 14-m solar panel and boom, carried 17 C-band transponders and a 1-m-resolution CCD camera and, for the first time on INSAT, a radiometer for assessing water vapor. This radiometer had 2-m resolution in the visible band and 8-m in infrared and such instruments have proved to be very accurate in forecasting the likely volume of rainfall. INSAT was also able to pick up and retransmit data from over 100 remote location weather platforms in uninhabited areas. The satellite was expected to operate until 2011.

REACHING THE VILLAGES

By 2000, 700 television and almost 200 radio stations used space-based signals. Ninety percent of the Indian population now received space-based television, compared to 20% before the first INSAT was launched. In the rural areas, 35,000 of India's 520,000 villages had satellite terminals. INSAT 2E's performance was such that the European INTELSAT consortium leased 11 transponders for 10 years for a cost of Rs8.86bn (€188m). The International Telecommunications Satellite Organization, Intelsat, which provided international lines to India, noted that its Indian revenues almost doubled in two years, from Rs752m (€16m) in 1996 to R1.372bn (€29.2m) in 1998.

Typical Indian satellite dishes are in the 10-m-diameter range, but the size is gradually being brought down to 3 m and later 2 m. The later INSAT 2 series operationalized a system first tested in 1987 whereby it could be commanded to trigger battery-powered systems in Indian villages to set off sirens to warn of impending cyclones. At one stage, with a cyclone heading in, the alarms triggered the evacuation of 170,000 people and many lives were saved. Typically, the siren klaxons loudly for a minute, followed by a warning in the local language, giving warning of the impending danger and the precautions to be observed. By 2000, 250 receivers were installed in the most cyclone-prone areas.

India's satellite systems have been used for video teleconferencing, training for farmers, programming for rural development workers, with networks set up under the Indira Gandhi Open University and the National Dairy Development Board. In

February 1995, Prime Minister P.V. Narasima Rao dedicated a training and development channel on INSAT. It was put to early use in the training of rural development workers. By 2000, about 120 hr of educational television were going out each month to 4,000 schools and colleges. The Training and Development Channel was used to transmit distance education in rural development, women's concerns and child development. Three hundred and seventy-five terminals were in use for this purpose, but with expansion set to reach the 2,000 mark. It is an interactive service – one that enables learners to phone back questions to the teachers (the system is called *talk-back*). Later, digital television will enable lessons to be downloaded and reused locally. By 2000, the INSAT system was providing 4,700 voice circuits via 430 Earth stations and 1,200 terminals. A unique experiment in development education through satellite began in 1996. Television was supplied to the poorly developed region of Padya Pradesh, evening educational television being provided in health care, literacy and the management of water resources to 150 sites. It received such a positive response that the program was extended to a further 200 sites in the adjacent districts of Dhar and Barwani.

PROMISE OF INSAT 3

Next in the series was INSAT 3. Work on defining the specifications for INSAT 3 began in 1992 and five satellites were ordered in 1998. The aim was to increase the numbers of transponders available from 70 in 1998 to 130 by 2002 and later to introduce digital television. INSAT 3 was so big that the enormous Ukranian Antonov 124 transports had to be used to lift the satellites from India to the European launch site in Kourou.

The first INSAT 3 was actually INSAT 3B, launched by Europe's new heavy launcher, the Ariane 5, on 21st March 2000. This was a communications-only satellite and was brought forward due to a high demand for channels. INSAT 3A followed on 10th April 2003 at a cost of Rs3,257m (€69.2m), used its 440-N liquid apogee motor to drift to station and began operations from 93°E the following month. INSAT 3A was a 2,950-kg multipurpose 24-transponder communications, weather and search-and-rescue satellite with a temperature sounder. By this stage, the INSAT system was providing 121 transponders not only for India, but for people of Indian origin throughout Asia.

INSAT 3E was lifted by Ariane 5 from Kourou on 28th September 2003 on V162 to 55°E, the Ariane also carrying in its cargo bay Europe's first Moon probe, SMART. INSAT 3C was a communications-only satellite, while 3D was primarily a weather satellite with a six-channel imager and a 19-channel sounder.

The next step forward was the INSAT 4 series, each carrying 12 kW of power and 24 transmitters (half C-band, half Ku-band), with the intention of dedicating channels to telemedicine between hospitals and the villages. The INSAT 4 satellites were box-shaped, with two solar panels 15.4 m across generating 6 kW of power, three lithium batteries and 32 thrusters for station-keeping. INSAT 4A and 4B were launched on Europe's Ariane (21st December 2005, 11th March 2007), with

INSAT 3A view of India

a view to later missions flying on India's own geostationary launch vehicle, the GSLV (see Chapter 6), completing the promise made 30 years earlier of India launching its own home-made communications satellites on its own home-made launcher.

The INSAT series made possible the extensive development of telemedicine, to the point that in March 2005, ISRO convened a three-day conference for 550 delegates on the topic in Bangalore. By the following year, ISRO had connected 33 specialized hospitals to 132 remote, rural or district hospitals in its telemedicine network, enabling specialists to provide advice to doctors working in the remote rural areas, with the prospect, over time, of robotic surgery carried out via satellite.

Using INSAT 3A, ISRO inaugurated its system of 25 interactive Village Resource Centers (VRCs), opened by Prime Minister Manmohan Singh in Sempatti, Tamil Nadu, in October 2004. The concept was that remote sensing satellite data relevant to the area around the village be made available in each VRC, offering information and advice to people in the area, covering such fields as land records, watersheds, harvest conditions, wasteland for reclamation, cropping patterns, disease warning, pest control, soil conditions, fishing and conservation, with information distributed from a central node in Chennai using a dedicated channel on INSAT 3A.

Telemedicine van

Following INSAT 4, ISRO developed, in cooperation with the leading European satellite designer and manufacturer, EADS Astrium, the W2M comsat, weighing over 3 tonnes and with 32 transponders, for launch on Europe's Ariane 5 in December 2008. The INSAT series is summarized in Table 5.2.

Table 5.2. INSAT series

10 Apr. 1982	INSAT 1A	Delta	Cape Canaveral
30 Aug. 1983	INSAT 1B	Delta	Cape Canaveral
21 Jul. 1988	INSAT 1C	Ariane	Kourou
12 Jun. 1990	INSAT 1D	Delta	Cape Canaveral
10 Jul. 1992	INSAT 2A	Ariane	Kourou
23 Jul. 1993	INSAT 2B	Ariane	Kourou
7 Dec. 1995	INSAT 2C	Ariane	Kourou
4 Jun. 1997	INSAT 2D	Ariane	Kourou
2 Apr. 1999	INSAT 2E	Ariane	Kourou
21 Mar. 2000	INSAT 3B	Ariane 5	Kourou
10 Apr. 2003	INSAT 3A	Ariane 5	Kourou
28 Sep. 2003	INSAT 3E	Ariane 5	Kourou
21 Dec. 2005	INSAT 4A	Ariane 5	Kourou
11 Mar. 2007	INSAT 4B	Ariane 5	Kourou

ASLV first stage

NEW LAUNCHERS: ASLV

India's first launcher, the SLV, could put only 40 kg into low Earth orbit and India wanted to go on to put much larger satellites into orbit. India would have liked to have proceeded straight to a much larger, polar launch vehicle capable of putting a tonne into orbit, but this could not be done for a number of years, so an intermediate rocket was decided on, the ASLV (Augmented SLV). Meantime, the ageing SLV was adapted as the first stage of the *Agni* intermediate-range ballistic missile.

India proceeded with the upgrading of the launcher in order to place larger, 150-

kg payloads into 150–300-km orbit. The ASLV was essentially the Scout-class SLV, but with the addition of strap-on boosters. The weight of the ASLV was almost 40 tonnes, the height 23.5 m. This time, the launcher was assembled vertically on the launch pad on a 40-m-tall mobile service structure with lifts, access platforms and clean room. The strap-on boosters were first tested in flight in November 1985, when they were attached to a *Rohini RH-300* sounding rocket. In place of the unguided SLV, the ASLV had new nozzles and a new guidance system.

ASLV was seen as an intermediate rocket whilst a much larger one was in design, so only four ASLVs were commissioned. The extra thrust increased payload from 40 to 150 kg, so the new satellites were called "stretched *Rohini*" (acronym SROSS) – an extension on the earlier small *Rohini* satellites. They typically carried gamma ray detectors, a monocular optical scanner, ionospheric monitor, two retro-reflectors for laser tracking and an X-ray instrument. These had a different shape, being "octagonal prismoids" – multisided cylinder shapes with solar panels in vanes behind. SROSS A had a laser tracker and a gamma burst detector. SROSS B had a stereo camera developed in Germany.

Like its predecessor, the ASLV was hard to tame. The ASLV failed on its first flight in March 1987: when the first set of strap-on rockets completed its burn, the next set failed to fire due to a short-circuit in the ignition system. Modifications of the safing and ignition systems were made. The second flight, on 13th July 1988, also failed. This time, the next stage ignited on time, but the rocket broke up at 50.4 sec at an altitude of 25 km, cartwheeled while still firing and impacted into the ocean after 257 sec aloft. Commands to separate the strap-ons failed. The unfortunate satellite was flung free. High winds that day contributed to the disaster, turning a minor yaw error into something that could not be corrected. The post-flight analysis revealed a

ASLV first-stage assembly

basic problem in the ASLV design: not only was the ASLV an unusually long and thin rocket, making it vulnerable to instability, but the crucial maneuver of dropping the strap-ons and igniting the first stage took place at the moment of maximum dynamic pressure, when the atmospheric forces on the rocket were at their most acute. Observers of the American Space Shuttle noted how the Shuttle engines were always throttled back at this stage.

Both failures had taken place at almost identical moments in the mission: 50 sec. As a result, two passive fins were added. The autopilot was given more authority to adjust the sequencing of the maneuver in the light of real, not planned, performance. The strap-ons were designed to burn more slowly, reducing the dynamic pressure. New rules were set down about when launches could take place, to avoid high winds at altitude.

These modifications took four years but paid off when the ASLV finally succeeded in orbiting SROSS C on 20th May 1992. The core stage and two strap-on rockets ignited at the zero-point in the countdown. The other four strap-ons fired at 50.5 sec, each running for 74 sec. First-stage burnout took place at 111 sec, followed by staging. The second stage burned for 150 sec before the third stage solid rocket motor took over. This brought the vehicle to 421 km some 380 sec after lift-off. The rocket coasted for 3.5 min before the two small fourth-stage liquid-fuelled engines ignited for 405 sec. Due to a problem in the fourth stage, the orbit achieved was much lower than planned – 267–433 km instead of circular at 475 km. SROSS had six small thrusters, using 4.8 kg of hydrazine fuel, designed for six months of orbital

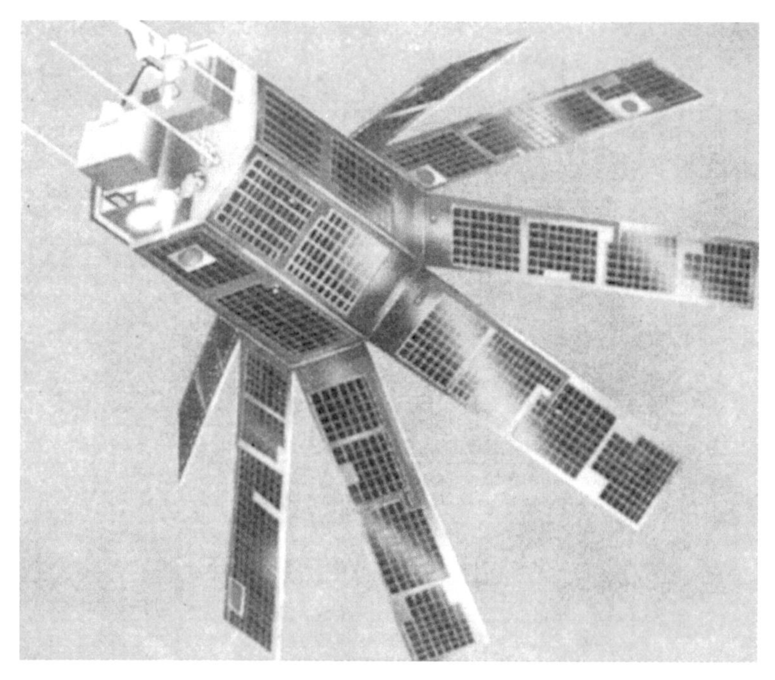

SROSS

Table 5.3. ASLV satellite launches

20 May 1992	SROSS C	ASLV	Sriharikota
4 May 1994	SROSS C2	ASLV	Sriharikota

operations, but it never lasted that long: because of the low perigee, the payload burned up after two months on 14th July.

SROSS was cylindrically shaped with eight solar panels generating 45 W of power. SROSS C carried a retarding potential analyzer for the National Physics Laboratory to measure density and temperature of electrons and ions in the ionosphere over equatorial and low altitudes and a gamma burst experiment developed by ISRO in Bangalore. The gamma burst detector was switched on during orbit 119 and after a few false alarms, when it passed through the radiation belts, recorded its first gamma ray burst on orbit 337. It was the first Indian satellite to have indigenously made batteries – 12 nickel cadmium cells able to complete up to 18,000 charge–discharge cycles.

Two years later, ASLV launched SROSS C2 at night on 4th May 1994, the solid rocket boosters turning the dark sky to light as it headed skyward. The orbit was not quite perfect, 437–938 km, but this time high enough for a full-length mission. SROSS C2 had a small motor with 5 kg fuel: in July, the satellite made a series of maneuvers to lower the height of its orbit to reach its intended destination of from 429 to 628 km, 46° and still had 1.5 kg fuel remaining. It was the best launch in the ASLV program.

SROSS was octagonally shaped, carrying solar cells on its body and on attached solar panels. Like its predecessor, the 113-kg satellite carried a gamma ray burst experiment and a retarding potential analyzer to investigate the ionosphere and thermosphere at equatorial and low latitudes. The detector found 12 gamma bursts in the first year in orbit and the analyzer made 600 sets of ionospheric data. It was still in operation in late 1999. This launch marked the conclusion of the ASLV series, which is summarized in Table 5.3.

SROSS C2

PSLV: INTO THE BIG LAUNCHER LEAGUE

The ASLV was always seen as an intermediate step toward India's second generation of launch vehicles, the Polar Satellite Launch Vehicle, the PSLV. The aim of the PSLV was to place 1,000-kg remote sensing satellites into 900-km Sun-synchronous polar orbits – a substantial advance. The PSLV was designed for the Indian Remote Sensing Satellite (IRS) series, which had thus far been launched by the USSR. In effect, the PSLV enabled India to join the "big launcher" club of the United States, Russia, Europe, China and Japan. The PSLV was almost 10 times bigger and heavier than the ASLV, being 275 tonnes in weight and 44.1 m tall. It was broadly comparable to the American Delta and its Japanese cousin, the N-II. The development costs were Rs4,136m (€88m) with the effective cost thereafter of an individual launching Rs658m (€14m). After domestic satellites had been launched, it was ISRO's intention that the PSLV would capture at least some of the lucrative world launcher market.

The size of the PSLV made it the third largest *solid*-fuel rocket in the world (after the boosters on the American Space Shuttle and the Titan). India was now a world leader in the development of solid-fuel rocket engines and had obtained global patents for its work in the development of castor oil-based fuels, India being the second largest producer of castor oil in the world. The intention was that the first stage should bring the rocket to 56 km, the second stage to 250 km, the third to 400 km and, after a long coasting phase of 273 sec, the fourth stage to orbital altitude of 904 km.

Rather than construct one giant solid rocket stage, the manufacturers made five smaller 2.8-m-diameter segments, each 25 tonnes in weight. The first stage generated 45 tonnes of thrust for 90 sec. The fuel used was hydroxyl-terminated ploybutadiene resin, with, as oxidizer, ammonium perchlorate, all domestically developed. The fuel was fortified by high explosive to give extra performance. Take-off of the PSLV was assisted by solid-propellant strap-on rockets, two of which ignited on the pad and four after 30 sec. The first stage weighed 128 tonnes and was made of maraging steel. The first stage of the PSLV was first tested in October 1989. Placed horizontally near the launch site at Sriharikota, the rocket gushed a blast of flame twice its own length, sending enormous dirty brown clouds billowing into the south-east Indian sky. The third stage was also solid-fuelled but was made of different material – polyamide kevlar. During ascent, course was maintained by two independent packages of thrusters. The third stage was India's most advanced solid propellant rocket, with a specific impulse of 293 sec.

The PSLV was an unusual mixture of solid and liquid-fuelled rocket (solid–liquid–solid–liquid). Until that point, rockets had been either liquid or solid-fuelled, or had been liquid-fuelled with relatively small solid strap-ons. The PSLV marked an unusual combination of the two. On top of the huge solid-fuel first stage was a liquid-fuel stage. India had tested its first liquid-fuelled rockets may years earlier – for the first time on 15th May 1974, when a small rocket stage, 25 cm in diameter and generating 600 kg thrust, had been put on the second stage of a sounding rocket.

The PSLV now marked the introduction of liquid-fuelled rockets in the Indian

PSLV on pad

PSLV second-stage engine

space program. The second and the fourth stages had storable liquid propellants – UDMH (unsymmetrical dimethyl methyl hydrazine) and nitrogen tetroxide. The second stage used a European motor, the SEP (the French Societé Européenne de Propulsion) Viking, the same one as used on the European Ariane program and which generated thrust for 145 sec. This required the construction of new test facilities, namely the Liquid Propulsion Center at Mahendragiri and the first Vikas engines were tested there in 1988.

The fourth stage, built indigenously, had two identical engines that burned for 425 sec, its propellant tanks being built of titanium, with engines that could be gimbaled in two planes. Here, the fuels were slightly different – mixed oxide of nitrogen (3% MON-3/97%N_2O_4) and monomethyl hydrazine (MMH).

LAUNCHING THE PSLV

The first PSLV launched on 20th September 1993. At first, all went well. The PSLV rose steadily, small items of débris flying free at lift-off, all as scheduled. But, at the end of the second-stage firing, there was an unplanned interval of three seconds before the third stage ignited, causing a significant loss of velocity. The small rocket designed to pull the second stage clear failed and instead nudged the third stage slightly off its course (collision would be too strong a word). Despite these events, the mission could still have been saved had the upper stages compensated for the loss of thrust and corrected the attitude, as the computer was programmed to do. Unfortunately, due to a programming flaw, it did not make the necessary adjustments. By the time of fourth-stage ignition, the top of the rocket had reached an altitude of only 340 km. The fourth stage lacked sufficient thrust to get the payload into orbit and it fell back to Earth after describing a long arc of 1,700 km. A failure analysis committee was set up and determined that the problem was due to "a software error in the pitch and control loop of the on-board guidance and control processor which occurs only when the control command exceeded the specified maximum limiting value" and an unintended contact between the second and third stages. Positively, the review board found that the engines, guidance and other systems had worked well and determined that the rocket's design was fundamentally sound. Still, the expensive and valuable IRS payload was lost: the satellite was in fact the refurbished engineering model of IRS-1A. It carried a German Monocular Electro-Optical Stereo Scanner, a LISS-1 scanner and a carbon dioxide monitor. First launch failures now seemed a well established tradition.

On the next occasion, there were no such problems and on 15th October 1994, the PSLV put into 825-km polar orbit an 870-kg remote sensing satellite, the IRS-P2. The rocket was painted dark red and white, with the Indian flag atop the shroud and the letters "India" painted on vertically below the ISRO logo. The previous time, the Mauritius ground station had waited in vain for the satellite to appear overhead, but this time, all went well and good signals were picked up. In order to avoid overflying Sri Lanka, achieving polar orbit required controllers to steer the rocket through a 55° yaw maneuver 100 sec after lift-off.

INTRODUCING IRS-POLAR

The P series of IRS was successfully introduced with the first "Polar" SLV on 15th October 1994. Centerpiece of the new satellite was the new LISS scanner, LISS-II, designed by the Space Applications Center in Ahmedabad, Gujarat, with a swath of 131 km. The CCD camera operated in visible light and infrared, with a resolution of 30 m across four spectral bands. The desk-sized spacecraft had two solar panels providing 510 W of electrical power. The mission focused on mapping water resources, checking for floods and monitoring agricultural yields. Under the government's Integrated Mission for Sustainable Development, 1992, the country was divided into 157 zones for water management and the satellite would help build

PSLV take-off

up a water picture for the different areas. The concept of "integrated" was that the program brought together space and ground-based sensing; diverse disciplines such as public administration, health, science and soil development; and was linked to programs of rural development. IRS-P2 returned 60,000 images in its first year of operation. The P2 and P3 followed similar orbits to IRS-1C and D, 101-min period, 98.6° inclination and orbital height of 817 km.

In its second successful launch, the PSLV put another remote sensing satellite, IRS-P3, into orbit on 21st March 1996. IRS-P3 concentrated on the assessment of

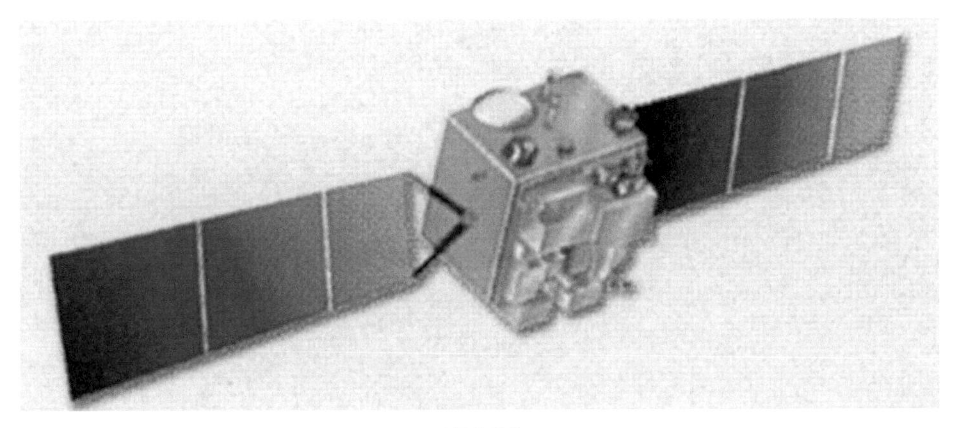

IRS-P3

vegetation, snow studies, geological mapping for minerals and the analysis of chlorophyll in the oceans and sediment entering the oceans. P3 carried an X-ray astronomy telescope, with three counters, a sky monitor, a pinhole camera and a German modular opto-electrical scanner called MOS. P3 carried a wide field sensor with an image area of over 800 km and a resolution of 190 m, able to provide broad-scale maps of India.

On its fourth flight, the PSLV introduced the C, or continuation, version of the launcher. This had 10 tonnes more propellant in the first stage (139 tonnes) and 2.5 tonnes more propellant in the second, between them sufficient to increase the payload from 930 to 1,250 kg.

IRS-1D IN TROUBLE – BUT SAVED

The next brown-and-white PSLV rose into the autumn sky of Sriharikota on 28th September 1997. The national flag was mounted on top, flanked by an emblem to mark 40 years of Indian independence. IRS-1D was a replacement for IRS-1C and carried a panchromatic camera, linear imaging self-scanner and wide-field sensor, fed by six solar panels and had experimental facilities to measure its orbit in reference to the Global Positioning System. It was aimed at a circular orbit at 820 km.

There was a small thrust shortfall in the fourth stage due to what appeared to be a fuel tank leak. Controllers in Sriharikota noted that the rocket was not achieving the intended velocity and commanded the engine to burn another 15 sec when the final 420-sec burn ended.

The 300-km perigee was too low for imaging and, worse, so low as to lead the IRS to spiral into the Earth's atmosphere and destruction in a few months. In an attempt to save the mission, the satellite used its thrusters to gradually raise its orbit, but in the knowledge that the satellite would not be able to carry out its full program. The setback came at a bad time, for Ground Control in Hassan was already struggling with power failures on the INSAT 2D. The chairman of ISRO, Dr K. Kasturirangan, went straight from Sriharikota to the ISRO tracking center in Bangalore to join his colleagues in emergency session to save IRS. The satellite had 84 kg of fuel, originally designed for station keeping over the three years of the satellite's normal operation. By 3rd October, ISTRAC had fired the 11-kN thruster engine five times and had raised the perigee to 375 km and planned to use small burns to raise the orbit to something in the range of 600–680 km. By mid-month, they had raised it to 700 km and the cameras on board the spacecraft were working perfectly. By December, ISRO was able to announce that the satellite would now be able to complete a five-year program after all.

EXPANDING THE EARTH OBSERVATION PROGRAM

The introduction of the PSLV enabled India to expand its Earth observation program significantly, with almost yearly launches. The Earth observation program became more specialized, with different satellites dedicated to the oceans and mapping. On a clear morning, cheered on by Indian Prime Minister Atal Bihari Vajpayee, India's first dedicated ocean observation satellite was put in orbit by the PSLV on 25th May 1999. IRS-P4, more popularly known as Oceansat 1, weight 1,050 kg, carried an ocean color monitor and a multi-frequency scanning radiometer able to penetrate clouds. Power came from two 9.6-m^2 solar arrays generating power of 750 W. Oceansat 1 entered a circular orbit at 727 km. Oceansat 1 was derived from the IRS series, but with ocean color monitor, scatterometer to measure wind speed and direction, altimeter to measure wave height and thermal infrared radiometer to measure temperatures and ocean currents. Its instruments were able to see down to 200 m deep.

By the 16th orbit, the main systems had been deployed and signals were received in Bangalore, Sriharikota, Lucknow, Mauritius, Biak (Indonesia) and Bear's Lake (Russia). By the 17th orbit, the camera had been activated, providing a rich color image of southern India, the blue Bay of Bengal and the north-west of Sri Lanka. Its orbit was fine-tuned in June with the firing of 11-N thrusters to settle the spacecraft in a perfect 727-km orbit crossing the equator at precisely noon local time every day. P-4's launch had been delayed by American sanctions following India's nuclear tests of 1998, which had prevented the fitting out of the ocean color monitor, but the equipment had, in the event, been obtained from Germany. The ocean color monitor operated in eight spectral bands in a 1,420-km swath and was used to detect chlorophyll, plankton, aerosols and sediments. The monitor was capable of covering the country every two days.

On 22nd October 2001, the PSLV put three satellites into orbit: a 1,108-kg Technical Experimental Satellite (TES), the main payload; a 94-kg Belgian satellite, PROBA; and a 92-kg German optical imaging satellite, *Bird*. The technical experimental satellite was intended to test out reaction wheels, controls, structures, recorders, an x-band phased array and panchromatic camera with 1-m resolution. The orbit was polar Sun-synchronous 567–572 km, 97.76°. Indian space officials ducked press questions as to whether photographs from TES could or would be used by the military. The mission had a postscript, for two months later, the PSLV upper stage blew up, scattering 300 pieces of débris into an orbit of 200–1,100 km. Because of the growing problem of orbital débris, most spacefaring nations had begun to "passivate" their upper stages by releasing all their residual propellants. It seems, though, that this did not happen and that the hypergolic fuels mixed and exploded.

Although the PSLV was designed to put satellites into polar orbit at around 900 km, it could be stretched to reach geosynchronous orbit by adding an extra half-tonne of propellant to the fourth stage. Metsat 1 was put into geosynchronous orbit by the PSLV on 12th September 2002. Unlike the INSAT satellites, which combined weather forecasting with telecommunications, Metsat was a dedicated weather satellite, which replaced the old INSAT 1D, which had been operating since 1990. Metsat reached its intended location of 74°E on the 25th. The 1,060-kg Metsat carried a very high-resolution infrared radiometer able to make hourly scans in the visible, infrared and water vapor bands and a data relay transponder to collect data from weather stations and then relay them to the Meteorological Data Utilization Center in New Delhi. The following year, its functional name gave way to a new name, *Kalpana*, as a memorial to India's most famous astronaut.

In October 2003, the PSLV put into orbit Resourcesat 1, or IRS-P6, a 1,360-kg payload costing €42m into an orbit of 821 km, 98.76°. It had three cameras and a memory of 120 GB to store images before downloading them. It became operational only a week after entering orbit. A small, 50-kg amateur mini-satellite also entered

Resourcesat 1

orbit. Resourcesat carried a three-band scanner with 5.8-m resolution and infrared scanners with 23.5 and 56-m resolution, able to fly over each site every five days.

The PSLV put into orbit the first dedicated mapping satellite, Cartosat 1, IRS-P5, on 5th May 2005. Cartosat 1, weight 1,560 kg, was a mapping satellite with 1-m resolution stereo cameras fore and aft, a wide-field camera, LISS 4 scanner with 6-m resolution and LISS-3 scanner with 23-m resolution, all designed to assist in the detailed mapping of India from 617-km polar orbit. The first Cartosat had to sit on the pad for two weeks, waiting for heavy rain, thunder and lightning to clear, and was accompanied into orbit by Hamsat, a 42.5-kg amateur radio satellite. The launch was watched by President Abdul Kallam. Cartosat 1 was to provide photographs for mapmakers, town planning, road and canal building, water resources management and disaster assessment. The satellite cost Rs2.48bn and India claimed it as the most advanced mapping satellite in the world – a claim hard to refute, as it was the only dedicated such satellite. It also marked the first launch from the second pad at Sriharikota.

It was succeeded by Cartosat 2 on 28th April 2008. This was a 824-kg mapping satellite with 1-m resolution and able to take 3D images, storing them in a 64-GB solid-state memory which can be downloaded at 336 MB/sec. This launch, the 13th of the PSLV, put an Indian record of satellites into orbit: Cartosat 2, Indian Mini Satellite 1 (IMS 1) and eight nano-satellites from Canada and Europe. Within days, high-quality pictures were received in Hyderabad. Cartosat 2 cost Rs1.2bn and IMS Rs220m. Cartosat 2, orbiting at 630 km, 97.4 min, 97.9°, had a panchromatic

Cartosat 1

camera making swathes 9.6 km across while IMS was pioneering new, miniaturized technologies with a multispectral camera of resolution 37 m and a hyperspectral camera with resolution 506 m. The PSLV was the third core-only, with no straps-ons, reached orbit after 14 min and popped each satellite out, one after the other into 635-km polar orbit. Cartosat's images were outstanding and illustrated some of the world's greatest cities, from Vienna to Mumbai, in astonishing clarity and detail. The early PSLV missions are summarized in Table 5.4.

Table 5.4. PSLV launches of Earth observations and applications satellites

15 Oct. 1994	IRS-P2	PSLV	Sriharikota
21 Mar. 1996	IRS-P3	PSLV	Sriharikota
28 Sep. 1997	IRS-1D	PSLV	Sriharikota
25 May 1999	IRS-P4 Oceansat	PSLV	Sriharikota
22 Oct. 2001	TES	PSLV	Sriharikota
12 Sep. 2002	Metsat 1 *Kalpana*	PSLV	Sriharikota
17 Oct. 2003	Resourcesat 1	PSLV	Sriharikota
5 May 2005	Cartosat 1	PSLV	Sriharikota
28 May 2008	Cartosat 2	PSLV	Sriharikota
			IMS 1

For the future, India has outlined an extensive program of Earth resources missions, as shown in Table 5.5.

Table 5.5. Future Indian Earth resources and related missions

Oceansat 2, 3
Cartosat 2B, 3
Radar Imaging Satellite (Risat) 1, 2, 3 (C-band radar)
INSAT 3D (weather only, 50-m resolution camera)
GEO High Imaging Orbiter
Resourcesat 2, 3
Megatropiques (with France)
Saral Altika Argos 3 (with France)
DMSART (C-band and X-band radar)
Science missions: I-stag (aerosols and gases), Aditya 1 (solar studies), Sense P, E (double mission for near-Earth magnetic environment), ITM (inner magnetosphere)
Two mini-satellites are also planned, like Youthsat (solar flares, to accompany Resourcesat 2), Saral (oceanography, with France)

India also plans its first astronomy mission, Astrosat. This was a project of the Tata Institute of Fundamental Science in Mumbai, the University Center for Astronomy and Astrophysics in Pune and the Rama Research Institute of the Indian Institute of Astrophysics in Bangalore, to develop a satellite to detect radio frequencies, X-ray and electromagnetic radiation from neutron stars and black holes. Astrosat will carry four X-ray instruments and an ultraviolet imaging

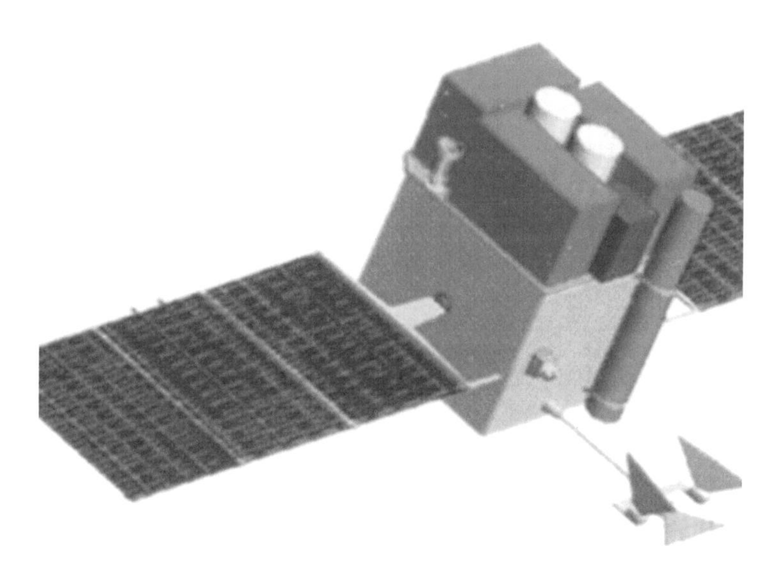

Astrosat

telescope. Hitherto, space science has been one of the least developed fields within the Indian space program, with only the early *Rohini* satellites carrying scientific instruments. This began to change with the Chandrayan lunar mission and, in addition, India joined the Russian solar science mission, *Koronas Foton*, launched in January 2009, supplying a 55-kg low-energy gamma ray telescope to detect solar and galactic radiation.

Megatropiques was an especially important project with France to examine tropical air currents and long-term climate patterns over the seas and continents. Set to orbit at 867 km, at 20°, it has three payloads: multi-frequency microwave scanning radiometer to provide information on rain and water vapor (MADRAS, India and France), multichannel microwave to make humidity profiles (SAPHIR) and multichannel radiation budget scanner (SCARAB, France). From them, it was hoped to construct a model of how tropical weather works and thus improve weather predictions.

Developing applications further, the Indian government decided in May 2006 to develop its own navigation satellite system, called Indian Regional Navigational Satellite System (IRNSS), with seven satellites launched on the PSLV in the early 2010s. An agreement signed in India in March 2006 by President Putin envisaged Indian access to the Russian GLONASS system in exchange for India launching two GLONASS for Russia. That summer, on 4th July, ISRO organized a national conference on navigation satellites in Bangalore, which announced the satellite locations – 34°E, 55°E, 83°E and 132°E° – and their designs (1,330 kg in weight, two solar panels with 1,400 W).

CONCLUSION: THE PROGRAM MATURES

In the late 1980s, India developed an uprated version of its small Scout-class SLV launcher to place new, heavier payloads in orbit. Although the Augmented Satellite Vehicle (ASLV) did not make a successful mission until 1992, it eventually placed two stretched *Rohini* satellites into orbit. The big leap forward took place with the Polar Satellite Launch Vehicle, which was 10 times bigger than the original SLV. The PSLV continued the regrettably well established Indian tradition of first-flight failures, but quickly went on to become a successful launcher. The PSLV enabled India to develop its own capacity to put large polar orbiting resource satellites weighing more than a tonne into pathways sweeping across the Indian subcontinent with a variety of ever more specialized missions making a significant contribution to India's economic, social and environmental development.

6

India: Manned and lunar flight

The early 21st century saw India strike out in new directions. Although India's first satellite (1980) followed a full 10 years behind Japan and China (1970), Indian space capabilities grew at such a pace that the country was able to challenge both its Asian rivals with a lunar program. India at last achieved the vision outlined in the 1960s of not only building its own geosynchronous communication satellites, but launching them itself through a new powerful launcher, the GSLV, making it fully self-sufficient in its space technology. This paved the way for India to send its own astronauts into space, making it one of the space superpowers and an Asian challenge to China.

CHANDRAYAN: TO THE MOON

First suggestions of an Indian Moon mission came in 2002, when a study group was established. The success of the Metsat mission showed that India's PSLV could successfully send a small satellite to 24-hr orbit, showing that a Moon mission was possible. The study group reported to a conference in Bangalore in January 2003 and the subsequent peer review to ISRO on 6th April 2003. The idea of such a project was strongly backed by the head of ISRO, Krishnaswamy Kasturgian. The initial concept was a mission called *Soyana* to put a 400-kg lunar orbiter into 100-km orbit to study the physical and chemical properties of the Moon. Approval of the project by the government was announced by Prime Minister Atal Behai Vajpajee on 15th August 2003, setting aside an initial €70m and giving it the formal name of *Chandrayan Prathim*. Appointed as mission director was Mylaswamy Annadurai.

The broad aim was to improve our knowledge of the Moon, its origin and its role in the formation of the solar system, with the sub-text of giving India its own first lunar dataset. Specifically, it was to compile a three-dimensional map of areas of the Moon of scientific interest, the chemical mapping of the lunar surface and the testing of new technologies (e.g. lithium batteries, gimbaled antennae star sensors, communication systems). Weight of the probe was 1,380 kg (fuelled), with the capacity for up to 11 instruments. India invited international participation in the

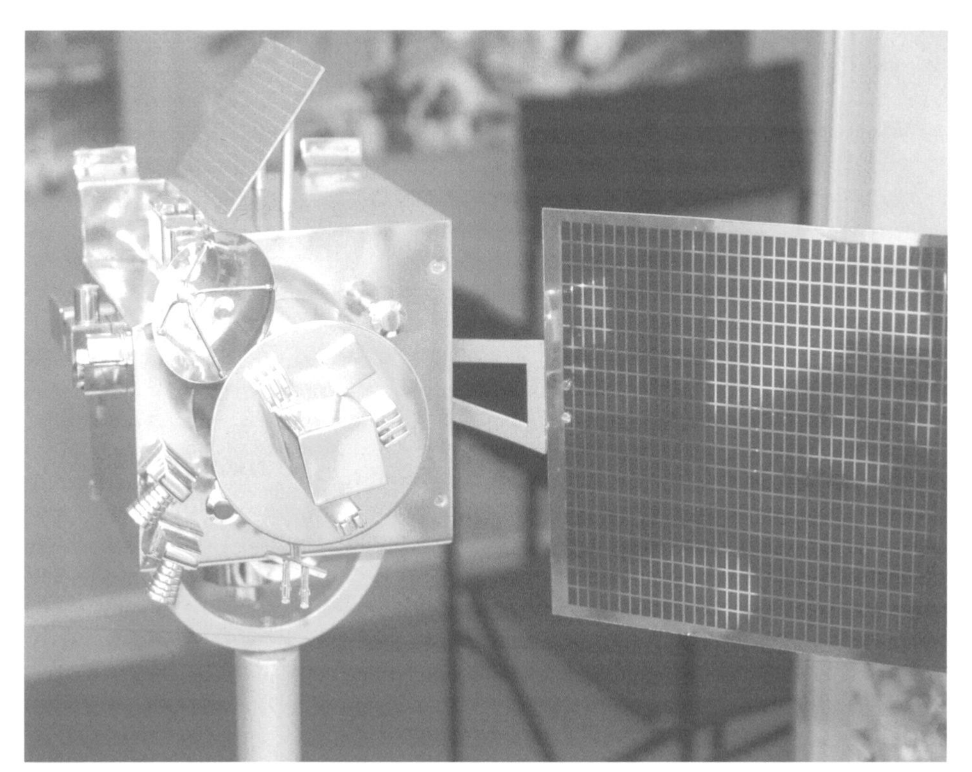

Chandrayan side view

mission and it was much the most international in flavor of the three Asian Moon probes of 2007–2008. There was a formal "announcement of opportunity" in 2004 to which a number of countries responded, including Britain and the United States. The Chandrayan instruments were:

- cartographic terrain mapping camera, 5-m resolution, 40-km swath (India);
- hyperspectral imager, 55 kg, resolution 15 m, swath 40 km, for mineralogical mapping (India);
- lunar laser altimeter topographic ranging instrument, 10-m accuracy (India);
- Moon Impact Probe (India);
- high-energy X-ray mapping camera, 20–250 keV, 20-km resolution (India);
- low-energy X-ray spectrometer, 1–10 keV (Britain);
- infrared mineral spectrometer (Germany);
- Moon mineralogy mapper (USA);
- atomic reflecting analyzer 10 V–2 keV (Sweden);
- radar mini-synthetic aperture radar s-band to detect water or ice (USA);
- radiation dose detector (Bulgaria).

The lunar mission required the construction of a deep space tracking network, the cost being Rs1bn (€18m). Put in charge was S.K. Khivakumar and two dishes were constructed in Byalalu, some 45 km from Bangalore. One dish was 32 m in diameter,

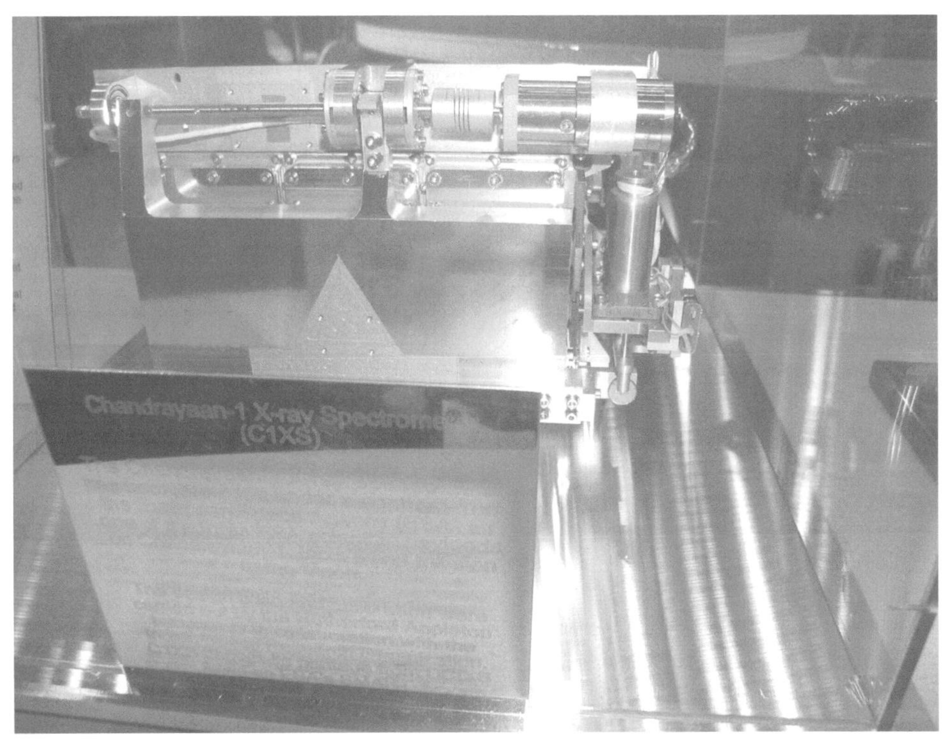

Chandrayan spectrometer

the other 18 m. Data were to be sent to the National Science Data Center to turn the raw data into user-friendly format.

Even as Chandrayan was prepared for launch, Indian space officials began to put together a second lunar mission, called Chandrayan 2. This was aimed at 2012 (coincidentally or not, the date of China's second planned Moon mission). The objective was to land a roving vehicle on the Moon. To do this, the Indians headed straight to the country with the only experience of automated lunar roving vehicles, namely Russia, leading to a first agreement signed in Moscow in November 2003 by the Indian Prime Minister Atal Behari Vaypayee and then a second by Manmohan Singh in November 2007. Under this, the Russian space agency Roscosmos took responsibility for the 100-kg rover while ISRO took charge of the launcher (GSLV), orbiter and lander. Encouraged by swift progress on the mission, India soon announced plans for Chandrayan 3 to bring back lunar samples in 2015, ahead of China and there was even some heady talk of an ultimate manned mission.

Even as Chandrayan was prepared for launch, Indian space officials began to speak of their desire for missions to Mars. At the International Astronautical Federation conference held in Bangalore in September 2007, ISRO chairperson, G. Madhavan Nair, spoke of ISROs interest in Mars and the outer solar system, indicating that a small Indian probe could soon fly there, following Japan's *Nozomi* and China's planned *Ying Yuo*.

PSLV Chandrayan night before launch

Chandrayan was launched on 22nd October 2008. The Sirharikota launch site was drenched in rain the day before the launch and when Chandrayan's Polar Satellite Launch Vehicle took off, it disappeared into clouds. The rocket momentarily reappeared about 20 sec later, trailing a long orange flame, before disappearing once more.

Like the other Asian Moon probes, Chandrayan took a long and slow route to reach the Moon. The PSLV put Chandrayan into 255–22,860-km, 6 hr 30 min, 17.9° Earth orbit and, on the following day, the 45-kg 440-N thrust Liquid Apogee Motor fired for the first time, for 18 min, to raise the apogee to 37,900 km. The motor burned on 29th October to set a new orbit of 465–267,000 km, 11 hr, and, after a 145-sec fifth and final burn bringing it out to 380,000 km, the probe was ready for capture by the Moon. This duly took place with an 805-sec burn, which placed Chandrayan into a perfect lunar polar orbit of 90°, 504–7,502 km.

Over the next four days, Chandrayan made three maneuvers totaling 16 min to reduce the orbit to its circular orbital height. The periselene was reduced to 200, 182 and then 100 km, while, at the same time, the apolune was brought down to 255 and then 100 km. This marked the 10th and final use of the Liquid Apogee Motor. Ground controllers began turning on the scientific instruments straight away. India had become the sixth country (or group of countries) to reach the Moon, after Russia, the United States, Japan, Europe and China.

The next milestone took place on 14th November, when the 35-kg Moon Impact Probe (MIP) was released. It carried three instruments: a video camera, radar altimeter and mass spectrometer to measure the Moon's extremely thin atmosphere

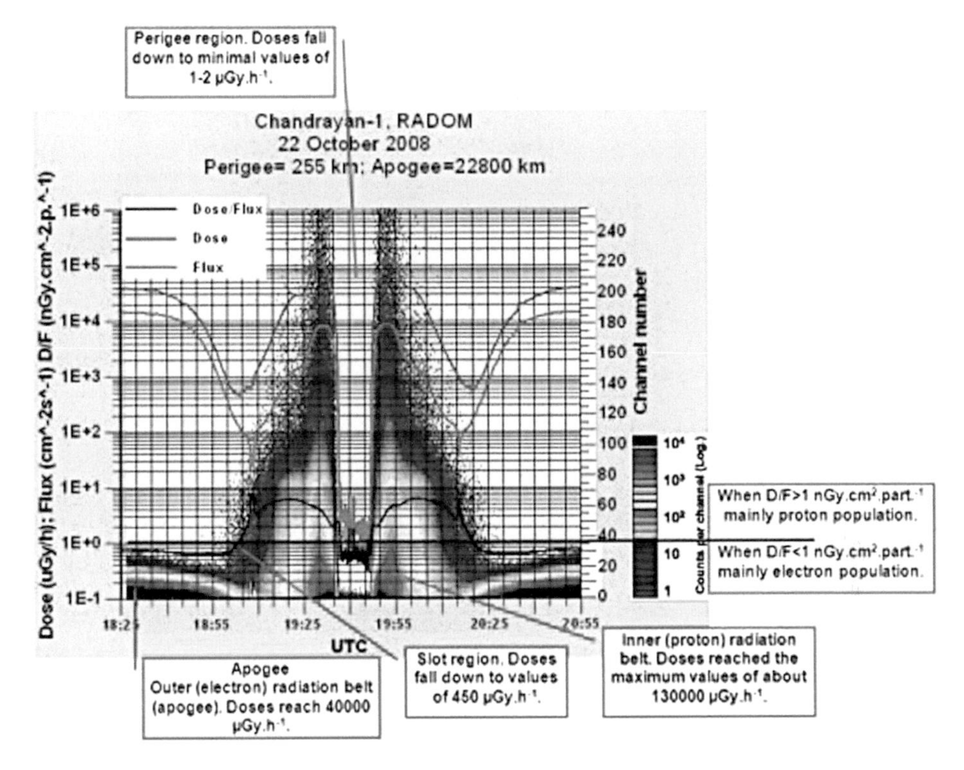

Chandrayan radiation results

as it descended. The MIP had a small motor that fired to achieve a new orbit with a low point of –200 km, one sufficient to make it crash within a half-orbit. MIP made a 20-min descent into Shackleton crater near the lunar south pole, its instruments taking pictures as it came down and barreled into the lunar surface, carrying miniature Indian flags on its side. The announcement of success was made by ISRO chairperson, G. Madhavan Nair, and with him, at his side, was former president, Abdul Kalam. India was now the fifth country (or group of countries) to hit the Moon, beating China to the surface.

By August 2009, Chandrayan had completed 3,000 orbits. In May, it had maneuvered from its 100-km circular orbit to a new one, also circular, at 200 km and 70,000 images had been sent back to Bangalore.

GRAMSAT TO THE VILLAGES: THE GSLV

From the 1960s, India's long-term goal had been to launch its own communications satellites to 24-hr orbit. For the first two decades, India had relied on other countries to launch its communications satellites. Of the first nine INSATs, three flew from Cape Canaveral – two on a Delta and one on the Space Shuttle – while six flew from

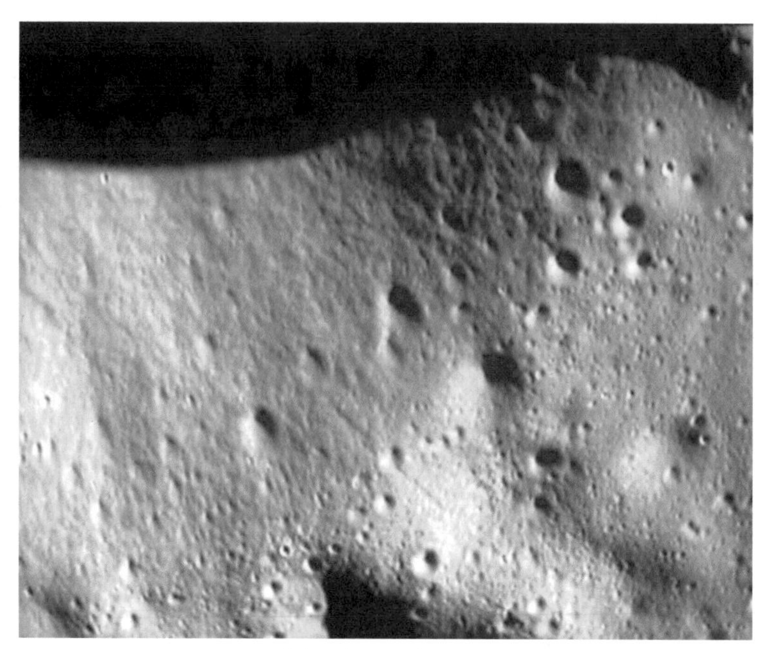

Chandrayan impact

Kourou, Guyana, on the European Ariane launcher. Although they more than repaid their investment, commercial foreign launches were expensive.

In 1987, India felt in a position to begin design and construction of its own geosynchronous launcher, named prosaically the Geostationary Satellite Launch Vehicle (GSLV). Not only would it give India launcher independence, but it would do so at much lower price overall, the aim being to reduce launch costs from Rs2.66bn (€56.6m) a go on Delta and Ariane to an effective cost of Rs883m (€18.8m). These individual launch prices did not take into account the substantial development costs of the GSLV, estimated at Rs13bn (€276m). One cost saving, though, was that in Sriharikota, India already had a launching range relatively near to the equator.

The first specific payload intended for the GSLV was *Gramsat*, an educational relay platform based on the INSAT 2 design (*gram* is the Hindi word for village). *Gramsat* was designed to transmit television, CD-quality sound, data and internet to 60-cm dishes in programming devoted to rural development, transmitting programs in health and hygiene, agricultural production, family planning, environmental awareness, vocational training and entertainment. *Gramsat* would have a national beam with six powerful transponders for educational television and two advanced spot beams to transmit sound in four different languages. Ion engines would be used for station-keeping.

For the bottom stages of the GSLV, India decided to take advantage of its existing knowledge of the behavior of rockets in the lower atmosphere derived from the PSLV. The new rocket would use the PSLV's solid-fuelled first stage, but instead

GSLV nighttime

of using solid-fuel strap-ons, it would use four PSLV second-stage liquid-fuelled engines as strap-ons, thereby generating enormous thrust at lift-off. The PSLV Vikas second stage would be used as the GSLV second stage as well. First ground tests of the GSLV's engines were made at Mahendragiri in June 1994. For the very final, fourth stage of placing payloads into geosynchronous orbit, the GSLV would use the Indian Liquid Apogee Motor or LAN, a 440-N liquid-propulsion system that had already accumulated 10,000 sec of testing. The GSLV became the first – indeed, it still is the only – geosynchronous rocket to use a solid-fuel first stage. During the late 1990s, the new GSLV stages were seen regularly at Sriharikota under test.

Where the launcher broke new ground was the use of a cryogenic, liquid-hydrogen-powered third stage. Cryogenic fuels, while enormously powerful, require

GSLV second stage

considerable handling skills because of their low boiling points. The turbopump must operate at up to 40,000 RPM. The concept of an Indian cryogenic upper stage was approved by DOS in late 1987. An engine of about 12 tonnes would be required. This was not only a big technical challenge, but became a large political one was well.

Indian engineers calculated that up to 15 years' work was involved to fully master the difficult cryogenic technology. Originally, India made enquiries with Japan about purchasing the LE-5, but nothing seems to have come of this. Getting wind of this, the American General Dynamics Corporation approached India, offering both a cryogenic engine and technology transfer. Europe's Arianespace also made an approach. In both cases, the cost was prohibitive and ISRO proposed, with the approval of the Space Commission, to reconsider its decision to shop abroad and to go it alone, despite the long time this would take. Just as the Cabinet was about to confirm the decision, the Soviet Union offered two engines and technology transfer for a relatively bargain price of Rs2.35bn (then about €188m). This came as quite a surprise, because the Soviet Union had never developed a cryogenic upper stage. Or so everyone thought.

Russia offered a hitherto unknown rocket engine, the KVD-1, industry code 11D56. It was actually an old engine, having been developed by the Isayev design bureau in 1964 for the Soviet Moon-landing program. Under the 1972 Moon-landing plan, the KVD-1 was to brake the Russian lander into Moon orbit and slow it gently to the surface, so that three cosmonauts could spend a full month on the Moon. The KVD-1 was first test-fired in June 1967. The engine was tested for 24,000 sec in six starts. Five motors were made and tested over 1974–1976 and just as they were completed, the Moon plan was canceled. The KVD-1 was never used and for subsequent missions to geosynchronous orbit and deep space, the Russians contented themselves with using a kerosene/liquid-oxygen-based engine, the "block D". The West wrongly presumed that the Russians had never been able to develop a cryogenic upper stage. In fact, the KVD-1 had unsurpassed thrust and capabilities that made it unmatched for years. Its turbopump engine generated 7,300 kg of thrust in a single chamber of 54.6 atmospheres with a specific impulse of 461 sec, still the highest in the world for an engine of its kind. It was a small motor, 2.146 m tall, 1.28 m in diameter and weighing 292 kg, the stage being 3.4 tonnes empty and 19 tonnes fuelled. Burn time was 800 sec, it had two nozzles and could be re-started five times.

In 1988, outline agreement was reached between ISRO and Glavcosmos, followed by a technical agreement in June 1991, whereby the USSR would deliver two engines and the associated technology by 1995 for the originally stipulated amount of Rs2.35bn. Then the trouble started. In May 1992, President George Bush condemned the agreement as a violation of the Missile Technology Control Régime, imposing sanctions on both ISRO and Glavcosmos for two years, taking the form of non-cooperation and sales embargoes.

India objected strongly to the American actions, pointing out that a high-powered hydrogen-fuelled upper stage that took a long time to prepare was of little military value in attacking a neighboring country with which they already had a land border.

The Americans had offered them the very same technology themselves and had

KVD-1 engine

not made any objection during the period 1988–1991, when the deal had been negotiated. The Americans, in response, tried to prove the military connection by publishing a list of seven ISRO scientists who had gone to work on missiles – as if American rocket scientists had never ever worked on missiles.

The following year, though, with the accession of Bill Clinton as president, the American attitude relented. He approved a reopening of cooperation with ISRO and Glavcosmos if the Russians transferred individual engines, but neither the blueprints nor the production technology that would enable India to design its own cryogenic engines. In July 1993, after negotiations in Washington, DC, Russia backed off its contract to transfer technology to India and suspended its agreement, invoking *force majeure* (circumstances beyond control), to the Indians' fury.

The KVD-1 had now become caught up in a much bigger game – the negotiations between America and Russia for the construction of the International Space Station. Russia suggested compensation for loss of the Indian contract and named a price of $400m. Soon after, the United States paid Russia for seven American flights to Russia's Mir space station, the price being – no surprise – $400m. With this deal under its belt, Russia tried to restore its relationship with India. The Indians struck a hard deal and in a revised agreement with India, made in January 1994, Russia agreed to transfer two engineering models and seven ready-to-fly KVD-1s for a modest €8.5m, but, in deference to the Americans, without the blueprints ("technology transfer"). India was formally required by the United States to agree to use the equipment purely for peaceful purposes, not to re-export it, nor to modernize it without Russia's consent.

The two engineering models were delivered between 1997 and 2000. The first of seven ready-to-fly KVD-1s arrived on 23rd September 1998 in Madras (Chennai), whence it was taken to Sriharikota. It is possible that the Russians transferred the blueprints anyhow, American intelligence following four suspicious shipments from

Moscow to Delhi on covert flights run by Ural Airlines in autumn 1994. As a cover, they used "legitimate" transhipments of Indian aircraft technology traveling the other way to Moscow for testing in Russian wind tunnels. The plot thickened when, at the same time, two senior scientists at the Liquid Propulsion Systems Center, S. Nambi Narayanan and P. Sasikumaran, were arrested for "spying for foreign countries". Eventually, the Central Bureau of Investigation admitted that the charges against S. Nambi Narayanan and P. Sasikumaran were false and baseless and they were freed. Later, the United States was accused of setting them up as part of a dirty-tricks campaign against the sale of the KVD-1. Although not properly recognized for his achievement, Nambi Narayanan went on to develop the Vikas engine that eventually sent Chadrayan to the Moon in 2008.

GSLV third stage

The American intervention was costly for both countries. India's acquisition of the cryogenic technology was delayed by five to ten years. The Russians lost over three-quarters of the price of their original deal. Few believe that the American move had much to do with missile proliferation – but had a lot to do with keeping India out of the global launcher market for communications satellites, where it was more than successful.

GSLV FLIES

Eventually, after 10 years in construction and at a cost of Rs14bn (€350m), the GSLV was completed with its Russian upper stage and the *Gramsat* educational satellite. The first attempt to launch the GSLV was on 28th March 2001, transmitted live on television. The GSLV stood 49 m tall on its pad and weighed a record

GSLV ignition

401 tonnes. Things did not go according to script. Ignition took place at 15.47 India time (10.17 Universal Time). Flames licked around the rocket at *zero!* but then flickered out. There was stunned silence in both Launch Control in Sriharikota and ISRO Mission Control in Bangalore. It transpired that a computer had detected a leak and shut the rocket down, just in time. There was no explosion. Embarrassment, but no one was hurt.

A fresh, 57-hr-long countdown began on 16th April, with the live television cameras kept away this time. Two days later, with only 16 min to go, the computer took over. At T–4.6 sec, the liquid fuel strap-on boosters ignited. This time, there were no show-stoppers and the main engines of the 125-tonne first stage roared into life. The second stage, 11.56 m long with 39 tonnes of propellant, took over at 162 sec, building up speed from 2.63 to 5.18 km/sec and up to 126 km. The Russian KVD-1 third stage brought *Gramsat* into an orbit of 181–32,051 km, 19° – its first mission in its 30-year history. The satellite's own motor then fired to raise the perigee to 3,000 km and adjust the inclination to 13.7°. On the 20th, the motor fired for 4 min to raise the apogee to the intended height of 35,880 km. On the 21st, the motor

Edusat

burned for 29 min to raise the perigee to 11,900 km and adjust the plane to 5.5°. A final burn was to bring the satellite up to geostationary orbit, but now there was a disappointment, for it depleted its fuel before reaching 24-hr orbit, leaving the satellite in a 23-hr orbit instead.

The second successful launching of the GSLV took place on 8th May 2003. This was called the "D2" mission (Development 2) and took off into a clear blue sky watched by ISRO head, Krishnaswamy Kasturiangan. The Russian upper stage put into 24-hr orbit an experimental 1,825-kg comsat called GSAT 2 with seven transponders (four C-band, two Ku-band, one mobile phone) and four experiments (radiation meter, surface charge potential meter, radio beacon, solar X-ray spectrometer).

The third GSLV, on 20th September 2004, was advertised as its first operational flight and it carried the 1,950-kg *Edusat*, or educational satellite, which carried six Ku, six C-band transmitters and five beams drawing off 3 kW of power from solar panels. The launch went smoothly, the first stage burning out at 150 sec, the second at 288 sec. The Russian third stage burned until 999 sec, leaving *Edusat* in a 180–35,985-km orbit with the 440-N liquid apogee motor bringing it to its station at 74°E. *Edusat* worked primarily through schools and colleges, each beam transmitting directly to 200 classrooms but ultimately reaching about 600,000 students. One of its first projects, formally launched in March 2005, was the primary education project providing an initial 100 animated science, maths and language programs to 885 villages in remote areas in Chamarajanagar in local languages. Each terminal was 1.2 m across, supplied with a 70-cm-screen television, set-up box, battery and solar panel. In addition, *Edusat* was used extensively for teacher training.

The fourth GSLV, its second operational flight, was lost on 10th July 2006, carrying INSAT 4C. About a minute into the mission, one of the GSLV's side rocket

GSLV with INSAT 4

motors lost thrust, causing it to go off course. The GSLV crashed into the Bay of Bengal. This was the first Indian launch failure for 13 years. It took over a year to organize a replacement mission, 2nd September 2007. The GSLV was loaded with its cargo, the INSAT 4CR ("R" for replacement) communications and meteorology satellite, equipped for digital news-gathering, television, high-data-rate transmission and weather forecasting. In attendance were new Indian president Pratibha Patil and Prime Minister Manmohan Singh. They had to be patient, for the countdown was halted with only 3 sec to go at 16.21. Launch was re-scheduled first to 17.40, then 18.10, each time being halted because the rocket's computer seemed unable to receive the launch command from the control center. Eventually, the GSLV soared into the Bengal sky at 18.21, two hours later than originally planned. INSAT 4CR had 12 C-band channels. The Russian upper stage put the satellite into its correct transfer orbit of 168–34,710 km, 20.7°. Now, the Indian apogee motor misburned, placing the INSAT in an orbit of 2,983–31,702 km, 11.1°, far below what was intended. Although it would be possible to reach geosynchronous orbit using its station-keeping fuel, the on-orbit duration of the mission would be much diminished. The early GSLV launches are summarized in Table 6.1.

Table 6.1. GSLV launches

18 Apr. 2001	GSAT-1	GSLV	Sriharikota
8 May 2003	GSAT-2	GSLV	Sriharikota
25 Sep. 2004	*Edusat*	GSLV	Sriharikota
2 Sep. 2007	INSAT 4CR	GSLV	Sriharikota

The GSLV demonstrated just how far India had come in 30 years, as Table 6.2 shows.

Table 6.2. Evolution of Indian launchers

Launcher	*Stages*	*Height (m)*	*Weight (kg)*	*Payload (kg)*
SLV	4	22.7	17,000	40
ASLV	4	23.56	60,170	150
PSLV	4	44.2	295,000	1,360
GSLV	4	51	414,000	2,000 (to GTO)

INDIA'S OWN UPPER STAGE

An interesting feature of these five GSLV launches was that publicity surrounding these launches rarely mentioned the use of a Russian upper stage, as if not wanting to draw attention to the murky intrigue that had gone on during the 1990s. Over time, though, this version of the GSLV came to be called the "mark I" and the one with the Indian upper stage, which was formally called the Cryogenic Upper Stage

GSLV in its hangar

(CUS), came to be called the "mark II". The Indians, though, did refer to and report on their own effort to develop their cryogenic upper stage, which they had to reverse-engineer from the KVD-1 hardware acquired from Russia and its blueprints, assuming that they had them (though this aspect was not mentioned either).

India had already tested a sub-scale engine for the first time on 21st July 1989. Three years later, a hydrogen manufacturing plant was opened and a thrust chamber manufactured. Ironically, India obtained its liquid hydrogen supplies from an American company that built a plant for ISRO at its test facility in Mahendragiri in Tamil Nadu. The first test engines were then built, but things went far from smoothly, for there was an explosion when India fired a 10-kN liquid-hydrogen engine in July 1993. There were two further explosions and another engine gave up after 10 sec. A prototype engine with 1 tonne of thrust was eventually tested at the Liquid Propulsion Systems Center in Mahendragiri in 1997 and, five years later, ran for 1,000 sec.

Even though the operating time required during a real mission was only 700 sec, the designers required 1,200 sec continuous running of a full-scale CUS engine before they were prepared to approve the design for flight. A 720-sec firing was achieved in November 2007, so the testing program was nearing completion. Once ready to fly, the GSLV mark II had a long manifest of educational and communications satellites to launch – at least one a year to 2015.

Indian officials had already begun to turn their minds to means of upgrading the GSLV. This would double its payload to 24-hr orbit to 5 tonnes, but also make possible the launch of a manned spaceship. A new version of the GSLV, the mark III, was approved by the Indian government on 17th August 2002, with a budget of Rs4,520.8m. The mark III was shorter (42.4 m) and fatter than the mark II, but much more powerful. Chief designer was S. Ramakrishnan. The first stage had a

GSLV mark III with designer

thrust of 629 tonnes, two-thirds of it coming from two S200 solid rocket boosters, each loaded with 207 tonnes of powder. The second stage used storable liquid fuels, UH25 and azote peroxide, and two Vikas engines of 160 tonnes' thrust. The third stage would be much larger, with 25 tonnes of propellant, double that of the KVD-1 or CUS. An early task of the mark III was to carry SRE 2 (Space Recovery Experiment 2) in advance of a manned flight in 2014, the next great leap forward in Indian space flight.

FIRST INDIAN IN SPACE

The idea of an Indian cosmonaut had been first suggested casually by the world's first space traveler, Yuri Gagarin, when he visited India on his triumphant world tour in 1961. In 1978, the Soviet government made the suggestion of a guest Indian cosmonaut to Prime Minister Moraji Desai, with a formal agreement subsequently being reached between Leonid Brezhnev and Indira Gandhi in 1980. The USSR was a natural partner, for India had launched Indian satellites. Recently, Moon samples taken from the Sea of Crises by Luna 24 had been sent for analysis to the Physical Research Laboratory in Ahmedabad. The offer was to fly an Indian cosmonaut up to the manned Soviet orbiting space station Salyut for a week-long visit. From 1976, cosmonauts from guest nations had been training in the Soviet Union, the first of them flying from 1978 onwards. Initially the guests were from the Soviet block countries, but the USSR widened the program – a French cosmonaut flying in 1982.

One hundred and fifty people applied for the mission and the Indian authorities cut the number to eight. Eventually, six were sent to the Soviet Union for evaluation the following year. Two were recommended by the Russians in September 1982 –

Sharma and Mulhotra

Wing Commander Ravish Mulhotra and Squadron Leader Rakesh Sharma. Mulhotra was born on 25th December 1943 in Lahore and Sharma on 13th January 1949 in Patiala, Punjab. They had 3,400 and 1,600 flying hours, respectively, to their credit. Ravish Mulhotra had an air force career, had flown in both India's wars against Pakistan (1965 and 1971) and was commander of Bangalore Test Pilot School. Sharma was a MiG jet commander in the Indian Air Force and became a test pilot in the Aircraft and Systems Design establishment in Bangalore.

Both went to Star Town for training. Indira Gandhi met them there during a visit to Moscow, when they were in the process of setting up home on the 13th floor of one of the cosmonaut accommodation blocks. Part of the training was done in the Air Force Institute of Aviation Medicine back in Bangalore, which developed a special cardiograph for the mission.

Their first task was to learn Russian and, within six months, they were speaking the language and taking their lecture notes in Russian. In June 1983, they were introduced to the Soyuz T spacecraft trainer and taken to splashdown training at Fedosia on the Black Sea. The trainee cosmonauts returned to India in July both for holidays and to visit the Indian research centers where the experiments for their mission were in preparation. The trainees were then assigned to their Soviet crew members in October. The prime crewman was Rakesh Sharma, assigned to fly with Yuri Malashev and Nikolai Rukhavishnikov. The backup crew was Ravish Mulhotra, Anatoli Berezovoi and Georgi Grechko. Although Sharma had less flying experience, he was better at Russian – indeed, he was quite a linguist, being fluent in English, Punjabi, Hindi and Telagu. In February, with the actual mission fast approaching, the Russians made a final crew change, pulling Rukhavishnikov due to illness and replacing him with Gennadiy Strekhalov. A feature of the missions was that the guest country could fly experiments to be operated by the visitor, so the Indians chose 43 experiments including:

- MKF-6M, KATE-140 and handheld cameras to observe the Indian continent during 11 overpasses;
- attempts to search for gas and oil (*Terra* experiment);
- Earth resources assessment of the Nicobar, Andaman and Laccadive islands;
- mapping of the Himalayas and Karakorn regions to assess ice melting;
- use of the cardiograph developed in Bangalore (called *Vektor*);
- study of blood flow to the cranial regions;
- study of the vestibular and visual systems (*Anketa* and *Optokinisis*);
- airlock smelting and coating of silver–germanium spheres (*Evaporator-M*).

In addition, the Indian cosmonaut would be examined three times a day by the doctor on board the Salyut 7 station, Dr Oleg Atkov. Atkov and his colleagues, Leonid Kizim and Vladimir Solovyov, were then in the midst of a record-breaking 237-day mission aboard the station.

The experiment that attracted the most public attention was a yoga exercise in which the cosmonaut would attempt to test the value of yoga in combating the decay of muscles in space. To complete the Indian theme to the mission, the cosmonaut would bring up typical Indian foods for his colleagues and the permanent crew –

Soyuz T-11 crew

principally curry, pineapple, mango juice, crisp bananas and fruit bars. They would surely be a welcome supplement to the notoriously dull Soviet space food.

The prime crew of Yuri Malashev, Gennadiy Strekhalov and Rakesh Sharma arrived at Baikonour cosmodrome for their flight on 23rd March 1984. Their spacecraft, Soyuz T-11, was rolled out to the launch pad on 1st April and erected into position. On 3rd April, seen simultaneously live on Indian and Soviet television, Soyuz T-11 soared into a clear blue sky to begin chasing Salyut 7 across 5,000 km of space. After several maneuvers by Soyuz, the two spacecraft were only meters apart by lunchtime the following day. Following docking, Malashev, Strekhalov and Sharma entered Salyut, Sharma bringing with him an Indian flag, pictures of the country's political leaders and a handful of soil from the Mahatma Gandhi memorial. There were now six men aboard Salyut 7, its largest ever crew. Sharma was the 139th person in space.

The week of joint Soviet–Indian experiments began the following day. There was a 15-min hookup between Sharma and Prime Minister Indira Gandhi, held both in English and Hindi, in the course of which Sharma described how beautiful India looked from orbit. Later, Sharma held a press conference in which he responded to questions from Indian journalists at the flight control center in Moscow. With his Soviet colleagues, he carried out medical and Earth resources experiments. The

Russian–Indian commemorative stamp

weather was good over India during his mission, making observations possible and on 8th April, the cosmonauts radioed a warning about a forest fire that they had spotted in nearby Burma. Blood samples were taken and analyzed on board. *Vektor* provided a record of the bio-electrical activity of Sharma's heart. Using the smelter, he carried out an experiment to make samples from mixing germanium and silver. Sharma used the multi-spectral camera in the course of 11 passes over India, the passes lasting between 2 and 11 min, though, unfortunately, there was a lot of cloud cover over the Bay of Bengal.

After a week, the cosmonauts loaded up the results of the week's joint mission. On 11th April, after bidding farewell to Kizim, Solovyov and Atkov, the visiting cosmonauts boarded the old Soyuz T-10 spacecraft (leaving the cosmonauts their own ship), undocked, fired retrorockets and plunged Earthwards through the atmosphere. The Soyuz cabin came to rest in a plowed field 56 km east of Arkalyk in the standard recovery zone. They were lifted out of the tiny cabin, seated in special chairs and then flown back to Moscow. Sharma's haul included 1,000 complete sets of MKF-6M photographs of India and 200 large-format KATE-140 photographs. He had been in space for 7 days 21 hr.

Sharma returned to test flying in Bombay. His Soviet-made spacesuit was put on exhibition. Sharma broke his leg five years after his famous mission when ejecting from a stalling aircraft but he made a full recovery.

SHUTTLE ASTRONAUTS WHO NEVER FLEW: BHAT AND RADHAKRISHNAN

As was the case with Japan, Indian astronauts first flew with the Russians, not the Americans, but there were plans for an Indian to fly on board the American Space Shuttle. NASA offered India the opportunity to fly an astronaut on board the Space Shuttle, suggesting that an engineer or scientist could assist in the launch of the INSAT 1C satellite from the payload bay of the Space Shuttle and, like Sharma on Salyut, would have the opportunity to bring a set of experimental payloads for testing during the mission. A formal agreement was signed in late 1984, with launch of INSAT 1C set for mid-1986.

Two Indian payload specialists were assigned to train to launch INSAT 1C, mission 61J, due to fly on the Space Shuttle *Challenger* on 27th September 1986. The candidates were Nagapathi C. Bhat and Paramaswaram Radhakrishnan. Their fellow crew members would have been Donald Williams, commander; Michael Smith, pilot; mission specialists James Bagian, Bonnie Dunbar and Sonny Carter; and a reporter on the first journalist-in-space mission. The final selection of who of the two would fly was never made. Both were training in Houston, Texas, on the very day that *Challenger* blew up. It was an ill-fated mission. Michael Smith died on the *Challenger* that day, the shuttle was grounded and then Sonny Carter was killed in a plane crash. Like a number of other payload specialists from other countries (e.g. Britain, Indonesia) hoping to fly at the time, their missions were effectively disbanded and they returned home. INSAT 1C was later launched on the European Ariane (Chapter 5).

Nagapathi Bhat was born on 1st January 1948 in Sirsi, north Kanara, and was a graduate from Karnataka University (bachelor of mechanical engineering, 1970) and the Indian Institute in Bangalore (engineering design, 1972). He joined the Indian Space Research Organisation in Bangalore in 1973 and worked on the *Aryabhata*, *Bhaskhara* and APPLE projects. Paramaswaram Radhakrishnan was born on 2nd October 1943 in Thiruvananthapuram, Kerala, and obtained degrees from University College Thiruvananthapuram (bachelor of science in physics and mathematics, 1963 and master of science in physics, 1965). He also joined ISRO, working at the Vikram Sarabhai center in Thiruvananthapuram on the *Aryabhata*, *Rohini*, APPLE and INSAT programs, becoming a member of the program planning and evaluation group of the center. Later, he became group director of electronics there and head of the test and evaluation division of the systems reliability group.

SHUTTLE ASTRONAUTS WHO FLEW: CHAWLA AND WILLIAMS

Two Indians did get to fly on the Shuttle before it concluded its operations. They did so as Indian-born Americans, but both women's achievements were closely followed in their home countries, especially Kalpana Chawla, who became the most famous. The Chawla family had been refugees from partition in 1947, trekking from Pakistan and not stopping until Karnal on the great trunk road between Delhi and Chandigarh. The family had been in the clothes business, but later moved into a more lucrative trade in car tires. Born in 1961, Kalpana was pushed by her family to achieve educationally at a time when few girls went to school. Her town had one of the few flying clubs in India and this inspired a lifetime's interest in aviation, which led her first to the Punjab Engineering College and then to the University of Texas, Arlington, not only renowned for aeronautical engineering, but already popular with Indian students. She married an American, Jean-Pierre Harrison, then moving to Colorado to study aerodynamics and, in turn, onto the NASA Ames Research Center in California to examine "ground-effect" and airflows around planes. A small, short, shy but determined and self-willed woman, she kept up her flying in the local club and was one of 19 successful candidates chosen from 2,962 applicants to NASA for the 1994 intake of astronauts. She quickly impressed NASA in Houston, for she received an early astronaut assignment as robotic arm operator on mission

Kalpana Chawla

STS-87. She flew into orbit in November 1997, being responsible for the deployment and recovery of the *Spartan* solar physics free flier spacecraft along with Japanese astronaut Takao Doi.

Only two years later, Kalpana Chawla was assigned to STS-107, the last pure science mission planned for the Shuttle fleet. Due to a series of snags, the 16-day mission was delayed for months, not lifting off until January 2003. She was mission specialist responsible for research studies into the atmosphere. Although very much part of the American science community, Kalpana Chawla kept up her links with home and India, staying a vegetarian, which was normal in India but less conventional in the United States, and bringing Indian music with her into orbit in her personal pack. On 1st February 2003, several hundred schoolchildren gathered in her old school in Karnal to watch on television her triumphant return to Earth. At Cape Canaveral, though, the Shuttle was late in landing. Apprehension grew quickly, because the Shuttle cannot be late: either it lands on time or it does not land at all. In no time, ground observers who got up early to watch the Shuttle streak across the dawn sky in the western mountain states saw the spaceship break up, fireballs crashing over Texas. Kalpana Chawla was mourned all over India, an international airport and the Metsat 1 satellite being quickly named after her. President Kalam applauded her remarkable story of courage and determination, poetically calling her a "citizen of the Milky Way".

A second American astronaut of Indian background is Sunita Williams, daughter of Deepak and Ursaline Pandya of Falmouth, Massachusetts, and a close relative of the home minister of Gujarat, Haren Pandya. In the United States, she became a helicopter pilot, US Navy test pilot and instructor and married Michael Williams, hence her name.

She joined the astronaut corps in 1998 and, eight years later, spent seven months on the International Space Station, becoming the longest-flying woman astronaut.

INDIAN MANNED FLIGHT

Sunita Williams

The idea of Indian manned spaceflight began to emerge soon after 15th October 2003. On that landmark day, China successfully put its first astronaut (*yuhangyuan* in Chinese) in orbit, Yang Liwei. Japan set up a Commission to review its space objectives.

India convened a three-day conference to discuss manned flight – a venture that later acquired the acronym of MSM or Manned Space Mission. Several study groups were set up, leading to a much bigger meeting, one prompted by discussion between Prime Minister Manmohan Singh and ISRO head G. Madhavan Nair. Eighty leading Indian space engineers were brought together in Bangalore on 7th November 2006 and they listened to the concept papers and background studies that had been carried out over the previous three years.

The conference agreed that the best way forward was to develop a two-person cabin that could be launched on the forthcoming GSLV mark III rocket (there would be three flights of the GSLV mark III before astronauts were launched on one). The first mission would be for a day, the second for a week, echoing the Chinese approach, with splashdown in the Bay of Bengal. The following year, discussions were held with Russia on spacecraft design and a joint working group established in March 2008. The overall cost of an Indian manned space project was estimated at Rs100–150bn. For a start, an initial Rs950m (€20m) was allocated and this was used for the development of simulation facilities and life-support systems (e.g. spacesuits).

Following the conference, the proposals were sent on to Manmohan Singh's government and he set up a commission to review the proposal. India had already made progress in the preliminary tests that would be necessary for a manned flight, especially the tricky phase of recovery of a payload from orbit. Two initial drop tests took place in June 2005 and the third and final one took place on 19th August 2004 when a cabin was dropped by helicopter from an altitude of 5,000 m and descended under parachute into Pulicat Lake near the Sriharkota launch center.

The next step was an orbital test. On 10th January 2007, the PSLV was launched from Sriharikota with a number of payloads. In this case, the most important one

SRE preparation

was a 550-kg cabin called SRE 1 (Space Recovery Experiment 1), which originally entered an orbit of 637 km, adjusted on 19th January to 485–639 km. SRE was a blunt cone shape with a base diameter of 2 m and a height of 1.6 m, made of mild steel and the cabin carried two microgravity experiments (metal melting and crystal synthesis). Retrofire took place on 22nd January, some 650 km over the Atlantic Ocean for a splashdown 46 min later. The engine burned for 9 min. SRE 1 was equipped with silicon tiles and carbon fiber ablative for its fiery descent at 1,500°C through the atmosphere. Following a two-minute blackout between 80 and 40 km, the parachute system worked perfectly, popping out when the cabin was 4 km high, with splashdown 140 km east of Sriharikota. On splashdown, a flotation system deployed and the cabin was then recovered by a coastguard helicopter. A second test, SRE 2, was scheduled for 2011.

Following the working group with Russia, agreement was reached between the two countries in December 2008 for the Indian manned spacecraft to be built on the basis of the trusted Russian Soyuz design, which dates back to 1960. The date for the first Indian manned mission was 2014–2016, with an Indian cosmonaut to gain experience on a Russian Soyuz flight in 2012. In the meantime, ISRO bought a 12-ha site near Bangalore airport for clearing for the construction of an astronaut training center.

Like Japan, India also carried out research into shuttles and lifting bodies, though the financial commitment was much smaller. By 2008, India had two projects under way to test out these difficult advanced technologies. One was a military hypersonic demonstrator intended to test out scramjets and while its focus was primarily on the

SRE helicopter test

development of cruise missiles, many of the lessons arising were applicable to shuttle technology. The second was an unconnected project in the Vikram Sarabhai Space Center, Thiruvananthapuram, to develop a Mach 8 scramjet called HEX (Hypersonic EXperiment). This was part of a 15-year project to develop a Mach 6 ramjet, a reusable lifting body demonstrator, and to test automatic landing systems, ultimately aiming at *Avatar*, a single-stage-to-orbit vehicle using a combination of turbofan, ramjet and scramjet.

ORGANIZATION

We conclude by examining the organization, infrastructure and budget of the Indian space program. The broad organizational lines of the Indian space program laid down during the time of Vikram Sarabhai remain intact: the Department of Space (DOS), under the Prime Minister's office; the advisory policy-making Space Commission; and ISRO.

The Indian space program is quite decentralized, with facilities in many different parts of the country. The spiritual original home of the space program was Ahmedabad, capital of Gujarat, where Vikram Sarabhai was born and lived. It is still the location of the Physical Research Laboratory that Sarabhai founded in 1947 and remains the primary research body for Indian space science. Ahmedabad is the site of the Space Applications Center (SAC), the main applications center for the work of ISRO, employing over 2,000 people in a leafy site – one where monkeys swing from trees. SAC developed the use of the Indian space program for telecommunications, television broadcasting, the survey of natural resources, space meteorology and geodesy as well as being responsible for the development of transponders for satellites, portable antennae, dishes, modems and the use of satellites for news gathering. The Development and Educational Communications Unit is also there, responsible for the on-the-ground application of television programs for education and literacy. Due to the problem of terrorism in the area, it must be well protected by walls and soldiers.

Flagship of the Indian space program is the Vikram Sarabhai Space Center (VSSC), located 16 km north of Thiruvananthapuram and not far from the site of the original sounding rocket launches in the 1960s. Geophysically, its location is important for being just south of the magnetic equator. It is the largest ISRO center, employing over 5,000 people, and is the main research and development center of the Indian space program, where its rockets are designed. VSSC has sections that deal with avionics, solid propulsion, materials, reliability, planning, computers and information systems. The center includes the Solid Propulsion Group, which has led the development of India's solid-fuel rockets – although the actual rockets are manufactured at Sriharikota, close to the launch site. Also located there is the Inertial Systems Unit for rocket guidance.

Building the Polar Satellite Launch Vehicle required a major expansion in the

Space center Thiruvananthapuram

Indian rocket test

facilities in Thiruvananthapuram and the construction of an 80-ha extension facility at Valiamala nearby. The Rs280m (€5.95m) center was responsible for computer-aided design, combustion studies, the integration and testing of the PSLV. Alwaye, between Hassan and Thiruvananthapuram, is the site of the Ammonium Perchlorate Experimental Plant where the fuel for the PSLV is made: it is one of only eight solid-fuel rocket factories worldwide and its initial production capacity was 125 tonnes of fuel a year. Liquid propulsion systems are tested in Mahendragiri, near the south coast at a site nestled under cloud-covered mountains. The center was called the Liquid Propulsion Test Facility and employs 1,500 people. The services there include a test stand able to fire engines for the full duration of their intended burn, including cryogenic engines; a high-altitude test facility to simulate high altitudes for upper stages; clean rooms; propellant storage facilities; and instrumentation services.

Indian rocket test facility

Solid fuel test

ISRO HQ Bangalore in construction

Mahendragiri has hydrogen and monomethyl hydrazine production facilities. Its main projects are the Vikas stage of the PSLV, the cryogenic upper-stage project and the liquid apogee motor.

The headquarters of the program are in Bangalore, one of the main centers for economic and scientific development in modern India, sometimes called India's silicon valley. Bangalore is the headquarters of the Space Commission, ISRO, the Department of Space, ISRO Telemetry, Tracking and Command and the ISRO Satellite Center (ISAC), where most Indian satellites have been designed and built. The satellite center alone has over 2,400 staff. The center includes the Large Space Simulation Chamber, one of only eight in the world, built in 1990 to simulate the heat, vacuum and solar radiation of Earth orbit. There, satellites are baked, radiated, frozen and vacuumed for periods of up to 20 days in a stainless-steel chamber. It is a sprawling everything-under-one-roof facility for the full range of satellite design and testing, using economies of scale typically associated with the Soviet space program.

Further west of Bangalore is Hassan in Karnataka, the Master Control Facility for the INSAT communication satellites. The center began life with two 14-m dishes for INSAT 1 and then added two 11-m dishes for INSAT 2. It is supplemented by large stations in Bombay, New Delhi, Chennai (Madras) and Calcutta, with smaller stations in Leh, Jaipur, Patna, Bhubanshwar, Lucknow, Hyderabad, Ernakalam, Kavoratti, Andaman, Nicobar and Shilong. Remote terminals were set up in the late 1970s in Srinagar, Jodhpur, Bhuj, Goa, Minicoy, Pondicherry, Aizawl, Imphal, Kohima, Gangtok and Itanagar. By the late 1990s, this network had expanded to the point that there were over 280 medium and large ground stations, 5,000 small-aperture terminals and 200 stations able to receive satellite weather data. Despite this, demand on the Hassan facility was so great – by 2003, it was controlling seven satellites simultaneously – that approval was given for the building of a second satellite control center 1,100 km away in Bhopal, capital of Madhya Pradesh, an ultra-modern, splendid, drum-shaped, state-of-the-art facility opened in April 2005.

Mission Control Hassan

For tracking other satellites, ISRO has developed ISTRAC. Almost 500 people are employed in the main tracking center in Bangalore, called the Spacecraft Control Center, which has a main mission control room, data and computer facilities, supplemented by stations at Thiruvananthapuram, Sriharikota, Nicobar and Lucknow, with an overseas station in Mauritius, each equipped with tracking dishes of between 8 and 10-m diameter. An optical and laser tracking station was built at Kavalur, Tamil Nadu, with Russian help: this can pick up satellites as dim as 13th magnitude in Earth orbit as far out as the 24-hr position, 36,000 km high.

Hyderabad is the headquarters of the National Remote Sensing Agency, with the main Earth station being 55 km distant at Shadnagar and the Institute for Remote Sensing at Dehradun. Shadnagar originally received data from American *Landsat* satellites until India's own *Bhaskhara* and IRS satellites got into orbit. The agency is responsible for training, interpretation and dissemination, holding the Data Center that archives the information from IRS and distributes products both photographically and digitally. From 1987, regional applications centers were set up, starting in Bangalore, Nagpur, Jodhpur, Kharagpur and Dehradun, and 23 were built, the latest being in Shillong, near Burma.

SANDBAR LAUNCH SITE: SRIHARIKOTA

The main launching center is the Satish Dhawan Center on Sriharikota Island, 90 km north-east of Chennai (Madras) in Nellore, Andhra Pradesh. Not only is it

Indira Gandhi at Sriharikota

the principal launching center, but it is also the main location for static tests. Other, but minor, launch sites are Thumba on the south-west coast and Belasore on the north-east, used only for meteorological and sounding rockets. There is a military launching site at Wheeler Island on the Orissa coast, which, in 1999, was used to test the *Agni II* missile during the stand-off with Pakistan.

By government decision on 5th September 2002, Sriharikota was called the Prof. Satish Dhawan Space Center, named after the second head of the Indian space program after Vikram Sarabhai and a statue was erected in his memory. The launch center was established because the sounding rocket site at Thumba on the west coast was not suitable for large rockets. Rocket engineers generally need an east coast site, so as to launch rockets eastward over sea, preferably in an uninhabited area, with good communication links and with ease of construction. Sriharikota was selected when the domestic rocket and satellite program got under way and the first launch from there was a *Rohini* sounding rocket on 9th October 1971. First director was the head of Thumba, H.G.S. Murthy. Born in 1918, he was a graduate in mechanical engineering from Balngalore and had worked in Hindustan Aeronatuics there, later spending time abroad with the Oerlikon missile company in Zurich and at NASA's Wallops Island launch base in Virginia, used to launch the Scout.

While close to the sea, like the Japanese launch sites, by contrast, the terrain at Sriharikota is flat, covered with eucalyptus, cashew, coconut and casuarina trees in forest, shrub and plantation. From October to November, the site is often lashed by monsoon rains. The island is shaped like a sand spit, not unlike parts of the Florida coast near the Kennedy Space Center, another similarity being the narrow channels that run down the middle of the island. As one arrives at the launch center, driving south, there is the security main gate. Soon thereafter is the control center. The road then divides. Seven kilometers to the right is the SLV, later the ASLV complex (now decommissioned); 5.6 km to the left is the PSLV complex and the clean rooms to receive spacecraft before launch. At the landward side of the island are storage centers, administration and accommodation and in the middle is a radio receiving station.

The island also houses a solid propellant plant, 5 km from the pad, and a static test and evaluation complex. The Solid Propellant Space Booster Plant was built in 1977 for the SLV but was later expanded to manage much larger rockets. The evaluation center provides facilities for horizontal firing of rockets, a vibration platform to shake equipment up to 100 G, and a high-altitude test facility simulating conditions of near space for the firing of upper stages. There was a serious fire on 23rd February 2004 in the solid-fuel plant, causing several injuries, much damage and a cost of Rs7m (€130,000).

The PSLV complex was commissioned in 1990. It has an enormous 3,000-tonne mobile service tower 76 m tall. It is a mammoth structure, made out of galvanized steel, with a 21 by 23-m cross-section, a 60-tonne crane at the top, able to withstand winds of up to 230 km/hr. It has to be repainted every five years.

The normal launch procedure is to stack the launch vehicle vertically at the pad area in the mobile service tower. Satellites are first checked out in two clean rooms, where they can be examined, balanced, tested for leaks, fuelled and checks made of the solar panels and electrical systems. The first clean room, where the spacecraft arrives, is 7 km from the launch pad, close to Mission Control. The purpose of this stage is to verify the health of the spacecraft, which is kept at a temperature of 22°C with 50% humidity and with less than one dirt particle per million. The second clean room is designed for propellant loading.

This mobile tower then brings the rocket down to the pad on twin rail tracks on four bogies before launch, crawling at 7.5 m per minute. The PSLV is placed on a 175-tonne pedestal ready for launch. This has a third clean room, on the pad, with an airlock to protect the integrity of the satellite. The tower is then rolled back 180 m for the launch itself. For the final stages of the countdown, the PSLV is connected to a 50-m-tall steel-made umbilical tower, which provides cooling and electrical power right up to the last moment. At take-off, deflector ducts take away the flames of the solid rocket motor exhausts and the rocket rises. The launch control center is a safe 5.6 km distant.

A second €50m launch pad was completed in November 2003, some 1,500 m from the PSLV pad, able to take either the PSLV or the GSLV and its successor versions and enabling India to conduct two launch campaigns simultaneously. Its first PSLV launch was Cartosat in May 2005 and its first GSLV launch was on 20th July 2006. The second pad has a vehicle assembly building so that the rocket is mainly stacked indoors, moved by rail to the pad and erected beside an umbilical tower with inspection platforms.

About 2,400 people work at the 145-km^2 site. Ascending rockets are tracked at Sriharikota by monopulse C-band precision radar, which is a dish antenna on a rotating structure that can follow the rocket up to a range of 3,200 km, marking its range, height and angle. Later, tracking stations take over in Nicobar in the Bay of Bengal.

GSLV and its launch tower

Like Japan, India continues to use small sounding rockets that are launched from the original launch site, Thumba, which has three launch pads, capable of taking sounding rockets of up to 560 mm in diameter. Three versions of the Rohini series are used, the numbers referring to their diameters, the 200 (two-stage), the 300 (single-stage) and the 560 (two-stage). They shoot into the atmosphere at enormous speed – indeed, the *Rohini 200* has a burn time of less than 8 sec. The *Rohini 560* is much the largest and heaviest and is a red-painted rocket fired at an angle from a crane-like platform. In 1993, there was a salvo of sounding rockets, with three Rohini launched to 300 km, each releasing barium clouds at altitude to measure the distribution of ions and electrons in the ionosphere. On 29th April 1998, a Rohini sent a 127-kg payload up to 432 km.

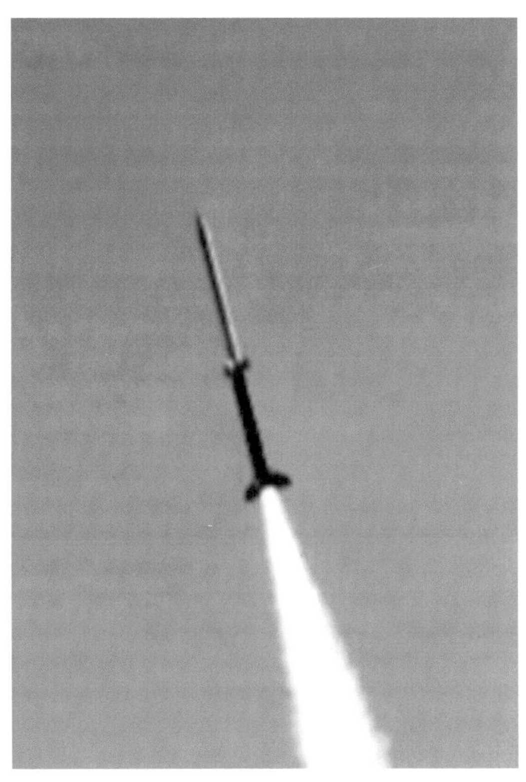

Rohini 200 rocket

SPREADING THE BENEFITS TO INDUSTRY

With its emphasis on applications, it is no surprise that India chose to emphasize the importance of spin-off and the transfer of the benefits of the space program to Indian industry. A technology transfer coordinator was appointed at ISRO headquarters in Bangalore, a manager of technology transfer was appointed at the Vikram Sarabhai Space Center in Thiruvananthapuram and other officials with responsibility to promote spin-off were put in place at the National Remote Sensing Agency in Hyderabad, at the Space Applications Center in Ahmedabad, at Sriharikota and the Liquid Propulsion Systems Center. By 1998, 231 distinct new technologies had been licensed to private industry.

The space industry has had an important effect in driving technology in India. More than 150 industries were involved in the development of the ASLV and PSLV and their associated satellites, spurring advances in such diverse areas as materials, electronics, fuels, optics, electro-mechanical systems and computers.

Much of the spin-off reflects the orientation of the Indian space program toward Earth resources, remote sensing, weather forecasting and telecommunications. Specific examples of spin-off technologies include mapping systems and cameras, communications terminals, color printers, film coatings, graphic displays, antennae, water-measuring meters and systems for news-gathering and dissemination.

In addition, there is a broad area of spin-off deriving from the construction of rockets, satellites and their related ground facilities. Examples are adhesives, glass and carbon composites, lubricants and chemicals deriving from solid rocket fuels. Some of the spin-off items are apparently mundane, but nonetheless important, such as crane controls (derived from building launch towers), domestic electric shock protectors and even (in the 1987 spin-off list) a "worm-driven mechanical jack". A new composite material developed for rockets called kevlar reinforced polymethacrylate was developed as a cheap, tough, non-toxic alternative to traditional dentures.

President Kalam in Sriharikota

Since much of the hardware used in the Indian space program is contracted out to private companies, there are considerable gains to know-how in Indian industry. About 500 companies participate in the space effort. The construction of rockets and satellites involves the manufacturing of products to unprecedentedly high levels of strength, precision and reliability. Even the making of such mundane items as batteries becomes a challenge when they must work unaided for years, otherwise the whole satellite will fail. The need for high standards is especially true in the making of solid rocket boosters, where India has developed world-class expertise in the making, mixing and use of solid rocket fuels and the forging of steel for casings, joints, rings and segments. This has involved the mastery of a range of materials – maraging steel, aluminum, copper, titanium, cobalt and magnesium. The recent building of a domestic liquid hydrogen engine involves the management not just of highly explosive fuels under great pressures, but control of hydrogen at temperatures of lower than –200°C – a feat achieved elsewhere by few other industrialized nations.

The government established the Antrix Corporation to promote the application of the space program both to domestic industry and abroad. Antrix earns over €30m a year in exports for India, selling spacecraft components such as valves, pressure transducers and solar panel hinges. In addition, an unknown amount of revenue is generated from leasing out lines on the INSAT system, where India is the largest provider for all south-east Asia. Antrix has promoted the PSLV as a launcher of small satellites for overseas customers. The breakthrough came in May 1999, when, with Oceansat, the PSLV launched two small satellites, Korea's KITSAT and Germany's TUBSAT (TUBSAT stands for Technical University of Berlin). KITSAT was a 107-kg remote sensing mission while TUBSAT was a 45-kg experimental remote sensing German satellite and, for them, India was paid Rs44m (€943,000). No less than eight nano-satellites were launched on the PSLV in April 2008, Antrix netting €500,000 for doing so.

AGILE

Under a Rs35m (€763,000) agreement reached in 1998 between Antrix and Europe's Arianespace, the PSLV launched the small 100-kg Belgian Earth observation technology demonstration satellite PROBA (Project for On-Board Astronomy). The January 2007 test of the SRE 1 carried a 56-kg Indonesian satellite, *Lapan*, and a 6-kg satellite for Argentina, called *Pehvensat*. A year later, on 21st January 2008, the PSLV put into orbit a 300-kg Israeli military satellite, TECSAR, but popularly known as Polaris, giving Israel a radar observation capacity to watch threatening military movements in neighboring states.

Perhaps the most substantial paid payload was AGILE, launched by the PSLV on 23rd April 2007, the first dedicated entirely to a foreign launch customer. AGILE was a small 100-kg Italian astronomical satellite standing for *Astrorivelatore Gamma ad Imagine Leggore* or light gamma ray imaging detector, which was to study black holes, gamma ray bursts, pulsars, supernova remnants and active galactic nuclei.

With the PSLV and the new GSLV, Antrix would like India to be able to break into the world commercial communications satellite market, where substantial money is to be made. This is dominated by Russia (Proton, Sea Launch, Land Launch), Europe (Ariane) and the United States (Delta, Titan) and a raft of American regulatory controls that have kept out newcomers, notably China. For the time being, India may have to settle for a more modest role with niche scientific, university and military payloads. Table 6.3 summarizes foreign payloads to date.

Table 6.3. Indian launches of foreign payloads

25 May 1999	TUBSAT (Germany) KITSAT (Rep. Korea)	PSLV	Sriharikota
22 Oct. 2001	PROBA (Belgium) Bird (Germany)	PSLV	Sriharikota
5 May 2005	Hamsat	PSLV	Sriharikota
10 Jan. 2007	Pehvensat (Argentina) Lapan (Indonesia)	PSLV	Sriharikota
23 Apr. 2007	AGILE (Italy)	PSLV	Sriharikota
21 Jan. 2008	Polaris (Israel)	PSLV	Sirharikota
28 May 2008	Rubin 8 (Germany) AAUSAT II (Denmark) Cute 1 (Japan) Can X-2 (Canada) Can X-6 (Canada) Compass 1 (Germany) Delfi C3 (Netherlands) SEEDS 2 (Japan)	PSLV	Sriharikota

INDIA'S SPACE BUDGET

India's space program can claim to be one of the most cost-effective in the world. It has been careful to limit its ambitions and focus on a narrow range of activities, principally those generating an economic or social benefit. Indian space spending has generally been the lowest of the three Asia space powers, but accelerated in recent years. India has obtained considerable and often generous help from abroad, either at nil or low cost, as may be seen from such examples as the use of the *Symphonie* satellite (France), ATS-6 (the United States), the APPLE mission (Europe) and several free or low-cost launches from the Soviet Union. The space program has never been politically controversial, having obtained the support of all the prime ministers from Nehru onward and has rarely been criticized in Parliament. It has attracted a high level of political support, prime ministers and presidents frequently visiting facilities, attending launches, opening conferences and sending congratulatory messages.

The Indian space budget was less than Rs5bn annually until 1993. Thereafter, it began to climb rapidly, passing the Rs10bn mark in 1996 and Rs17.598bn (€374m) in 2000. There was a rapid hike to Rs27bn in 2005, Rs36bn in 2007 and Rs40bn in 2008 (€665m), with a projected increase to €672m in 2009. A substantial part of the increase was taken by the development of the GSLV launcher. ISRO has a staff of 16,000, with a further 20,000 people working directly in the space industry and many more indirectly in contracting companies.

CONCLUSIONS: INDIA

By the 21st century, at a time when Japan found that it had to restrict its space ambitions, India had, by contrast, developed the self-confidence to make further steps forward, mounting its first deep space mission, Chandrayan, to the Moon and then carrying out the first steps preparatory to manned flight. India already had a preliminary but limited experience in manned flight, directly through Rakesh Sharma's mission to Salyut 7 and indirectly through the experience of two Indian-born American astronauts, Kalpana Chawla and Sunita Williams. What was of more immediate value to the Indian space program was the slow but ultimately successful development of the Geostationery Satellite Launch Vehicle, achieved with Russian help and despite international intrigue. The GSLV fulfilled a 30-year vision of self-sufficiency in rocketry, while the satellites launched by the GSLV enabled India to continue to be world leader in educational television and exciting applications in such areas as telemedicine.

7

Iran: Origins – the road to space

It was a long road, from a cold December day in New York at the United Nations (UN) in 1958 to the nighttime February launch of the Omid satellite from the Semnan launch site in Iran in 2009.

It did not take long for the General Assembly of the UN, in its 792nd plenary meeting, to establish an *ad hoc* Committee on the Peaceful Uses of Outer Space (COPUOS) after the first artificial satellite, Sputnik, started orbiting the Earth. Resolution 1348 (XIII), adopted on 13th December 1958, tasked this committee, composed of the representatives of 18 countries, to report to the General Assembly at its 14th session on the space-related aspects that could be foreseen at that time.[1] In its 856th plenary meeting, on 12th December 1959, the General Assembly adopted Resolution 1472 (XIV) in which it formally established COPUOS, now with 24 members, to "review, as appropriate, the area of international cooperation, and to study practical and feasible means for giving effect to programs in the peaceful uses of outer space which could appropriately be undertaken under United Nations auspices".

INTERNATIONAL CONTEXT

As one of the founding members of COPUOS, Iran set out to pursue a vision about using space and its technologies for peaceful purposes and for the nation's welfare. In each epoch of continuous evolution of Iran's space program, the driving attitudes and visions of the authorities have, of course, influenced its progress and development. This often resulted in space application being first, with space science as second priority. The strict division between civil and military (applications) was one of the main reasons that the application of space technology spread in Iran; different organizations and institutions had their own vision on space matters and there was not much cooperation, let alone combining of resources. It took until 2004, with the establishment of the Iranian Space Agency (ISA), for the civilian application of space science and technology to become a coordinated effort in Iran.

Historical facts about Iran's journey to becoming a country capable of indigenously launching a satellite into space are not easily forthcoming. Available information mostly comes from sources where Iran works together with international institutions, but not much information has been preserved from the early days. With the help of the UN Office for Outer Space Affairs (UNOOSA), it could be determined only from verbatim records of the 14th session of the General Assembly that Iran was actively present there. It shows that Iran was involved in international space decisions in the early days, but it does not provide much insight into Iranian's space developments. From those records, it could be ascertained that a Dr Mehdi Vakil was the permanent representative of Iran at the General Assembly.[2] Dr Vakil appears also in other documents and at those times seems to be the voice of Iran in space debates.[3, 4]

Another space-related organization of which Iran became a member is the Committee on Space Research (COSPAR) from the International Council of Scientific Unions (ICSU), which is now the International Council for Science (ICS). COSPAR organized its first Space Science Symposium in 1960. A further logical step for Iran was to sign the Treaty on Principles Governing the Activities of States in the Exploration and Use of Outer Space, including the Moon and Other Celestial Bodies (Outer Space Treaty) in 1967.[5] COPUOS, COSPAR and the Outer Space Treaty all have the common denominator of the peaceful use of space and this denominator served as a guideline for the space endeavors of Iran.

SPACE APPLICATIONS AS DRIVERS: COMMUNICATIONS

In Iran, the need for good communications swiftly became clear. The country tried to modernize at a fast pace and imported a significant amount of foreign capital and knowledge in all areas of commerce and space applications; for example, when the Iran Telecommunication Manufacturing Company (ITMC) in 1967 was established in Shiraz, the Siemens Company of Germany had to fund 20% of its capital and is still one of the big stockholders of ITMC.[6] In 1970, the Iran Telecommunication Research Center (ITRC) and, in 1971, the Telecommunication Company of Iran (TCI) were founded and these institutions ran the main of all telecommunication affairs of the country. From 1972 onwards, all telecommunication became the responsibility of the Ministry of Post, Telegraph and Telephone (PTT). Iran joined the International Telecommunications Satellite Organization (Intelsat) in 1970 after, in 1969, having constructed and commissioned the installation of two Standard A and one Standard B antennae (30-m-diameter antennae) at the Shahid Dr. Ghandi satellite communication center in Asadabad, Hamedan. Through these antennae, Iran became connected to the Pacific Intelsat system for international communications.[7]

At that time, also the need for a state-owned satellite communication system for national use became apparent, but it was well into the 1990s before plans for a national satellite system took shape. The planned geostationary satellites were called Zohreh (Venus). At UNISPACE III (1999), in its national paper, Iran stated that "The Telecommunication Company of Iran has also announced a tender for the construction and launch of two Ku-band geostationary satellites to be placed at 34° and 47° E. The project is called Zohreh and is intended to carry out the domestic traffic. The project is expected to be implemented within the next two years". Countries that were reportedly in the race for building geostationary communication satellites for Iran are India (1970s), Russia (1980s), France (1990s) and Russia again (2000s). Initial plans for Zohreh were to provide telecommunications facilities to remote areas in Iran, support terrestrial telephony, provide military and data communications and develop Iran's broadcasting capacity, with two spacecraft stationed at 26° and 34° E to provide both L-band (Inmarsat-compatible) and Ku-band links. The Shah's government perceptively reserved three positions (26°, 34° and 47° E) in geosynchronous orbit for Iran. To fill the need in satellite geostationary communications, Iran leased a Russian Gorizont communication satellite (1989-081A, Gorizont-19), which had been launched in September 1989. Gorizont satellites only had life spans of about three years and so the advantage of a geostationary communication satellite for Iran was short-lived.

In 1987, the Boomehen Earth Satellite Ground Station, located some 35 km north of Tehran, was established and equipped with one Intelsat Standard A and one Standard B antenna. Moveable antenna terminals of Standard A and C were set up to provide an international data communication network with different parts of the world. Nowadays, through this center with three Standard A antennae and three Standard B antennae, Iran maintains international communications with Europe, Africa and Asia. Since 1991, with one antenna, Inmarsat A (telephone, telex and

Main communication antennae of the Boomehen Earth Satellite Ground Station. [Boomehen ESGS]

data) and Inmarsat C (telex) services for the Indian Ocean Region are also rendered through the Boomehen station. The refurbished station was inaugurated by President Rafsanjani on 20th May 1995, and was billed as the largest satellite telecommunication center in the Middle East at the time.[8] In 1987, another satellite Earth station was built in Mobarakeh, Isfahan, with one Standard B antenna for the Intelsat satellite on 62° E in the geostationary orbit. Together with Asadabad and Boomehen, they provide direct international services with 48 countries and through transit with 182 countries via more than 3,300 circuits. In the future, Mobarakeh will be equipped with a new Standard A antenna for the Intelsat satellite on 64° E.

For regions in Iran where terrestrial communications services are not easily feasible, satellite communication was the proposed solution, for the time being via leased transponders on international satellites. The national Domsat system was put into effect in 1990 by implementing its phase 1, which consisted of seven hubs and 61 terminals configured in seven star sub-networks. Use was made of access through transponders of the Ku-band east spot of the Intelsat 63° E satellite. The ground segment was later augmented by the installation of two star networks comprising two hubs and 900 very-small-aperture terminals (VSAT) accessing the same satellite. In addition, a separate nationwide network consisting of two hubs and some 1,700 VSATs owned and operated by the Central Bank of the Islamic Republic of Iran is now in service. A tender was issued by the Telecommunication Company of Iran (TCI) for the acquisition of nine gateway hubs and 300 demand-assigned multiple access Earth stations, all using the 14/11-GHz band.[9]

Radio and TV broadcasting through satellites was developed by Iran, starting during the 1970s. The Islamic Republic of Iran Broadcasting (IRIB) organization implemented many projects making effective use of three 72-MHz bandwidth Ku-band transponders on the 63° E Intelsat satellite. Four national television channels now broadcast nationwide, making use of 2,600 television receive-only terminals, thus providing almost complete national television coverage. IRIB has also begun a Ku-band television broadcast over Europe and the Middle East via a European Telecommunications Satellite Organization (Eutelsat) satellite. In addition, IRIB owns two C-band Earth stations relaying news items to Asiavision and also internationally through Intelsat. Two transportable ground stations are also available for satellite news-gathering transmission from any point around the country and neighboring countries. Initially (and still), use was made of Intelsat satellites; later, other (commercial) satellite transponders were added, such as Eutelsat, HotBird and Telestar-5, currently providing Iran with some 25 TV channels and 30 radio channels.[9, 10]

OBSERVATION OF THE EARTH

Although meteorological observations have taken place in Iran since 1919 (establishment of the Barzegaran School and Meteorological Station), meteorological activities on a more professional level date from after World War II. In 1958, a general office responsible for all meteorological activities was established and in

Iran as seen by a meteorological satellite from NOAA with AVHRR sensor received by the Mahdasht Satellite Receiving Station and processed by IRSC (currently the Remote Sensing Administration of ISA) [IRSC]

1959, Iran became the 130th member of the World Meteorological Organization (WMO). From 1960 onwards, stations studying the atmosphere (Tehran, Shiraz and Mashad) were equipped with land-based radars and other new technologies.[11] Antennae and satellite receivers were acquired for NOAA (USA), Meteosat (Eumetsat), GOMS (Russia) and FY-2 (China) meteorological satellites when they became available. In 2007, the UniScan ground station was installed in the Mahdasht Space Center. The ground station was formerly located at Saadat Abad headquarters of ISA in Tehran but, after moving headquarters to the Sayeh Building in Tehran in 2006, the facility was moved to Mahdasht Space Center.

The weather satellite receiving system for Meteosat data user stations and NOAA automatic picture transmission were installed at the Iranian Meteorological Organization (IRIMO) headquarters in Tehran in early 1992. IRIMO expanded the receiving station with high-resolution picture transmission and meteorological data distribution units in 1998. Data from meteorological satellites are used by the

IRIMO forecasting center not only for weather-forecasting purposes, but also for atmospheric disaster mitigation objectives. NOAA receiving facilities are also installed in the Iranian National Center for Oceanography (INCO) and the Mahdasht Space Center. The Advanced Very-High Resolution Radiometer (AVHRR) data received by the former IRSC (currently the Remote Sensing Administration of ISA) acquisition system are used for Earth resource monitoring and studies as well as dissemination of results and documents in the public domain; data received by the National Cartographic Center (NCC) and INCO agencies are used for their own studies and research projects. In addition to atmospheric data, the National Committee on Natural Disaster Reduction (NCNDR), within the framework of a joint research project, is using space-based positioning systems to monitor plate movements along major active faults in Khorasan Province (in the north-eastern part of the country) and the Tehran region, both of which have historical and recent earthquake records and reactivation potential. The project is carried out through a trilateral endeavor including the Geological and Mineral Research Survey (GMRS) of Iran and the NCC.[9]

Satellite remote sensing data were recognized as an efficient and modern tool for studying and monitoring the environment and resources of Iran and have been used since such data were made available commercially. The launch of the American ERTS satellite in 1972 – which later was renamed Landsat-1 – further developed interest in satellite remote sensing and related space technologies. The use of remote sensing products actually started with the "Satellite Applications Plan" in the Iranian Planning and Budget Organization. A feasibility study pointed at the current site of the Mahdasht receiving station, some 65 km west of Tehran (formerly known as Mard Abad, Karaj) as the best place to establish a station for direct satellite data reception. The facility was build at Mahdasht to not only receive remote sensing imagery from satellites, but also to process and distribute relevant imagery products to users throughout Iran for resource planning and management. Iran became the fourth country to install such a ground-receiving station and, for quite some time, Mahdasht Satellite Receiving Station was one of only a few major sites in the world for data acquisition from Landsat. Over the years, Mahdasht has supplied data to assist in identifying areas for development. It enabled scientists to identify areas prone to earthquakes, floods, landslides and other natural disasters and threats. The center has also been used to investigate greenhouse gas emissions and air pollution in large urban areas and to monitor wetlands, inland water basins and the environment of the Caspian Sea and the Persian Gulf.

The Iranian Remote Sensing Center (IRSC), established in 1973, operated the Mahdasht facility. The United States and Iran, in their first cooperation program in space technology, agreed that Mahdasht directly should acquire satellite data from Landsat satellites. It was agreed also that Iran would supply these data to the 33 countries that fell under the antenna coverage of Mahdasht. In 1974, a contract was signed between Iran and the American General Electric Company to install the receiving station and after starting building in 1976, the station became operational in 1978. The advent of the revolution in 1978 caused General Electric to leave the country and suspension of the project. IRSC, once the only legal governmental body

۵ - ۱۲ ماهواره‌های هواشناسی زمین آهنگ

ماهواره‌های هواشناسی زمین آهنگ تحت پروژهٔ پاییدن آب و هوای جهان، WWW، که توسط سازمان هواشناسی جهانی، WMO، سازمان یافته به فضا پرتاب می‌شوند و همان طور که در شکل ۵ - ۱۲ - ۱ نشان داده شده است، کل کرهٔ زمین را با پنج ماهواره پوشش می‌دهند.

این پنج ماهواره عبارتند از: مِتئوسَت، METEOSAT (آژانس فضایی اروپا، ESA)، اینسَت (ماهوارهٔ ملی هند)، INSAT، ماهوارهٔ هواشناسی زمین آهنگ، GMS (ژاپن)، ماهوارهٔ زیستمحیطی عملیاتی زمین آهنگ شرقی، GOES-E (ایالتهای متحد امریکا)، ماهوارهٔ زیستمحیطی عملیاتی زمین آهنگ غربی، GOES-W (ایالتهای متحد امریکا). برنامهٔ این ماهواره‌ها در پیوست ۱ آمده است.

ماهواره‌های متئوست - ۵، اینست - ۱D، ۴ - GMS و ۷ - GOES از سال ۱۹۹۱ در حال فعالیتند.

ماهوارهٔ ۴ - GMS سنجنده‌ای به‌نام VISSR (تابشسنج اسکن‌کنندهٔ چرخشی مرئی و فروسرخ نزدیک) با دو باند مرئی و فروسرخ گرمایی دارد. سنجندهٔ VISSR چهار خط برای باند مرئی و یک خط برای باند گرمایی را به طور همزمان از شمال به جنوب اسکن می‌کند، که آن طور که در شکل ۵ - ۱۲ - ۲ نشان داده شده، برای پوشش دادن کامل یک نیمکره ۲۵ دقیقه زمان لازم است.

تعداد کل خطهای اسکن ماهواره‌ها ۱۰۰۰۰ خط برای باند مرئی و ۲۵۰۰ خط برای باند گرمایی است.

آن طور که در شکل ۵ - ۱۲ - ۳ نشان داده شده، ماهوارهٔ هواشناسی زمین آهنگ، GMS، دارای یک سیستم سکوی گردآوری داده (DCP) برای گردآوری اطلاعات مختلف، نه فقط از ایستگاههای زمینی، بلکه از ایستگاههای دریایی است.

داده‌های تصویری در یک مُد با توانِ تفکیک بالای سیگنالهای S-VISSR (تابشسنج اسکن‌کنندهٔ چرخشی مرئی و فروسرخ نزدیک) و یک مد با توانِ تفکیک پایین WEFAX به ایستگاههای زمینی منتقل می‌شوند، که با تجهیزات ساده‌تر و ارزانتری می‌توان آنها را دریافت کرد. بعضی داده‌های آماری مثل هیستوگرامها یا نمودارهای ستونی، حجم توده‌های ابر، دمای سطح دریا، توزیع باد و غیره در آرشیوهایی ثبت می‌شوند که شامل مجموعه‌های ISCCP (پروژهٔ ماهوارهٔ بین‌المللی اقلیم شناسی ابرها) است.

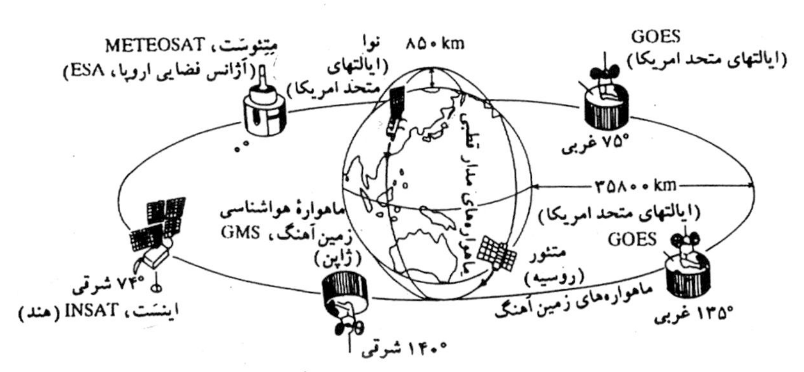

شکل ۵ - ۱۲ - ۱ مکان ماهواره‌های زمین آهنگ گرد زمین

۱۱۶

Page 116 of the book *Remote Sensing Notes* from the Japan Association on Remote Sensing, translated in 1996 into Farsi for the IRSC. The title of the chapter reads: "5-12 Geostationary Meteorological Satellites" for providing remote sensing data in Iran, was involved in the application of Earth space data for about 25 years. In addition to data-providing, IRSC controlled the development and implementation of different applications as well as research projects and programs requested by different organizations and institutes. Furthermore, IRSC pursued research in the development of remote sensing technology focused on the development of a CCD sensor.[12] The plans for the development of Mahdasht were put on hold and the future was uncertain. People who were directly involved, such as experts, engineers and managers, tried to save the station and to convince the decision makers and high-level officials that remote sensing and other space technologies played an important role in the management and control of the development of Iran. In these changing times, the people at the station managed to maintain and operate the facilities in such a way that not only direct data acquisition from meteorological satellites became possible, but training and complete management of data were institutionalized. In the early 1990s, responsibility for remote sensing, in all its aspects, transferred to the state-run IRSC that became affiliated to the Ministry of PTT. The transfer of responsibility to IRSC created legal problems for the Mahdasht station in terms of securing financial resources and the inability to develop the necessary plans and programs. It was then decided that IRSC should be downsized and the activities of Mahdasht should be minimized. This led to the temporary suspension of the activities of the station. In the early 2000s, there were organizational and administrative changes in the Ministry of PTT and this ministry changed in December 2003 into the Ministry of Communications and Information Technology (CIT). In 2004, ISA was created and this agency was tasked with all remote sensing activities throughout the country. As a direct result, the Mahdasht receiving station was revived with the cautious commencement of reconstruction and operationalization of the station in 2003. The receiving facilities (antennae) at Saadat Abad north of Tehran were gradually transferred to Mahdasht, but the antenna for receiving Landsat data was abandoned.

The Mahdasht Satellite Receiving Station had to grow into the Mahdasht Space Center and comprises comprehensive and multi-task ground space complexes and laboratories, as well as work, living and leisure facilities for Iran's space science and technology specialists, scientists and officials.[13] Products from newer-generation satellites such as SPOT, IRS and Radarsat were used. Nowadays, data acquired from commercial remote sensing satellites are widely used for research and development programs by a variety of organizations and institutions in Iran. The center obtained (and stores) more than 10 years of NOAA-AVHRR (Advanced Very-High Resolution Radiometer) data on magnetic tape. The center is capable of receiving data in both S- and X-band frequencies that are used by existing and future satellites. Not only NOAA-AVHRR data were acquired, but also data from the NASA TERRA-MODIS (Moderate Resolution Imaging Spectroradiometer) and Russian OKEAN satellites have been collected. The receiving station for data acquisition from the MODIS sensor was put in service in October 2001. In September 2002, the station was made capable of receiving data from the Indian Remote Sensing (IRS) satellite. When, in February 2004, the Iranian Space Agency

Antennae at the Mahdasht Satellite Receiving Station [IRSC]

Mahdasht Satellite Receiving Station's early mission was to receive data from the US Landsat satellite. The 11-m antenna for receiving from Landsat is seen at the top. The main building and control building are seen in the (lower) middle [IRSC]

was established, all the functions and duties of the IRSC were transferred to this agency under the Remote Sensing Administration. The Mahdasht Satellite Receiving Station is now officially called Mahdasht Space Center.

Lut Desert in center/eastern Iran as seen by Landsat-TM, received by the Mahdasht Satellite Receiving Station and processed by IRSC. It stretches about 320 km from north-west to south-east and is about 160 km wide. In its lowest, salt-filled depression – less than 300 m above sea level – the summer heat and low humidity are believed to be unsurpassed anywhere in the world [ISA/Geographic Atlas of I.R. Iran]

SPACE APPLICATIONS ORGANIZED

The main agencies involved in space application activities in Iran, such as communications and Earth resources remote sensing, include ministries that deal with these applications. To mention a few: ITMC, ITRS, TCI and the Ministry of PTT for communications; for remote sensing: ISA/IRSC, GMRS, the Forest and Range Organization, the Soil Conservation and Watershed Management Research Center, the Ministry of Jahad-e-Agricultural, the Iranian National Center for Oceanography, the Ministry of Energy, the Ministry of Petroleum and the Ministry of Science, Research and Technology.

The history of Iran's space efforts and its drive to pursue independent space projects began during the shah's reign. The main goal in 1977 was to establish an Iranian communications satellite system (Zohreh). In addition, several Iranian organizations were involved in plans to send small research satellites into space.

Those satellites would pave the way for launching a military intelligence-gathering system.[14] Iran could not achieve these goals purely indigenously and needed foreign assistance in the field of technology and knowledge. Iran turned to Russia, China and India, being established spacefaring nations, but also to North Korea for rocket technology and, later, to Italy for satellite technology. In 1977, the first step to establish a space agency was taken. The director of the Iranian National Organization of Radio and TV presented to the Director of the Planning and Budget Organization a plan to create the Iranian Space Agency (ISA), but the revolution (1979) and the war with Iraq (1980–1988) halted all efforts in the institutionalization of space activities in Iran.[15]

The Iranian revolution transformed Iran from a pro-Western constitutional monarchy into an Islamic, theocratic republic. This changed the country in every aspect internally and space research was put on the backburner. Space applications, such as communications and remote sensing, were barely allowed to continue. Contacts with the West decreased rapidly as the country turned in on itself and accepted only Islam as leading its endeavors.

Iran needed to reform its space efforts to be able to deal with civil space on the one hand (communications and remote sensing) and military space on the other hand (command/control, observation of the Earth and rocketry). A number of agencies and industries were involved in space development, but in an uncoordinated way. In 1987, the Prime Minister of Iran, Mr Mir Hussein Mousavi, proposed a Commission for the Coordination of Space Issues. The importance of space applications was again emphasized, but it would take until 2003 for the Bill for the establishment of the Iranian Space Agency to be approved in the Iranian Parliament.[15] In 2004, ISA was finally established with a mandate for all civilian applications of space science and technology. The agency's assignment from the Iranian Supreme Space Council (SSC) was to generate programs for the peaceful applications of space technologies. This included a satellite and launch vehicle program, as well as promotion of the ground segment.[16]

According to Dr Parviz Tarikhi, Head of the Microwave Remote Sensing Department at the Mahdasht Space Center, the creation of ISA did not happen overnight. The Iranian Space Agency was established on 1st February 2004 according to Article 9 of the Law for Tasks and Authorizations of the Ministry of Communications and Information Technology, passed on 10th December 2003 by the Parliament of the Islamic Republic of Iran. This law received approval of the Guardian Council of the Constitution of Iran on 18th June 2005. The Guardian Council's main task is to monitor and control the consistency of laws passed by the Parliament with the Islamic laws and regulations. Based on the approved statute, ISA got a mandate to cover and support all the activities in Iran concerning the peaceful applications of space science and technology under leadership of a Supreme Space Council (SSC) chaired by the President of Iran. This Council's main goals included policy making for the application of space technologies aiming at the peaceful use of outer space; manufacturing, launching and use of national research satellites; approving space-related state and private sector programs; promoting the partnership of private and cooperative sectors in efficient uses of space; and identifying guidelines concerning the regional and international cooperation in space issues. To be able to follow and implement the strategies set by the Council, ISA affiliated with the Ministry of CIT in the form of an autonomous organization. The President of ISA held the position of the Vice-Minister of CIT and the secretariat of the SSC at the same time. ISA continued implementing its tasks and duties under supervision of the SSC until the government decided to merge the supreme councils according to the approval of Administrational Supreme Council in August 2007 and in line with the implementation of the Fourth Development Program of the country. The Supreme Council of Education, Research and Technology was established by merging the Council of Sciences, Research and Technology and the Supreme Councils of Information Technology, Communications, Space, Atomic Energy, Communication Media Security, Education and Training, Educational Revolution Logistics, Informatics, Science Applications, Biotechnology, and Standards. However, this new Supreme Council of Education, Research and Technology was dissolved soon after, in February 2008, and its functions were transferred to the newly set-up Science, Research and Technology Commission under the Cabinet of Iran. The change in the status of the SSC necessitated the revision of the statute of ISA to allow it to function according to legislation, approved laws and regulations. In this connection, the Council of Ministers of Iran, on 15th June 2008, approved the amendments to the statute of ISA set down earlier in June 2005. This followed the investigations of the Guardian Council that led to final approval on 2nd July 2008. The most important change in the new statute is that the supervision of the SSC by the leadership of the President of Iran has been annulled. As a result, ISA is only an administration under the Ministry of CIT that reports to the related minister. It is an indication of the limitation and confinement of ISA, but the new statute provides ISA with more financial authorization to focus on and to regulate its efforts in institutionalization of space activities and to benefit from potentials and available sources to reach its goals. Moreover, the new statute authorizes ISA to proceed with establishing space research centers and firms with the endorsement of the Council for Development of Higher Education. This last task was not included in the older statute of ISA approved in June 2005. The new statute also authorizes ISA to receive the approved tariffs for offering space services based on the rates approved by the Cabinet and maintain funds at the state public revenue account. Furthermore, in line with Article 68 of the Law for Management of Country Service, approved in 2007, ISA, in coordination with the Presidential Deputyship of Management and Human Assets Development, is authorized to make necessary payments with the endorsement of the Cabinet to draw and retain appropriate human resources for the specialized and managerial positions.[17]

THE SUPREME SPACE COUNCIL

The Supreme Space Council (SSC) is not an organization as such, but a board of people on ministerial level. Its secretariat is performed by the Iranian Space Agency (ISA), the president of ISA being the secretary of the SSC. The main members of the SSC are:

The President of Iran, who is the president of SSC at the same time
Minister of Communications and Information Technology
Minister of Science, Research and Technology
Minister of Defense and Armed Forces Logistics
Minister of Foreign Affairs
Minister of Industries and Mines
Minister of Road and Transportation
Director of the I. R. Iran Broadcasting [Organization]
President of ISA who is the secretary of SSC

The SSC has other members as observers and advisors.

STATUS OF THE SSC

Beginning from February 2004, ISA continued implementing its tasks and duties under supervision of the SSC until the state's decision to merge the supreme councils according to the approval of Administrational Supreme Council in August 2007 and in line with the implementation of the IV Development Program of the country. The Supreme Council of Education, Research and Technology (SCERT) was established by merging 12 supreme councils, including the SSC. However, the new SCERT was dissolved soon after, in February 2008, and its functions were taken over by the new Science, Research and Technology Commission (SRTC) subordinated to the Cabinet of Iran. The change in the status of the SSC necessitated the revision of the statute of ISA to allow it to function according to the legislation and approved laws and regulations. In this connection, Iran's Council of Ministers, on 15th June 2008, approved the amendments to the statute of ISA, approved in June 2005, which ultimo led to final approval on 2nd July 2008.

The most important change in the new statute in comparison to the former one is that the supervision of SSC, by the leadership of the President of Iran, has been cancelled. As a result, ISA now is only an administration under the Ministry of CIT that reports to the related Minister. ISA now has somewhat limited power to execute policy but the new statute provides it with more financial authorization to focus upon and regulate its efforts in institutionalization of space activities. Moreover, ISA can benefit from potential and available sources to reach its goals. For example, the new statute authorizes ISA to establish space research centers and firms with the endorsement of the

Council for Development of Higher Education. This task was not included in the former statute of ISA. ISA now also is entitled to receive the approved tariffs for offering space services based on the rates approved by the Cabinet and settle the funds to the state public revenue account. Furthermore, according to the new statute, ISA receives additional financial authorizations.

The Iranian Parliament, though, considered the dissolution of the 12 supreme councils as illegal and decided to revive the dissolved councils by the approval of the "Assembly for Distinguishing the Prudence of the Regime" on 27th September 2008. The state has a mandate to revive the dissolved councils after eight months since their dissolution. By reviving the SSC, the need for again changing the statute of ISA became mandatory. The new statute ratifies the relation of the revived SSC with ISA and redefines the functions and duties of the Agency in the new configuration based on the goals and mandates of the SSC.

[Parviz Tarikhi: Statutes of the Iranian Space Agency. *Journal of Space Law*, The University of Mississippi School of Law, USA, No. 2, Vol. 34, Winter 2008]

AEROSPACE RESEARCH INSTITUTE (ARI)

The Aerospace Research Institute (ARI), an academic organization affiliated with the Ministry of Science, Research and Technology, was founded in 2000 with the aim of conducting fundamental and applied aerospace research. Since then, ARI has been actively involved and instrumental in meeting aerospace research requirements at the national level as well as in establishing links and working relationships with relevant industries. The principal objective of ARI is the identification and introduction of state-of-the-art aerospace and related technologies, and collaboration with organizations in conducting innovative research. ARI is involved in research and analysis of booster rockets, re-entry vehicles, rocket engines and satellites/payloads, as well as in other aerospace-related topics. ARI is organized into different departments: Aeronautical Sciences and Technology, Space Science and Technology and Aerospace Law, Standards and Management. Furthermore, there is an Aerospace Physiology Research Group and a Strategic Aerospace Studies and Future Planning Think Tank. Currently, ARI also concentrates on the aerodynamic design and analysis of launch vehicles. It is assessed to be capable of estimating aerodynamic coefficients and determination of flow patterns around launch vehicles with various levels of accuracy that are required in different phases of a design process. Planning and conducting wind tunnel tests for validation of analytical and numerical results is also among the capabilities of ARI. A group that deals with sounding rockets works on sub-orbital rockets and their payloads. It has carried out several study programs in the field of sounding rockets' capabilities and applications, their payloads and experiments. This group is capable of planning sounding rocket

Main building of the Aerospace Research Institute in Tehran [MvanEijkeren]

experiments, as well as selecting and/or designing the required payload and equipment. The orbital débris team of ARI, as a part of the Department of Aerospace Law, Standards and Management, is working on a variety of subjects such as categorization, characteristics, tracking and laws on orbital débris. Mathematical simulation and collision probability functions as well as hazard analysis are the prospective topics of the department's studies.[18, 19]

In 2008, ARI employed some 65 researchers, 33 organizational members and 13 Ph.D. students under the scholarship of the institute studying abroad and in the country. It conducts and carries out several research projects in collaboration with the industry and indigenously. A number of space-related projects have been carried out in the form of contracts with aerospace industries of Iran. One of those projects was the space lab project. This involved the design and prototyping of the Kavoshgar rocket payload based on a contract between the Ministry of Science, Research and Technology, and the Ministry of Defense and Armed Forces' Logistics. Another project was the evaluation of the standards employed in the design, manufacturing and testing of aerospace systems, funded by the Aerospace Industries Organization. Some other space-related projects are studies on propulsion fundamentals and performance of microsatellites propulsion systems; satellite dynamics software; studies on the cooperation of private companies in space activities such as space commercialization and space tourism; the effects of microgravity on bone health of astronauts; analysis and investigation of satellite navigation systems and devising strategies for the development in this area and publication of a series of books on subjects such as space débris and medicine and space physiology.[20]

EDUCATION AS A FOUNDATION FOR MASTERING SPACE

Iran aims at the acquisition and mastering of space science and technology to support the development of space applications and industrial activities. Space education is considered to be indissolubly connected to the future of the country. Currently, there are tens of universities (see Annexes) offering space-related education, postgraduate space courses and degree programs, amongst others in space remote sensing and geographic information systems (GIS). In addition to those universities, other administrative bodies such as the National Cartographic Center (NCC), ISA/IRSC and the Soil Conservation and Watershed Management Research Center (SCWMRI) provide discipline-oriented or special courses on new space technologies. The NCC, which is affiliated to the Management and Planning Organization, is the national body responsible for topographic base maps and data production. NCC has more than 50 years' history in map production and is responsible for planning, directing, standardization and supervision of map and spatial information production. National technical capabilities are utilized in production and updating of this spatial information. In addition to the activities mentioned here, NCC is using satellite navigation for projects including the Triangulation Networking and the National Leveling Project and its subsequent linkage with regional and international GPS networks. It also uses GPS in the national 1:25,000 scale Topographic Mapping Project, geodesic surveying projects, accurate leveling projects, and the "Determination of the Geoid of Iran" project. The SCWMRI of the Agricultural Research and Education Organization is the focal point for soil conservation, watershed management, flood management and exploitation, river engineering and training, coastal protection, hydrology and water resources development in the Ministry of Jahad-e-Agriculture, in Iran. SCWMRI works in cooperation with universities, particularly Tehran University, and focuses on research topics on the mentioned areas.

To extend existing knowledge, Iranian space specialists regularly participate in courses supported by the United Nations Economic and Social Commission for Asia and the Pacific (ESCAP) or offered by other regional or international bodies such as the United Nations regional Center for Space Science and Technology Education in Asia and the Pacific (CSSTE-AP), the Inter-Islamic Network on Space Sciences and Technology (ISNET) and the Japan International Cooperation Agency (JICA). Attending various seminars, symposiums, conferences and workshops also plays an important role in promoting the existing expertise of Iranian scientists.[9]

INTERNATIONAL EXPERIENCE TO SUPPORT NATIONAL POLICY

After the war with Iraq, Iran started to rely on strict Five-year Development Plans. The fourth Five-year Development Plan concludes in 2009. High priority was given to space technology applications as the effective tool for the sustainable development of the country. In the meantime, according to the 20-year Vision Decree of the country, issued on 4th November 2002 by the leader of the Islamic Republic of Iran, potentials and capacities throughout the country should be focused to increase Iran's contribution to global scientific production. In this connection, Iran should gain access, in particular, to new technologies including nanotechnology, biotechnology, information and communication technology, environmental technology, and aerospace and nuclear technology: "... at the end of the twenty-year Development Plan, in 2025, it is expected that Iran will become the number one country in applications and development of space technology in the Middle East region."[21]

As a part of this Development Plan, Iran's policy is to acquire (space) technology and knowledge and show its engagement with the world community. This is partly done through membership of foreign and international organizations. One of the organizations it turned to was the Inter-Islamic Network on Space Sciences and Technology (ISNET), an interstate, nonpolitical and nonprofit agency. It is an independent, autonomous and self-governing institution under the umbrella of the Organization of Islamic Conference (OIC) Standing Committee on Scientific and Technological Cooperation (COMSTECH). ISNET was founded by nine OIC Member States – Bangladesh, Iraq, Indonesia, Morocco, Niger, Pakistan, Saudi Arabia, Tunisia and Turkey – while Syria joined in 1997 and Iran and Sudan joined in 2004, thus increasing the Membership of ISNET to 12 Islamic countries. SUPARCO, the national space agency of Pakistan, is the host organization of ISNET.[22]

Another organization of which Iran is a member is the Asia Pacific Space Cooperation Organization (APSCO), which evolved from Asia Pacific Multilateral Cooperation in Space Technology and Applications (AP-MCSTA) (1992). The objectives of APSCO are to focus on space science and technology, its applications, education and training and cooperative research to promote peaceful uses of outer space in the Asian region. Seven workshops and international conferences were organized by AP-MCSTA from 1994 to 2003 in Thailand, Pakistan, Republic of Korea, Bahrain, Iran, China and Thailand, respectively. As a result of these conferences, APSCO was formalized and Bangladesh, China, Indonesia, Iran, Magnolia, Pakistan, Peru and Thailand signed the APSCO Convention.[23]

At regional level, Iran actively cooperates with ESCAP, the United Nations Economic and Social Commission for Asia and the Pacific. This commission is the regional development arm of the United Nations for the Asia-Pacific region. With a membership of 62 governments, 58 of which are in the region and a geographical scope that stretches from Turkey in the west to the Pacific island nation of Kiribati in the east and from the Russian Federation in the north to New Zealand in the south, ESCAP is the most comprehensive of the United Nations five regional commissions. It is also the largest United Nations body. Part of ESCAP is RESAP, the Regional Space Applications Program for Sustainable Development, in which Iran contributes actively. ESCAP and ISA have advanced plans for the establishment of a Center for Informed Space-based Disaster Management and an affiliated research center in Iran. Here, a five-day workshop on the Use of Space Technology for Environmental Security, Disaster Rehabilitation and Sustainable Development took place in Tehran in May 2004. The United Nations Office for Outer Space Affairs organized the workshop jointly with ISA. Another organization, of which Iran has been a member since its founding in 1981, is the Asian Association on Remote Sensing (AARS). This association is an organization with a specific Asian identity and attracts remote sensing scientists, technologists, applications specialists, industrialists and entrepreneurs, as well as government decision makers and planners throughout the region, Europe and North America. Working hand in hand with governments, academic societies and education institutions, AARS is an important partner in developing capabilities in space applications in Asia and the Pacific, in particular, in close cooperation with ESCAP and RESAP. At least nine members of AARS have been involved in national multi-mission small satellite projects.[24]

On a global level, Iran obtained great success by chairing an action team on the development of a worldwide comprehensive strategy for environmental monitoring at UNISPACE-III. This third United Nations conference on the exploration and peaceful uses of outer space was held in Vienna in 1999. UNISPACE III aims to foster a greater understanding and better use of space science and technology to assist and stimulate economic and social growth, particularly in developing countries.[25]

Although Iranian officials have declared that Iranian space endeavors are for peaceful purposes and Iranian diplomats have supported efforts within the United Nations to prevent an arms race in outer space, it would be naïve to assume that Iran would not use space applications for military purposes. Analysts worry that Tehran could use its two largest types of space assets – satellites and rockets – to enhance its strategic and/or military capabilities. Iran has launched satellites, one of them indigenously, and is reported to have other programs under development. The satellites are officially being developed for civilian purposes but since most of these satellites are still being built and are shrouded in secrecy, it is generally too early to conclude how these satellites will be used. Some of these satellites could potentially be used by Iran for strategic gains. Such potential uses could include the use of satellite imagery to observe the territory of neighboring states for reconnaissance purposes. The first Iranian satellite, Sina-1, is owned by "the non-civilian sector and its data/imagery has not yet been made accessible, even to ISA".[21] According to official announcements, Sina-1 is a research satellite whose mission is remote sensing and communications and it offers 50-m resolution in panchromatic mode with a 50-km-width swath. The multi-spectral scanning mode has a resolution of 250 m with a 500-km swath width. The current status of the satellite is kept secret. One should consider the strategic potential of each Iranian satellite individually before classifying it as military or civil. Due to the multilateral framework in which the Small Multi Mission Satellite (SMMS) is being developed, it is unlikely that the SMMS is used for military purposes (see Chapter 8). In addition, the Sina-1 satellite

Mockup of the Safir IRILV at the ISIG center in Tehran [FARS News Agency/ H. Ghaedi]

that was launched on 28th October 2005 does not seem to have the imaging capability that would allow it to be used for reconnaissance/surveillance purposes. Similarly, Dr John B. Sheldon, a leading expert in the Space Security program at the Center for Defense and International Security Studies (CDISS), doubts that Iran intends to use the Zohreh satellite for non-civilian purposes, since the satellite does not appear to be jam-proof. Iran would need to make the satellite less vulnerable to attack if it were meant to serve strategic functions. There has also been significant speculation among foreign experts that Iran could use its space launch vehicle (SLV) program as a cover for developing ballistic missiles with longer ranges. Sheldon estimates that Iran could use space launch vehicle technologies to extend the range of its ballistic missiles beyond current capabilities, allowing Tehran to strike the entire Middle East with its improved ballistic missiles. SLVs and ballistic missiles are built with essentially the same technology, but to accomplish this, Iran would need to make several modifications to its space launch vehicles. This would include adjusting the stages it uses in its rockets, adapting the engines used in each system and fitting the missiles with warheads, advanced guidance systems and improved command-and-control mechanisms. As with its satellite programs, Iran's intended uses for its SLV program are presently unclear. Analysts hold conflicting estimates regarding the military utility of Iran's space launch technology. Nonetheless, after the launch of Omid, this discussion broke loose again.[26, 27]

PLANS LEADING TO HARDWARE

Because of the secrecy around Iran's missile program and its affiliation with the space launcher program, it is difficult to acquire information on the space launcher development. Also, the information on Iranian satellites is scarce and often conflicting. Trying to get a clear picture of Iranian industry related to their space developments is therefore very difficult. After Iran launched the Omid satellite, the official state news agency IRNA stated that the launch was part of the country's effort to build a space industry.[28] Most of this space industry (satellites and launch vehicles) is considered to be located in south-east Tehran, namely in the Khojir region. Reportedly, most space-related development and production – especially of launch vehicles – falls under the umbrella of the Aerospace Industries Organization (AIO) of the Ministry of Defense of Iran. This organization is also named the Air and Space Organization (ASO). Civilian bodies involved in space-related development and production are the Ministry of Science, Research and Technology, and the Ministry of Communications and Information Technology. Space-related information tends to be overshadowed by information on the nuclear industry, which also falls under the AIO/ASO. The author, visiting Tehran three times, was refused visits to people who could shed light on the case. High-level industries that are capable of building satellites and launch vehicles are undoubtedly linked to military assets that use these industries, too.

For satellites, there must be industries that can build the hardware, electronics and apparatus and sufficient software developers. SAIran is the main Iranian electronic industrial body, also named Iran Electronics Industries (IEI). SAIran/IEI, a non-civilian organization, affiliated with the Ministry of Defense and Armed Forces Logistics (MODAFL), was founded in 1972 from a mix of different manufacturing plants and companies in different parts of the country with long time experience in electronics, optics, electro-optics, communications, semiconductors and computers. Presently, it is the major producer of electronic systems and products in Iran. IEI *de facto* is a holding company with eight subsidiaries, over 100 different kinds of electronic products and more than 5,000 experienced personnel, including some 700 qualified and highly trained engineers in different disciplines. IEI is considered the largest Iranian electronic conglomerate and a major producer of (military) electronic systems and products. Currently, it cooperates with well known companies across the world. With a reputable and renowned brand name, well managed nationwide distribution network and service centers, IEI has the biggest share of the domestic electronics market and started venturing into the regional market. The major fields of activities of this company are audio visual products, communications, information technology, security of ICT, satellites, avionics, consumer electronics, test and measurement, optics, electro-optics and laser, micro-electronics, security systems, training aids and electro-medical equipment and products, electronic warfare, radar tube manufacturing and refurbishment and missile launchers. IEI has developed an extensive potential of R&D, which is the technological backbone of the company. A subdivision of IEI is the Iran Space Industries Group (ISIG) that deals with satellites.[21, 29] At the location of ISIG in Tehran, a mockup of the Safir IRILV (Islamic Republic of Iran Launch Vehicle) has been placed and used for photography of dignitaries after the *Kavoshgar-1* sounding rocket launch on 4th February 2008.[30] There have also been photographs published in which people can be seen wearing dustcoats with the ISIG logo for the first time. Unfortunately, there is no official website for ISIG that could shed more light on their space activities.

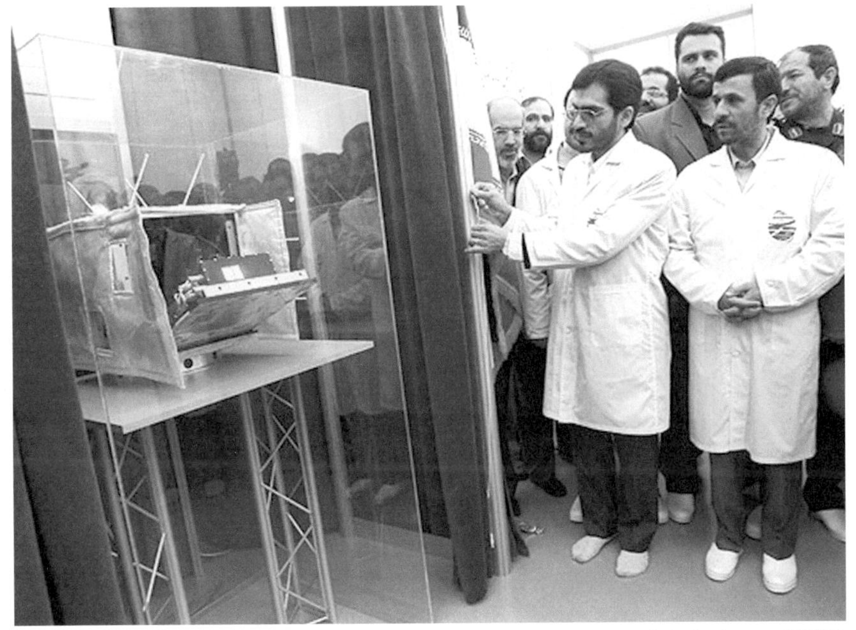

President of Iran, Mahmoud Ahmadinejad, wearing a dustcoat with the logo of ISIG at the presentation of the Omid satellite (February 2008) [FARS News Agency/H. Ghaedi]

President Ahmadinejad leaves the ISIG building (February 2008) [FARS News Agency/H. Ghaedi]

Iran is aware that educational and technological development is needed to take fundamental steps towards advancing space technology, especially in the field of satellite design and manufacturing. To meet this goal, Mesbah, a small research satellite project, has been defined for design and development purposes as a microsatellite for launch to low Earth orbit. The main task of the project is to train Iranian specialists and to support Iranian research centers and universities with satellite manufacturing technologies. Objectives of this project include designing and developing a microsatellite in the amateur radio frequency band with the aim of research, e-mail and store and forward data communication. It should foster scientific research work and training to gain experience and potential for developing satellite communications systems of the store-and-forward type. Technological goals involved in these areas include hardware establishment, definition of steps required for space research, improvement of domestic industries for space activities and familiarization with remote sensing, Earth observation and related technologies.

Exploration in the outer atmosphere is another basic activity of space-related sciences within Iran. A variety of sounding rockets of low, medium and high-altitude capability were planned in 2002. Ionosphere studies, upper atmospheric winds, microgravity, atmospheric composition and atmospheric structure (including pressure and density) were selected topics for further investigation and to meet the defined objectives.[9]

SPACE INFRASTRUCTURE

To meet all its space goals, Iran had to invest in infrastructure. A launch site was needed as well as a satellite control center, assembly buildings, factories to build the rockets, manufacture propellants and so on. Of course, there was already military-related infrastructure – especially concerning the launch vehicle – and Iran made use of this. Because of the military relation and the common use of infrastructure, not much is known in detail. Luckily, official television broadcasts and photography on the internet reveal some information. On internet blogs, every piece of information is thoroughly analyzed: from these sources, the space infrastructure of Iran can be determined.

Iran operates ballistic missile test ranges and that provides (at least the military) experience to handle (space) launch vehicles. Space launch sites in Iran, given its restricted geography, are relatively limited. When trying to find those ranges with *Google Earth* or other commercially available space imagery, one looks for related infrastructure, such as storage bunkers for propellants, assembly buildings, revetments and so on. Also, now that Iran has orbited the Omid satellite, it is possible to calculate from where the launch originated.

Iran news media reported several times that Omid was launched from a launch pad in the Semnan area. They also showed on television how the launch pad looked from the ground. From this information, one can determine what the layout of the launch site is and one could try to find it. It is not possible yet to find this launch site with *Google Earth* because used space imagery is some four to five years old.

The launch site from which the *Kavoshgar-1* was launched on 4th February 2008 [Iran Military Forum]

Much publicity was given to the launch of the *Kavoshgar-1* on 4th February 2008 [Islamic Republic of Iran News Network]

The erector service/fueling gantry with six service levels [Iran Military Forum]

Location of a ballistic missile site in the Semnan province (35°13′19″N 53°53′34″E) [Image (probably 2005) Digital Globe by Google Earth]

Same location of a ballistic missile site in the Semnan province (35°13′19″N 53°53′34″E) [Image (2008) Digital Globe by http://www.timesonline.co.uk]

Commercial and more currently available space imagery shows relevant changes that mark the infrastructure.

Analysis of photography taken by the Digital Globe QuickBird satellite led *Jane's Intelligence Review* to the conclusion that the Semnan missile test site is at about 35°14′00″ N, 53°55′15″ E and must be the place from where Iranian space-related launches took place.

Movie clips were released by the Iranian Military Forum and the Safir-e-Omid Advertisement Center, which is affiliated with AIO and promotes the Omid satellite and the Safir launcher. They released a special Safir/Omid VCD. These videos show

This slab of concrete/asphalt is the place (35°14′00″N 53°55′15″E) where the Iranian space-related launches took place [Image (probably 2005) Digital Globe by Google Earth]

a circular, 50-m-diameter concrete launch platform that accommodates a transporter erector launcher and an erector service/fueling gantry with six service levels. The launch pad features four lightning towers with flood lights on the outer edge of the circular launch platform area. Trucks used for fueling, high-pressure gases and power are part of the operations facility support elements. Also shown are command and communications control trucks with radar and telemetry antennae. The site from which Iran launched its sounding rockets (Kavoshgar) and space launch vehicle (Safir) is very rudimentary compared with Western launch sites. It is of a temporary nature and does not have permanent fixtures that make regular launches easier, such as propellant storage tanks, bunkers, etc. It is therefore assumed that the use of this site is temporary for as long as the indigenous launching of satellites into orbit is not a routine event. The video footage contains a mixture of real footage and good animation. It shows, among other things, the launch procedure in which the Safir launch vehicle is loaded from the Transporter Erector Launcher (TEL) vehicle to the service gantry. All kinds of support vehicles can be seen. Before launch, the gantry is lowered, hinged away at the foot. Then the TEL also releases the support cradle for the Safir and the vehicle is launched.

When Iran launched the *Kavoshgar-1* sounding rocket on 4th February 2008, government spokesman Gholam-Hossein Elman made remarks on the sidelines of the inauguration and said that "Iran's first space terminal is an underground control station and a launch pad at a space center southeast of Tehran in the province of Semnan".[31] No further information on this underground control station has been released since. In (the province of) Semnan is located the site of a ballistic missile test range and production facility, built with Chinese assistance. According to reports published in Russia, apparently based on information developed by the Russian Federal Security Service, the Semnan test site, located 170 km east of Tehran, serves as a test range for ballistic missiles.[32] On Iranian Military Forum video footage, an above-ground control center can be located at Charmshar.

The infrastructure and support that were used for the launch of Omid consisted (apart from the services needed on the launch pad and the launch pad itself) of a Central Flight Control station, three mobile TT&C stations, four ranging stations and ground receiving stations and terminals.[33] At different (aerospace) exhibitions in Iran small telemetry, tracking and control (TT&C) vehicles were shown. These vehicles can be forwarded to convenient places in Iran to support launch and in-orbit operations. Furthermore, on video footage, ranging stations were located in four places in western Iran.

The latest addition to the Iranian space infrastructure is a ground site in the south of Tehran which belongs to the National Geographic Organization, a non-civilian body affiliated with the Ministry of Defense and Armed Forces Logistics

Telemetry, tracking and control truck [ISA]

View on the Semnan Safir launch pad where the preparations for the launch of the Omid satellite are in full swing [jamejamonline.ir]

(MODAFL). In January 2009, the Iranian military reported the construction and completion of its first satellite image ground station and said it would receive images from Iranian satellites. "Through the construction of such an indigenous station, we can receive several simultaneous images from various satellites on a constant basis," Amir Nami, head of the military's Geographic Unit, said. "Also, we can build as many of these stations anywhere we would like."[34]

REFERENCES

1. United Nations Committee on the Peaceful Uses of Outer Space, www.oosa.unvienna.org/oosa/en/COPUOS/members.html (accessed 11 March 2009).
2. E-mail exchange between author and Sirgiy Negoda, legal officer of Committee Services and Research Section, UN Office for Outer Space Affairs (UNOOSA) at Vienna, 2 July 2008.
3. *The New York Times*, 24 September 1960.
4. *Flight International*, 11 January 1962, p. 62.
5. United Nations Office for Outer Space Affairs (UNOOSA), www.oosa.unvienna.org/oosa/SpaceLaw/outerspt.html (accessed 11 March 2009).
6. Iran Telecommunication Manufacturing Company, www.irantelecom.ir/eng.asp?page = 7&code = 1&sm = 7 (accessed 11 March 2009).
7. Telecommunication Company of IRAN, Technical Report, 9 February 2008.
8. Iran's Telecom and Internet Sector – a comprehensive survey from Open Research Network, 1999, www.science-arts.org/internet/node39.html (accessed 11 March 2009).
9. International cooperation in the peaceful uses of outer space – activities of Member States in 2002, www.unoosa.org/oosa/natact/2002/iran.html (accessed 30 March 2009).
10. www.lyngsat.com (accessed 11 March 2009).
11. A history of the meteorology of Iran, www.weather.ir/english/about/history.asp (accessed 11 March 2009).
12. Tarikhi, Parviz: *Applications of Remote Sensing Data in Iran*. Iranian Remote Sensing Center – Tehran, 19 January 2002, www.geocities.com/parviz_tarikhi/OOSA/Presentation_stsCOPUOS39.htm (accessed 11 March 2009).
13. Tarikhi, Parviz: Mahdasht satellite receiving station verging into a space center. *Res Communis*., 13 October 2008, http://rescommunis.wordpress.com/2008/10/13/guest-blogger-parviz-tarikhi-mahdasht-satellite-receiving-station-verging-into-a-space-center/ (accessed 11 March 2009).
14. Shapir, Yiftah: Iran's effort to counter space – strategic assessment. *The Institute for National Security Studies*, 8, no. 3, November 2005, www.inss.org.il/publications.php?cat = 21&incat = &read = 160&print = 1 (accessed 11 March 2009).
15. Tarikhi, Parviz: *Milestones of the Establishment of the Iranian Space Agency*. Iranian Remote Sensing Center – Tehran, 2003.

16. Tarikhi, Parviz: Iran's ambitions in space. *Position Magazine*, Issue 35, June–July 2008.
17. Tarikhi, Parviz: *New Statute for ISA – More Confinement or More Freedom?* Tehran, 11 September 2008, http://rescommunis.wordpress.com/2008/09/11/guest-blogger-parviz-tarikhi-new-statute-for-isa-more-confinement-or-more-freedom/ (accessed 11 March 2009).
18. Brochure Aerospace Research Institute. Tehran, February 2009.
19. Behnam, Farjam: *Iran Almanac and the Book of Facts 2007–2008*, p. 273, ISBN 978-964-2865-16-1.
20. Aerospace Research Institute, www.ari.ac.ir/images/stories/introduction.pdf (accessed 28 March 2009).
21. Private conversation between author and source. Tehran, February 2009.
22. Inter-Islamic Network on Space Sciences and Technology, www.isnet.org.pk/ (accessed 11 March 2009).
23. Asia Pacific Space Cooperation Organization, www.suparco.gov.pk/pages/apsco.asp (accessed 11 March 2009).
24. Regional Space Applications Program for Sustainable Development, www.unescap.org/ (accessed 11 March 2009).
25. UNISPACE III, www.un.org/events/unispace3/ (accessed 11 March 2009).
26. Sheldon, John B.: A really hard case – Iranian space ambitions and the prospects for U.S. engagement. *Astropolitics*, Vol. 4, No. 2, Summer 2006.
27. Iranian military space activities. Secure World Foundation, www.secureworldfoundation.org/index.php?id=117&page=Iran_Military (accessed 30 March 2009).
28. Anonymous: Iran launches satellite. *Washington Jewish Week*, 2009, *HighBeam Research*, www.highbeam.com/doc/1P3-1646152251.html (accessed 1 April 2009).
29. Iran Electronics Industries (IEI), Promotional Item in the *Iran Daily*, February 10, 2008, www.iran-daily.com/1386/3060/pdf/i10.pdf (accessed 1 April 2009).
30. Iran's first space launch vehicle Safir IRILV. Norbert Brügge, Germany, www.b14643.de/Spacerockets_1/Diverse/Safir-IRILV/Safir.htm (accessed 1 April 2009).
31. *Iran News Daily*, 5 February 2008.
32. www.fas.org/nuke/guide/iran/facility/semnan.htm (accessed 30 April 2009).
33. 46th Meeting of the Scientific and Technical Subcommittee of the Committee on the Peaceful Uses of Outer Space (COPUOS), 9–20 February 2009, Vienna, Iranian Space Agency, *Omid Satellite Launch Report*, www.oosa.unvienna.org/pdf/pres/stsc2009/tech-15.pdf (accessed 30 April 2009).
34. Middle East News Line, *Iran Builds First Satellite Ground Station*, 14 January 2009.

8

Iran: Development – space launch systems and satellites

The National Space Science Data Center/World Data Center for Satellite Information recorded the launch of the Iranian Omid satellite in its monthly publication, *Spacewarn Bulletin*, as follows: "Omid, a small communications satellite, was launched by Iran on a Safir 2 rocket on 2nd February 2009 at 18:34 UT. The satellite, whose name means hope in Farsi, consists of an experimental control system, communications equipment, and a small remote sensing payload. Initial orbital parameters are: period 90.7 minutes, apogee 364km, perigee 258km and inclination 55.5°."[1] This was by no means the only reference to this remarkable feat. Every self-respecting news agency and other publishing body on the internet mentioned the successful launch. Most of them speculated also on its importance and consequences. Not only the building of a satellite, but actually getting it into space was important news. Ever since Iran publicly made clear in the late 1990s that it eventually would launch a satellite into orbit indigenously, it was monitored, not to say scrutinized, for information on their progress. Iran, on its part, did its utmost to conceal from the world its development and production of space-related hardware, how it would proceed and what its timetable would be. Since so many different bodies published findings on the internet, to assess what information is correct and what is just hearsay is difficult. Sources refer to each other, obscuring the initial one. Taking this into account, this chapter attempts to describe the development in Iran of the most important assets for spacefaring: launch systems and satellites.

LAUNCH SYSTEMS

Rocketry in Iran has always been a part of its military strategy, from small, rocket-propelled grenades and surface-to-air missiles, most of which were imported from the United States of America and Europe, to later, after the revolution, larger ballistic missiles. The space launch capability and technical knowhow that led to the launch of the Omid satellite are largely based on North Korean missile technology.

The Shahab missiles – the name "Shahab" means meteor or shooting star in Farsi – are the bases for long-range ballistic missiles and ultimately for the space launcher. The Shahab-3 is a medium-range ballistic missile and from this missile, different versions have been developed, of which one has become the space launcher. This "space launcher" missile has been dubbed IRIS, or sometimes Shahab-3C and/or Shahab-3D. The Shahab-3 is a military, single-stage, liquid-fueled, road-mobile, medium-range ballistic missile with a range of approximately 1,200 km. The Shahab is a derivative from the North Korean No-Dong missile. Following official documents and publications, it is possible to construe how Iran developed its space launch capability.

Iranian interest in ballistic missile acquisition can be traced to the war with Iraq in the 1980s. Because Iraq's modified Scud missiles outnumbered and outranged those of Iran, Iran turned to North Korea to supply it with ballistic missiles. North Korea obliged, sending Iran Scud-Bs, 77 of which were fired against targets in Iraq during the second "War of the Cities" in 1988. The missiles provided by North Korea had been reverse engineered from Scuds it had obtained from Egypt in the early 1980s. In the early 1990s, Iran turned again to North Korea to acquire ballistic missiles. Some analysts believe that Iran was involved in North Korea's No-Dong program from its outset in the late 1980s and that it provided substantial funding for its development. By the mid-1990s, Iran had as many as 10 No-Dongs, either in component form or as completed missiles. Over the same period, Iran had also begun to establish the infrastructure that would permit it to produce ballistic missiles within the country, ending its dependence on assistance from abroad.[2] Since mid-1980, there have been reports of the development of intermediate-range ballistic missiles in Iran and the program has been given various names, including Shahab (also translated as Shihab, Shehob) and Zelzal. A Turkish intelligence report said that Iran would purchase and/or manufacture a total of 150 No-Dong missiles.[3]

In 1997, other reports suggested that Iran was conducting a series of motor tests for a missile program and that the missile was called Shahab-3. North Korea supposedly delivered to Iran a small number of missile assemblies and transporter-erector-launcher (TEL) vehicles. In 2002, Iran successfully tested the Shahab-3 and although several tests of the missile appear to have failed, a ceremony was held on 30th July 2003, when the Revolutionary Guards were equipped with it. In 2004, Iran said that it could mass produce the Shahab-3.[4] After the first flight test of the Shahab-3 in July 1998 and the display of two further missiles in September of that year, it took two years for the Shahab-3D to be tested. Iran described this missile as a prototype satellite launch vehicle (SLV) using solid and liquid propellants. This could indicate that the first stage of the SLV was a standard Shahab-3 liquid-propellant motor, with a solid-propellant second stage. Analysts at that time publicly were disagreeing about whether the Shahab-3 could be developed into an SLV or not.

The Aerospace Industries Organization (AIO) produces the Shahab-3 and the final assembly and testing are done by the Shahid Hemmat Industrial Group (SHIG) in Tehran (see Annexes). Most of the component production (missile motors, fuel tanks, etc.) takes place in large underground facilities in the Khojir region. Although

Shahab-3 military, single-stage, liquid-fueled, road-mobile, medium-range ballistic missile [A. Kholoosi]

the Iranian government has not identified specific missile production plants or sites, it has publicly acknowledged the existence of such facilities. In September 2002, an Iranian news organization reported on a public ceremony inaugurating three facilities to produce missiles and that the AIO manages these facilities. Reportedly, the AIO manages a number of factories and research centers, including: the Missile Center of Saltanat-Abad, the Vanak Missile Center, the Parchin Missile Industries factories, the Baqeri base factories Numbers 1–3, the Tabriz Bakeri base factory, the Bakeri Missile Industries factory, the Hemmat Missile Industries factory, the Bagh Shian (Almehdi) Missile Industries, the Shah-Abadi Industrial Complex, the Khojir

Shahab-3 medium-range ballistic missile in military parade in Tehran [Reuters]

Complex, the Baqerololum Missile Research Center, the Mostafa Khomeini base factory, and the Quadiri Base factory.[5]

The first space-related launch occurred on 4th February 2008. With much media attention, the *Kavoshgar-1* was launched from a missile site in the Semnan province on a suborbital trajectory. Some analysts and observers suggested that a launch on 25th February 2007 was to be earmarked as Iran's first space-related launch. That day, Iranian TV (Shia News Agency) broke the news of this launch, saying "the first space rocket has been successfully launched into space". It quoted the head of Iran's Aerospace Research Institute, Mohsen Bahrami, as saying that "the rocket, dubbed Kavoshgar, carried material intended for research created by the ministries of science and defense". However, Ali Akbar Golroo, executive director of the same institute, was later quoted by the FARS news agency as saying the craft launched was a suborbital rocket for scientific research. "What was announced by the head of the research center was the news of launching this sounding rocket," Mr Golroo said. It would not remain in orbit but could rise to about 150 km before a parachute-assisted descent to Earth. The controversy whether it was a space launch or another missile test arose because no pictures of the reported launch have been shown on Iranian state TV. Furthermore, the Russian news agency, Novosti, quoted a source in the Russian Ministry of Defense as saying that the Russian radar station in Gabala, Azerbaijan, did not detect any signal of a space rocket launch from Iranian territory. American intelligence sources also strongly implied that there had been no Iranian launch.[6, 7]

Kayhan News (Feb. 5, 2008)

کیهان

تولید گندم از مرز ۱۵میلیون تن گذشت

اعتراض مردم بحرین به دستگیری مخالفان سیاسی

با حضور رئیس جمهور و پرتاب موشک کاوشگر یک

اولین سفیر ایران به فضا رفت

فهرست جبهه متحد اصولگرایان برای تهران نهایی شد

انهدام کاروان حامل ۵تن مواد مخدر در خراسان جنوبی

افشای دیدار وزیر خارجه قطر با وزیر جنگ رژیم صهیونیستی

Resalat (Feb. 5, 2008)

All Iranian newspapers carried the news about the *Kavosgar-1* launch on 4th February 2008. [Kayhan News, Resalat]

The launch on 4th February 2008, on the other hand, received immense media attention on television as well as in the written press. Even the homepage of the Presidency of the Islamic Republic of Iran had tens of photographs. Not only the Western press picked up the story rapidly, but also the Chinese and Russian. The launching of the rocket came during the celebrations of the 29th anniversary of the Islamic Revolution (February 1979) and the event was attended by President Mahmoud Ahmadinejad, several ministers and a number of senior officials. Government spokesman Gholam-Hossein Elhan referred to the launch as a "great success" and an indication of the self-belief of the Iranian nation.[8] PRESSTV showed an image of a launch vehicle with a blue base and a blue pointed nose and called it the *Kavoshgar-1* (Explorer-1) sounding rocket or suborbital research rocket. This launch vehicle/missile was very much like the Shahab-3B ballistic missile and there was also much similarity between the Transporter Erector Launcher vehicle of the *Kavoshgar-1* and a Shahab-3B. However, FARS News Agency and the *Iran Daily* showed pictures of a rocket with a white base and with a blunter-tipped blue top while referring to the Kavoshgar launch. The launch vehicle showed a supporting cradle that did not reach all the way to the top of the launch vehicle as is the case with a Shahab-3B type of missile, so therefore it must be longer. This launch vehicle was in fact a mockup of the Safir space launch vehicle believed to be erected at the ISIG center in Tehran.

The *Kavoshgar-1* sounding rocket is considered to be a derivative of the Shahab-3 and to be a liquid-propellant rocket. On 19th February 2008, Iran announced that the space probe it had sent on the back of the rocket reached an altitude of 200 km and returned to Earth after some minutes. Contrary to earlier Iranian news agencies' reports that suggested that the *Kavoshgar-1* was a three-stage rocket, "the launch vehicle had two sections of which the first separated after 100sec and returned to earth with a parachute. The second section continued to an altitude of 200km," said the head of Iran's space agency, Ahmad Talebzadeh. He added that the second section of the rocket received data on the atmosphere and the electromagnetic waves on its path and simultaneously made contact with the base and returned to Earth with a parachute after five to six minutes.[9, 10] The *Kavoshgar-1* sounding rocket bore the logo of the Aerospace Industries Organization, manufacturer of the Shahab ballistic missiles.

On 16th August 2008, Iran test-fired a Safir launch vehicle, which, state television said, was capable of putting a light satellite into low Earth orbit between 250 and 500 km. The nighttime launch of the two-stage *Safir-e Omid* (*Ambassador of Hope*) rocket was claimed to be a resounding success, be it that the launch vehicle was launched in a sounding rocket mode. Television showed footage of the rocket launch, saying that the Safir was about 22 m long, with a diameter of 1.25 m, and that it weighed more than 26 tonnes. After analyzing the television images, it is assessed that the first stage of this Safir launch vehicle was a liquid-propellant Shahab-3 (derivative). The transition from the first stage to the second stage appeared to have been successful. "The Safir satellite carrier was launched today and for the first time we successfully launched a dummy satellite into orbit," Reza Taghipour, head of the Iran Space Agency, told state television.[11] This premature

The Safir Islamic Republic of Iran Launch Vehicle (IRILV) at the ISIG center in Tehran is longer than a Shahab-3. This launch vehicle is, confusingly, also named Kavoshgar by some media [*Iran Daily*]

This television grab shows the *Kavoshgar-1* sounding rocket [IRINN]

These television grabs show the *Kavoshgar-1* sounding rocket [IRIB Daily News][10]

media claim was later withdrawn. Actually, this launch was the first time Iran (successfully) launched a two-stage SLV and this could be said to be a leap in technology. But, again, there was disagreement as to whether the launch was

This television grab shows the Safir-1 launch vehicle, launched on 16th August 2008. Mark the blunt nose section that differs from the nose sections of ballistic missiles [Al Alam News Channel]

successful. Iran state television aired only a few seconds of the launch[12] and there is no further information to support the statement. On the other hand, an unnamed US official stated that "The attempted launch failed. The vehicle failed shortly after liftoff and in no way reached its intended position. It could be characterized as a dramatic failure".[13] However, the official failed to give information about how the United States could know the launch was a failure. If this launch was a failure, then certainly the Safir launch on 2nd February 2009, with the Omid satellite on board, was a successful two-stage SLV launch.

A second Kavoshgar sounding rocket was launched on 26th November 2008. The *Kavoshgar-2* payload landed 40 min later with a special parachute. A reporter on Iranian state television showed a graphic of the rocket followed by around 10 sec of footage of it blasting off .[14] The rocket looked much like the 17-m-tall Shahab-3 ballistic missile and not like the 22-m-tall Safir space launch vehicle. According to the media, the main goals of the launch were to carry out space experiments on the atmosphere, to test the retrieval system (parachute) and coordination between the scientists in an academic project. It is assumed that the scientific experiments were developed by the Aerospace Research Institute because their logo was on the payload section. State news agency IRNA said the launch was in line with Iran's strategic space program and prepared the grounds for further scientific and technological progress.

With the launch of the *Omid* research satellite on 2nd February 2009, Iran joined the select group of countries or agencies that were able to launch satellites into space indigenously, marking the 30th anniversary that month of the 1979 Islamic revolution. The two-stage Safir space launch vehicle lifted off from the Semnan launch pad at about 18.30 UTC in a south-easterly direction over the Indian Ocean

The Safir space launch vehicle being readied for its launch on 16th August 2008 [ISNA/V. Alaee]

The Safir space launch vehicle being readied for its launch on 16th August 2008 [ISNA/V. Alaee]

to avoid flying over neighboring countries. It brought the satellite in a low Earth orbit (258–364 km) with an inclination of 55.5°. Many video clips and photographs were released and Western analysts had a field day in scrutinizing them. Luckily, President Ahmadinejad liked to have his photo taken with used and exhibited hardware. From such photography, we could deduce that the first stage of the Safir space launch vehicle was indeed derived from Scud/No-Dong technology. This first stage was capable of bringing the second stage and payload to an altitude of about 68 km. The second stage might have been developed indigenously and propelled by two smaller, gimbaled engines. This stage performed angle corrections and inserted the 27-kg payload into a low Earth orbit. The total length of the launch vehicle was about 22 m, the overall diameter 1.25 m and the mass some 26 tonnes.

The payload section of the *Kavoshgar-2* sounding rocket, bearing the logo of the Aerospace Research Institute [Unknown]

This television grab shows the payload section of the *Kavoshgar-2* sounding rocket after retrieval with a parachute [Al Alam News Channel]

The Safir-2 space launch vehicle being readied for its launch with the Omid satellite on 2nd February 2009 [jamejamonline.ir]

The Safir space launch vehicle layout [N. Brügge]

SATELLITES

On 27th October 2005, Iran became the 43rd country to have a satellite in orbit when Sina-1, together with at least seven other satellites, was launched on a Cosmos-3M rocket from the Russian Plesetsk space launch center and achieved orbit. Apart from the early plans (1977) to set up an Iranian satellite communications system (Zohreh), there were apparently no records of the design and production of satellites at Iranian ministries prior to 1996. During that year, a number of researchers from the Iranian Research Organization for Science and Technology (IROST) – affiliated to the Ministry of Science, Research and Technology – were given the task of preparing a

Safir-1 in storage/assembly hall [FARS/V.R. Alaei]

space project. A plan to design and manufacture a research satellite, Mesbah, was drawn up and related research projects began: Iran was to pursue space technology to be able to jointly develop (originally with Russia, later with Italy) a satellite for research purposes. The name for the project, Mesbah, was not announced until 1999. From the start, there was the discussion of whether this satellite would be military or civil, forgetting that almost all satellites are dual-use and that military advantage only would emerge when rudimentary satellites improved considerably. Of course, one should not neglect to note that in Iran, the amount of finance that is devoted to the non-civilian space industries is considerably more in comparison with the investment in civilian space industries that is mainly allocated by the Ministry of Science, Research and Technology, the Ministry of Communications and Information Technology and the Islamic Republic of Iran Broadcasting (IRIB) organization.

It is also clear that the non-civilian sector puts its efforts and forces mainly on developing space transportation systems, while the main focus of the civilian sector is on developing satellite and communication systems as well as systems for remote sensing and navigation.[15] Now follows a description of Iranian satellites.

SINA-1

The first Iranian satellite, Sina-1 (Explorer of knowledge) or Z-S.4, is owned by the non-civilian sector and its data and imagery have not yet been made accessible, even to ISA.[15] According to official announcements, Sina-1 is a research satellite whose mission is remote sensing (on natural disasters) and communications. On 1st May 2009, its orbit parameters were perigee 681 km, apogee 704 km, inclination 98.06° and orbital period 98.63 min (a Sun-synchronous, near-polar orbit). The other payloads in this launch were Mozhayets (Russia), TopSat (Britain) and Tsinghua (China). The SSETI Express (ESA) released three 1-kg satellites: CubeSat XI-V (Japan), Ncube 2 (Norway) and UWE-1 (Germany).

The approximately 160-kg satellite was launched on 27th October 2005 at 06.52 UTC from the Plesetsk cosmodrome in northern Russia with a Cosmos-3M space launch vehicle (it received the catalog codes 28893 and 2005-043D). The box-type satellite with four folding solar panels had two cameras and communication equipment on board. The one camera offered 50-m resolution in panchromatic mode, with a 50-km swath width. The other in a multispectral scanning mode has a resolution of 250 m, with a 500-km swath width.

Sina-1 was developed by the Russian Polyot design bureau that belongs to the Polyot Production Association from Omsk, which also builds the Cosmos-3M space launch vehicles. The development is based on the design of the Sterkh satellite bus. The rocket's launch was first scheduled for 30th September 2005 but repeatedly delayed because of production difficulties with Sina-1. Because of the hectic rush around this launch, technicians forgot to re-code the communications module and left it with an RS-25 call sign, which can be received by amateur communications systems. The downlink of Sina-1 is on 435.325 or 435.345 MHz (CW beacon) and FM data. The telemetry gives information on housekeeping, different temperature readings and power consumption.[16] The current status of the satellite is kept secret, but as the Sterkh satellite bus has an average lifetime in space of about three years, it is considered that the Sina-1 satellite ended or is at the end of its operational life.

On 10th April 2006, Ebrahim Mahmoudzadeh, director of Iran Electronics Industries, said that Sina-2, also named Pars, would be launched in two years.[17] "We reached an agreement with Iran's Aerospace Organization to produce Iran's second satellite," he said.

SMALL MULTI-MISSION SATELLITE (SMMS)

On 8th September 2008, telecommunications Minister Muhammad Suleimani said on Iranian television that Thailand has helped Iran to put a joint research satellite

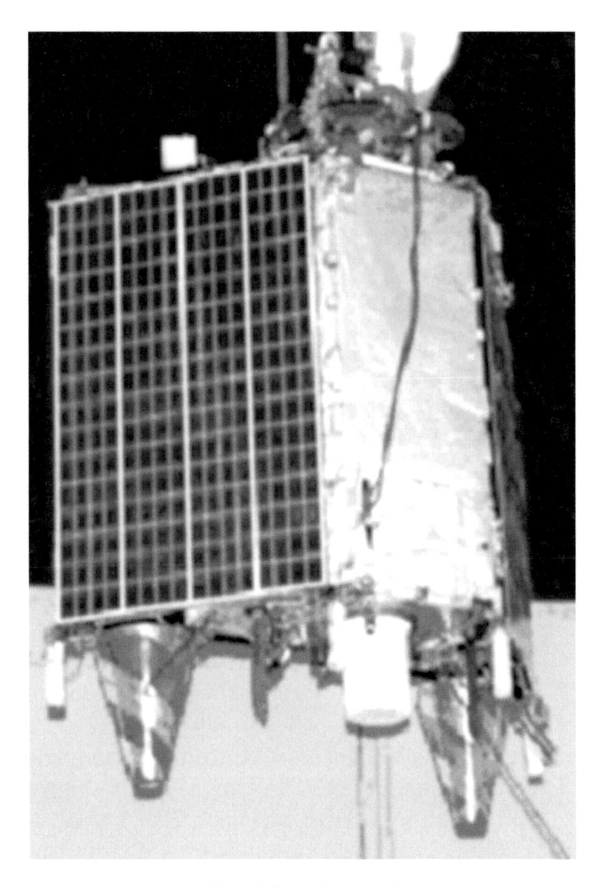

Sina-1 [Roskosmos]

into space, piggyback atop a Chinese SLV. "Several countries contributed to the design and construction of the rocket, but Iran, China and Thailand completed the project and are entitled to use the data provided by the satellite," Suleimani said.[18] However, the quoted names Small Multi-Mission Satellite, SMMS or SM2S are not in the SpaceTrack database. To confuse things further, the Asia-Pacific Multilateral Cooperation in Space Technology and Applications (AP-MCSTA) called the SMMS in the beginning the Small Multilateral Mission Satellite. Since the SMMS is designed and manufactured in a multilateral framework as the AP-MCSTA, it is considered to be a civilian satellite with which Iran has been involved. SMMS is an international joint venture in cooperation with China and Thailand mainly aimed at disaster and environmental monitoring, civilian remote sensing and communications experiments. It is a medium-class, low Earth polar orbit (650 km), Sun-synchronous satellite weighing 490 kg. Out of the US$44m cost of the satellite, Iran's share for manufacturing and launching it is US$6.5m. Access to an advanced Earth observation satellite could greatly help Iran, especially after a natural disaster. The satellite carries a low-resolution charge-coupled device (CCD) camera and an experimental telecommunications system. Iran contributed to building the CCD sensor. In an undated paper from the Malek Ashtar University of Technology in Tehran, it was said that Iran has been also responsible for the S-band TT&C subsystem of the SMMS. Some of the technologies used to develop the device are expected to enhance Iran's long-term sensor design and manufacturing capabilities. The AP-MCSTA Joint Technical Coordination Meeting of the SMMS project scheduled the launch for 2004 atop a Chinese booster rocket, but later the launch was postponed until 2006. The satellite bus would be a CAST968 from the Chinese Academy of Space Technology.[19]

In his televised comments, minister Suleimani must have hinted at the Chinese

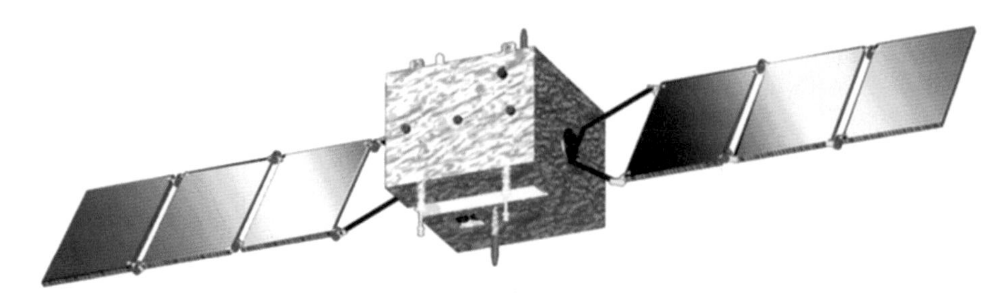

A CAST968 satellite bus [CAST]

launch of two satellites two days earlier, because no other Chinese launches qualify. On 6th September 2008, at 03.25 UTC, China launched two Huan Jing (Environment) satellites (HJ-1A and HJ-1B) with a Long March 2C rocket from the Taiyuan launch site in northern China (they received the catalog codes 33320/33321 and 2008-041A&B, respectively). These satellites – based on the CAST968 satellite bus from the China Academy of Space Technology – carried optical and infrared cameras to monitor natural disasters. The initial orbital parameters of both satellites were similar and were perigee 627 km, apogee 670 km, inclination 98.01° and orbital period 97.6 min.[20] China stated, when these two satellites were commissioned into service on 30th March 2009, that they were part of a constellation of four optical and four radar satellites being launched for all-weather, 24-hr monitoring and environment and natural disasters forecast, forming a complete image of China once every 12 hr. The multi-mission CAST968 bus is a box-shaped satellite with overall dimensions of 0.83 by 0.85 by 1.33 m, weighing 200–300 kg. It can accommodate a payload of up to 300 kg. Electrical power is provided by two solar arrays of 5.67 m^2. The Chinese Ministry of Civil Affairs, in charge of disaster relief and reduction, and the Ministry of Environmental Protection are the users of the two satellites, which both have CCD cameras for 30-m resolution imagery at a 720-km swath width. The satellite lifetime is three years or more.[21, 22] As China did not name one of those two satellites SMMS and it did not even refer to this project when launching those satellites, it must be assumed that the SMMS project as such no longer exists.

MESBAH

Although the Mesbah project was jointly implemented by IROST and the Ministry of CIT at first, the responsibility for conducting and advancing the project was laid down with ISA the moment this agency was established in 2004. The Space Technology Group of the ECEDEP carried out the design and development of the Mesbah satellite in cooperation with the Ministry of CIT. Reportedly, this group currently follows the plan for system engineering design for the follow-on Mesbah for ISA.[15] Because of the project's high cost, which was beyond the budget of IROST, no significant breakthroughs were initially made. In 1997, it was decided that the project should be pursued in a joint venture with the Ministry of CIT. On 27th June 1998, an agreement was reached and the Mesbah project entered a new phase. Finally, in the summer of 2005, the project to build this satellite at a price of some $10m was completed and on 3rd August 2006, the satellite was displayed publicly for the first time. The satellite was then sent to Italy for final tests and in December 2006, it reportedly was transported to Russia to be launched into space aboard a Cosmos launch vehicle. For unknown reasons, Mesbah was, by the time of writing, never launched.

The Mesbah (Lantern) micro-satellite is a store-and-forward communication satellite basically aimed at attaining know-how in the process of design, assembly and expansion of international cooperation. The system initially was said also to be intended to obtain images for a variety of civilian purposes, to include larger data collection and distribution, assisting in efforts to find natural resources and predict the weather. The project was implemented in cooperation with the Italian Carlo Gavazzi Space Company (CGS). In September 2005, the managing director of CGS, L. Zucconi, held a presentation in Shiraz in which much more became clear about this satellite: "Iran set up the Mesbah satellite project with the following main objectives:

- Enable Iranian access to space by developing, launching and operating a small low Earth orbit communication satellite for data collection/distribution, e-mails and store & forward services;
- Reinforce the Iranian telecommunication infrastructures by exploiting the capability of the Mesbah satellite of connecting remote villages and rural areas;
- Enhance the Iranian educational and scientific knowledge through practical involvement in space activities especially in the field of satellite communications."

There was no mentioning (anymore) of remote sensing. CGS worked in close cooperation with ITRC/IROST in the design, development and manufacturing of the Mesbah system, which is composed of a space segment, a ground segment and user terminals.

With dimensions of 0.7 by 0.5 by 0.5 m, the satellite weighed 65 kg. It had gravity gradient stabilization and a VHF/UHF TT&C package. The payload was a UHF receiver/transmitter.

Mesbah would orbit at 900-km altitude (Sun-synchronous) and be controlled from a ground station located in the Iran Telecommunications Research Center (ITRC) in Tehran, operated by ITRC/IROST personnel; the backup station would be operated by the CGS in Milan, Italy. When launched, Mesbah would orbit the Earth 14 times a day. Four times within 24 hr, there would be line-of-sight with ground stations in Iran. A three-year lifetime was expected, but it would be capable of continuing operation for up to five years.[23]

Mesbah at the Carlo Gavazzi Space Company [CGS]

Mesbah at the Carlo Gavazzi Space Company [CGS]

ZOHREH

Zohreh (Venus) is a satellite planned to meet Iran's telecommunication needs since the 1970s. Although its manufacturing has not been completed, it is still an active concern of the Iranian authorities, particularly in the telecommunications domain. They want to manufacture and launch into geostationary orbit a satellite capable of providing numerous services, including television and radio broadcasts, telephone and fax, and data transmission and internet/e-mail. France and Germany, as well as China, have been party to different aborted contracts with Iran to manufacture the satellite. In 1998, Iran entered into negotiations with Russian organizations. These contacts were suspended in 2003 and renewed in late 2004 – probably with different organizations in Russia. In January 2005, it was announced that a contract had been signed between the Russian Avia Export Company and Rosaviakosmos (the Russian space agency), and the Iran Telecommunication Company [24]. In the new version of the Zohreh contract, the project is reported to cost US$132m – much less than the US$300–350m price of the previous transaction. Delivery of the satellite in orbit was to be by mid-2007, but in March 2006, a report by the Russian news agency, Itar-Tass, indicated that the launch of the Iranian Zohreh satellite could be delayed owing to the unavailability of certain Western-made spacecraft components. The head of Russia's Federal Atomic Energy Agency, Sergei Kiriyenko, said there had so

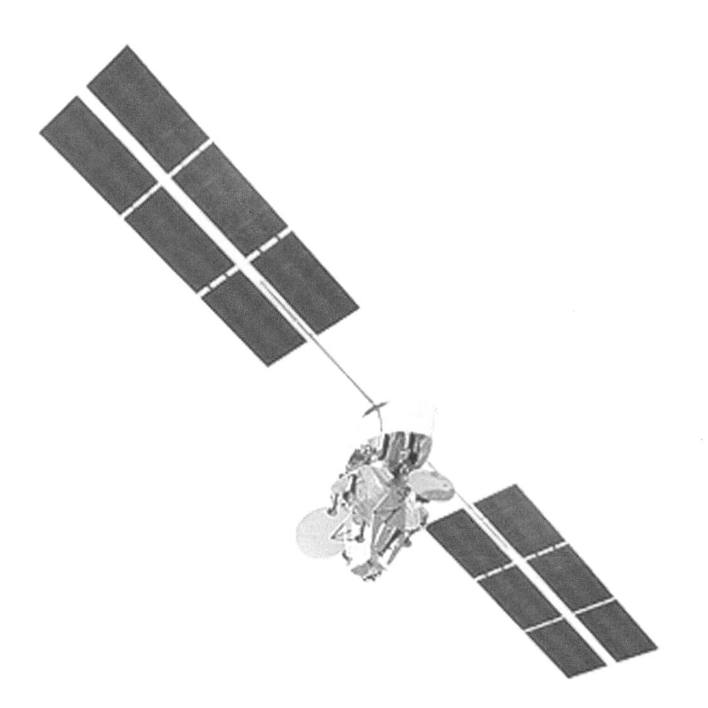

Ekspress-AM-type geostationary communications satellite that possibly will be used for the Zohreh satellite [NPO PM]

far been "no news from France and Germany", which were supposed to supply an onboard computer and a star sensor for the satellite. Mr Kiriyenko told Iranian Minister of Economics and Finance, Davoud Danesh-Ja'fari, that Russian specialists "are ready to make this equipment, but Iran must make a decision". The satellite would be based on the Ekspress bus, with a lifetime expectancy of 12 years. Subcontractor for the 12 transponder payload module was Alcatel (Alenia) Space. Launch services were to be provided by Russia's TsENKI (Center for Operation of Space-related Ground Infrastructure Facilities with Roskosmos).[25] Then, on 11th December 2006, a small article in *Iran News* reported that Zohreh could be launched no earlier than 2009 and not in 2007. "The contract for the satellite's launch is signed, but it has not taken effect yet. Under the agreement, the satellite can begin to be produced only when the contract is effective," the deputy chief of Russia's space agency, Roskosmos, Yuri Nosenko, told Itar-Tass in an interview.[26] If this contract is carried out, the launch date still could be set for later this decade.

OMID

After more than 1,300 orbits around the Earth, the Omid satellite burned up in the atmosphere on Saturday 25th April 2009. It is estimated that the satellite decayed at

46.4°S, 324.9°E. ISA's official website (Farsi version) announced on 30th March 2009, quoting Mr Asghar Ebrahimi, the Omid project's coordinator, that the satellite's mission was ended and it would remain in orbit for about one more month. Moreover, Fars News Agency (in Farsi), quoting the words of Mr Fathullah Ommi, the observer of the Omid project, announced that the 50-day mission of Omid would conclude on 24th March 2009. Presumably, Omid ceased operations after 50 days in orbit because its batteries ran out of power. The mission of Omid was to act as a store-and-forward telecommunication satellite and, according to Iranian reports, it was successful.[15]

The plan to launch Omid was announced at the inauguration of Iran's space complex on 4th February 2008, when the Kavoshgar-1 sounding rocket was launched. FARS News Agency published numerous photographs of President Ahmadinejad and other dignitaries being shown around. It was reported that Omid would be launched in the "next Iranian calendar year", which started on 20th March 2008. Omid was eventually launched on 2nd February 2009 at 18.34 UTC with a Safir Space Launch Vehicle (SLV) in a southern direction, leaving Iran airspace on a southbound trajectory, taking it over the Arabian Sea and Indian Ocean. It received the catalog codes 33506 and 2009-004A. The initial orbital parameters of the satellite were perigee 258 km, apogee 364 km, inclination 55.5° and orbital period 90.7 min. SpaceTrack also recorded the orbital parameters of the second (final) stage of the Safir SLV.

The news of the successful launch of Omid led to many different reactions and the news agencies had a field day. Trying to listen in to Omid were the amateur space observers and amateur radio operators. For the downlink frequency, Iran had registered 464.9875–465.0125 MHz, for the command uplink 401.000–401.025 MHz. It was the downlink that amateur radio operators would try to receive and did. At Sven's Space Place[27] and Robert Christy's Zarya,[28] excellent analysis can be found.

Omid is a square box with 0.4-m sides and it weighs 27 kg. From two faces of the

Folded-out view of Omid [ISA]

Omid satellite in launch configuration [ISA]

box, two sets of four antennae protrude. This secures communications at all times with a spinning box. Around the periphery of the box, there are patch aerials for the reception of GPS signals at all times. The box is covered with some kind of insulating material for passive thermal control and does not have solar cells. The Omid satellite box is constructed in an isogrid way, probably from aluminum. Inside, there are three packs of batteries, a power distribution unit, an onboard computer, a GPS receiver and control system, a receiver and a transmitter.[29]

IRAN: PLANS – THE ULTIMATE GOAL

Iran never failed in its plans for a space program, but history often caught up with those plans and longtime satellite projects such as Zohreh, SMMS and Mesbah have not yet seen a successful launch. Now that the main goal has been achieved, becoming a member of that select club of countries that have launched homemade satellites with indigenously built launch vehicles, it is time to look ahead. A follow-on satellite has to have a purpose other than to "prove you can do it". There has to be an application for it, be it commercial, scientific or military. Building and launching satellites are not cheap and the country is in no position to waste money on daydreaming. Government officials often make public announcements of satellite projects, but the details are kept so secret, for no obvious reasons, that they almost disappear from memory. But if they are so secret, why mention them in the first place?

BESHARAT

At the inauguration of the Iranian space terminal (underground control station and launch pad) on 4th February 2008, Iran's Minister of Defense, Mostafa Mohammad-Najjar, said that Iran planned to build a satellite that will be called Besharat, in collaboration with Islamic countries. On 21st August 2008, the head of ISA, Reza Taghipour, confirmed this, saying that the country had plans to design and build the satellite in collaboration with the Organization of Islamic Conference (OIC).[30] On 18th February 2009, the *Iran Daily* reported that the "Head of Iranian Space Agency Reza Taghipour on Tuesday said feasibility studies for design and production of 'Besharat' satellite have been completed". "If the specifications of the satellite are approved by the Organization of Islamic Conference (OIC) states, we will start its construction," Taghipour told Mehr News Agency. He recalled that the satellite was meant for remote sensing and will be used for missions such as assisting management of natural resources, managing natural disasters, meteorology and mapping. The need to implement the project within the framework of OIC was re-emphasized by the official. "Since aerospace technology is expensive, most countries design and build satellites with the participation of other countries. As information from space can be made available to many countries, in design and production of 'Besharat' we will use cooperation from Muslim states so that at the end of the project its information and data can be made accessible easily to OIC members," he concluded.[31] The OIC is an international organization grouping 57 states that have "decided to pool their resources together, combine their efforts and speak with one voice to safeguard the interests and secure the progress and well-being of their peoples and of all Muslims in the world". When asked for a reaction from the OIC, the author received no answer.

OTHER SATELLITES IN THE MAKING

Iran's Minister of Communications and Information Technology, Muhammad Suleimani, has said that Iran was constructing four more satellites, the Mehr News Agency reported on 8th February 2009. "The details of the satellites will be disclosed step by step at their final stage of preparation," Suleimani said, adding that "With the capabilities attained, we are trying to raise the weight and the altitude of the satellites to be launched". Four days later, Suleimani said that Iran was building seven new satellites, including three for altitudes lower than 36,000 km. Four low-orbit satellites weighing less than 100 kg were under design and construction and all seven were being built by Iranian scientists.[32] Asked for comments, it was said "that details of Iran's new satellites announced by the Minister of CIT were not yet declared. However, the light-weight, small satellites were going to be designed and produced by universities for research and exploration purposes. The heavy-weight satellites were the ones which being designed by the state organizations and bodies and will be used for Earth observation, environmental monitoring, telecommunications, broadcasting and data transmission".[33]

MANNED SPACE

For decades, manned space missions used to be the monopoly of American and Russian astronauts, with astronauts from other countries occasionally taking part. China now has its own manned space program as well, but there are no other countries able to launch a human in space and bring him safely back to Earth.

Iran has kicked off a 12-year project to send an astronaut into space, just days after putting its first home-built satellite into orbit, Iran's English-language satellite news channel Press TV reported on 12th February 2009. One day earlier, Reza Taghipour said "The program's preliminary needs, assessments and feasibility studies have been carried out. The Aerospace Organization [ISA] had drawn up a comprehensive plan for the project and various academic and research institutions must play to carry out a successful space mission by 2021. China and India managed to send an astronaut to space in a 15-year program. We see ourselves taking the same path, but we hope to reach that goal in a shorter period".[34]

Of course, there is always another way to get into space. Anousheh Ansari spent her childhood in Iran dreaming of venturing into space and that dream came true when this businesswoman with American citizenship boarded a Russian Soyuz spacecraft. After immigrating to the United States in her teens, she earned a bachelor's degree in electronics and computer engineering and a master's degree in electrical engineering. The telecommunications tycoon from Texas became the fourth person – and the first woman – to pay an estimated US$20m to travel into space. She went on a 10-day journey to the International Space Station as part of a crew exchange flight. Ms Ansari was wearing both the American and Iranian flags on her spacesuit in training. She said she wanted to recognize both countries' contributions to her life.[35]

Anousheh Ansari became the first Muslim woman in space and the first woman to pay for her own space trip [http://alianamirza.files.wordpress.com/2007/10/ansari.jpg]

A MORE POWERFUL SPACE LAUNCH VEHICLE

Building heavier satellites is not a problem; getting them into space is. Iran's President Mahmoud Ahmadinejad said, on 14th April 2009, that his country was constructing a new rocket that could carry heavier satellites. Iran was constructing "a rocket that can go 700km to 1,500km up which will carry heavier satellites", Ahmadinejad was quoted as saying.[36] There is quite a difference between launching a 27 or 500-kg satellite into low Earth orbit, let alone into a 740-km Sun-synchronous orbit (especially from inside Iran). Iran will need much more powerful rockets than the current Safir type and it is doubtful whether Iran will be able to master the technologies needed and will have the resources to build such space launch vehicles within the next 10 years.

- According to Reza Taghipour, President of the Iranian Space Agency, "entering to the club of countries which use their own launching and satellite-building capabilities has been the goal that Iran invested US$ 500m in the course of its fourth Five-year Development Plan".
- According to item 9 of the main policies of Iran's fourth Five-year Development in cultural, scientific and technological issues Plan (issued on 2nd December 2003), it was necessary that the facilities and capabilities of the country be organized and mobilized to increase Iran's contribution to

the world scientific production: Iran should commit to the software revolution and promote research. In the meantime, the country should reach new technologies, including nanotechnologies, biotechnologies, information and communication technologies, environmental technologies, as well as aerospace and nuclear technologies. At the end of the Twenty-year Vision Decree, in 2025, it is expected that Iran will become the number one country in applications and development of space technology in South-West Asia (including Central Asia, Caucasus, the Middle East and neighboring countries).

- On Tuesday 27th February 2007, Fars News Agency, quoting the Minster of Communications and Information Technology, Muhammad Suleimani, reported "Investment in space is very serious and requires time, but we are trying to speed this up". According to Suleimani, Iran hoped to launch four more satellites by 2010 and to increase the number of land and mobile telephone lines to from 22m to 80m. It also hoped to expand its satellite capabilities to increase internet users from 5.5m to 35m in the next five years.

PLANNING SPACE GOALS

Iran's fourth Five-year Development Plan ends in 2009. High priority has been given to space technology applications as the effective tool for the sustainable development of the country. In the meantime, according to the Twenty-year Vision Decree of the country that was issued on 4th November 2002, by the leader of the Islamic Republic of Iran, potentials and capacities throughout the country should be focused to increase Iran's contribution to global scientific production. In this connection, Iran should gain access, in particular, to new technologies, including nanotechnology, biotechnology, information and communication technology, environmental technology, and aerospace and nuclear technology. At the end of this Decree, in 2025, it is expected that Iran will have become the number one country in applications and development of space technology in the Middle East region. Satellite-based remote sensing is one of Iran's top priorities. The idea of having self-owned satellites to secure the needs of the country for remote sensing data in addition to other demands, including communications and broadcasting, is also considered important. International professional cooperation and exchange are the key factors in developing satellites. This always has been regarded as important and vital, while it is hoped that there will be follow-up in the future.[37]

REFERENCES

1. National Space Science Data Center/World Data Center for Satellite Information, *Spacewarn Bulletin 664*, http://nssdc.gsfc.nasa.gov/spacewarn/spx664.html (accessed 14 April 2009).
2. Iran's ballistic missile and weapons of mass destruction programs. Hearing before the International Security, Proliferation, and Federal Services Subcommittee of the Committee on Governmental Affairs, United States Senate, 21 September 2000.
3. United Press International Article. 15 May 2002, HighBeam Research, www.highbeam.com/doc/1G1-86216670.html (accessed 16 April 2009).
4. Dareini, Ali Akbar: Iran says it's capable of mass producing Shahab-3 missile. Associated Press, 9 November 2004.
5. www.ncr-iran.org/ (accessed 16 April 2009).
6. BBC News: Iran rocket claim raises tension, http://news.bbc.co.uk/1/hi/world/middle_east/6394387.stm (accessed 30 April 2009).
7. Hilal, Khalid: Iran's announcement of a space rocket test: fact or fiction? http://cns.miis.edu/pubs/other/wmdi070404.htm (accessed 30 April 2009).
8. *Iran News Daily*, 5 February 2008.
9. Press TV: Iran provides space launch info, www.presstv.com/detail.aspx?id=43849§ionid=351020101 (accessed 30 April 2009).
10. *www.youtube.com/watch?v=glujfPnD4OE&NR=1* (accessed 30 April 2009).
11. Hafesi, Parisa: Iran says it has put first dummy satellite in orbit, www.reuters.com/article/worldNews/idUSHAF75296620080817 (accessed 30 April 2009).
12. www.youtube.com/watch?v=7kt0bEAoARw (accessed 30 April 2009).
13. Reuters: Iran satellite launch a failure – U.S. official, www.reuters.com/article/topNews/idUSN1935578420080819 (accessed 30 April 2009).
14. www.youtube.com/watch?v=A6NCXUlQaKg (accessed 30 April 2009).
15. Private conversation between author and source. Tehran, February 2009.
16. Sina-1 mit RS-25 Modul, www.dk3wn.info/sat/afu/sat_sinah.shtml (accessed 30 April 2009).
17. Reuters: Sina-2 set for early 2008 launch, www.iran-daily.com/1385/2534/html/economy.htm (accessed 30 April 2009).
18. Bangkok Post Article: Thailand, Iran, China launch research satellite for non-military purposes, 8 September 2008, www.bangkokpost.co.th (accessed 20 October 2008).
19. Detail Design of TT&C Transponder for SMMS satellite, www.dlr.de/iaa.symp/Portaldata/49/Resources/dokumente/archiv5/1311P_Payamifard.pdf (accessed 30 April 2009).
20. ribs SC&I/DB&C Databases, Breda, The Netherlands, 2008.
21. www.cast.cn/CastCn/Show.asp?ArticleID=15050 (accessed 30 April 2009).
22. www.cast.cn/CastEn/Show.asp?ArticleID=30801 (accessed 30 April 2009).
23. Zucconi, L.: *The Mesbah Project*, www.itrc.ac.ir/ist2005/Keynote/K5/MESBAH_Conf.pdf (accessed 30 April 2009).

24. Bawaba, Al: *Iran and Russia Sign "Zohreh" Satellite Deal*, 31 January 2005, *www.highbeam.com* (accessed 30 April 2009).
25. Itar-Tass: No Western components for Iran's Zohreh? 1 March 2006.
26. Iran News: Iran's Zohreh satellite to be launched in 2009, 11 December 2006, www.iranian.ws/cgi-bin/iran_news/exec/view.cgi/24/19451 (accessed 30 April 2009).
27. www.svengrahn.pp.se/trackind/OMID/OMID_signals.htm#AntennasBox (accessed 30 April 2009).
28. www.zarya.info/Tracking/Omid/Omid.php (accessed 30 April 2009).
29. 46th Meeting of the Scientific and Technical Subcommittee of the Committee on the Peaceful Uses of Outer Space (COPUOS), 9–20 February 2009, Vienna, Iranian Space Agency, *Omid Satellite Launch Report*, www.oosa.unvienna.org/pdf/pres/stsc2009/tech-15.pdf (accessed 30 April 2009).
30. Taghipour, Reza: Iran, OIC to build satellite "Besharat", Iranian Space Agency, 21 August 2008, www.payvand.com/news/08/aug/1223.html (accessed 5 May 2009), www.presstv.com/detail.aspx?id = 67190 (accessed 5 May 2009).
31. *Iran Daily*: "Besharat" studies complete, 18 February 2009, www.iran-daily.com/1387/3344/pdf/i1.pdf (accessed 5 May 2009).
32. Fars News Agency: Iran building 7 more satellites, 12 February 2009, http://english.farsnews.com/newstext.php?nn = 8711240734 (accessed 5 May 2009).
33. Private conversation between author and source. Breda-Tehran, May 2009.
34. Middle East News: Iran to send astronaut to space, build 7 new satellites: defying odds, 13 February 2009, www.esinislam.com/News200902/MiddleEastNews/MiddleEastNews_0213.htm (accessed 5 May 2009).
35. BBC News: Anousheh Ansari: a passion for space travel, 15 September 2006, http://news.bbc.co.uk/1/hi/sci/tech/5345872.stm (accessed 5 May 2009).
36. IRNA: Iran constructing rocket carrying heavier satellites, 14 February 2009, http://satellite.tmcnet.com/news/2009/04/14/4131203.htm (accessed 5 May 2009).
37. Unofficial translation by Dr Parviz Tarikhi of: Iran's twenty-year vision decree, issued by the leader of the Islamic Republic of Iran on 4th November 2002, 14 May 2009.

9

Brazil: Origins – the road to space

In 1957, the International Geophysical Year, two students from the Technical Institute of Aeronautics (ITA) in São José dos Campos, in the Brazilian province of São Paulo, wrote a letter to the US Naval Research Laboratory. Fernando de Mendonça and Júlio Alberto de Morais Coutinho wanted to install a device to monitor the signals from the Vanguard Project satellites that were developed at that time. The Naval Research Laboratory accepted their proposal and a Minitrack station was set up in São José dos Campos with the aid of the Institute of Research and Development (IPD), which, like ITA, formed a part of the Brazilian Aeronautics Technical Center (CTA). The students had the Minitrack station ready when the Soviet Union astonished the world by launching the first artificial satellite, Sputnik, on 4th October 1957. Within a week, Mendonça and Coutinho had adapted the Minitrack station to monitor Sputnik's transmissions. In January 1958, they were also able to receive the signals from the first American artificial satellite, Explorer-1.[1] Mendonça and Coutinho can so be associated with all the people who developed Brazil's national space research, from the very beginning.

Almost three years later, in November 1960, Professor Luiz de Gonzaga Bevilacqua, an astronautics enthusiast, attended the first Inter-American meeting on space research in Buenos Aires, Argentina. He was the president of the Brazilian Interplanetary Society (SIB) that was affiliated to the International Astronautical Federation. The meeting took place within a symposium organized by the Argentinean National Committee on Space Research, where the Inter-American Committee on Space Research was created. Here, the following was adopted as one of its main guidelines: "Each local group shall incentivize the formation of National State Committees, or seeks government support for increased activity in space research." This guideline was supported heartily by Professor Bevilacqua and, because of his doings, State President Jânio Quadros nominated a commission to "study and suggest a policy for a Brazilian space research program and to propose measures for implementing research in this field" on 17th May 1961. The commission consisted of Admiral Octacílio Cunha, president of the National Research Council (CNPq), Colonel Aldo Vieira da Rosa, director of IPD/CTA, and the presidents of the SIB, Luiz de Gonzaga Bevilacqua and Thomas Pedro Bun. The

The Minitrack station that received signals from the first artificial satellites, set up by Fernando de Mendonça and Júlio Alberto de Morais Coutinho in 1957 [Archive Fernando de Mendonça]

commission proposed the creation of an Organizing Group for the National Commission on Space Activities (GOCNAE), with the purpose of forming the core of trained personnel and developing activities in the areas of astronomy, radio astronomy, optical satellite tracking and satellite communications.

Of course, at that time, new developments in space activities had major effects in the world. On 12th April 1961, Yuri Gagarin became the first man to fly in space. In Brazil, President Quadros, who had already shown his enthusiasm for space-related subjects and for the nationalism of socialist countries, decorated Gagarin when he visited the country in July of that year. Only days later, he signed the decree that created GOCNAE as a unit subordinate to CNPq. This group was tasked with, amongst other things, the execution of space research projects and the coordination, stimulation and support of space-related activities. With the establishment of GOCNAE, Brazil was one of the first countries to officially include space activities within its government program. It is worth noting that GOCNAE received formal support from the Ministry of Aeronautics that provided a site at São José dos Campos and personnel to form part of the initial staff. The formal inauguration of GOCNAE's first directorate took place on 22nd January 1962. Although GOCNAE was a civilian institution, the military involvement in the development of Brazilian space was evident: the president and executive members, all were military; scientists were only nominated to the council. GOCNAE, mostly referred to as the National Commission for Space Activities (CNAE), lasted until 1971. In that year, a decree replaced GOCNAE with the National Institute of Space Research (INPE). In 1964, the Ministry of Aeronautics established the Executive Group for Space Studies and Projects (GETEPE), which, in 1969, created the Institute of Space Activities (IAE). IAE was one of the five technical divisions of the Ministry of Aeronautics' CTA that was responsible for developing the SONDA sounding rockets and VLS orbital launcher for the national space program and for operating the Alcantara launch site. These activities were consolidated in 1971 under COBAE, the Brazilian Commission of Space Activities – a coordinating inter-ministerial body under the head of the Joint Chiefs of Staff of the Armed Forces. This Commission was responsible for the development of a national Brazilian space program and was replaced more than a couple of political decades later by the civilian Brazilian Space Agency (AEB). It could be said that in those times, Brazil's military primarily led the country's space program, but changes in Brazil's space scene were imminent.

São José dos Campos became the seat of CNAE. This was mainly because it was part of the industrial area of São Paulo and close to CTA and its Technical Institute of Aeronautics. All the initial facilities were provided by the military and in 1963, CNAE moved to more permanent buildings next to CTA on ground provided by the Air Ministry. Contacts were established with NASA. The plan for space research in the areas of ionosphere, geomagnetism and meteorology was largely inspired by NASA projects. Fernando de Mendonça, at that time a lieutenant of the Brazilian Air Force studying at the Stanford University in California, played an important role in these contacts. After finishing his Ph.D. in radio sciences, he brought back equipment, compliments from NASA, with which a complete station for receiving signals from ionospheric satellites was assembled. This station was the beginning of the CNAE's Space Physics Laboratory. Around 1963, Brazil was part of a small group of countries that was on the outset of space sciences, of course led by the United States and the Soviet Union, who were in fierce competition to prove themselves the best. In 1963, at the Municipal Planetarium of São Paulo, the International Space and Aeronautics Exhibition was held, where Soviet and American projects were displayed.

GETEPE was to build the Barreira do Inferno Rocket Range (CLFBI) at Natal in the state of Rio Grande do Norte.[1] Ultimately, GETEPE had to stop dependency on foreign launch capacity. It was CNAE that drew up the initial plans for this rocket range and proposed its construction in the north-east of the country, close to the equator. The first launches from CLFBI were part of Brazil's contribution to the so-called International Quiet Sun Years (1964–1965), during which space physics research was conducted under conditions of minimum solar activity. The CLFBI was inaugurated on 15th December 1965, with the launch and tracking of an American Nike-Apache rocket, a joint operation between CNAE, the Air Ministry and NASA. The launches programmed for CLFBI included a series of meteorological rockets in the Meteorologia por Satélite (MESA) program – reception of meteorological images – as part of the Experimental Interamerican Meteorological Rocket Network (EXAMETNET) in which Brazil, Argentina and the United States cooperated. The launches formed part of CNAE's Aeronomical Rocket Sounding Project (SAFO) that was made possible by NASA's cooperation in training personnel and supplying equipment. CLFBI became one of the most active rocket ranges in the world. Although the rockets launched from CLFBI were mostly foreign, some of the payloads were built by CNAE. The firm Avibras, from São José dos Campos, under contract of GETEPE and later of IAE/CTA, started the development of the

Aerial view of the construction of the Barreira do Inferno Rocket Range, Natal, 15th December 1965 [Archive Adauto Gouveia Motta]

SONDA rocket family. Although CLFBI was Brazil's primary rocket range, most launches of sounding rockets were performed from the Cassino Beach Range in the south of the country, which was constructed in the early 1960s. Here, as was the case at CLFBI, the Air Force was responsible for the actual rocket launches.

The National Institute for Space Research, INPE, succeeded the National Commission for Space Activities, CNAE, and became responsible for the ground and space segments of Brazil's application satellite programs. As such, it developed research and management areas, including satellite tracking and control, atmospheric and space science, space technology, Earth observation and spacecraft integration and testing. Furthermore, INPE's interests include propulsion, space materials, solar cells, sensors, computer sciences, meteorology, remote sensing and science scanning technology development. INPE formed many cooperative partnerships with space agencies and organizations of various countries, including the United States, Russia, China, Japan and India, several European countries, and

Nike-Apache rocket ready for launch at the Barreira do Inferno Rocket Range, Natal, 1965 [Archive INPE]

countries in Central and South America. It also works with the Brazilian Space Agency and numerous Brazilian institutions.[2]

During the 1970s, several government organizations embarked on space-related, sometimes overlapping, projects. INPE (meteorology), the Ministries of the Army (missile and guidance technology), Aeronautics (launch vehicles) and Navy (satellite navigation and geodesy), the Ministry of Communications (domestic communications satellites) and even the Ministry for Mines and Energy (natural resources) all had space plans in one way or another. Within COBAE, plans took form to coordinate all these efforts into one space program, with a common goal with Brazil's technological development as the prime objective.

THE BRAZILIAN COMPLETE SPACE MISSION – MECB[3]

The scope, main objectives and directions of the Brazilian space program were established in a project called the Brazilian Complete Space Mission (MECB). This mission was approved in 1979. The MECB initially involved civilian (INPE) and military institutions (Ministry of Aeronautics) separately. These institutions were formed, on the civilian side, by the Secretary of Science and Technology (this secretariat was later replaced by the Ministry of Science and Technology). The INPE space segment was directly subordinated to that ministry. On the military side of the Brazilian space program, the military were still responsible for the space launch vehicle VLS and the necessary launch sites. The main goals stated in the MECB were the design, development, launching and operation of two small low Earth orbit Data Collection satellites (SCD) and two remote sensing satellites (SSR). The space program included the ground facilities, a laboratory for satellite integration and test, and the design, construction and implementation of the launch site at Alcantara. The main MECB objectives were to:

- develop human resources and related infrastructure to support space activities in Brazil;
- call for partnerships with industry for developing space technologies;
- develop satellites with applications related to specific Brazilian needs (including those of interest to low-latitude regions worldwide);
- arrange for Brazil to participate in international space programs.

The MECB has since been extended through the years to include other satellites and programs, such as the Scientific Satellites Program and the joint China–Brazil Earth Resources Satellites (CBERS) program.

Political changes within Brazil – marked by the demise of the military government that had ruled the nation for 20 years between 1964 and 1985 – combined with external pressures coming largely from the United States to demilitarize the space program led to the creation of the Brazilian Space Agency (AEB) to replace the inter-ministerial COBAE in 1994. In name and deed, the agency emphasized the civilian role of Brazil's space activities.[4] The AEB, reporting directly to the President, became linked to the Ministry of Science and Technology with the purpose of developing space activities of national interest. Since then, while satellite development is overseen by INPE, AEB has become responsible for managing Brazil's space policy in general. AEB also establishes the National Program of Space Activities, which summarizes Brazilian's space activities with a horizon of 10 years, and reviews it every two years.

THE NATIONAL PROGRAM OF SPACE ACTIVITIES

According to the National Program of Space Activities (PNAE), this program is strategic for the sovereign development of Brazil.[5] The importance of capacity building in the domain of space technology, which, in a broader sense, includes launch centers, launch vehicles, satellites and payloads, arises from its relevance for the nation's future: "... because no strategic technologies will be made available by third parties, these must be developed with domestic resources, in a widespread and integrated manner, in order to address the challenges posed by the era of Earth observation and telecommunications. Only those countries that master space

AGÊNCIA ESPACIAL BRASILEIRA

pnae

PROGRAMA NACIONAL DE
ATIVIDADES ESPACIAIS
2005 - 2014

National Program of Space Activities (PNAE) [AEB]

technology will have the autonomy to develop global evolution scenarios, which consider both the impact of human action, as well as of natural phenomena. These countries will be able to state their positions and hold their ground at diplomatic negotiating tables." The third review of the PNAE, which covers the period between 2005 and 2014, pursues the above guidelines and forms the basis for the following overview.

The program states that space activity significantly contributes to the development of Brazil, by providing imagery and data acquired over the national territory and by stimulating innovation in the effort towards the acquisition and development of technologies and knowledge considered critical to meet the needs of the PNAE, benefiting both industry and society. Satellite, sub-orbital payload and balloon missions for Earth observation, meteorology, space science and technology, and telecommunications meet government needs to implement efficient and effective public policies and address national problems in those areas. The development of launch vehicles – another important component of the nation's autonomy regarding access to space – also allows the commercial exploitation of launch services. The entire infrastructure overseen by PNAE is extremely important in supporting all space vehicle production, integration, testing, launch and control activities, without which autonomous development would not be possible: "... in order to face the technological challenges involved in large scale projects, PNAE has become, by way of R&D activities with the support of the academic community, a strong innovation fostering agent, which plays a fundamental role towards the leveraging of national industrial capacity and competitiveness, through the acquisition of strategic capacities and technology and of new work processes and methodologies in compliance with international quality standards."

It is essential for Brazil to select aspects that are technologically advantageous to the PNAE and effectively meet the needs of its society. Toward this end, the following activities are selected: Earth observation and meteorology, technological and scientific missions, and telecommunications. The efforts of PNAE toward the development of sounding rockets and launch vehicles, as well as their associated technologies, aim at assuring access to space, and are fundamental to enable orbital and sub-orbital missions, as established in the program: "... satellites and suborbital and orbital payloads are fundamental in providing the Brazilian government with important information, essential for the effective implementation of public policies aimed at the sustainable use and preservation of natural resources, to meet the needs of the productive sector and society at large in terms of information on climatic and environmental changes and furthermore, information supporting scientific research towards the production of precise knowledge of the planet and the universe. The PNAE has consolidated a scientific community consisting of competent professionals in space engineering and technology as well as researchers in space applications such as remote sensing and meteorology, and other space sciences."

Space activities in Brazil have been developed within the framework of the National System for the Development of Space (SINDAE) that includes, as executive branches, the Institute of Aeronautics and Space, the Alcantara Launch Center and the Barreira do Inferno Launch Center. All three are now part of the recently reshaped CTA, now called Commando-Geral de Tecnologia Aerospacial (General Command for Aerospace Technology). The Research and Development Department of the Air Force Command has been abolished and its role transferred to the new CTA. Within SINDAE, the PNAE now defines a group of various institutions that play distinct roles:

- the AEB, acting as a central coordination agency, reports to the Ministry of Science and Technology (S&T) and ministries and other governmental agencies, as well as entities of the civil society, that are represented in the Superior Council of the AEB;

- INPE, as an executive agency, reports to the under-secretariat for the Control of Research Institutes of the Ministry of S&T;
- participating organizations and bodies within the industrial sector and Brazilian universities, which conduct space research.

The development of sounding rockets and a space launch vehicle has allowed Brazil to consolidate knowledge regarding propulsion, materials and processes, guidance and control and scientific experiments, which have significantly increased the participation of industry and scientific research. A large support infrastructure comprises the Alcantara Launch Center, the Barreira do Inferno Launch Center, the Testing and Integration Laboratory (LIT), the Satellite Tracking and Control Center (CRC), the Coronel Abner Propellant Plant (UCA), as well as observatories and research laboratories, and technological niches, mostly in the private sector. Industry has contributed significantly towards the development of the PNAE projects, demonstrating the highly qualified technical capacity of their staff, similar to that of the SINDAE research institutes. Despite this favorable scenario, when the program has grown in recent years, serious annual budget restrictions created obstacles in the maintenance and re-staffing of technical teams, as well as in project procurement.

SPACE SCIENCE AND TECHNOLOGY

PNAE considers a space mission scientific when it involves scientific experiments on board satellites, sounding rockets, satellite launch vehicles or recoverable orbital platforms. A space mission is considered technological when it tests the performance of new space components or systems. The objective of the Brazilian scientific and technological missions is to ensure the means to conduct successful orbital and sub-orbital experiments. To be able to conduct such experiments, international cooperation with exchange of scientific and technological knowledge is mandatory. The following projects are part of PNAE and are described in more detail in Chapter 10:

- **Lattes.** The Lattes program/satellite is named after one of the most important Brazilian scientists, Cesar Lattes, and is the continuation of the Equatorial Atmosphere Research Satellite (EQUARS) and X Ray Imager and Monitor (MIRAX) programs. The EQUARS Project called for the development of a small scientific satellite, to be launched into a low-altitude equatorial orbit. The satellite was planned to observe atmospheric phenomena in the equatorial region, such as luminescence, electrical discharges, particle fluxes and the abundance of atmospheric constituents. MIRAX was planned to be a small satellite for X-ray astronomy. This first astronomical satellite had to bring Brazil in the group of nations that developed and launched satellites to observe the universe.
- **Sub-orbital platforms.** The sub-orbital platforms launched by sounding rockets are extremely useful and cost-effective in research of the atmosphere, ionosphere and gravitational or magnetic fields, as well as of new processes

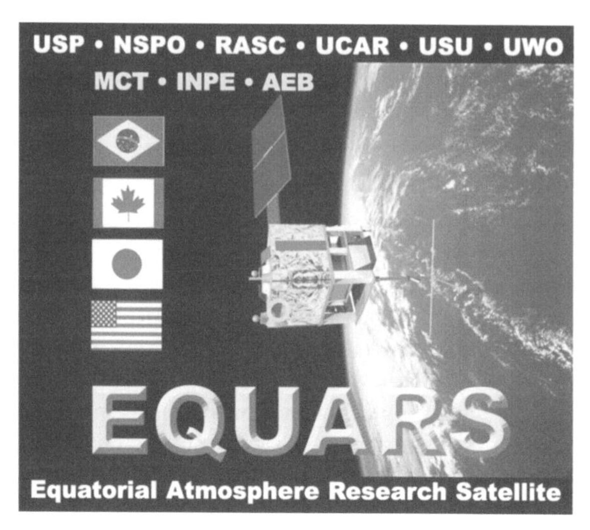

The former Equatorial Atmosphere Research Satellite (EQUARS) project [INPE]

and materials in a microgravity environment. Sounding rockets, a technology that Brazil masters very well, are routinely used to conduct scientific and technological experiments of interest to the Brazilian industrial and academic communities.

- **Recoverable orbital platforms.** Amongst many of the innovative alternatives used in microgravity experiments, recoverable orbital platforms are of particular interest. In the near future, the Atmospheric Re-entry Satellite Project (SARA) will be implemented, in which a reusable platform will be developed that will be able to carry a payload of up to 55 kg for 10 days of microgravity.
- **International Space Station (ISS).** The participation of Brazil in the ISS provides the country with the possibility to conduct manned space operations and long-duration scientific experimentation in microgravity.
- **The Microgravity Program.** Microgravity conditions are created by orbital and sub-orbital flights, such as those by sounding rockets, recoverable orbital platforms and the International Space Station. These conditions allow experiments in the areas of biotechnology, pharmaceuticals, human physiology, combustion and materials science. The program is made possible through the collaborative efforts of the AEB, INPE, IAE/CTA and the Brazilian Academy of Sciences as well as Brazilian universities.

EARTH OBSERVATION

Given its continental dimensions, Brazil is a country that, for the greatest part, can only be effectively observed by aerial and space remote sensing. Control by observation satellites is, in practice, the only fully effective way of monitoring very large areas. The Amazon region, for example, the largest tropical forest in the world, covers about 5m km^2. The use of satellite imagery is also fundamental when there is a need for information on events whose location and time are difficult to predict, such as natural disasters, human-induced events or crisis management. There are several Earth observation programs operational and under development in Brazil:

- **The Remote Sensing Satellite Program (SSR).** For Brazil, as far as ecological, economic and strategic aspects are considered, environmental protection and adequate control of land use and natural resources are extremely important. The issue here concerns the need to control the rapid and sudden changes frequently brought about by natural events, such as floods and droughts, but also by human intervention, such as fires, farming, deforestation, lumber extraction and mining. These and other similar events require the availability of an efficient monitoring system, which provides rapid access to relevant information on a frequent basis. Toward this end, the SSR program was established. With equatorial or polar orbits, satellites of SSR use the Multi-Mission Platform (PMM), which actually is a multipurpose satellite bus.
- **The Sino-Brazilian Program.** The initial China–Brazil Earth Resources Satellite (CBERS) program called for the production and launch of two remote sensing satellites. The first satellite was launched in October 1999 and operated successfully until August 2003, surpassing its designed operational life by almost two years. While in operation, the satellite produced images that were routinely used in both China and Brazil. More satellites in this Program have been produced and launched and based on the success of these satellites, Brazil and China initiated a second phase in their partnership for the production of more satellites.
- **The Data Collecting Program.** The Brazilian Satélite de Coleta de Dados (Environmental Data Collecting System – SCD) started with the launch of the SCD-1 satellite in 1993. Although the SCD-1 had a design life of only one year, it has been operational for more than 11 years. The system's operational capacity was expanded with the launch of SCD-2 in 1998 and the CBERS satellites. The system provides environmental data on the various regions of the country through ground-based Data Collecting Platforms. Local environmental data are collected by the platforms from various sensors, which then are retransmitted by the system's satellites to the receiving stations at Cuiaba and Alcantara. When SCD-1 was launched, the initial network of ground-based data collection platforms comprised about 60 platforms. By the end of 2003, the number of such platforms reached 600.
- **Synthetic Aperture Radar Program (RADAR).** The aim of the RADAR satellite is to provide Brazilian users with high spatial resolution imagery, regardless of weather conditions, through Synthetic Aperture Radar (SAR) technology. The satellite will allow the scanning of the entire national territory, producing imagery that may be used in thematic and cartographic mapping, as well as other applications that depend on permanent monitoring.

Local environmental data are collected by Data Collecting Platforms and routed via SCD satellites [INPE/F. Nolla]

METEOROLOGY

Climate has a direct impact upon practically all sectors of the Brazilian economy and no country is safe from natural disasters. In Brazil, serious social and economic problems can be caused either by prolonged droughts in the semi-arid regions of the northeast or by heavy rains in densely populated areas. In addition, cyclones occur along the southern coast. Therefore, quick and correct weather reports are very important to anticipate or to mitigate the effects of extreme natural phenomena. Meteorological information, obtained from satellite data and images, is of fundamental importance for efficient and precise global weather forecasts. Brazilian users of meteorological data from foreign geostationary satellites cannot fully utilize these resources, due to operational routines and data distribution policies of the countries that own the geostationary meteorological satellites covering this region of the globe:

- **Meteorological Payload on-board Geostationary Satellite.** The main objective of this mission is the high temporal frequency imaging of the national territory, both visible and infrared, to support weather forecasting, climate monitoring and severe storm warnings, in order to eliminate dependency on the operational routines and meteorological information dissemination policies established by foreign institutions. The main users of these data will be weather-forecasting centers, civil and military aviation support systems, river and maritime navigation support systems, as well as public agencies in the areas of agriculture, civil defense, environment and water resources. The orbital position that meets the needs of the majority of these potential users is ideally at 55°W.
- **Global Precipitation Monitoring (GPM).** Atmospheric precipitation is one of the main climatic variables, and information regarding its spatial distribution is of fundamental importance for climate monitoring. Precise and reasonably frequent precipitation measurements are difficult to obtain, since they depend essentially on an adequate surface and high-altitude observation network. The southern hemisphere comprises an extensive oceanic area, as well as vast uninhabitable areas that are difficult to access. Therefore, satellite data are essential to complement those of the existing network. The most adequate electromagnetic spectrum region to measure precipitation is that of microwaves. Therefore, a satellite equipped with a sensor in such range will significantly contribute toward this end. This satellite may also be part of the Brazilian contribution towards the GPM program, increasing the national capacity to monitor precipitation. Brazil already possesses microwave radiometer technology, having taken part in the development of the Humidity Sounder for Brazil (HSB) on board NASA's EOS PM-1 satellite and may also develop, through partnership, specific technology for GPM. Moreover, a satellite of this type may use wide-field cameras, similar to the CBERS Wide Field Imager, and a small-scale platform such as the PMM.

TELECOMMUNICATIONS

The growth of the international telecommunications market has generated new opportunities for the development of the Brazilian industry. With the privatization of the Brazilian market and given the vast numbers of applications currently in use in the telecommunications area, the growth of domestic demand for these systems has led to intensive use of capacity available on foreign satellites. These considerations justify a national capacity-building effort in this area so that Brazilian industry may participate in this market. The development of a complete Brazilian geostationary telecommunications system, which could also meet the needs of a meteorological mission, is a direct response to the PNAE objectives of capacity building and increasing competitiveness of the national industrial sector.

The aim of the telecommunications mission was the development in Brazil of geostationary satellites to meet the objectives and needs of the government in areas of

The launch of the first Brasilsat (SBTS 1) telecommunications satellite was performed on 8th February 1985, with an Ariane 3 from the Kourou space launch site in French Guyana [Arianespace]

secure communications, meteorology and air traffic management. Given the wide coverage of the PNAE, any decision regarding the use of geostationary orbits was of interest to the AEB. Despite the AEB's potential interest in the decisions regarding the use of geostationary orbits, it is the National Telecommunications Agency (ANATEL) that decides, monitors and controls the use of the Brazilian orbital positions for commercial telecom services.

SOUNDING ROCKETS AND SPACE LAUNCH VEHICLES

The development of orbital and suborbital launch vehicles is of strategic importance to Brazil because it is the only way to ensure a country's independent access to space. Through the Institute of Space Activities, the former Aerospace Technical Center and the Brazilian aerospace industry, Brazil has conceived and produced a successful range of sounding rockets. These rockets have allowed Brazilian universities and research centers to develop a series of scientific and technological experiments and have attracted the attention of foreign users. The acquired sounding vehicle technology has contributed towards the development of a small, four-stage, solid-propellant satellite launcher. The next stage of the program calls for the development of liquid-fuel technology, which will allow the development of mid-to-large-scale launch vehicles:

VLS-1 [AEB]

- **Sounding rockets.** These are small launch vehicles used for sub-orbital space exploration missions and able to carry payloads for scientific and technological experiments. Brazil has operational sounding rockets that meet current needs, with a successful and impressive launch record. The close involvement of universities and research centers in the space program will increase the demand for this type of rocket, justifying their continued production that is estimated to be at least two rockets per year. The production of the sounding rockets will gradually be transferred to the national industrial sector, to meet the needs of scientific and technological missions.

- **Small space launch vehicles (VLS).** The first-generation Veículo Lançador de Satélites (VLS-1) is a small launch vehicle able to inject payloads ranging from 100 to 350 kg, into orbits between 200 and 1,000 km. The flight qualification of this vehicle began with the launch attempt of the first prototype in November 1997, followed by two other prototypes between 1999 and 2003, when a catastrophic accident occurred on the launch pad during the third launch campaign. Currently, the VLS-1 project is undergoing a critical review, and the fourth prototype still has to be launched. An upgraded version of the VLS-1, named VLS-1B, is in a preliminary study phase. This rocket will have a liquid-fuel third stage, which will enable injection of 850-kg satellites into 750-km low Earth orbits.
- **Cruzeiro do Sul.** The Cruzeiro do Sul (Southern Cross) program is to succeed the VLS-1 space launch vehicle program and to provide for a family of small, medium and heavy lift launch vehicles. The proposed space missions are composed of micro, mini and medium-weight satellites into low inclination as well as polar orbits that cannot be fulfilled by the VLS-1. The possibility of being able to launch geostationary missions would complement the range of launch vehicles.

INFRASTRUCTURE AND GROUND SUPPORT

To support its space activities – development, testing and operation of satellites and launch vehicles, including sounding operations – Brazil developed and built the necessary infrastructure. The first mention of space infrastructure was in 1956, when Americans used a station in Fernando de Noronha to follow rockets launched from Cape Canaveral.[6] Nowadays, the infrastructure consists of launch centers, telemetry stations, laboratories, a propellant plant and so on. Furthermore, universities, industrial plants and manufacturers support space development:

- **Alcantara Launch Center (CLA)/Alcantara Space Center (CEA).** CLA, built on a site on the Atlantic coast across the Rio Mearim from São Luiz, where it is isolated in the interior of Maranhão, provides near-equatorial launch services for sub-orbital and orbital space missions. Rockets can be launched safely (impact area) from almost due east up to about 100° inclination. Since February 1990, CLA has provided a total of five launch pads for the VLS space launcher, SONDA sounding rockets and meteorological rockets. CLA was also home to foreign (NASA) launches, such as VIPER, NIKE-ORION, NIKE-TOMAHAWK and BLACK BRANT. All activities comprising space vehicle launching and tracking are conducted by the CLA under the responsibility of the Research and Development Department of the Air Force Command. Until the 2003 VLS-1 accident, the Air Force was also responsible for managing the base. In order to provide support for commercial launch activities as anticipated in the PNAE, CLA is directly subordinated to the Brazilian Space Agency.

Alcantara Launch Center [EIJournal.com]

Changing the name from Alcantara Launch Center to Alcantara Space Center is part of an attempt to give Alcantara a civilian profile. The center will not only provide launch capabilities, but also host a university campus and a space museum. The entire base area totals some 620 km^2 and the plan to set aside more than half of this area for the people of Alcantara will contribute to this. The military area will be restricted to about 10%. The establishment of CEA is essential for the accomplishment of the commercial launch operations foreseen with the Ukraine (Tsyklon-4) and other interested countries. The CEA installations will provide the necessary means and utilities for the launch sites for national and international technical teams involved in the launch operations. Installations will also be provided to other governmental agencies directly involved in operations and activities, including the reception, inspection and preparation of rocket and satellite parts and components. Advanced university centers, laboratories, biomass and bio-diesel plants will be established at CEA, as well as companies associated with the space sector and other technological activities.

Following his inauguration ceremony on 25th March 2008, the new president of the AEB, Carlos Ganem, said the work on the launch tower that was lost in the 2003 accident at the CLA would be resumed shortly.[7] He also made clear that he did not intend to create a new Brazilian space program but

Barreira do Inferno Launch Center (CLBI) [brazilianspace.blogspot.com]

rather keep existing projects: "... those projects are a new launch platform at the Alcantara Launch Center, the partnership with Alcantara Cyclone Space (a bi-national mission center made up by Brazil and Ukraine) and the creation of an Alcantara Space Center which is expected to guarantee work for area residents."

Alcantara is also proposed as the world's first spaceport for tourism. During a national conference in Brasilia in 2007, it was recommended that developing space tourism should be one of the space goals in the future.

- **Barreira do Inferno Launch Center (CLBI).** CLBI is a smaller launch site than Alcantara and is located at the Atlantic coast near Natal, almost at the most easterly point of Brazil. The Ministry of Aeronautics set it up in 1964 primarily for sounding rocket launches and the development of Brazil's research capability in space and atmospheric sciences. The place of the center was not only chosen for the proximity to the equator and the ocean for impact area, but also for conditions such as favorable winds. It has six concrete launch pads with launch rails that suit different types of rockets. Launches at CLBI include, apart from the Brazilian SONDA sounding rockets, mainly foreign rockets, such as the already mentioned BLACK BRANT, as well as AEROBEE, ARCAS, LOKI, NIKE-APACHE, -CAJUN, -IROQUOIS and -ORION, SERGEANT and SKYLARK.

  SONDA launches took place from 1965 until 28th April 1989. Although all launches have now been transferred to CLA, the capacity to operate sounding

Launch site at Praia do Cassino [INPE]

rockets remains. CLBI made more than 230 rocket launchings.[6] In mid-1970, the center was upgraded with telemetry reception and processing facilities for Ariane space launch vehicles launched from the Kourou space launch center in French Guyana. CLBI, subordinated to the Research and Development Department of the Air Force Command, is now the main support station for the monitoring and control of launch vehicles launched from the CLA. Improvements are incorporated to support other foreign space agencies.

- **Cassino Beach Range.** The Cassino Beach Range (CBR), or Praia do Cassino, was located close to the City of Rio Grande in the south of Brazil. In 1966, hundreds of scientists, researchers and technicians from many different countries participated in cooperative programs for atmospheric research during the total eclipse of the sun on 12th November. That day, seven SONDA-1D rockets were launched to support the Brazilian eclipse mission. The results of this mission were presented at an international symposium held at CNAE in February 1968. In all, in 1966, some 27 rocket launches were performed from CBR, including SONDA and NIKE-APACHE, -HYDAC, -JAVALIN and -TOMAHAWK.
- **Satellite Tracking and Control Center (CRC).** The CRC at São José dos Campos is responsible for the planning, management and execution of all INPE satellite operations activities. Since 1989, the CRC enables INPE to monitor and control satellites in orbit, as well as to provide support services to foreign satellites. The center is composed of the Satellite Control Center (CCS) and two ground stations, Cuiabá and Alcantara, which are linked by a data communication network, called RECDAS.[8] In Alcantara, there will be an additional station for the control of scientific satellites. A new ground station will provide satellite control for the CBERS-3 and CBERS-4 satellites. The two S-band ground stations of Cuiabá and Alcantara are almost identical in terms of hardware configuration. Cuiabá, however, is the only one equipped with a receiver of Data Collecting Platforms (DCP), being the data receiving station for the SCD1 and SCD2 satellites payload.

Antennae at the Cuiabá Data Reception Station (ETC) [INPE/E. Girão]

- **Cuiabá Data Reception Station (ETC).** In 1972, the remote sensing satellite data reception station ETC, strategically located due west of Brasilia near the geographical center of South America, became operational. SPOT, LANDSAT, ERS and CBERS data stations are operated at this site (tracking and/or reception of imagery). In May 1973, with this station, Brazil became the third country in the world (after the United States and Canada) to have an operational system for receiving data from civil remote sensing satellites (ERTS-1/LANDSAT-1). In the Brazilian Environmental Data Collecting System, local environmental data, collected by ground-based platforms with various sensors, are retransmitted by the system's satellites to the ETC. For INPE, the Cuiabá station functions as a backup station for the CRC.

  In 1991, INPE received from the Canadian government – as part of the contract for the purchase by Brazil of the first generation of Brasilsat communication satellites – a station for receiving and processing data from microwave remote sensing satellites. For a country that has much of its territory spending the major part of the year covered by clouds, this was a major advantage. The receiving equipment was installed at the ETC and the image processing installations at Cachoeira Paulista.

- **Cachoeira Paulista Mission Center (CPMC).** INPE's Image Processing Laboratory (IPL) for remote sensing products was installed in Cachoeira Paulista and became operational in September 1974. CPMC is now the mission center for the SCD satellite data collection. SCD data that are received at the ETC are processed into digital and photographic products at the CPMC and made available to users through the internet. The main applications are in the areas of meteorology and hydrology. In 1974, INPE also created a number of laboratories at CPMC to fulfill its needs in the area of basic and applied research, such as in ionosphere sounding, geomagnetism, atmospheric airglow and satellite meteorology and satellite data processing. IPL distributes meteorological images to users such as the Brazilian (national) meteorological system, institutions interested in climate and weather studies, and the media. On 3rd September 1989, INPE's SARSAT Local User Terminal installed in Cachoeira Paulista helped locate a Varig Airlines Boeing 737 that made a forced landing in the middle of the Amazon forest. The station had been donated by Canada and is part of the international search and rescue system COSPAR-SARSAT.
- **Fortaleza Telemetry Station.** In 1968, Brazil and France (on behalf of the European Organization for the Development and Construction of Space Vehicle Launchers, ELDO) agreed to build a telemetry station at Fortaleza to monitor and track space vehicle launchings from French Guyana.[9] The station, built on a site within the boundaries of land that was allotted for the construction of CLA, comprised, *inter alia*, fixed radio receiving and transmitting facilities for maintaining direct contact with the Guyana Space Center and other stations. It also was agreed that the station could be used for the Brazilian government's own scientific activities if those were not detrimental to the station's programs. Under a new contract between Brazil and the European Space Agency (ESA), the telemetry station was moved to CLBI.[10]
- **Colonel Abner Propellant plant (UCA).** The Colonel Abner plant for the production of solid and composite propellants was planned to meet the demand for propellants for the vehicles envisaged at the outset of the space program. UCA is subordinated to the Air Force Command. The original design foresaw the possibility of expanding the plant, to produce boosters with greater diameters or at a higher production rate. This expansion will eventually take place to meet the needs of the planned launch vehicles. Currently, scientists of UCA are, in cooperation with Russia, also researching and developing liquid propellants.[5]
- **Combustion and Propulsion Laboratory (LCP).** The LCP is subordinated to INPE. It was founded in 1968 and relocated from São José dos Campos to Cachoeira Paulista in 1976. LCP is engaged in research activities and

Satellite testing in the Thermal/Vacuum Chamber at LIT [INPE]

developments in combustion, propulsion of satellites, auxiliary propulsion and catalysis with applications in combustion and propulsion. The Brazilian Government considers LCP as the core of excellence in satellite propulsion. An important part of the LCP is the Satellite Thrusters Test Facility with Altitude Simulation (BTSA), which is the only one of its kind in South America, and a test bank for research under atmospheric conditions (BTCA).

- **Integration and Test Laboratory (LIT).** During the 1980s, INPE established the Laboratório de Integração e Testes (LIT) that started with developing highly specialized activities essential to the Brazilian's space program in 1988. It was the largest such facility in the southern hemisphere (10,000 m^2) and the

CBERS CCD image of Fortaleza [CBERS/INPE]

only one of its type in Latin America. It was established in São José dos Campos to develop and implement assembly, integration, functional and qualification testing, validation activities for satellites and other orbital systems, as well as to perform space parts fault analysis. Besides meeting the needs of the PNAE space projects, the laboratory has other important objectives such as technology transfer through test and analysis, on all levels, from parts to integrated space systems, promotion of the national industry participation in space activities and participating in international development programs aimed at intensifying the exchange of technology. LIT activities include parts and material engineering, procurement, parts qualification and acceptance tests, as well as qualification and acceptance of complete integrated systems. LIT is also responsible for the promotion of ongoing training and capacity-building efforts aimed at the adequate implementation of the proposed activities.[5] The laboratory includes a 1,600-m^2 hall, a 450-m^2 integration room and a control room. It has a 13-kN and a 80-kN shaker, a 22-m^3 thermal vacuum chamber and electromagnetic test chambers. In 1997, the French aerospace and defense company Intespace added an acoustic chamber. With a growth in demand for its services, INPE's Integration and Test Laboratory is undergoing further expansion.[11]

REFERENCES

1. Oliveira, F. de: *Caminhos Para O Espaço/Pathways to Space*. Instituto Nacional de Pesquisas Espaciais (INPE) – São José dos Campos, 1991.
2. *Jane's Space Directories*, 1986–2008.
3. Fonseca, Ijar M. and Bainum, Peter M.: *The Brazilian Satellite Program – A Survey*. AAS 05-479, 2005.
4. *TWAS Newsletter*, Vol. 18, No. 2, 2006.
5. National Program of Space Activities 2005–2014: PNAE/Brazilian Space Agency. Brasília: Ministry of Science and Technology, 2005.
6. ribs SC&I/DB&C Databases, Breda, The Netherlands, 2008.
7. Ligia Formenti in Sao Paulo O Estado de Sao Paulo (Internet), Sao Paulo, 26 March 2008.
8. Orlando, V. and Kuga, H.K.: *Flight Dynamics Operations of INPE's Satellite Control Center*. São José dos Campos, SP, Brazil, INPE, 2001.
9. Exchange of letters constituting an agreement between France and Brazil concerning the installation of a telemetry station at Fortaleza. Rio de Janeiro, 20 June 1968.
10. Agreement between the Government of the Federative Republic of Brazil and the European Space Agency on the establishment and use of tracking and telemetry equipment to be installed in Brazilian territory. Brasilia, 20 June 1977.
11. Feller, G.: *Brazil Seeks Larger Place in Global Space Industry*, www.satellite-today.com/enterprise/applications/03/14961.html (accessed 8 August 2008).

10

Brazil: Development – space launch systems, space probes and satellites

On 17th October 1969, the Institute of Aeronautics and Space (IAE), an organization subordinated to the national military Aerospace Technical Center (CTA), succeeded GETEPE as the main space institute in Brazil and became responsible for the research and development of space launch vehicles. IAE has since pursued a sounding rocket development program whose rockets have been widely used by national and foreign research institutions. Nowadays, it is effectively the executive branch of the Brazilian space agency AEB for rocket development.

Even before GETEPE was founded, the Ministry of Aeronautics initiated space activities as early as 1963, resulting in a small meteorological sounding rocket called SOMFA. In 1961, Avibrás Aerospace Industry (Avibrás Indústria Aeroespacial S.A.), a company producing military rocketry and missiles, had been founded by engineers associated with the CTA. In the late 1960s, Avibrás was awarded the development and production of the series of sounding rockets named SONDA (Table 10.1). The development, building and operating of these small, multistage rockets, with assistance of the United States of America, provided much experience and technology for future launch vehicles. Since then, IAE has been the major player in Brazil's efforts to develop a satellite launch vehicle to bring it into the exclusive group of countries that are able to launch satellites indigenously.

SONDA FAMILY OF SOUNDING ROCKETS

Sounding rockets are rockets used for upper atmospheric research as well as for sub-orbital space exploration missions and carry payloads for scientific and technological experiments. As part of programs to determine electron density in the low ionosphere – a question of practical importance for aircraft navigation – Brazil has operated sounding rockets since the first test launch in 1965 of the SONDA I rocket. In the accompanying tables, details of the different families of developed (sounding) rockets are given. Not all rockets of the same basic type always have exactly the same parameters and they differ, such as in launch weight (different payloads) and total

SONDA I sounding rocket [FAB-CLBI]

length (different fairings). Also, there are differences in the reports about the number of rockets that has been launched during the years. One official report[1] quotes quite different numbers (more than 200) of SONDA launchings than are documented in other open sources (fewer than 50).[2, 3] See Annexe 1 in Chapter 15 for all space-related Brazilian launches.

The two-stage solid-propellant SONDA I rocket was capable of carrying a 4-kg instrument package up to 60–75-km altitude. Most notable was the successive launch of seven SONDA ID rockets on 12th November 1966. On that day, hundreds of scientists, researchers and technicians from many different countries gathered in Brazil and participated in cooperative programs for atmospheric research during a total eclipse of the sun. Together with the seven SONDA launches that day, some 15 NIKE launch vehicles in different sounding configurations – such as NIKE-APACHE, -HYDAC, -JAVALIN AND -TOMAHAWK-9 – were also launched to investigate the eclipse. The first launch of the SONDA I took place from Barreira do Inferno Launch Center (CLBI), all other launches from the Cassino Beach Range.

The SONDA II rocket was a small, single-stage solid-propellant rocket for sounding operations (ionosphere and aeronomy research). Dolinsky states that there were 61 flights in total of this version, starting with the first launch in April 1970.[1] Launches were performed from Alcantara Launch Center (CLA) and from CLBI. SONDA II's maiden launch, as documented in open sources, was in 1990 with an ionosphere research mission for INPE. After six more launches, this sounding rocket was decommissioned in 1996. The names of the missions were taken, in general, from a city or other geographical name in the Brazilian state in which the launch happened. The last six launches have flight numbers ranging from XV53 (1990) through XV58 (1996), suggesting at least 58 launches or launch efforts for SONDA II.

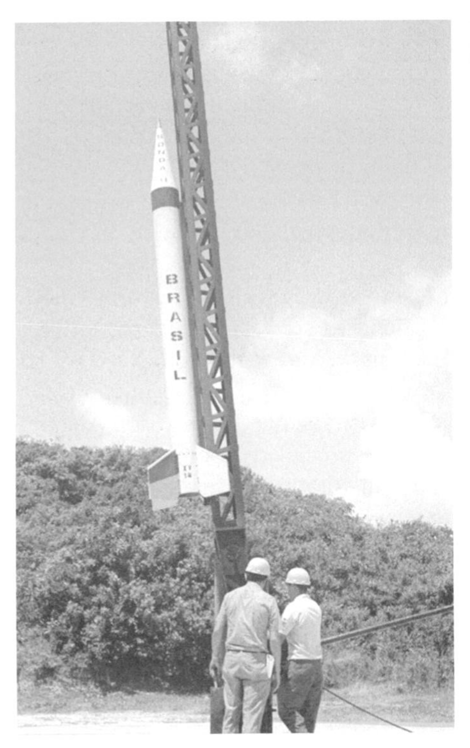

SONDA II sounding rocket [FAB-CLBI]

SONDA III sounding rocket [FAB-CLBI]

In 1976, SONDA III, a new two-stage, unguided sounding rocket, was launched for the first time in a test that was called Serido, reaching an altitude of only 47 km: this was considered a launch failure. The following launches, with the basic SONDA III, all reached a mean altitude of 550 km, with the exception of the two Colored Bubbles Ionosphere (CBI) missions in 1982 that reached 335-km altitude. These CBI missions were launched 24 hr apart (17/18th September 1982) and were part of plasma diagnostic experiments. The CBI missions were carried out in cooperation with the German Max-Planck Institute. The two Brazil Ionospheric Modification Experiments (BIME) missions, launched earlier that month, were done in cooperation with the geophysics laboratory of the American Air Force. Part of the BIME experiment was the launch of two Black-Brant rockets, also from CLBI. These ionospheric modification experiments were conducted to test theories for generating plasma instabilities in the equatorial ionosphere.

SONDA III rockets, for the first time in Brazil's rocket development, carried a complete instrumentation system. Furthermore, they carried a tele-destruct system, a rudimentary telemetry system, three-axis attitude control for the payload that could be recovered from the sea and a newly developed electronic apparatus to monitor the flight. Because of different mission requirements, SONDA III was developed in two basic versions. The second stage of the SONDA III was boosted by an S20 motor

(0.300 m in diameter, 3 m in length and 237 kg of propellant). The second stage of the SONDA III M1 was boosted by an S23 motor (0.300 m in diameter, 1.63 m in length and 114 kg of propellant). This was done to accommodate longer payloads from the Air Force Geophysics Laboratory (AFGL) and others, but keeping the original aerodynamics. There was also just one flight of the SONDA III M2 with an even shorter S24 in the second stage. The SONDA III M was launched four times between 1978 and 1980 and reached mean altitudes of 240 km with payloads weighing between 50 and 80 kg.

Between 1976 and 2002, 28 SONDA III rockets were launched: 24 from CLBI, two from CLA and two from Wallops Island. The last two were considered failures because only altitudes of respectively 81 and 55 km were reached. All SONDA III missions were ionosphere sounding missions except for the year 2000 mission. On 9th December of that year, a microgravity mission with the name Operation Alecrim was launched from CLBI to test and qualify the so-called PSO platform. The platform was developed by INPE and consisted of a stabilized experiment housing, onboard data computation and processing and telemetry transmission. The platform provided short-duration microgravity conditions.

The last SONDA III mission was Operation Parnamirim on 12th May 2002. The 28th SONDA III was launched from CLBI and carried an ionosphere instrument package to measure the concentration of ionization between 150 and 500-km altitude. The flight lasted some 13 min and reached a maximum altitude of 593 km.

SONDA IV sounding rocket [FAB-CLBI]

In 1979, Brazil established three goals under the rule of the MECB (Brazilian Complete Space Mission). One of these goals was the design, development and construction of the VLS satellite launch vehicle. To achieve this goal, the two-stage SONDA IV rocket was developed as a technological building block and qualification and technology demonstrator. Its main objective was the testing of rocket three-axis attitude control, which was an explicit requirement for the VLS. Between 1984 and 1989, four launches were conducted with the SONDA IV in a sounding rocket mode of operations. The third launch failed due to a malfunction in the second stage. The main testing was on control systems, payload fairing ejection and other technologies necessary for a satellite launch vehicle. However, the capacity of the SONDA IV rocket (600 km, 500 kg) brought it under the regime of the MTCR (Missile Technology Control Regime) and this put Brazil at odds with the United States. Although Brazil adhered to the MTCR (signed in 1995), it was not permitted to import technologies that would help Brazil to develop its own space launch vehicle and that put the development of the VLS seriously on the back burner.

Table 10.1. SONDA family of sounding rockets: characteristics

	SONDA I	SONDA II	SONDA III	SONDA IV
Development	IAE	IAE	IAE	IAE
First launch	19650401	19900221	19760226	19841121
Last launch	19661112	19960828	T.b.d.	19890328
Number of launches	9	7	28 [20020512]	4
Configuration	2-stage rocket	1-stage rocket	2-stage rocket	2-stage rocket
Mass at launch	65 kg	368 kg	1.548 kg	7,417 kg
Length at launch	3.950 m	4.100 m	8 m	11.000 m
Thrust at launch	27 kN	32 kN	102 kN	203 kN
Max attainable altitude	65 km	100 km	< 650 km	< 1,000 km
Max payload	4 kg	< 70 kg	< 150 kg	< 500 kg
Stage 1: Denotation	S-10-1	S-20	S-30	S-43
Stage 1: Mass	30 kg	250 kg	1,212 kg	5,579 kg
Stage 1: Propellant	Solid HTPB	Solid HTPB	Solid HTPB	Solid HTPB
Stage 1: Thrust (vac)	27 kN	32 kN	95 kN	234 kN
Stage 1: Isp	? s	257 s	266 s	265 s
Stage 1: Number of engines	1	1	1	1
Stage 1: Burn time	? s	18 s	29 s	43 s
Stage 1: Length	1.350 m	4.2 m	3.837 m	5.348 m
Stage 1: Diameter	0.127 m	0.300 m	0.557 m	1.008 m
Stage 2: Denotation	S-10-2	n/a	S-20	S-30
Stage 2: Mass	30 kg	n/a	332 kg	1,212 kg
Stage 2: Propellant	Solid HTPB	n/a	Solid HTPB	Solid HTPB
Stage 2: Thrust (vac)	4.2 kN	n/a	32 kN	95 kN
Stage 2: Isp	? s	n/a	257 s	266 s
Stage 2: Number of engines	1	n/a	1	1
Stage 2: Burn time	32 s	n/a	22 s	21 s
Stage 2: Length	1.790 m	n/a	4.2 m	3.837 m
Stage 2: Diameter	0.114 m	n/a	0.300 m	0.557 m

VS FAMILY OF SUB-ORBITAL ROCKETS

Based on the SONDA experiences, Brazil developed new sub-orbital rockets for many kinds of research. SONDA IV tested propulsion components and motors that were used in these new rockets.[4, 5] See Table 10.2 for the characteristics of the VS family of launch vehicles.

The first of these sub-orbital vehicles that was launched was the VS-40, an unguided, two-stage, aerodynamically stabilized, solid-propellant sounding rocket. The VS-40 was initially conceived to carry out tests of the fourth stage of the VLS-1 in a vacuum environment. This was necessary because Brazil did not have vacuum chambers for simulations. The first stage of the VS-40, the S-40TM, was developed for the second stage of the VLS-1 and was tested in two launches. The second stage, the S-44, was to become the fourth stage of the VLS-1. The VS-40 was flown from CLA two times. The first mission (2nd April 1993) was Operation Santa Maria (PT-01) in which the S-44 motor was tested and qualified in vacuum. The VS-40 was tested to be able to bring payloads to high altitudes. An altitude of 950 km was reached with a 500-kg payload, creating 760 sec of microgravity. The second (PT-02) mission, called Operation Livramento (21st March 1998), involved a 236-kg payload (VAP-1) of the Dutch space firm Fokker Space. The mission monitored propulsion, pressure, acceleration and vibration of the VS-40.

VS-40 sounding vehicle [www.iae.cta.br]

The IAE is preparing a revision launch of this two-stage rocket in 2010. This mission is denominated EN-03/SHEFEX 2 and will be launched from CLA. The main objective of this mission is to test the improved first stage of the rocket; also tested will be standardized components that will be used in the third stage of the VLS-1. The mission will be carried out in cooperation with DLR (Deutsches Zentrum für Luft- und Raumfahrt) from Germany.

Another future (2010) mission for the VS-40 is the SARA (Satélite de Reentrada Atmosférica) sub-orbital mission to flight-test the SARA recovery system, electronic systems and the experiment module. The SARA sub-orbital payload will weigh about

VS-30 sounding vehicle [IAE/D. dos Santos]

350 kg and, launched by a modified VS-40 from CLA, it should achieve eight minutes of microgravity.

The VS-30 rocket is basically an unguided, single-stage, solid-propellant rocket. It has been used in its basic configuration but also with foreign upper stages and launched both from CLA and the Andøya range in Norway. Furthermore, it is used as the second stage of the VSB-30 rocket. DLR had been interested in a suitable rocket for its microgravity experiments that could be launched from a range in Europe and stimulated the development of the VS-30.

The first launch of a VS-30, under control of CTA/IAE, was on 28th April 1997 from CLA and the rocket carried a test from DLR. This mission was called Operation Santana and consisted of technological experiments to test, among other things, the feasibility of launching the VS-30 from the Andøya Test Range. The data were transmitted to CLA by means of an onboard telemetry transmitter. This flight qualified the VS-30 for scientific and commercial sounding operations for microgravity and aeronomy missions.

The Andøya launches occurred on 12th October 1997 and 31st January 1998 and both carried the rocket-borne Optical Neutral Gas Analyzer with Laser Diodes (RONALD and RONALD 2) experiments. RONALD was an experiment developed at the University of Stuttgart in Germany in cooperation with scientists from Norway.

Three more basic VS-30 rockets were launched from CLA carrying microgravity missions called San Marcos (15th March 1999), Lençóis Maranhenses (6th February 2000) and Cumã (1st December 2002). Another one, the Angicos mission (16th December 2007), was launched from CLBI.

The Operation San Marcos consisted of three experiments from Brazilian universities in which the effect of microgravity on worms and the formation of biomedical crystals for antibiotics were studied. Operation Lençóis Maranhenses experimented with recovering the payload on land. The payload was a platform of DLR carrying two Brazilian experiments and one experiment from DASA. The payload stayed in space long enough to create about 200 sec of microgravity, sufficient for the experiments on board. Operation Cumã must be considered a failure because 29 sec after launch, the payload module separated from the rocket prematurely. This separation resulted in only 30 sec of microgravity, only just enough time to transmit the telemetry of the start of the data flow.

Operation Angicos was a result of Argentinean and Brazilian cooperation that started in 1998. This first joint space mission by two South American countries involved the seventh flight of a VS-30 rocket launched from CLBI, two Argentinean microgravity experiments, a Brazilian Global Positioning System (GPS) receiver built by scientists from the Grande do Norte Federal University (UFRN) and an attitude control system using cold gas nozzles. The results of the GPS tests will, when analyzed by the UFRN scientists, be integrated into future satellites. It was a perfect mission in which telemetry and data transmission worked without a hitch. It took the rocket seven minutes to reach the operational altitude of 121 km, providing about four minutes of microgravity. The payload splashed down some 120 km off the Brazilian coast and was recovered by helicopters of the Brazilian Navy. The results of the Argentinean experiments were taken to Buenos Aires to be examined.

The basic VS-30 was fitted out with an Orion upper stage four times, creating a two-stage rocket capable of carrying a payload of some 150 kg to an altitude of more than 300 km, resulting in about seven minutes of microgravity conditions. The Orion upper stage, an American–Italian design from the early 1980s, was supplied by the German Moraba (Mobile rocket base) organization. Moraba, part of the Space Operations and Astronaut Training Department, is an organization for planning, preparing and implementing scientific sounding rocket campaigns in the fields of aeronomy, magnetosphere, astronomy and microgravity research. The Germans opted for the VS-30 rocket because of its gentle acceleration compared to other vehicles of this kind. The VS-30/Orion was to replace the Skylark rocket on certain missions and to fly microgravity payloads complementary to current launch systems.

The maiden flight of the VS-30/Orion was on 21st August 2000 from CLA in

Angicos payload before attachment to the VS-30 [www.geo.net]

which the AEB cooperated with DLR to test the effectiveness of the combination of VS-30 and Orion. It carried the Baronesa microgravity mission (160 kg) up to 330 km. The second flight (23rd November 2002), also from CLA, carried the Pirapema ionosphere mission. The payload (two instruments from Germany, one from INPE) reached an altitude of 434 km and experienced 11 min of microgravity.

The two following launches of the VS-30/Orion combination took place from the Andøya rocket range in Norway. On 27th October 2005, the rocket flew the SHEFEX (Sharp Edge Flight Experiment) mission. The main goal of SHEFEX was the examination of a new thermal protection system concept based on sharp-edged plane panels and to validate design tools. The SHEFEX payload reached an altitude of 212 km. All measuring data of this part of the flight, as well as live pictures of an onboard camera, were transmitted directly to the ground station. However, during the activation of the parachute system, it failed and the flying unit of SHEFEX was lost, but evaluation of the measuring data led SHEFEX to become a success.

The second VS-30/Orion flight from Andøya was on 31st January 2008. It carried the HotPay-2 payload that involved nine instruments from eight countries. The scientific instruments were divided into three groups, studying phenomena in the upper mesosphere and lower thermosphere, auroral science and cosmic ray flux. At the time of launch, Earth was inside a solar wind stream and the payload flew into an auroral arc. Another successful auroral research mission was flown with a VS-30/ Orion on 5th December 2008. The ICI-2 payload provided during a 10-min flight data on electron plasma structuring in an electron cloud. Instruments were provided by the University of Oslo, DLR and JAXA.

Brazilian–German cooperation improved remarkably and became of an unprecedented level with the development of a new two-stage, unguided sounding rocket called the VSB-30. To maintain the European sounding rocket microgravity program, there was a need to replace the aging (1957, 441 launches) Skylark sounding rocket design. The motor for this rocket would no longer be commercially available from 2005 on. DLR took part in the design and development of parts of the VSB-30. The rocket is basically a VS-30 with the well proven S30 motor powering the second stage, while the first stage, placed in tandem, uses a new S31 solid-propellant motor. The combination can be fitted with a TEXUS upper stage. In this configuration, the rocket flew four times from the Esrange Space Center in Sweden. In its basic configuration, it flew twice from CLA.

The successful maiden flight of the VSB-30 rocket, called Operation Cajuana, took place on 23rd October 2004 from CLA, and qualified it not only for sounding operations, but also for TEXUS launches from Esrange. It was the first launch since the VLS-1 accident in August of 2003. In this VSB-30 launch, the 392-kg payload contained no experiments but was ballasted, and was carried to an altitude of about 240 km. The payload service module was equipped with three-axis gyros, accelerometers, vibration accelerometers, a GPS receiver and temperature sensors that provided a wealth of information on the flight dynamics, trajectory parameters and the mechanical environment. A telemetry system transmitted data on housekeeping, GPS and other test results.[6]

The second VSB-30 flight from CLA was on 19th July 2007. This flight, with experiments from Brazilian and German universities, aimed at studying how gravity (or the absence of it) affected the speed of chemical reactions in enzymes and the quality of DNA repair following exposure to radiation. The payload was carried to about 280 km and the resulting microgravity time was some seven minutes. Because of faulty telemetry data, the payload could not be located in the Atlantic Ocean by Brazilian search teams and was lost. Only some of the transmitted data on experiments were received during the descent of the payload.

From the Esrange Space Center in Sweden – the Aerospace Operation Facility of the Swedish Space Corporation – four VSB-30-TEXUS missions were flown between 2005 and 2008. The TEXUS project is a sounding rocket program with the primary aim to investigate the properties and behavior of materials, chemicals and biological substances in a microgravity environment; it gives around six minutes of microgravity. The TEXUS program started in 1977 and is carried out at the Esrange Space Center jointly by DLR, EADS Astrium and the Swedish Space Corporation. The campaign is funded either by ESA (European Space Agency) and/or DLR.[7]

TEXUS 42, the first VSB-30-TEXUS launch, was jointly funded by the ESA and DLR. It was launched on 1st December 2005, reached an altitude of 263 km and provided 397 sec of microgravity time. The experiments on board performed nominally. During the flight, the scientists were able to interact with their experiments through telecommand in real time. Housekeeping data, scientific data and video were downlinked to Esrange. Two new technical systems, the service module and the recovery system, were operated for the first time. Also for the first time, the EML (Electro Magnetic Levitator) was used. The EML was developed for precision measurements of thermophysical properties of electrically conducting materials in a molten state. In the experiment chamber of the EML, an electrically conducting molten sample can be levitated by electromagnetic fields while surface oscillations are induced. In this way, interfacial tension and viscosity data can be derived. During the flight, the sample can be molted and solidified for testing purposes. The findings will contribute to the development of new metals for special applications.

VSB-30 sounding vehicle at Alcantara (23rd October 2004) [DLR/CTA/IAE]

TEXUS 43 was launched on 10th May 2006. The 407-kg payload reached an altitude of 237 km and five experiments were exposed to microgravity for 347 sec. The flight was nominal and the payload was recovered undamaged, but while three of the experiments worked flawlessly, two failed.

TEXUS 44 carried the second EML mission to 273-km altitude on 7th February 2008. There were four experiments on board, which had 376 sec of microgravity time. All experiments worked well and the video downlink operated without a hitch. Also, the newly developed recovery system worked flawlessly and supported the safe landing of the payload. TEXUS 45 was launched only 14 days after its predecessor (21st February 2008) and carried a 367-kg payload with three experiments to an altitude of 270 km for 398 sec of microgravity time. One experiment involved 72 cichlid fishes to investigate motion sickness when they entered weightlessness (why do certain people get motion sickness and others don't?). Another experiment researched

Table 10.2. VS family of launch vehicles: characteristics

	VS-40	VS-30	VS-30–Orion	VSB-30
Development	IAE	IAE	IAE	IAE
First launch	19930402	19970428	20000821	20041023
Last launch	19980321	20071216	20051027	T.b.d.
Number of launches	2	7	4	2 [20080221]
Configuration	2-stage rocket	1-stage rocket	2-stage rocket	2-stage rocket
Mass at launch	6,737 kg	1,508 kg	1,800 kg	2,626 kg
Length at launch	7.390 m	5 -7.8 m	9.6 m	12.640 m
Thrust at launch	? kN	102 kN	102 kN	240 kN
Max attainable altitude	640 km	160 km	320 km	270 km
Max payload	500 kg	< 300 kg	< 300 kg	392 kg
Stage 1: Denotation	S-40TM	S-30	S-30	S-31
Stage 1: Mass	5,664 kg	1,208 kg	1,208 kg	998 kg
Stage 1: Propellant	Solid HTPB	Solid HTPB	Solid HTPB	Solid HTPB
Stage 1: Thrust (vac)	208.39 kN	95 kN	95 kN	240 kN
Stage 1: Isp	275 s	266 s	266 s	? s
Stage 1: Number of engines	1	1	1	1
Stage 1: Burn time	56 s	28 s	28 s	11 s
Stage 1: Length	5,800 m	3.293 m	3.293 m	3.293 m
Stage 1: Diameter	1,000m	0.557 m	0.557 m	0.557 m
Stage 2: Denotation	S-44	n/a	Orion 2	S-30
Stage 2: Mass	1,025 kg	n/a	400 kg	1236 kg
Stage 2: Propellant	Solid HTPB	n/a	Solid HTPB	Solid HTPB
Stage 2: Thrust (vac)	33.24 Kn	n/a	13 kN	102 kN
Stage 2: Isp	282 s	n/a	? s	266 s
Stage 2: Number of engines	1	n/a	1	1
Stage 2: Burn time	68 s	n/a	32 s	20 s
Stage 2: Length	1.800 m	n/a	6.315 m	3.293 m
Stage 2: Diameter	1.000 m	n/a	0.355 m	0.557 m
Stage 3: Denotation	n/a	n/a	n/a	Texus

spray cooling methods for industrial processes and the third experiment studied flows in capillary channels to get answers on fundamental questions in fluid mechanics.

VEÍCULO LANÇADOR DE SATÉLITES (VLS)

After component and stage testing and qualification with the SONDA IV, the development of a space launch vehicle for small satellites commenced in earnest in 1984. See Table 10.3 for the characteristics of the VLS family of launch vehicles. The VLS-1 is a space launch vehicle comprising a core rocket and four strap-on boosters. The four boosters, forming the first stage, each have a solid propellant motor and are strapped to the core of the launch vehicle. This core comprises three rocket stages.

The first test of this configuration took place on 1st December 1985 with the VLS-R1. This launch vehicle had the same configuration as the intended VLS-1 but failed (propulsion problems) and reached an altitude of only 10 km. The second launch of this test vehicle was on 18th May 1989. This time, the vehicle also encountered problems, but reached an altitude of 50 km and was considered a success.

On 2nd November 1997, a real launch with a satellite on board was hoped to make Brazil a spacefaring nation. The mission was to carry a satellite (SCD-2A) to a 750-km circular equatorial orbit. At the ignition of the first stage, one of its four S43 solid rocket booster motors failed to ignite because a "Safe and Arm Device" failed to transmit the pyrotechnic detonation. The launch vehicle took off with strong asymmetric trust and mass, due to the inert motor. The control system used all its capacity to make the initial maneuvers and to keep the vehicle on top of the trajectory. Given that the launch vehicle was aerodynamically unstable, at a speed of 700 km/hr, the control system was unable to stabilize the flight and an increasing angle of attack developed until the structural load was unbearable. At 26 sec, the structure broke between the second and third stages and the launch vehicle fell safely into the sea. The command destruct system was activated later to check its performance. The SCD-2A, an environmental data relay satellite, was lost.

A second attempt took place on 11th December 1999. A scientific satellite (SACI-2) was on board. The launch vehicle followed the nominal trajectory up to the ignition of the second stage, when this motor blew up. The investigation committee concluded that there was a hot gas penetration between the propellant block and the flexible thermal insulation. The burning surface grew fast as a sub-chamber. It was ascertained that the local pressure might have increased up to some 1,100 MPA, causing the burst of the motor case front dome. The third and fourth stages continued a ballistic flight without control and, 200 sec after launch, were destroyed by safety officers.

VLS-1-3 space launch vehicle at Alcantara [CTA/IAE]

The third flight of a VLS-1 was planned for 25th August 2003. The launch had already been delayed from October 2002, 7th May and 20th June 2003. On 22nd August 2003, the launch vehicle was integrated on the launch pad and a team was working on the trimming of a harness duct on the fourth-stage motor. Suddenly, the first-stage motor D began to fire. The investigation committee could not prove the existence of an electric command of ignition. There was a suspicion that an electrostatic potential developed a spark that initiated the pyro. A good number of tests could not show the susceptibility of the electro-pyrotechnic initiator to electrostatic sparks. In the débris, the investigators found the two electro-pyrotechnics installed on the motor D. One of them was burned by the fire and the other showed evidence that it was somehow initiated. The ignition of the strap-on booster destroyed the launch vehicle and the launch pad, killing 21 people in the resulting blast.

Two satellites, developed by undergraduate students, were lost in the accident. SATEC was a technological satellite planned to test the technological equipment embarked in the VLS. It was a 60-kg box-shaped satellite to orbit at 750-km altitude. The second satellite, UNOSAT, was a nanosatellite comprising an FM transmitter, batteries, four solar panels, antenna and a computer. The satellite was to transmit at regular intervals a voice message and a packet of telemetry. The satellite had a mass of 8.8 kg and measured 46 by 25 by 8.5 cm.

Table 10.3. VLS family of launch vehicles: characteristics

	VLS	VLM
Development	IAE	IAE
First launch	19851201	TBD
Last launch	TBD	TBD
Number of launches	4 [19991211]	0
Configuration	4-stage	4-stage
Mass at launch	49,600 kg	15,786 kg
Length at launch	19.700 m	20.800 m
Thrust at launch	1,030 kN	260 kN
Max attainable altitude	750 km	200 km (5°)/800 km (98°)
Max payload	150 kg	100 kg (5°)/18 kg (98°)
Stage 0: Denotation	S-43 (4x)	n/a
Stage 0: Mass	8,550 kg (4x)	n/a
Stage 0: Propellant	Solid HTPB	n/a
Stage 0: Thrust (vac)	303 kN (4x)	n/a
Stage 0: Isp	260 s	n/a
Stage 0: Number of engines	1 (4x)	n/a
Stage 0: Burn time	62 s	n/a
Stage 0: Length	9,000 m	n/a
Stage 0: Diameter	1.007 m	n/a
Stage 1: Denotation	S-43TM	S-43TM
Stage 1: Mass	8,720 kg	8,720 kg
Stage 1: Propellant	Solid HTPB	Solid HTPB
Stage 1: Thrust (vac)	320.7 kN	359.1 kN
Stage 1: Isp	277 s	277 s
Stage 1: Number of engines	1	1

Table 10.3. *cont.*

	VLS	VLM
Stage 1: Burn time	61 s	61 s
Stage 1: Length	8.100 m	8.100 m
Stage 1: Diameter	1.007 m	1.007 m
Stage 2: Denotation	S-40TM	S-40TM
Stage 2: Mass	5,664 kg	5,664 kg
Stage 2: Propellant	Solid HTPB	Solid HTPB
Stage 2: Thrust (vac)	207.9 kN	202.1 kN
Stage 2: Isp	275 s	275 s
Stage 2: Number of engines	1	1
Stage 2: Burn time	57 s	57 s
Stage 2: Length	5.800 m	5.800 m
Stage 2: Diameter	1.007 m	1.007 m
Stage 3: Denotation	S-44	S-44
Stage 3: Mass	1,025 kg	1,025 kg
Stage 3: Propellant	Solid HTPB	Solid HTPB
Stage 3: Thrust (vac)	33.0 kN	33.0 kN
Stage 3: Isp	275 s	275 s
Stage 3: Number of engines	1	1
Stage 3: Burn time	58 s	58 s
Stage 3: Length	5.800 m	5.800 m
Stage 3: Diameter	1.200 m	1.007 m
Stage 4: Denotation	n/a	VLM-4 (S-33 motor)
Stage 4: Mass	n/a	377 kg
Stage 4: Propellant	n/a	Solid HTPB
Stage 4: Thrust (vac)	n/a	20.5 kN
Stage 4: Isp	n/a	275 s
Stage 4: Number of engines	n/a	1
Stage 4: Burn time	n/a	42 s
Stage 4: Length	n/a	1.120 m
Stage 4: Diameter	n/a	0.660 m

The VLM is a four-stage vehicle based on the core of the VLS-1. It lacks the four strap-on boosters, but a fourth, spin-stabilized stage has been added. The VLM is intended to launch microsatellites from CLA into LEO.

CRUZEIRO DO SUL

The Cruzeiro do Sul (Southern Cross) program is to succeed the VLS-1 space launch vehicle program and to provide a family of small, medium and heavy lift vehicles.[8] The space missions proposed in the PNAE (National Program of Space Activities) are composed of micro, mini and medium-weight satellites into low inclination as well as polar orbits that cannot be fulfilled by the VLS-1. The possibility of being able to launch geostationary missions would complement the range of launch vehicles. Currently, Brazilian sounding rockets and launch vehicles are all equipped with solid-propellant rocket motors and the performance in launch capability cannot be extended anymore. Therefore, a change into liquid propellant motors is mandatory. On the other hand, developing liquid-propellant motors is not consistent with the specific goal of the Cruzeiro do Sul program to reduce development and operational cost of the launch vehicles. Current technology in solid-propellant motors should provide part of the answer.

The first vehicle to be designed in this program (VLS Alfa, formerly also referred to as VLS-2) will have the four strap-on boosters with the proven S-43 motor and the first stage of the core rocket with the altitude-adapted S-43TM motor now used in the VLS-1. The upper stage will have a new liquid-propellant motor with about 75 kN of thrust. It is planned that the development of VLS Alfa will be carried out in parallel with the qualification of VLS-1. VLS Alfa must be capable of launching 400-kg payloads into LEO (400 km).

VLS Beta is the second launch vehicle in the Cruzeiro do Sul program and is a new concept compared with the VLS-1. The configuration is based on a medium-sized, solid-propellant motor for the first stage and a 300-kN-class liquid-propellant motor for the second stage. A third stage, similar to the VLS Alfa third stage, completes this launch vehicle. VLS Beta must be capable of launching 800-kg payloads into LEO and micro satellites into polar orbits.

A heavy satellite into polar orbits is the next step in which VLS Gama should provide. Keeping the three-stage concept, for this launch vehicle, a new 150-kN-class liquid-propellant motor is planned for the first stage, while the upper segment should

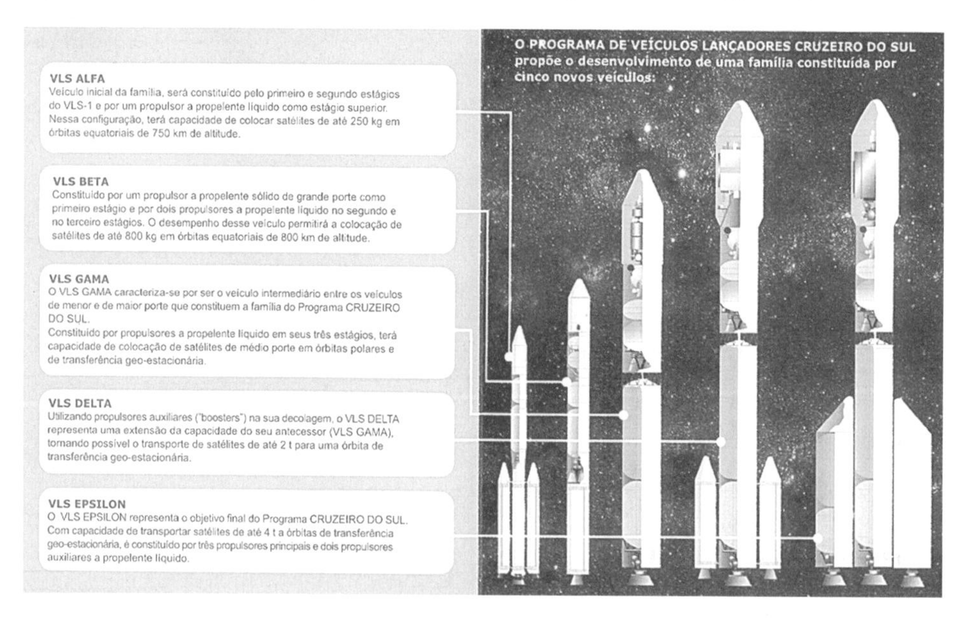

Cruzeiro do Sul program [AEB/IAE]

be similar to that of VLS Beta. Placing 1,000-kg-class Earth observation satellites into 1,000-km polar orbit is the ultimate target for this launch vehicle.

Two more launch vehicles are proposed in the Cruzeiro do Sul program: VLS Delta and VLS Epsilon. These launch vehicles are derivatives and complements of the previous VLS vehicles and will only be developed if a demand for the launch of heavier satellites arises.

The Cruzeiro do Sul launch vehicle program was proposed by the IAE to the AEB in October 2005, the same year as the IAE started to carry out mission analysis and preliminary studies on the VLS Alfa. Due to managerial and priority changes at the IAE and AEB, this action was interrupted in the beginning of 2007. Now, the program is again under discussion at both organizations and should be re-started in 2009. Due to the fact that the VLS Alfa will use the lower part (first and second stages) of VLS-1 and that qualification of the VLS-1 launch vehicle is delayed, the VLS Alfa will not be launched before 2014.[5]

SPACE PROBES AND SATELLITES

Brazilian space probes and satellites can be roughly classified into three categories: science, communications and Earth observation. Brazil used space probes on sounding rockets for scientific purposes from the 1960s. Sub-orbital sounding rockets provide the only means possible to take direct measurements in some regions of the Earth's atmosphere (e.g. sounding rockets can investigate the electrodynamics and irregularities in the ionosphere and mesosphere along the magnetic equator and study their relationship with the neutral atmosphere and winds). The early sounding rockets helped aeronomy scientists to better understand the unique properties of the Earth's ionosphere. Later, starting in 1999 with the use of the VS-30 launch vehicle, Brazil was able to fly microgravity missions for scientific purposes. The SARA mission, planned for 2010, is a prelude to orbital recoverable microgravity research satellites.

Brazil is aware that despite innovation of terrestrial technologies, communications satellites still hold a number of advantages over terrestrial counterparts. The benefits include, among others, ubiquitous coverage, reliability (few outages), bandwidth flexibility and scalability. Of course, there are drawbacks, too, particularly when compared to fiber-optic technology, but these are mostly satellite-related and can be resolved by adapting specific technologies on the Earth side of the process. In the early 1980s, Brazil turned to commercially available communication satellites to mitigate its big communication problems in the vast country. Satellites could be used primarily for broadcasting (reaching even the outback of Brazil) but there were many more applications in the future, such as point-to-(multi)-point communications and mobile services.

The same vast country was the reason that Brazil, from early on, turned to Earth observation by satellite. Although satellite observations can be less accurate than conventional observations, such as by aircraft, their great advantage is their broad geographical coverage in a relatively short time and (automated) change detection.

On the other hand, when a satellite is launched, it is almost impossible to customize its payload to meet later requirements. Brazil turned to observation of the Earth by satellite as early as 1972, when the Cuiabá Data Reception Station became operational. Data reception from foreign civil Earth observation satellites (LANDSAT, SPOT, ERS and CBERS) became possible.

MULTI MISSION PLATFORM (PMM)

The Brazilian Plataforma Multimissão (PMM) is a modern generic satellite architectural concept that provides support for low Earth (600–1,200-km) equatorial and polar orbit missions, which need an attitude control system incorporating a three-axis stabilized platform. With a mass of some 185 kg, the PMM can handle payloads of up to 280 kg and can be launched by 500–600-kg-payload-class launch vehicles. Amazônia, Lattes, MAPSAR and GPM are examples of satellites that will use the Multi Mission Platform. PMM has been developed by INPE and almost all (80%) of its subsystems were devised and developed by Brazilian companies.[9]

SCIENCE

PNAE's activities in atmospheric sciences concern basic and applied research related to phenomena that occur in the atmosphere and in outer space, with primary

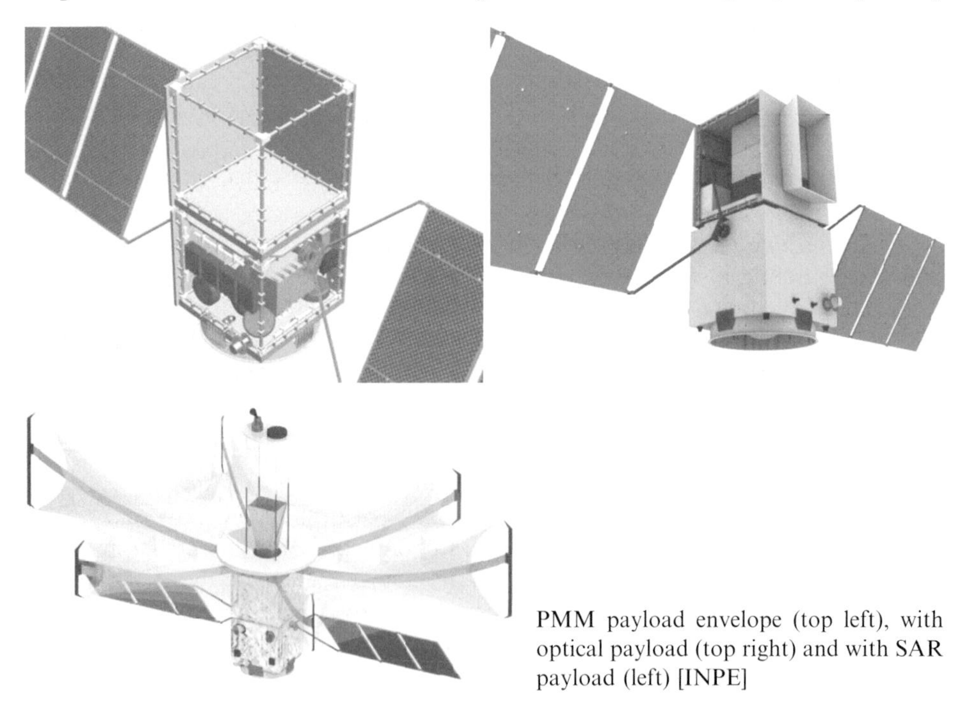

PMM payload envelope (top left), with optical payload (top right) and with SAR payload (left) [INPE]

SONDA I sounding rocket at Barreira do Inferno Launch Center [IAE]

emphasis on the areas of space aeronomy, astrophysics and geophysics. Design, development, qualification and launching of scientific payloads on board stratospheric balloons, atmospheric sounding rockets and satellites are carried out to support these research activities, along with the development of specialized scientific instruments and the establishment and maintenance of laboratories, observatories and other support facilities. The important contribution of the Southern Regional Center for Space Research of INPE, with headquarters in Santa Maria (RS), is considered essential to the accomplishment of these activities. The center's main objectives include research in the areas of aeronomy, space geophysics and radio-astronomy, in addition to fostering the expansion of space cooperation with other countries. Since the launch of the first SONDA sounding rocket in 1965, more than 45 aeronomy/ionospheric experiment payloads on sounding rockets have been launched (see section on sounding rockets in this chapter for more information on the science payloads).

- **Microgravity.** The Brazilian Microgravity Program (BMP), established in 1998 by the AEB, aims to provide a regular sequence of flight opportunities for the Brazilian scientific community to perform experiments and to test hardware facilities in a microgravity environment. Microgravity has become a very important environment for research, covering a wide range of activities such as fundamental and solidification physics, fluid science, biology, biotechnology, human physiology and medicine. The BMP supports the Brazilian microgravity activities by selecting and following the experiments from their initial design phases up to the flight mission and research after recovery. It also looks for feasible and appropriate worldwide microgravity flight opportunities to be used in the program. The objective is to provide the Brazilian scientific community with a range of different facilities and flight conditions in which their experiments can be performed. Among these are sounding rockets, space shuttle flights and the International Space Station. The BMP has its activities technically assisted by governmental institutions such as INPE and the IAE. Until 1998, a few microgravity experiments were done using international and Brazilian facilities, and those rare opportunities were based on private agreements between Brazilian researchers and foreign agencies and companies, as well as on VS-30 and VS-30/Orion sounding rocket flights done by the IAE and DLR/Moraba.[10]

The first documented dedicated Brazilian microgravity launch seems to be the Operação San Marcos Microgravity Mission on 15th March 1999. A VS-30 single-stage sounding rocket, launched from Alcantara, reached an altitude of 128 km, creating microgravity on its return back to Earth. Other microgravity missions were: Lençóis Maranhenses (VS-30, CLA, 6th February 2000), Baronesa (VS-30/Orion, CLA, 21st August 2000), Alecrim (SONDA III, CLBI, 9th December 2000), Cumã I (VS-30, CLA, 1st December 2002) and Cumã II (VSB-30, CLA, 19th July 2007). Between 2005 and 2008, four TEXUS microgravity missions were flown with VSB-30 sounding rockets from the Esrange Space Center in cooperation with the German DLR.

- **Centenary Mission.** In 2006, space shuttle STS-121/Expedition 13 mission restored the ISS to a three-person crew and focused on, amongst others, microgravity science. Expedition 13 began on 29th March 2006, with the launch of crew members commander Pavel Vinogradov and science officer and flight engineer Jeffrey Williams, on the Russian Soyuz TMA-8 from Baikonur, Kazakhstan. Spaceflight participant Marcos César Pontes, a lieutenant colonel from the Brazilian Air Force, also joined the crew. Marcos Pontes had been in astronaut training for eight years preparing for this mission, which lasted 10 days there and back. While on board the ISS, he conducted eight experiments – most of them taking advantage of the lack of gravity – in order to study its effects on enzymes, proteins, DNA and various types of seeds. Although the Pontes mission created much positive media coverage, popularized space science and boosted space morale in Brazil – which had much declined after the VLS-1 accident in August 2003 – the price tag of US$10m met with a lot of criticism.
- **SARA.** The SAtélite de Reentrada Atmosférica was developed by the IAE with the aim to launch microgravity experiments into space. As part of the development plan, first, two sub-orbital flights are planned. SARA is a small, 300-kg platform, designed to perform microgravity experiments lasting a maximum of 600 sec when launched on sub-orbital mission. The experimental sub-orbital module will be a sealed canister housing a 25-kg scientific payload in a fully monitored environment. The platform will be launched by a

Lieutenant Colonel Marcos César Pontes [NASA]

In 1997, Brazil became involved in the ISS and a year later, Marcos Pontes was selected by the AEB and NASA to represent his country in space. The cost for Brazil would be US$10m – only half the normal price according to the AEB because of the space partnership between Russia and Brazil.

Astronaut Marcos Pontes was born in Bauru, SP, on 11th March 1963. His life history is a source of inspiration for a lot of young people. He was born in a humble family and started to work at 14 to help to provide for his family. He dreamed of becoming a pilot and entered the Air Force. Pontes graduated as a military pilot. He has logged over 1,700 flight hours in more than 20 different aircraft, including F-15, F-16, F-18 and MiG-29. Pontes served Brazil in military functions until 1998, when he was selected and began to represent his country as a civilian astronaut in the ISS program at NASA, where he qualified as Space Shuttle mission specialist in December 2000. In 2005, he was invited to participate in Mission Centenary that was created, defined, contracted and managed by the AEB. The goals of this mission were to conduct national microgravity experiments to stimulate this kind of research in Brazil and to promote the Brazilian space program. On 2nd September 2005, Pontes started to train at Star City, near Moscow, to learn about the Soyuz space capsule's operational and life-support systems and to fly to the ISS.

Commander Pavel Vinogradov and flight engineer Jeff Williams lived in space for 183 days. They began their mission on 29th March 2006, when they were launched from the Baikonur Cosmodrome in a Soyuz TMA spacecraft. They were accompanied by Marcos Pontes, who returned after 10 days. Pontes flew to the Space Station as part of a commercial agreement with the Russian Federal Space Agency Roscosmos. In the Brazilian Centenary Mission, eight experiments, covering technological, biological and educational areas, were developed. During his time on the ISS, Pontes not only conducted microgravity experiments, but also completed a ham radio contact with a school in Brazil. Pontes came back to Earth with the Soyuz TMA-7.[11]

Marcos Pontes works in the Destiny Laboratory of the ISS during his eight days in space [NASA]

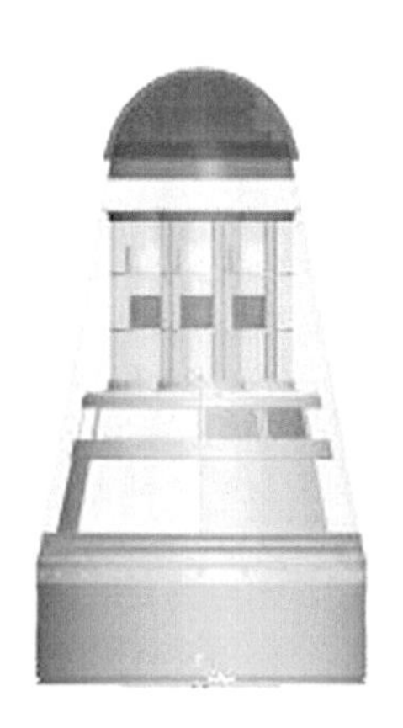

SARA payload [IAE]

modified version of the VS-40 sounding rocket that will provide a range of 300 km and an apogee of 400 km.[12]

When used operationally in Earth orbit, the 55-kg payload module of SARA will perform microgravity experiments for a maximum of 10 days. After the orbital phase, the spacecraft will initiate its return to Earth and land at Lençóis Maranhenses, a dune region near the Alcantara Launch Site.

- **SACI.** The Satélite Avançado de Comunicações Interdisciplinares (Advanced Interdisciplinary Communications Satellite) project was meant to be low-cost, accomplished in a short period of time and reliable. The philosophy of the project was to involve Brazilian universities and space industries. SACI-1 was a small 60-kg spin-stabilized satellite developed by INPE and launched piggyback on flight 1999-057A (CBERS-1, Long March 4B, 15th October 1999, 750-km Sun-synchronous orbit). The experimental scientific/technologic microsatellite housed a magnetometer, particle detectors and an atmospheric experiment. The intended lifetime of SACI-1 was 1.5 years but INPE lost contact with the satellite almost immediately after launch. The payload (28 kg) of SACI-1 was built around four scientific experiments. The main objective was to investigate the generation, development and decay of plasma bubbles, particularly in the Brazilian region of the ionosphere (PLASMEX). This investigation intended to elucidate the strong influence of such bubbles and associated plasma turbulence in several space application systems (remote sensing with radar, space geodesy, trans-ionosphere telecommunication, etc.). The discovery of such phenomena in the Brazilian ionosphere was reported by measurements with sounding rocket photometers and ground-based sounders. Another experiment had the objective of measuring the intensity of the global distribution of terrestrial airglow emissions (FOTSAT). A third experiment was called Solar and Anomalous Cosmic Rays Observation in the Magnetosphere (ORCAS), which was to measure anomalous cosmic rays. The fourth experiment dealt with geomagnetism (MAGNEX). The geomagnetic field controls the movement of charged particles in the space around the Earth and protects the planet from the solar wind. Simultaneous measurements of the geomagnetic field on the Earth's surface and in the near space plasma are essential to the study of geomagnetic phenomena.[13]

Although the shape and configuration of the SACI-2 satellite were different from the SACI-1, the purposes of its mission was scientifically the same as for SACI-1 and it had a function to collect environmental data such as temperature, air humidity, wind direction and atmospheric pressure from ground platforms located all over Brazilian territory. SACI-2 was an 85-kg satellite with a central body like a cube and the solar arrays formed something like a hat brim. For practical purposes, the solar array was designed and made as rigid as possible so that the spacecraft would be injected into its orbit in that configuration. It meant that no solar panel deployment operation would be necessary. On the other hand, the spacecraft comprised four mechanical arms that would be deployed in orbit. The main application of the arms was to keep the sensors away from the satellite electronics.[14] SACI-2 was installed on board and scheduled to be launched with the second VLS-1 launch vehicle. The VLS rocket failed to launch properly on 11th December 1999, and had to be destroyed by safety officers. INPE then cancelled the SACI program.

- **Lattes.** Initially, two 150-kg scientific satellites were based on an agreement between INPE and CNES. The motivation was to develop a low-cost platform that could be launched by the VLS space launch vehicle, but CNES abandoned this concept in 2003 in favor of the Proteus platform (satellites such as Jason-1 and -2 (topography of the oceans), Calypso (aerosol), Corot (exoplanets) and SMOS (sea salt) are based on this platform). With the exit of CNES, INPE initially assumed on its own the development of the MIRAX and EQUARS missions. During the review of the PNAE for 2008–2011, INPE proposed that the two scientific missions be consolidated into a single satellite, based on the PMM, that could be launched into an equatorial orbit.[15] For completeness, the EQUARS and MIRAX missions are described below.

- **EQUARS.** The scientific mission of the Equatorial Atmosphere Research Satellite was defined by INPE's Aeronomy division with the objective to understand atmospheric coupling with dynamical, electrical, photochemical and ionospheric processes and to apply the data to atmospheric, space weather and climate studies. The study is carried out in the equatorial low, middle and upper atmosphere, with special emphasis on vertical energy transport, propagation of gravity, tidal and planetary scale waves, and the generation and development of plasma bubbles in the ionosphere. EQUARS was to house for these purposes eight scientific instruments with a payload mass of 30 kg and would have been launched in a 20° inclination, 750-km equatorial orbit.[16]
- **MIRAX.** The X-Ray Monitor and Imager is an X-ray astronomy observatory that was proposed by the high-energy astrophysics group at INPE as a small-scale satellite weighing approximately 200 kg. This first Brazilian astronomical satellite hoped to bring the country into the select group of nations that developed and launched satellites to observe the universe. The objective of the mission was to observe regions of the universe that emit X-rays. The satellite was expected to be launched in a low-altitude (550–600-km) circular equatorial orbit around 2011. MIRAX would have been able to make unique contributions to the study of energetic transient phenomena in astrophysics by virtue of its observing strategy, which departs significantly from traditional pointed programs and scanning monitors. MIRAX would detect, localize, identify and study unpredictable phenomena that last on the timescales of minutes to days, which would otherwise be missed by traditional observing strategies.

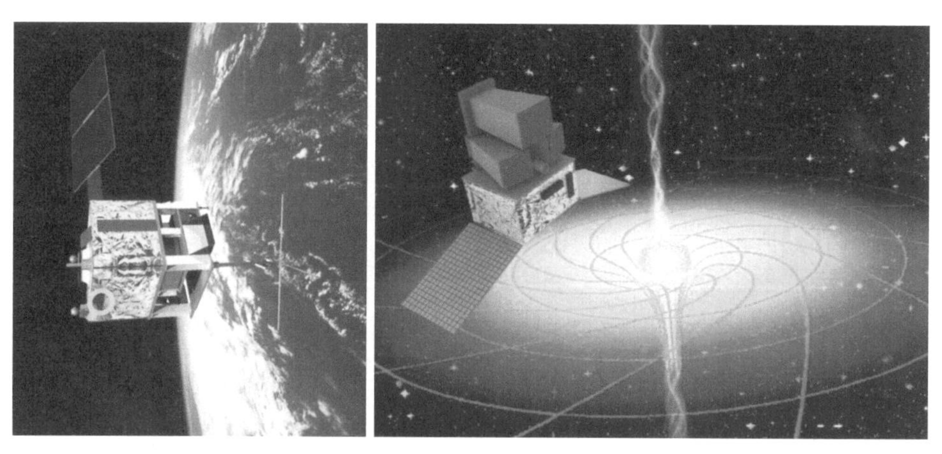

The EQUARS (left) and MIRAX (right) scientific satellites that precluded the Lattes scientific satellite [INPE]

COMMUNICATIONS

The Sistema Barasilero de Comunicação por Satélites (SBTS) or Brazilian Telecommunications Satellite System was put into operation early in 1985, with the launch of Brasilsat A1 or SBTS A1. This communication satellite was produced for Telebrás, based on the Hughes HS-376 space bus. Telebrás was a holding company, controlled by the government, with an interest in various telecommunications companies in Brazil. During the National Privatization Program, in May 1998, Telebrás was broken up into 12 holding companies, each one responsible for some of the Telebrás ownership interest, and these separate companies were auctioned to private bidders. The newly formed companies were the long-distance operator Embratel, three fixed-line regional telephone companies and eight cellular telephone companies. On 29th July 1998, the separation became effective and Empresa Brasileira de Telecomunicacões (Embratel) was turned into a state-controlled agency in charge of the SBTS, owning and operating the Brasilsat communications satellites and responsible for the domestic Brazilian communications network and links through the Intelsat system via its Tangua ground station. It is also Brazil's national Inmarsat signatory. The communication satellites are operated from a Satellite System Operational Center located in Guaratiba near Rio de Janeiro. This center comprises a Spacecraft Control Center (SCC) and a Communications Operations Control Center (COCC). The SCC provides tracking, telemetry and control through 14.2 and 6-m dish antennae; the COCC primary traffic routing is through a 16.2-m dish antenna.

- **Brasilsat.** The Brasilsat satellite was the standard built by Spar Aerospace Ltd of Canada under license from Hughes Space and Communications (now the Boeing Company). The Hughes HS-376 space bus was equipped by Spar Aerospace with 24 active and six backup C-band transponders and clocked some 1,200 kg at beginning of life when launched. The HS-376 spin-stabilized spacecraft is one of the world's most purchased commercial communications satellite models. These HS-376 models have two telescoping cylindrical solar panels and antennae that fold for compactness during launch. The basic satellite bus can accommodate all kinds of communication payloads and can be launched by most of the world's major space launch vehicles.

  In 1990, Telebrás purchased two more 376 models for the Brasilsat B series. These were the more powerful 376-W model spacecraft. In 1995, Brasilsat B3 was ordered, followed by Brasilsat B4 in 1998. Like the first-generation Brasilsat satellites, the Brasilsat B spacecraft provide basic telecommunications services such as telephone, television, facsimile and data transmission, and provide for business networks. The 376-W is wider and taller and provides 60% more radiated power than Brasilsat A. Brasilsat A provided service throughout the nation and the Brasilsat B satellites provided higher performance for VSAT customers and businesses using their own small antennae. Where the Brasilsat A spacecraft carried 24 C-band active transponders and had eight-year mission lives, Brasilsat B1 and B2 have 28

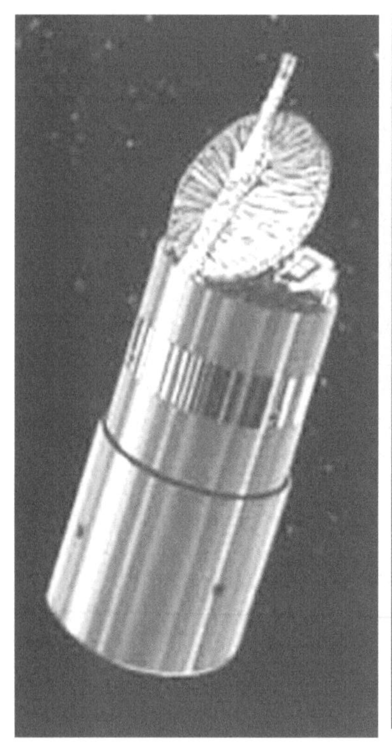

Brasilsat A series/HS 376 space bus [Boeing]

Brasilsat B series/HS 376W space bus [Boeing]

C-band active transponders and one military X-band. Brasilsat B3 and B4 do not have X-band transponders. Telebrás, and later Embratel, chose Ariane space launch vehicles to launch its satellites (see Annexes for actual launch data).

- **Estrela do Sul.** Estrela do Sul/Telstar 14 is a commercial, FS 1300 space bus-type satellite with coverage in North and South America and the North Atlantic Ocean Region. The satellite offers focused coverage in South America covering Brazil, the Andean region including Colombia and Panama and the Mercosul region including Argentina and Chile. The satellite offers connectivity options both within a region and between regions. The satellite was launched into space on 10th January 2004 with a Boeing Sea Launch Zenit-3SL space launch vehicle from the Odyssey Launch Platform, positioned on the equator in the Pacific Ocean and has been positioned at 63°W on the geostationary orbit. The satellite is operated by Loral Skynet do Brasil, equipped with 41 Ku-band transponders for video and internet services. The communications satellite failed to fully deploy one of its two power-generating solar panels, but is still considered operational, although, since May 2007, it started to drift between 61° and 65°W. Loral Skynet do Brasil is the first private Brazilian satellite company to offer Ku-band satellite

Estrela do Sul commercial Ku-band geostationary communications satellite [Space Systems Loral]

services and was formed primarily to address opportunities in the fixed satellite services market in Brazil and South America. The company has a ground station in Rio de Janeiro that will provide tracking, telemetry, and control and access management from in-country technical staff.[4, 17]

- **Amazonas.** The telecommunications satellite Amazonas was built by France's EADS Astrium for the telecommunications firm Hispamar, a Brazilian subsidiary of HISPASAT in Spain. The spacecraft has a mass of about 4,550 kg, is based on the Eurostar E3000 satellite platform and will serve customers in Latin America – as well as North American and European markets. Atop a Russian Proton-M launch vehicle that lifted off from the Baikonur launch site in Kazakhstan on 4th August 2004, the satellite was launched into a geosynchronous transfer orbit. Amazonas has 36 Ku-band and 27 C-band transponders on board for basic television broadcasting, internet, corporate communications and broadband applications. The expected lifetime of the satellite is 15 years. HISPASAT has awarded Arianespace the launching of their new communications satellite, the Amazonas 2, which will also be situated in the orbital position 61°W. With this new launch, planned for 2009, the group takes one more step in the process of growth and consolidation of this orbital position. Amazonas 2, whose construction was awarded in June 2007 also to EADS Astrium, will have an operational life of 15 years and will give fixed communications and radio broadcasting services by means of a total of 64 simultaneous

Star One C1 is shown being installed atop the Ariane 5 core stage [ESA - CNES - Arianespace]

transponders, of which 54 will operate in Ku-band and 10 in C-band. The satellite will have a mass at launch of 5.5 tonnes.[4, 17]

- **Star One C.** The first of this series of satellites, also known as Brasilsat C1, was launched into orbit on 14th November 2007 on an Ariane-5ECA launch vehicle from the Kourou space launch site in French Guyana. The satellite is based on an Alcatel Space 3000B3 space bus and has been positioned at 65°W on the geostationary orbit. It will provide communications and broadband internet services over the entire South American continent for the Brazilian satellite operator Star One, which is a joint venture between several South American countries under Embratel leadership. Star One C1 has a payload of 28 C-band, 16 Ku-band and one X-band transponders and has a total launch mass of some 4,100 kg. The estimated lifetime is 15 years. A duplicate satellite, Star One C2, was launched on 18th April 2008 and also positioned at 65°W. Star One C satellites will replace ageing Brasilsat B satellites.[4, 17]

- **SGB – Brazilian Geostationary Satellite.** Satélite Geoestacionário Brasileiro is a project of the federal government, run by the Brazilian Air Force Command. The first SGB is tentatively scheduled for launch in 2009. SGB is seen as fundamental for the aerospace sector and for the country's air traffic. The satellites will handle the needs of air traffic services in accordance with current international agreements and technology. It will operate in conjunction with other mechanisms, such as meteorological sensors, to be used in communications related to national security and civil defense. SGB will implement the Communications Navigation & Surveillance (CSN)/Air Traffic Management (ATM) concept. One of the objectives of SGB is that with this kind of satellite, the government could control state security/military communications and be independent of international suppliers. A second objective is that it can improve the monitoring of the Amazônia, maritime monitoring for meteorological purposes and air traffic control – all essential for the protection and sovereignty of the country. The SGB concept comprises three satellites in geostationary positions, operated by Brazil's telecommunications regulator ANATEL, of which two satellites will have C-band, L-band and X-band transponders and the other will have Ku-band transponders and meteorological sensors.[18]

  In 2008, CNES signed an agreement with Brazil in which CNES will help define a mission specification for the SGB. Based on initial data supplied by Brazil, CNES will analyze service requirements and programming constraints, and make a preliminary analysis of various mission and system scenarios and a risk analysis of the SGB mission. In addition, CNES will supply an instrument to be integrated into the satellite that will allow CNES to study propagation of high-frequency waves (Ka-band), used for future satellite telecommunications.[19]

EARTH OBSERVATION

Remote sensing by satellites is used for observation of the Earth and is of paramount importance for the monitoring of the vast Brazilian territory. Brazil has an Earth observation program comprising an indigenous Data Collecting Program, the Sino-Brazilian remote sensing program, planned Earth observation satellites and the use of civil foreign satellite imagery.

- **SCD.** The Satelite de Coleta de Dados Program started in earnest with the launch of the SCD-1 satellite on 9th February 1993, with a Pegasus launch vehicle, carried by a United States B-52 aircraft. The SCD-1 was the first satellite totally designed, manufactured, integrated and tested in Brazil. With a design life of one year, the SCD-1 has been operational since then. The SCD system provides environmental data on the various regions of the country, through the Data Collecting Platforms on the ground. Local environmental data are collected by the platforms from various sensors, which are retransmitted by the system's SCD satellites to the receiving stations at Cuiabá and Alcantara. The data are then processed by the Cachoeira Paulista SCD Mission Center and made available to users via the internet. The main applications are in the areas of meteorology and hydrology. By the time SCD-1 was launched, the initial network of data collection platforms comprised approximately 60 installed units. By 2003, the number of platforms had reached 600, reflecting the growing and continued interest of users. This interest derives mainly from the operational and economic advantages offered by the system for the collection of data on a national basis. The satellites were part of the MECB program that called for, amongst others, developing and building four such satellites. The second satellite, SCD-2A, was lost in the failed launch of the first VLS-1 launch vehicle on 2nd November 1997. On 23rd October 1998, the SCD-2 was successfully launched with a Pegasus launch vehicle, providing continuity and improvement to the SCD program. CBERS satellites are also equipped with an SCD transponder.

SCD-1 [INPE]

The SCD satellite is an octagonal-based prism (1 m in diameter and 1 m in height) with a mass of 110 kg. The satellite is spin-stabilized and is covered with solar panels. The data collection transponder works on UHF and S-band. The SCD-2 is very similar to the SCD-1 but has an additional reaction wheel experiment. The estimated lifetime of SCD-2 was two years but this already has been exceeded. The satellites fly in circular orbits (750 km) inclined 25° to the equator.[14, 20]

- **CBERS.** Initially, the China–Brazil Earth Resources Satellite or CBERS program (Table 10.4) called for the production and launch of two remote sensing satellites. The program is coordinated by the China National Space Administration and the AEB and carried out by the China Academy of Space Technology and INPE. Two identical satellites with the Chinese name Zi

CBERS-1 [China Academy of Space Technology]

Yuan 1 were launched on 14th October 1999 and 21st October 2003, respectively. Based on the success of these satellites, China and Brazil decided to produce two more CBERS satellites, but more advanced ones, to be launched in 2010 and 2013. In the meantime, it was necessary for the continuation of the supply of images that a new CBERS was launched. This satellite was almost identical to the first two but differed in the infrared camera that was replaced by an improved camera in the visible spectrum. Additional improvements were a new onboard recording system, incorporation of GPS (Global Positioning System) and a star sensor. A further two CBERS satellites are planned for 2016 and 2019. CBERS-7 will be a synthetic aperture radar satellite.

The second-generation CBERS-3 and CBERS 4 will also be identical. They will carry a Chinese four-band panchromatic (5-m resolution) and multispectral (10-m resolution) camera, a Brazilian CCD (20-m resolution in four bands), a Chinese infrared multispectral (40–80-m resolution in four bands) scanner and a four-band Advanced Wide Field Imager (AWFI-1) with 60-m resolution from Brazil. The satellites will have a mass of about 2,000 kg and a design lifetime of three years.[14, 21]

Table 10.4. First-generation CBERS parameters

	CBERS-1	CBERS-2A	CBERS-2B
Chinese name	Zi Yuan-1-1	Zi Yuan-1-2A	Zi Yuan-1-2
Launch date	1999–10–14	2003–10–21	2007–11–14
Launch vehicle	Long March 4B	Long March 4B	Long March 4B
Catalog number	1999-057A	2003-049A	2007-042A
Mass	1450 kg	1450 kg	1450 kg
Design lifetime	2 years	> 2 years	> 2 years
Orbit	780 km SSO	780 km SSO	780 km SSO
Sensor 1	5-band CCD Camera	5-band CCD Camera	5-band CCD Camera
Resolution	20 m	20 m	20 m
Swath width	113 km	113 km	113 km
Sensor 2	4-band IR multi-spectral scanner	4-band IR multi-spectral scanner	High-resolution camera
Resolution	80–160 m	80–160 m	2.5 m
Swath width	120 km	120 km	27 km
Sensor 3	2-band wild field imager	2-band wild field imager	2-band wild field imager
Resolution	260 m	240 m	240 m
Swath width	890 km	890 km	890 km
Aux. payload 1	Data collection system	Data collection system	Data collection system
Aux. payload 2	Space environment monitor	Space environment monitor	Space environment monitor
Status	Decommissioned 2003	Operational	Operational

- **Amazônia.** The first satellite built with the PMM satellite bus is Amazônia, an optical Earth observation satellite. Amazônia is scheduled for launch in 2011 and will be proof of concept for the PMM. The satellite will carry a UK-made high-resolution camera and two Brazilian cameras to monitor the Amazon region for deforestation and urban expansion from a polar, 780-km, circular orbit using an Advanced Wide Field Imager (AWFI-2) with 40-m resolution and a swath width of 700 km in four spectral bands. It will complement CBERS-3 and the two AFWIs together will have global coverage every two days. The design lifetime of Amazônia is four years.
- **MAPSAR.** The Multi Application Purpose SAR is the

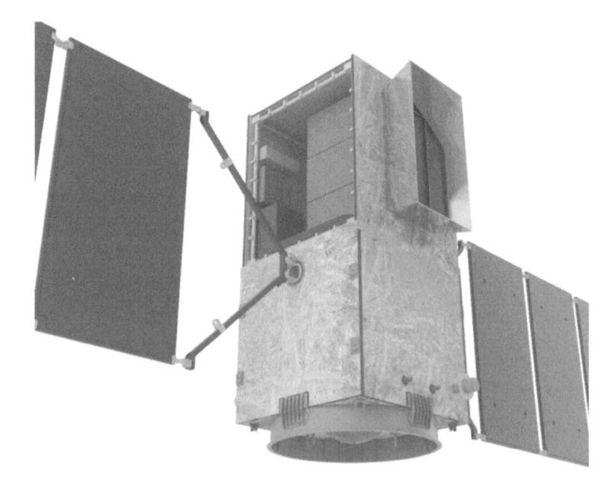

Amazônia [INPE]

result of an initiative for cooperation between INPE and the Deutsches Zentrum für Luft- un Raumfahrt (DLR), for the development of a satellite with a Synthetic Aperture Radar (SAR). In 2001, a feasibility study commenced for the MAPSAR. INPE would be responsible for the platform (PMM) and the integration of the satellite and DLR would design and manufacture the payload. MAPSAR had to fulfill requirements for German and Brazilian users for applications such as agriculture, cartography, disaster management, forestry, geology, geomorphology, hydrology, oceanography, urban studies and defense/intelligence. As all these applications required different parameters in frequency, resolution, orbit, etc., a choice had to be made. MAPSAR eventually will have a light and innovative multi-polarization L-band SAR sensor and the main mission objectives are the global assessment, management and monitoring of natural resources from a Sun-synchronous orbit. The resolution will be 3–20 m with a swath width of 20–55 km. During its four years' lifetime, different mission scenarios with different coverage sequences are proposed. The possibility of repeat pass SAR interferometry as well as SAR stereoscopy with the help of a slowly varying orbit will be one of the great advantages of this mission. MAPSAR is scheduled for launch in 2013.[22, 23]

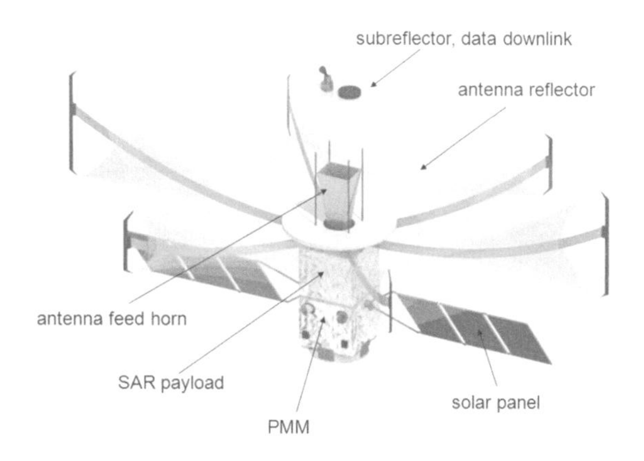

MAPSAR [INPE]

- **GPM.** The GPM-Br program implements an equatorial orbit satellite used for precipitation estimates and participation in the Global Precipitation Measurement International Program. It is a joint effort by the American NASA, Japan's JAXA, other participant agencies like the AEB and the GEOSS (Global Earth Observation System of Systems), which includes more than 50 countries. The program intends to monitor global precipitation every three hours in areas of 25 by 25 km. GPM will comprise a core satellite and a constellation of secondary small satellites. The core satellite's payload is composed of an active microwave sensor to observe precipitation's vertical structure, which will provide the calibration for similar sensors on board the secondary satellites. The small satellite constellation (around eight) will be equipped with passive microwave sensors and will provide three-hourly time interval measurements of global precipitation fields. The program also comprises calibration and validation measurement sites with instruments and

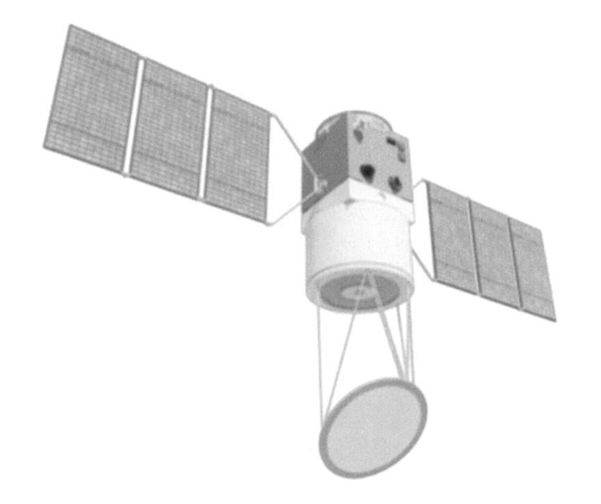

GPM-Br [INPE]

sensors able to measure different precipitation characteristics such as precipitation rate, raindrop size and distribution, hydrometeor characteristics, cloud type, etc. The GPM-Br will fly in a circular 600-km high orbit, inclined at 30° to the equator. It will feature two radiometers (three frequencies) with a 15-km resolution.[24]

REFERENCES

1. Dolinsky, Mauro Melo: IAE – presença Brasileira no espaço (IAE – The Brazilian presence in space). 21st August 1989 (in Portuguese).
2. www.planet4589.org/space/lvdb/ (accessed 2 March 2009).
3. www.astronautix.com (accessed 2 March 2009).
4. ribs SC&I/DB&C Databases, Breda, The Netherlands, 2008.
5. E-mail exchanges between author and officials of IAE, December 2008 and March 2009.
6. Ariovaldo, Felix Palmerio, et al.: Results from the first flight of the VSB sounding rocket. August 2005, ESA SP-590.
7. www.ssc.se (accessed 2 March 2009).
8. Moraes Jr, Paulo: An overview of the Brazilian launch vehicle program Cruzeiro do Sul. CTA Space Directorate, IAF, 2 October 2006, São José dos Campos, Brazil.
9. The multi mission platform & applications, Power Point presentation, INPE, Brazil, 31 October 2007.
10. Bandeira, I., Bogossian, O. and Corrêa, F.: Centenary Mission – first Brazilian microgravity experiments at ISS. *Microgravity – Science and Technology Journal*, Vol. 19, Numbers 5–6, September 2007.
11. www.marcospontes.com/ (accessed 1 July 2009); www.nasa.gov/ (accessed 1 July 2009).
12. Loures da Costa, Luis E. and Villas Boas, Danton J.F.: *SARA Suborbital Development Status*. IAC-06-A2.3.8, 2006.
13. Neri, José Ângelo C.F. and Fonseca, Ijar M.: *The Brazilian Scientific Microsatellite – SACI-1*. INPE, 1998.

14. Fonseca, Ijar M. and Bainum, Peter M.: *The Brazilian Satellite Program – A Survey*. AAS 05-479, 2005.
15. INPE Annual Report 2008.
16. www.laser.inpe.br/equars/eng/workshop.shtml (accessed 1 July 2009).
17. www.lyngsat.com/ (accessed 1 July 2009).
18. www.defesanet.com.br/md/satelite_geo.htm (accessed 1 July 2009).
19. www.cnes.fr/web/CNES-en/7134-press-releases.php?item = 1308 (accessed 1 July 2009).
20. National Program of Space Activities 2005–2014: PNAE/Brazilian Space Agency. Brasília, Ministry of Science and Technology, 2005.
21. Machado e Silva, A., Liporace, F., Dr, and Santos, M.: *CBERS: A Reference in the Brazilian Space Program*. Gisplan, 2007.
22. INPE: *MAPSAR (Multi-Application Objet SAR)*, www.obt.inpe.br/satelite/mapsar (accessed 1 July 2009).
23. Schröder, R., et al.: *The MAPSAR Mission: Objectives, Design and Status*. Germany, DLR/Brazil, INPE, April 2005.
24. AEB: *GPM-Br Mission*, http://pindara.cptec.inpe.br/gpm/implementacao_ing.html (accessed 8 July 2009).

11

Brazil: Plans – the ultimate goal

The ultimate goal for Brazil is independence in the space arena: independent access to space by means of indigenous satellite launch vehicles; independent satellites with beneficial applications. What started as a presumed military need in the 1960s changed into a realization that Brazil could become an environmental power and one of the first developed nations in the tropics.

The definite turning point seems to be the aftermath of the tragic accident on 22nd August 2003, when the VLS-1 launch vehicle exploded on the launch pad during its launch campaign. The sudden loss of all 21 people on the launch platform shocked the nation. "On behalf of the government I want to express deep regret over the death of these workers who were labouring in such a noble cause," Defense Minister Jose Viegas said. Brazilian President Luiz Inacio Lula da Silva "deplored" the accident and "paid homage to the victims and their families" his spokesman said, adding that he had "reaffirmed" Brazil's determination to pursue space technology.[1] The direct consequences of the accident were that Brazilian space activities declined to virtually nothing. No indigenous satellite was launched, although the first launch of the VSB-30 (23rd October 2004) successfully took place, carrying out the Cajuana microgravity test. It would take until 19th July 2007 before the next launch (also a VSB-30) occurred from Brazilian soil.

This did not mean that no space-related activities took place at all. By the end of 2007, three telecommunications satellites belonging to private Brazilian companies were launched and went into operation and two CBERS satellites were launched by China. International cooperation also continued. In January 2004, during an official visit to India, an agreement was signed for the installation of a receiving station for the Resourcesat-1 in Brazil and the launch of domestic microsatellites as secondary payload on board Indian rockets. Reportedly, the European Union was negotiating with Brazil on participation in the Galileo satellite navigation project and Marcos César Pontes flew on board a Russian Soyuz space capsule to the International Space Station. At the same time, the marketing of CLA and the VLS, and the joint development of satellites, were high on the agenda. Notably, an agreement between Brazil and Ukraine for use of CLA for launches of the Cyclone-4 launch vehicle was signed. Brazil and China signed follow-on agreements that enabled the launch of the

Launch of the Cajuana microgravity mission with a VSB-30 on 23rd October, 2004 [IAE/CTA]

CBERS-2B satellite to ensure that the supply of images was not interrupted and regulated marketing of images from CBERS satellites to other countries. See Annexe 3 in Chapter 15 for a compilation of Brazilian space-related institutions and industries.

A review of the National Space Activities Program (PNAE) was undertaken with vigor. The final document for the period 2005–2014 not only highlighted the space strategy for Brazil, but also laid out the guidelines to make that strategy come true. Earth observation, scientific and technological missions, telecommunications and meteorology, all, for Brazil, necessary applications, were addressed. The development of orbital and sub-orbital access to space was declared of strategic importance to ensure Brazil's autonomy and had to be done autonomously or through international partnerships. First of all, the small VLS-1 launch vehicle had to become operational. Second, the development of liquid-fuel technology for the development of medium and large space launch vehicles had to be embarked upon. And last but not least, subjects such as infrastructure, R&D, human resources, industrial policies, etc. were addressed.[2]

SPACE ACCESS

When PNAE 2005–2014 was published, it stated that sounding vehicle technology had contributed toward the development of the small satellite launch vehicle VLS-1 and that the next stage called for the development of liquid-fuel technology to allow for the development of mid and large-scale launch vehicles. Brazil has successfully operated sounding rockets for sub-orbital space exploration missions and scientific and technological experiments. The closer involvement of universities and research centers in the space program will increase the demand for this type of vehicle, therefore justifying their continued production, estimated to be at least two vehicles per year. The production of the sounding vehicles will gradually be transferred to the national industrial sector, to meet the needs of scientific and technological missions. Under development is the VS-43, a one-stage vehicle with attitude control using the S43 booster motor. In addition, to meet domestic needs, it is expected that the sounding rocket market has interest in this

The ill-fated VLS-1 on the launch pad at CLA [IAE]

vehicle. As the S43 motor is also used in the four boosters and second stage of the VLS, the development is probably affected by the VLS-1 accident. It took until 20th October 2008 for the IAE to accomplish a first test of the S43 motor at the UCA. The test was successful. The objective of the test was to evaluate the modifications of the S43 suggested by the Makayev Bureau.[3]

Flight qualification of the VLS-1 started with the first (failed) launch on 2nd November 1997. The second, also failed, launch on 11th December 1999 and the accident on 22nd August 2003 justifiably set the development on the back burner. A critical review was called for and a new qualification flight still has to be scheduled.

The Cruzeiro do Sul (Southern Cross) program is to succeed the VLS-1 space launch vehicle program and to provide for a family of small, medium and heavy lift vehicles. This is necessary because the space missions proposed in the PNAE comprise micro, mini and medium-weight satellites into low inclination as well as polar orbits that cannot be fulfilled by the VLS-1. The possibility of launching geostationary missions would complement this range of launch vehicles (see Chapter 10).

On 19th September 2008, RiaNovosty wrote in an unconfirmed press bulletin that in early 2008, Russia and Brazil had concluded an agreement to develop a family of launch vehicles as part of the Cruzeiro do Sul program: "... this technological alliance must develop an all-purpose rocket based on Russia's Angara vehicle plus three types of power units. Two of them are under development with Brazil's full participation and are based on Russian liquid-propellant engines. The first stages of Brazil's Gamma, Delta and Epsilon launchers will be powered by a unit based on the RD-191 engine developed for Russia's new Angara rocket. The upper stage, which will be the same for all Cruzeiro do Sul rockets, will be driven by an engine which is currently part of Russia's Molniya launch vehicle. The third stage will be a solid-propellant booster. It has been developed by Brazilian engineers for an upgraded version of the VLS-1 and its modifications. The Gamma launcher is part of the light-weight class, but using the near-equatorial position of the spaceport, it can place almost a ton of payload into a geostationary orbit. The Delta launcher (medium-weight class) differs from the Gamma by having four solid-propellant boosters attached to the first stage. Its payload deliverable to a geostationary orbit is 1.7 tonnes. The heavy Epsilon launcher (its first stage consists of three identical units, the same as on the Gamma and Delta) can place a four-tonne spacecraft in orbit, if it is launched from Alcantara. These three almost fully unified rockets will cater to all ranges of payload needed now and likely to be needed in the future. Helped by Russia, Brazil will not only gain a chance to enter space on her own, but also to make commercial launches for other countries. The Brazilian government is planning to allocate US$1bn for the project in the next six years. It has already set aside US$650m for the construction of five launch pads able to handle up to 12 launches per year. Its commercial launches can bring Brazil between US$60m and US$100m million. To pool its financial resources, the country has decided to cancel some of its cooperation agreements with other countries. Ukraine is a likely candidate for cancellation."[4]

A foreign space launch from Brazilian soil. Brazil has expressed an interest in

The Angara family is a new generation of launch vehicles built around the generic rocket module equipped with an RD 191 engine using kerosene and liquid oxygen (LOX) as propellant [Khrunichev Space Center]

using the Ukrainian launch vehicle Cyclone-4 for orbiting satellites, said Eduard Kuznetsov, Deputy General Director of Ukraine's National Space Agency. He said in September 2008 that protocols were already drawn up for using the launch vehicle in 2011–2014. The Cyclone-4 project is crucial for Ukraine's (launch vehicle) future and if the project proceeds, it will benefit the Ukrainian space industry considerably. To implement the project, the Ukrainian–Brazilian joint venture Alcantara Cyclone Space was set up. Brazil's greatest space asset is the Alcantara Launch Center (CLA) on the Atlantic coast, close to the equator. Advantages are that such closeness creates the best conditions for launching satellites into geostationary transfer orbits and that in the case of an accident, the fragments of a rocket would fall into the ocean.

The technology provides for up to 12 launches per year from CLA. But the possible figure is four to five launches: neither Ukraine nor Brazil has a niche for more in their space programs. To break even financially, the minimum number is six launches. In that case, all costs could be recouped within 10–12 years and profits would start coming in. In 2005, Ukrainian authorities said the rocket engineering specifications had been handed over to the Pivdenmash manufacturing facility in Dnepropetrovsk, which began making its parts. The first launch of the Cyclone-4 from Alcantara was scheduled for 2006. But recurring political crises in the Ukraine stalled financing. It is now clear that a quantity production of the Cyclone-4 at Pivdenmash will not begin until 2009, and launch facilities for it will be built in Brazil

The Cyclone-4 launch vehicle from the Ukraine [NSAU]

by 2011 at best. And these are not the only snags. Most important is unstable Russian–Ukrainian relations. Russia is Ukraine's contractor. Most of the space infrastructure is constructed by Russian industrial enterprises. And first among them is the Design Bureau of Transport Engineering (KBTM), which built launch facilities for Cyclone-2 and Cyclone-3 at Plesetsk. The first-stage engines of the Ukrainian rocket were developed by Russia's Energomash. Russia is also responsible for fuel components. Furthermore, the American aerospace lobby looks askance at Ukraine's efforts to enter the world space market. Another factor to consider is that Russia has proposed to Brazil they develop an environmentally friendly launch vehicle, because Cyclone-4 is not ideal in that respect. It was, incidentally, one of the reasons why Kazakhstan banned the launches of its forerunners from Baikonur: they were using the same toxic fuel.[4] Again, this press bulletin could not be confirmed by Brazilian officials.

APPLICATION SATELLITES

The aim of the PNAE is to enable Brazil to develop and use space technology in order to address national problems and thereby contribute toward the improvement of the quality of life by generating wealth and creating jobs through scientific research and by increasing awareness of issues regarding their territory and environment. Toward this end, the following applications were selected to have top priority: Earth observation, technological and scientific missions, telecommunications and meteorology. Some of the plans in the 2005–2014 version of the PNAE do not exist anymore. They were either canceled for a number of reasons or have been replaced by new plans. The following assessment has been based on INPE's *Annual Report 2008*[5] and INPE's 2009 presentation, "A vision for the future of INPE in the 21st century".[6]

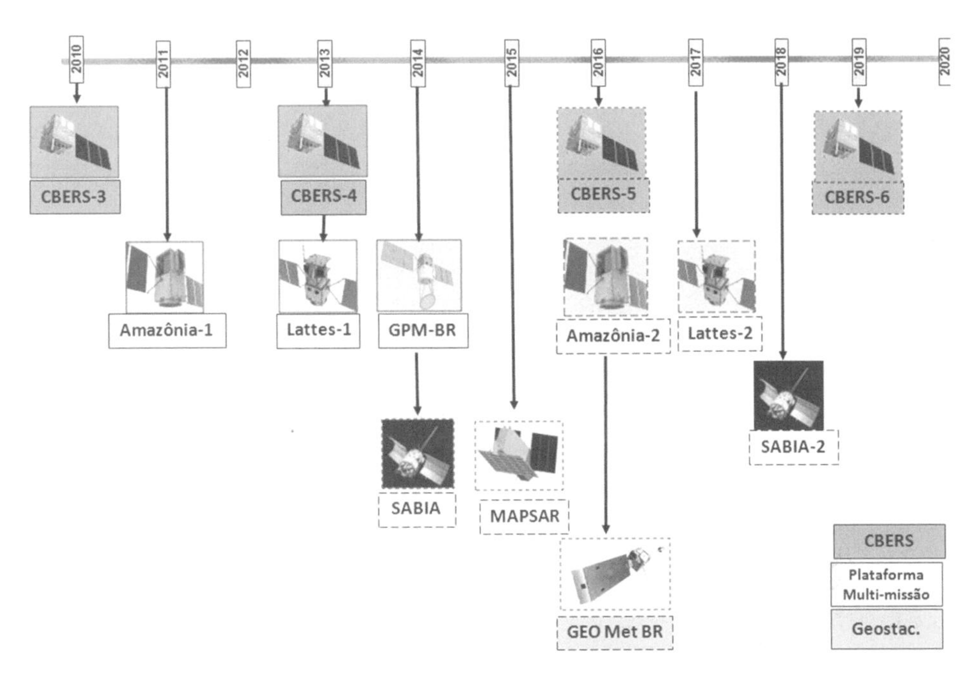

A vision for the future of INPE in the 21st Century (Version June 2009) [INPE/G. Câmara]

Nationally, INPE has to carry out the directives of the PNAE. As for satellites, INPE follows two paths. On the one side, the CBERS satellites that have been developed jointly with China and, on the other side, the development of the Multi Mission Platform (PMM) to be used as a basis for other Brazilian application satellites.

The Sino-Brazilian program. The program of CBERS remote sensing satellites is the result of space cooperation between Brazil and China since 1988. It resulted in the construction and launch of three satellites (CBERS-1 in 1999, CBERS-2A in 2003 and CBERS-2B in 2007). A new agreement has been signed for the construction of CBERS 3 (launch in 2010/2011) and CBERS-4 (launch in 2013/2014). The development of these last two satellites started in 2004, but a delay occurred attributed to restrictions in the supply of space components in the United States, budgetary constraints to meet the obligations of INPE at the appropriate time and technical difficulties of national companies hired to meet the schedule. Still, the CBERS program meets very well with Brazilian needs and the appropriation of two more satellites (to be launched in 2016 and 2019, respectively) is under consideration.

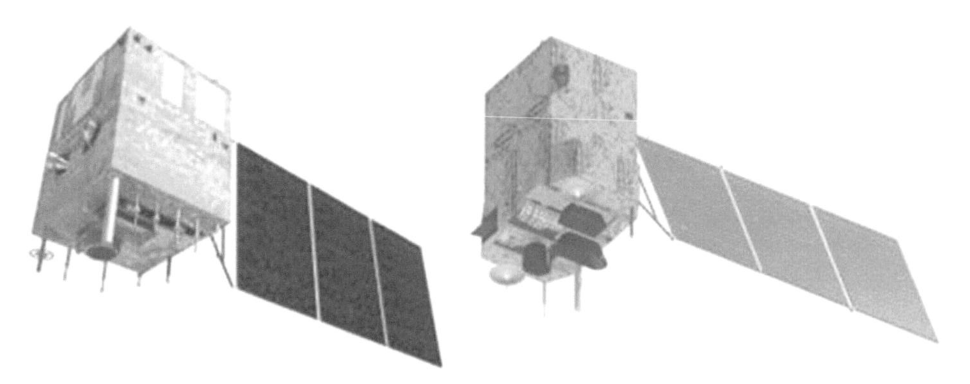

There is a payload difference in the first-generation (left) and second-generation (right) CBERS satellites resulting in the second being some 500 kg heavier [INPE]

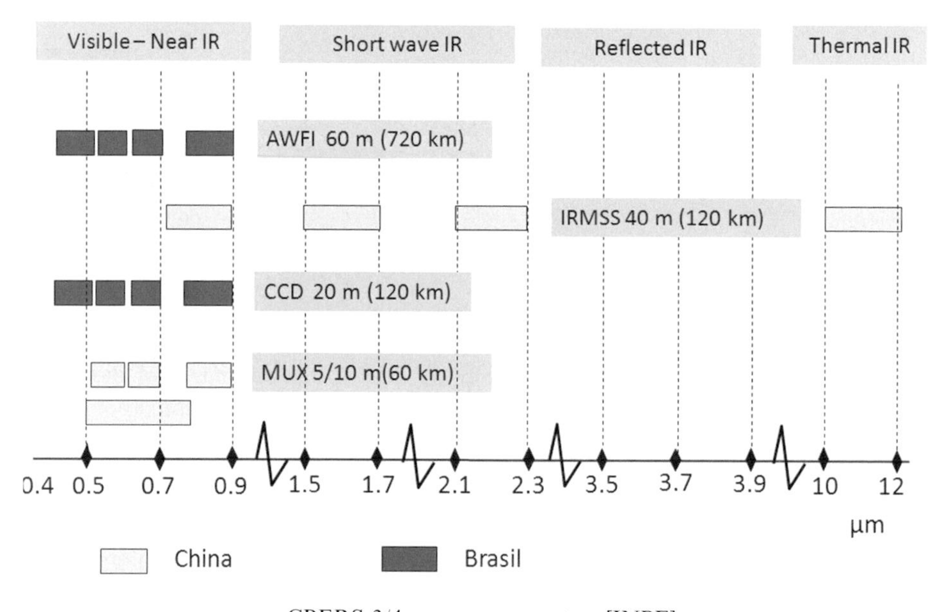

CBERS-3/4 camera parameters [INPE]

CBSAR (China Brazil Synthetic Aperture Radar). To ensure the continuity of the CBERS program, Brazil intends to sign an agreement for the construction of CBSAR, also known as CBERS-7 or CBERS-SAR.[5] CBSAR is an L-band synthetic aperture radar on a CBERS satellite bus. It is to be launched in 2015 and will have a design lifetime of about five years. In a Sun-synchronous dawn–dusk orbit, the satellite will be able to map the whole of Amazonia in five days (30-m resolution) or 25 days (10-m resolution). CBSAR will feature differential interferometric and multi (four) polarization SAR.[7]

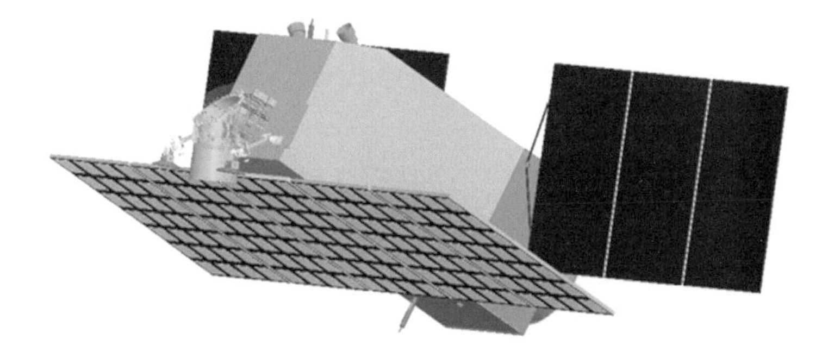

The proposed CBSAR satellite which is under consideration [INPE]

MULTI MISSION PLATFORM (PMM) SATELLITE PROGRAMS

The PMM supports a variety of different Brazilian space missions:

Amazônia. The first satellite that will be launched and is based on the PMM is the Amazônia-1 remote sensing satellite. This satellite will be capable of imaging the Earth from a 750-km circular orbit with a resolution of 40 m. The launch is scheduled for 2011/2012 and it has a design lifetime of four years. Its mission is to provide Brazil with images of its territory, particularly the Amazon region, with a revisit time of five days. The combination of CBERS and the Amazônia satellites will allow full coverage of the Earth every three days. In 2008, INPE signed an agreement with the Rutherford Appleton Laboratory in the UK that offered to provide its RALCam-3 camera with 10-m resolution and 80-km swath width from the intended orbit.

Lattes. As described in Chapter 10, the initial planning of two scientific satellites was based on 150-kg satellites. Two missions were to be conducted: EQUARS and MIRAX. EQUARS was aimed to study the dynamic and photochemical processes in the low, middle and upper atmosphere and ionosphere in the equatorial region. MIRAX was developed as a small astronomical X-ray satellite with a mission of continuously observing the central region of the galactic plane and to make broadband spectroscopic studies. In 2008, INPE decided to join these two missions into the Lattes program and base the satellite on the PMM.

GPM-Br. The Global Precipitation Mission is designed to frequently observe tropical rain, extending the observation areas to higher latitudes. The GPM program consists of a main reference satellite and a constellation of eight smaller satellites. The main satellite is equipped with a double frequency precipitation radar and a microwave radiometer. The smaller satellites carry a microwave radiometer that will be referenced to the one in the main satellite. The main satellite is developed by JAXA (Japan) with participation of NASA. Brazil participates in the GPM program with one of the small satellites: GPM-Br. INPE will provide the PMM. The microwave radiometer payload will be supplied by an international partner.

The proposed SABIA-MAR satellite [INPE]

SABIA-MAR. The latest satellite to be added to INPE's wish list is an ocean color measurement satellite called SABIA-MAR, a Brazil–Argentina cooperation development. This satellite will operate in 16 bands from 350 to 2,130 nm with a resolution of 1 km and a swath width of 2,800 km. The SABIA-MAR program is awaiting the Brazilian and Argentinean governments' financial approval.

MAPSAR. A Brazilian SAR satellite based on the PMM, named MAPSAR (see Chapter 10) is, according to the latest INPE information, scaled down from a definitive development and launch in 2013. At this point, it is not clear whether INPE will have two SAR satellites (MAPSAR and CBSAR) or has to choose one of them. It is not clear what the consequences will be: MAPSAR is an L-band SAR and the result of cooperation between INPE and DLR, while CBSAR will be a Sino-Brazilian joint undertaking.

BRMET. Brazil needs images and meteorological data with operational coverage but American (GOES) and European (Meteosat) satellites do not meet the needs of Brazil due to operational routines and data distribution policies. Therefore, Brazil is developing a meteorological payload on board a geostationary satellite called BRMET or GEO-Met-Br. The main objective of this mission is the high temporal frequency (15-min) imaging of the national territory, both visible and infrared, to support weather forecasting, climate monitoring and severe storm warnings, in order to eliminate the dependence on the operational routines and meteorological information dissemination policies established by foreign institutions.

Integration and Test Laboratory (LIT) [INPE]

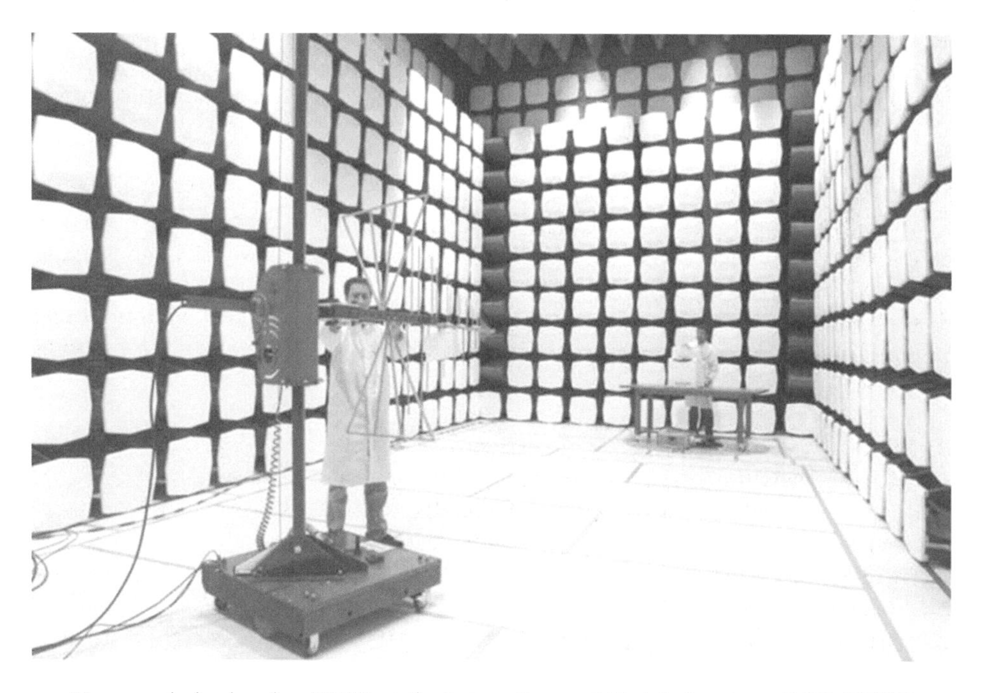

New anechoic chamber (2009) at the Integration and Test Laboratory (LIT) [INPE]

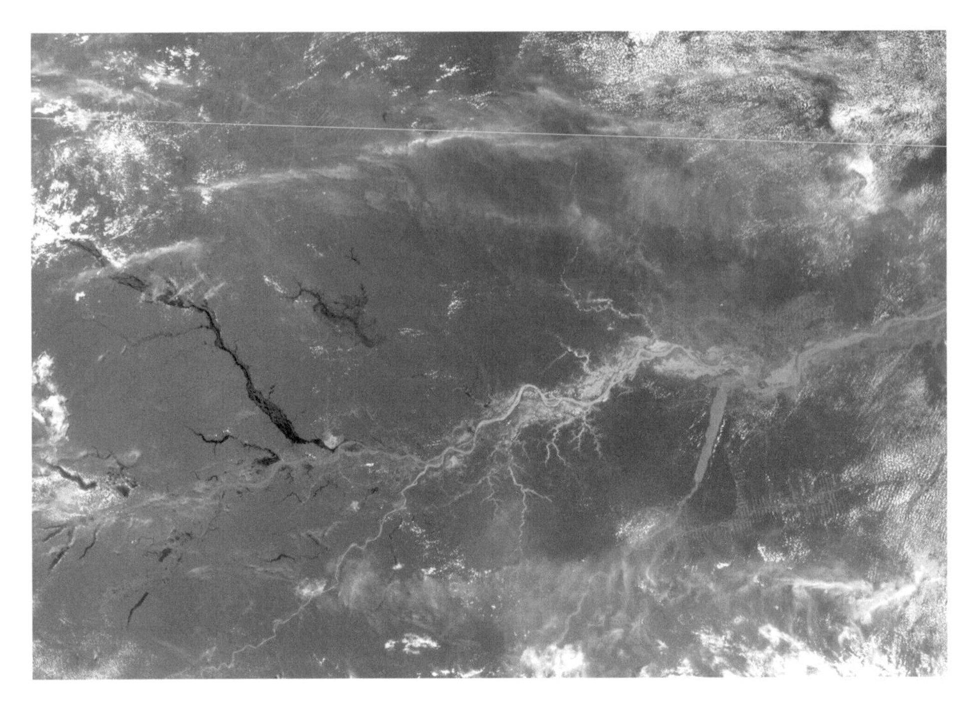

The Amazon River, which carries 20% of all river water discharged into the planet's oceans, is featured in this Terra MODIS image from 8th September 2002. Surrounding the river, which runs for approximately 6,400 km, is the Amazonian rainforest, one of the most ecologically diverse regions on the planet [NASA: http://www.visibleearth.nasa.gov/view_rec.php?id = 4565]

CBERS-2B High Resolution Camera (PAN – 2.7 m) + CCD (multispectral, 20 m) image of Guarulhos Airport, Sao Paulo, March 2008 [INPE]

> The National Program of Space Activities – PNAE – is strategic for the sovereign development of Brazil. The importance of capacity building in the domain of space technology, which, in a broader sense, includes launch centers, launch vehicles, satellites and payloads, arises from its relevance for the nation's future. No strategic technologies will be made available by third parties. These must be developed with domestic resources, in a widespread and integrated manner, in order to address the challenges posed by the era of satellite telecommunications and imaging. Only those countries that master space technology will have the autonomy to develop global evolution scenarios, which consider both the impact of human action, as well as of natural phenomena. These countries will be able to state their positions and hold their ground at diplomatic negotiating tables.
>
> National Program of Space Activities 2005–2014
> Eduardo Campos
> Minister of Science and Technology

REFERENCES

1. www.spacedaily.com/2003/030823072050.384ukdim.html (accessed 1 July 2009).
2. National Program of Space Activities 2005–2014: PNAE/Brazilian Space Agency. Brasília, Ministry of Science and Technology, 2005.
3. www.fab.mil.br/portal/acontecefab/mostra_conversapol.php?id = 143 (accessed 1 July 2009).
4. *Russia Begins Elbowing Ukraine Out from Brazil's Space Program*. RiaNovosty, 19 September 2008, http://en.rian.ru/analysis/20080917/116874710.html (accessed 1 July 2009).
5. Câmara, Gilberto: *Annual Report to INPE's Scientific and Technological Advisory Council for 2008*. São José dos Campos, INPE, May 2009.
6. A vision for the future of INPE in the 21st century. Gilberto Câmara, Director General National Institute for Space Research (INPE), Brazil, June 2009 (Portuguese version), www.dpi.inpe.br/gilberto/inpe_seculo21_eng.ppt (accessed 1 July 2009).
7. The CBERS program – an overview as of April 2009. Gilberto Câmara, Director General National Institute for Space Research (INPE), Brazil, April 2009, www.dpi.inpe.br/gilberto/cbers_overview2009.ppt (accessed 1 July 2009).

12

Israel: Small but efficient actor in space

Israel's space industry is the result of high-level R&D activities for defense and for business during the 1980s, when it became crucial to be independent for Earth observations, communications and broadcasting. Israel is autonomously developing spacecraft (with some cooperation from the European satellite industry) and launchers (with some partnership with South Africa and the United States). On 19th September 1988, Israel became the eighth country in the world to successfully launch a satellite (*Ofeq-1*) with its own vehicle (Shavit). The Ofeq series consists of compact mini-satellites with optical systems for high-resolution observations. Israeli Aircraft Industries (IAI) also developed the TechSAR radar satellite for military all-weather observations and proposed its use to the Pentagon.

The Israeli Space Agency (ISA) was established in 1983 by the Ministry of Science and Technology, in order to formulate and to coordinate the national space program. The ISA is mainly involved in scientific missions through bilateral cooperation with NASA (Space Shuttle and first spaceflight of Israeli astronaut with a dramatic end), ISRO in India (telescope on Gsat-4, Earth observation satellites) and CNES in France (atmospheric studies, vegetation monitoring). IAI is the main shareholder of Spacecom Ltd, which operates the geosynchronous AMOS communications and broadcasting satellites, and of Imagesat International, which markets high-resolution imagery services and the products of the EROS spacecraft.

ISRAEL, A SMALL BUT EFFICIENT ACTOR IN SPACE

On 19th September 1988 – the day of the fifth anniversary of the ISA (Israel Space Agency) – Israel joined the club of countries able to have access to space. For its 40-year birthday, Israel became the eighth nation to launch an indigenous satellite, when a three-stage solid-propellant booster, named Shavit, placed the *Ofeq-1* ("horizon" in Hebrew) demonstration satellite of 156 kg into low orbit (250–1,149 km). The launch took place from a mobile platform at the Palmachim Air Force Base on the Mediterranean Sea coast. The *Ofeq-1* launch showed the maturity of the Israeli industry in designing, building, launching and operating a satellite. The

Shavit launcher – the Hebrew name means "comet" – is believed to have been developed from the Jericho long-range missile. It was a very particular launch: the spacecraft was launched in the north-western direction over the Mediterranean Sea, passing over the Straits of Gibraltar and placed on a retrograde elliptical orbit: Israel had to prevent the over-flight of neighboring Arab countries. This procedure significantly reduces the payload capacity of the launch vehicle and severely limits potential operational trajectories, such as polar and equatorial orbits.[1]

The Israeli space program started in the 1960s around the National Committee for Space Research (NCSR). This Committee was geared more towards developing research and education than establishing an Israeli planning of long-term activities in space. It pushed ahead the fundamental basis for space science in

Israel used its missile-derived Shavit launch vehicle to place satellites in retrograde orbit after flight over the Mediterranean Sea

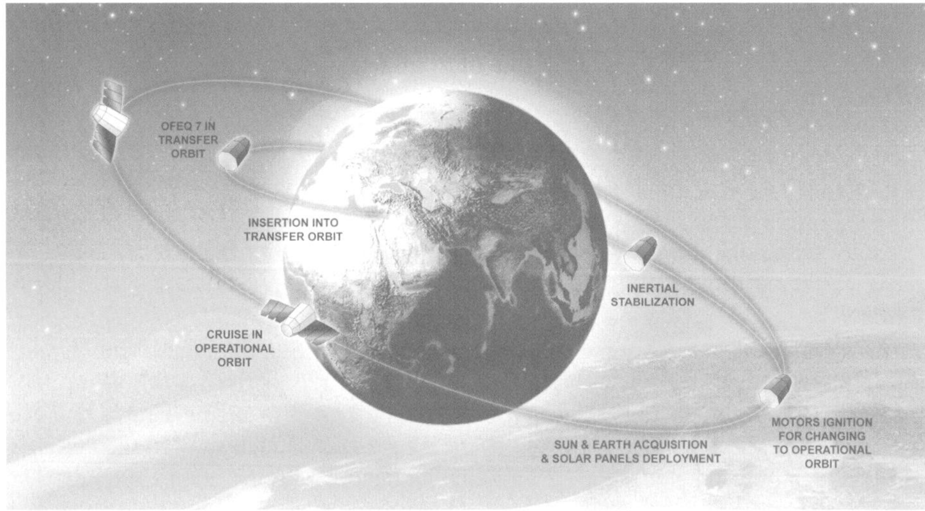

Israel. At the same time, on 5th July 1961, a solid two-stage sounding rocket was tested with a meteorological payload up to 80-km altitude. It is described as the first official step of Israel in space technology. Built by Rafael Armament Development Authority, this 3.76-m-high booster was intended to show the superiority of the Israeli rocketry compared to the Egyptian rocketry that was developed with the assistance of German engineers.[2]

In the 1970s and 1980s, Israel developed the infrastructure needed for space science and exploration. In August 1982, at the Second United Nations Conference on the Exploration and Peaceful Uses of Outer Space in Vienna, a national paper was presented by the National Committee for Space Research of the Israel Academy of Sciences. This committee was described as pursuing "routine activities as an intermediary and coordinating body".[3] It described experiments in manned spaceflight (with Skylab missions), remote sensing of Earth resources, meteorology, communications, physics of the solar wind, geophysics and planetary sciences, astrophysics and astronomy. Tel Aviv University and Ben Gurion University in Beer-Sheva, Technion with the Israel Institute of Technology in Haifa were cited as leading institutions for space research. The Israeli report stated that "in almost all categories of research, international cooperation is a prime ingredient".

LET'S GO INTO SPACE FOR INTELLIGENCE AND SECURITY!

"The Israeli space program was a by-product of the 1979 peace treaty with Egypt", according to a statement in a commemorative publication of the Israeli Ministry of Foreign Affairs, *Israel at Sixty – From Modest Beginning to a Vibrant State 1948–2008* (p. 101). "Spy planes could not fly over Sinai to make sure Egypt, now a friendly country, was not moving missiles into the demilitarized peninsula. Spy satellites were the answer. Israel didn't want to rely on anyone else to collect and convey the intelligence." At the Ministry of Defense in 1983, Moshe Arens was the right man in the right place when Prime Minister Menahem Begin decided on the creation of the ISA (Israeli Space Agency) (Sokhnut HaH'alal HaYisraelit). With the development of ballistic missiles, it became feasible for Israel to launch satellites. "Once we had the capability, we were on our way," explains Arens. "Then we saw the advantage of a satellite for intelligence purposes. We wanted independent capability because information is not always made available to you, even by your best friends." Security was the trigger for space research, yet, at first, the military were unenthusiastic. Arens notes: "When I was Defense Minister, some of the senior army people thought they had better use for the funds. But looking back, you won't find anybody today who would argue that the space program was a waste of money."[4]

The ISA was established on 19th September 1983 by Professor Yuval Ne'eman (1925–2006), who was the Minister of Science and Technology, under the aegis of the Israel Defense Force of Science and Technology. The first satellite was launched five years later, with success. The Agency developed space activities jointly with the Interdisciplinary Center for Technological Analysts and Forecasting of Tel Aviv

University, the NCSR (Israel Academy of Sciences and Humanities) and IAI (now Israel Aerospace Industries). Nowadays, this governmental body functions with an annual budget of some $70m, of which 90% is related to the technological programs for the Ministry of Defense. It coordinated all Israeli space efforts with scientific and commercial purposes, also through international cooperation, mainly with France (CNES), Europe (ESA), the United States (NASA) and India (ISRO). It is a unit under the authority of the Ministry of Science and Technology, linked to the Ministry of Defense.

Also involved in the ISA as co-founder is retired Brigadier General Haïm Eshed, who had, from 1980, a great influence on the development of Israel's space program, first as head of research and development for Israeli intelligence, then as director of space programs for the Ministry of Defense. Eshed, who will retire in 2009, is professor of aeronautics and astronautics: as director of the Faculty of Aerospace Engineering at Technion Institute, he pushed ahead the technology of microsatellites through the student Techsat-Gurwin program of the 1990s. Constantly battling back from launch failures and from chronic underfunding of space activities, he is looking ahead to the next milestone of Israel in orbital operations, such as the creation of constellations of very small and compact satellites.

PROFESSOR YUVAL NE'EMAN, PIONEERING "FATHER" OF THE ISA

Nuclear physicist and politician, Dr Yuval Ne'eman (born on 14th May 1925 in British Palestine, died 26th April 2006, in Tel Aviv) was one of Israel's most prominent scientists. As the founder and the first chairman (until his death) of the ISA, he could be considered as the "father" of the Israeli space program. He studied engineering at the Technion Institute of Technology in Haifa and at the Imperial College in London. He also was the leader of the nuclear program in Israel, serving on the Atomic Energy Commission (1965–1984) and director of one of Israel's nuclear plants.

Dr Yuval Ne'eman, the great pioneer of advanced research and high technology in Israel

Concerned by the political independence of his country, Dr Ne'eman was elected three times to the Knesset as head of the right-wing Tehiya party. He was twice minister: responsible for science and technology from 1982 to 1984, and energy from 1990 to 1992. He left politics after the elections of 1992. One of his greatest achievements in physics was his discovery in 1962 of the classification of

The first satellite of Israel, used for technological experiments

hadrons through what is now known as the quark model. Through his contribution to the deciphering of the atomic nucleus and its components, he was the pioneer of the subatomic physics in Israel. He died at the age of 80, due to an acute stroke.

The ISA is the child of Professor Ne'eman. Pushing ahead the indigenous development of space systems, it has the role of supporting industrial and academic projects, coordinating their efforts, initiating and developing international relations and cooperative missions, integrating projects involving different bodies, and creating public awareness of the importance of space science and technology in a small country such as Israel. When the Shavit launcher, for its first flight on 19th September 1988, was successful in placing the *Ofeq-1* mini-satellite in low Earth orbit, it was the result of high-tech development in Israel in electronics, computers, electro-optics and imaging techniques. The Ofeq satellite program began in 1982 with parallel efforts in research and development, construction of the necessary infrastructure, training of hundreds of engineers and technicians, designing, building, testing and launching satellites. The Israeli engineers had to make progress in the miniaturization of light and compact satellites in order to reduce the launch mass and the cost of their missions. The Israeli satellites – Ofeq and TecSAR – are developed under the funding and control of DRDD (Defense Research and Development Directorate) of the Ministry of Defense.

COMPACT OFEQ SATELLITES AND SMALL SHAVIT LAUNCHERS AS DEFENSE SYSTEMS

All Ofeq satellites are unusual in having a retrograde orbit. This is due to the geographical location of Israel, which gives only an obstructed launch path over the Mediterranean Sea. The national satellite launch vehicle Shavit is derived from the solid two-stage Jericho-2 ballistic missile and developed under the general management of

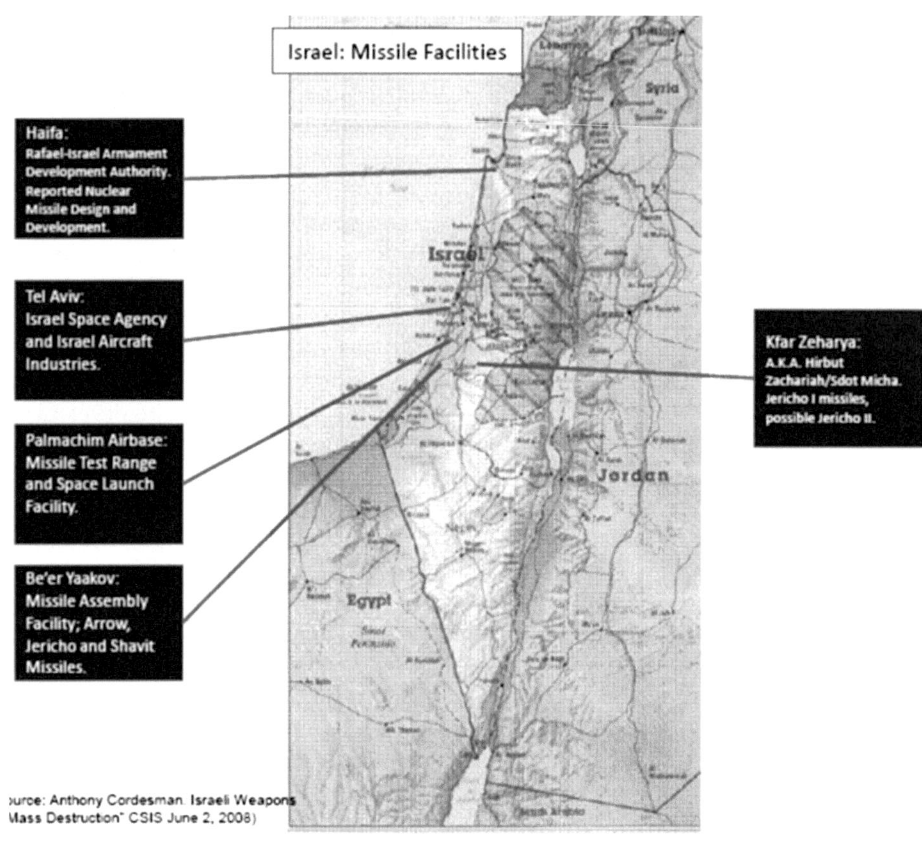
Israel: Missile Facilities
Haifa:
Rafael-Israel Armament Development Authority. Reported Nuclear Missile Design and Development.
Tel Aviv:
Israel Space Agency and Israel Aircraft Industries.
Palmachim Airbase:
Missile Test Range and Space Launch Facility.
Be'er Yaakov:
Missile Assembly Facility; Arrow, Jericho and Shavit Missiles.
Kfar Zeharya:
A.K.A. Hirbut Zachariah/Sdot Micha. Jericho I missiles, possible Jericho II.
Israel
Syria
Jordan
Egypt
urce: Anthony Cordesman. Israeli Weapons
Mass Destruction" CSIS June 2, 2008)

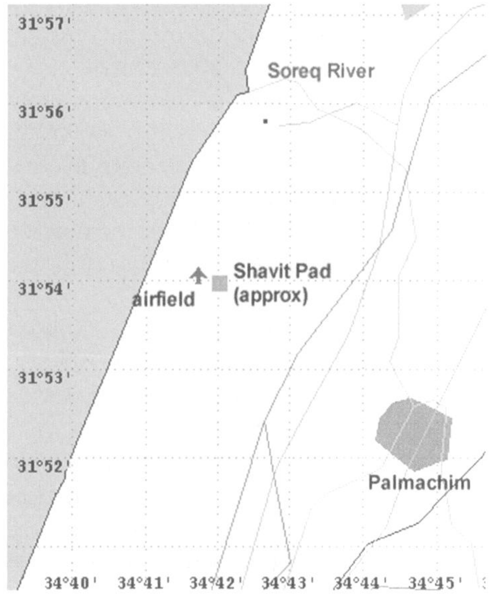
31°57'
31°56'
31°55'
31°54'
31°53'
31°52'
Soreq River
Shavit Pad
(approx)
airfield
Palmachim
34°40'
34°41'
34°42'
34°43'
34°44'
34°45'

The location of manufacturing and launching facilities in Israel

IAI through its MBT Systems Missiles and Space subsidiary. Jericho-2 is the first ballistic missile indigenously developed by Israel during the 1980s.[5] This solid two-stage rocket of 6.5 tonnes at launch is able to deliver an up to 1-tonne payload to a range of 1,500 km. IMI (Israel Military Industries) is responsible for the production of the first and second-stage motors, while Rafael Advanced Defense Systems is responsible for the third-stage motor, which is designated AUS-51 (advanced upper stage). IAI studied a more powerful upper stage, called the CTM (Cryogenic Transfer Module), to fly up to 2.1 tonnes to GTO (Geostationary Transfer Orbit). Its development was stopped by the lack of business for this stage.

In 1993–1994, an upgraded Shavit launcher version, named Next, was marketed by IAI to the international community. The commercial three-stage Next booster would be capable of delivering up to 400 kg to polar orbits from launch sites outside Israel. There were negotiations to launch it from the United States and also from Brazil (Alcantara). Later, it was used by Israel as Shavit-1, with a stretched first stage, and as Shavit-2, with both stretched first and second stages. The ISA and IAI came into a partnership with the US Coleman Research corporation to develop and to market the LK family of small launch vehicles: LK-A (basic Shavit), LK-1 (Shavit-2), with motors manufactured in the United States, and LK-2 (Shavit-3), with a plan to use Thiokol (now ATK) Castor 120 booster as first stage. All these attempts at commercial Shavit versions remained unsuccessful, mainly because of the constraints on classified rocketry systems of the Israeli Ministry of Defense and of US-import regulatory rules. An air-launched Shavit small satellite launcher, without the first stage, was studied by IAI for being dropped from a Hercules C-130 aircraft.

The first Shavit version, with a launch mass of 23.4 tonnes and 15.43 m high, was used for the first two Ofeq technological missions. *Ofeq-2*, 160 kg, launched on 3rd April 1990, was put into a very low orbit of 150–251 km. It increased its perigee, so that its orbital lifetime could be 40 days! It was described as a demonstration satellite to test communications systems, but no information about a possible military mission in space.

In the late 1980s, there was some connection between Israel and South Africa (still under the regime of apartheid). In exchange of uranium for nuclear activities and the use of the Overberg Test Range, Israel helped South Africa to develop its national space program with a three-stage launch vehicle and with small remote sensing satellites. There were some evident similarities between the Israeli Shavit and the RSA-3 launch vehicles using solid motors and between the Israeli Ofeq and South African Greensat spacecraft for Earth observations. An improved RSA-4 with the fourth liquid stage was also in the development phase. In June 1994, South Africa, which had other social priorities, decided to cancel its cooperative projects of space program and joint technological efforts with Israel.

The improved Shavit-1, with a launch mass of 27.25 tonnes and a height of 17.21 m, had a stretched TAAS booster as the first stage, with 12,750 kg of solid propellant instead of 10,215 kg. It was used to launch *Ofeq-3*, an operational satellite of a new generation, carrying the ERMS (Earth Resources Monitoring System) CCD developed by El-Op for reconnaissance observations (resolution of up to 2.5 m). According to Wikipedia, there was a first attempt scheduled for 15th

Successful launch of the improved Shavit-1 carrying the *Ofeq-5* "spy satellite" into orbit

September 1994, but no launch happened. Finally, *Ofeq-3* (225 kg) was launched on 14th April 1995, with success, to reach a retrograde orbit of 366–694 km (143.4° inclination). Using the three-axis stabilized Optsat-2000 platform with a high level of autonomy and good pointing accuracy, it was believed to be able to deliver imagery in the visible and ultraviolet bands with resolutions of about 1.5 m. It served as the test bed for the EROS satellites (Earth Resources Observation Satellite) for commercial operations of the Imagesat International company.

In a CSI/CIA report about Israel's quest for satellite intelligence,[6] E.L. Zorn, senior analyst with the US Department of Defense, stated that "The successful orbiting of *Ofeq-3* represented the initial satisfaction of a longstanding Israeli desire: an independent space reconnaissance capability. For more than 20 years before they began receiving imagery from the satellite, Israeli defense officials had recognized that spacecraft offered unique capabilities to intensify information-gathering in adjacent countries while extending their intelligence reach to more distant lands. After the nation was almost overwhelmed in 1973, Israeli intelligence officers further focused their efforts in preventing future surprise attacks. Satellite photography was seen as a vital tool, able to provide unprecedented warning about the movement of enemy troops and equipment in preparation for war, as well as the movement of enemy forces once hostilities were underway. ... The Israeli desire for a photoreconnaissance satellite did indeed originate with the October 1973 war. ... The Israeli need was not a consequence of having timely satellite photography from

The third Ofeq satellite of Israel Aerospace Industries represents a major step for Israel in the development of high-resolution remote sensing spacecraft

the United States, but rather the result of the failure of the United States to provide satellite intelligence to Israel just when it was required most".

Ofeq-4 was launched on 22nd January 1998 but did not reach its retrograde orbit because of a malfunction with the second stage of the Shavit-1 launch vehicle. The rocket with the satellite crashed into the Mediterranean Sea. Israel did not disclose the specific purpose or capabilities of this operational intelligence imaging satellite. This failure brought the space program of Israel to an unexpected halt and then there were budgetary limitations. The loss of this $50m spacecraft, which was planned to replace *Ofeq-4*, set back Israel's space reconnaissance program by several years.

TOWARDS MINIATURIZED SPACECRAFT FOR HIGH-RESOLUTION IMAGING

In July 1999, Israel took part to the Unispace III conference in Vienna. Colonel Aby Har-Even, then Director General of the Israel Space Agency (and head of the Shavit satellite launch program at IAI), made an interesting statement about the role and the efforts of his country in space: "Israel's emphasis continues to be on building an infrastructure geared to achieving optimal economic outcomes by making use of Israel's technological advantages in selected niches, notably small satellites and remote sensing. Major other areas of activity, in addition to the Ofeq and its Shavit launcher, are the development of applications of remote sensing by users in Israel."[7] He referred to some achievements and projects: the Techsat-1 microsatellite of Tecnion (Israel Institute of Technology), the TAUVEX telescope, the participation of the ISA in the scientific Sloshsat (microgravity mission) of the Netherlands, the development of space electrical thrusters and the MEIDEX (Mediterranean Israeli Dust Experiment) for the flight of an Israeli astronaut as payload specialist on the Space Shuttle.

It took more than four years between the *Ofeq-4* launch failure and *Ofeq-5*'s success in orbit. On 28th May 2002, from Palmachim air base, the third Shavit-1 had

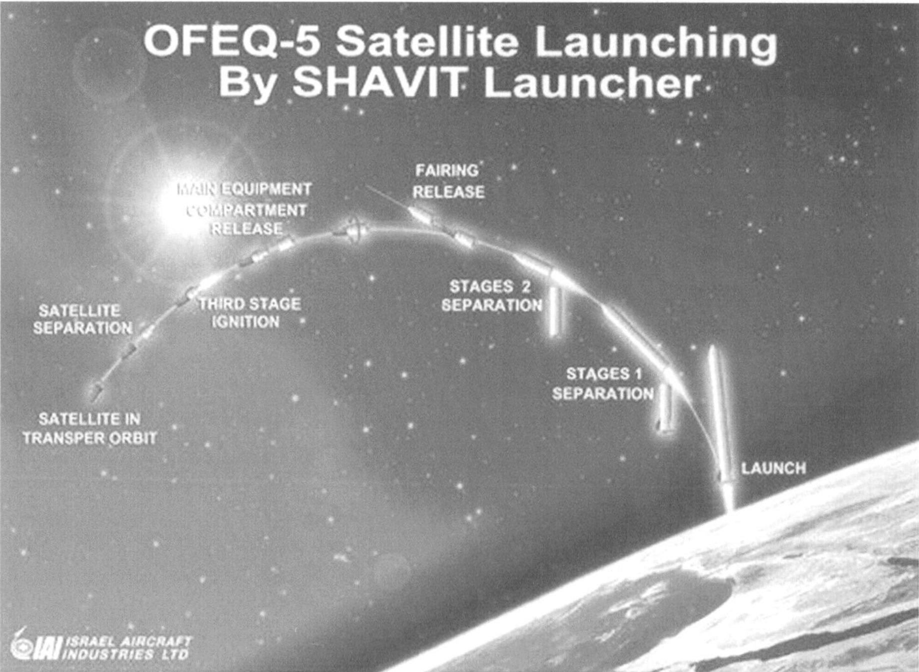

Success for the Shavit-1 launch of the *Ofeq-5* "spy satellite"

a perfect flight, placing *Ofeq-5* into a retrograde orbit of 370–600 km. The ISA website, in a long statement, described this launch of an "advanced spy satellite", "supposed to relay high resolution pictures of the entire Middle-East": "The countdown for the Ofeq satellite started in the early morning, under the supervision of the engineering teams of the Mabat and Malam companies of IAI, the developers of the satellites and launchers. The countdown was postponed several times and it reinitiated time and time again. The final preparations were completed and at 1825, Shavit broke loose and soared in the sky with Ofeq-5, heading west. ... Both rocket engines, a development of the Israeli Military Industry's Giv'on factory, brought the launcher to an altitude of 110km and positioned itself for canopy release. After the canopy separation, the main instrument pod was separated and the third phase engine (developed by Rafael) ignited. When the engine's fuel was depleted, it separated and the satellite remained at an altitude of 260km. When the Shavit launcher completed its mission, the Ofeq ground station, located in the Mabat factory in Yahud, came into play. This station was the center for monitoring the satellite's activity and the data stream from the satellite implied it was fully functional."

Capable of delivering color images with a very high resolution (less than 1 m), *Ofeq-5* had a projected lifetime of some four years. It had a mass of some 300 kg at launch, a height of 2.3 m and a diameter of 1.2 m. According to an ISA statement, "the satellite carries a telescopic camera developed by El-Op owned by Elbit Systems. The camera has a remote sensing capability which allows high resolution space observation. The pictures are transmitted via a data system developed by TadiranSpectrelink. The satellite's main mission is to monitor, identify, and provide reliable information to the Intelligence Department and the defense community regarding hostile foreign country army movements and other additional intelligence information".

The third stage of Shavit-1 uses a solid rocket motor of Rafael

Ofeq-6 was prepared for launch on 6th September 2004. It did not reach orbital velocity and disappeared into the Mediterranean Sea. The third stage of the Shavit-1 did not ignite or malfunctioned, so Israel faced another dramatic failure in its national space program.

Lift-off of the new Shavit-2, which launched *Ofeq-7*

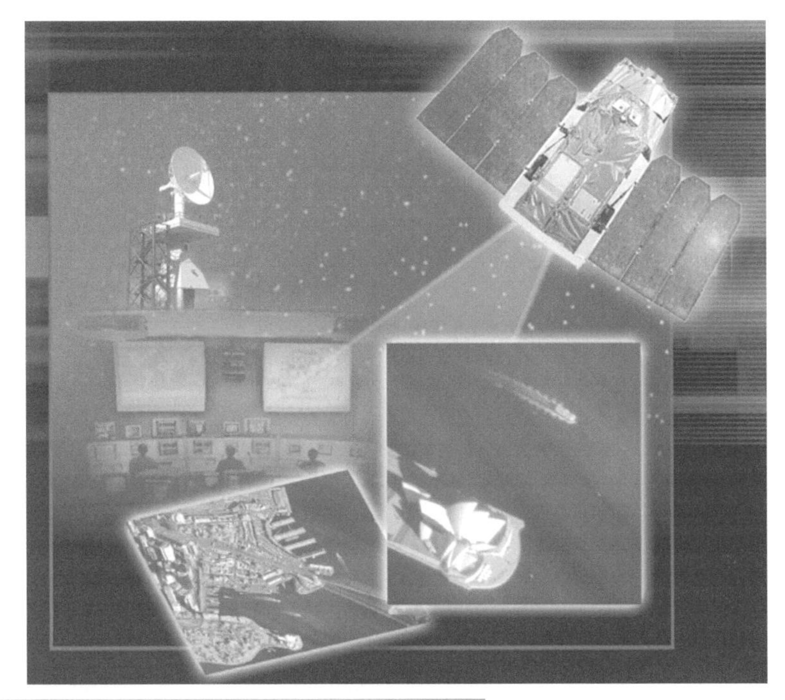

The Optsat 2000 is a commercial spacecraft proposed by the MBT Space Division of IAI (Israel Aerospace Industries) for the worldwide market of remote sensing satellites

The Israeli media estimated the financial loss at around $100m. This shock had no impact on the development of the new Shavit-2 or LK-1 with higher performance: compared to the basic Shavit, the upgraded launcher used stretched solid TAAS motors for the first two stages. With a launch mass of 31.15 tonnes and a height of 19.51 m, it was able to place up to 500 kg in retrograde low orbit. On 10th June 2007, Shavit-2 was used to place *Ofeq-7*, of some 300 kg, into an orbit of 340–576 km, inclined at 141.8°. The Israeli government did not release information about the payload and performances of the *Ofeq-7* spy satellite, which used the multi-purpose Optsat 2000 bus. This new satellite, with a design lifetime of four years, is able to store images taken during its flight and download them when flying over the IAI ground station.

The next *Ofeq-8* (Ofeq-B4 project) could be based on the Optsat 3000 concept that uses the IMPS II (Improved Multi-Purpose Satellite – second generation) bus. It would have a heavier payload with an impressive telescope system to take very high-resolution pictures in panchromatic and multispectral mode (possibly with infrared imaging capability). The compact 400-kg spacecraft will have a higher agility, enabling it to obtain a very great number of images, widely spread, in one satellite pass. Designed for more than a six-year lifetime, it offers high geolocational accuracy. Optsat 3000 is a commercial product of IAI, which exhibited a mockup at Bourget 2007 and Bourget 2009 Air Shows. The first satellite for military observations is planned for launch with a Shavit-2 rocket for Tsahal through the Israel Air and Space Force (IASF) in 2009–2010. According to a statement of the IAI MBT Space Division, the prime contractor, it can be controlled by a single ground control station and may serve multiple domiciliary uses.

A SPACE INDUSTRY SPECIALIZED IN LOW-COST, LOW-MASS SPACECRAFT

There is a project of Ofeq–Next related to a new type of optical satellite mission. Haïm Eshed, Director of space programs for the Israeli Ministry of Defense, aims to accelerate the advent of a networked, 3D constellation of multi-mission micro-satellites that are instantly accessible to multiple users around the world. In an interview published by *Defense News* (22nd September 2008), he admitted the difficulties in getting public money for satellites for tactical intelligence applications: "... we need something like $150m annually to advance ourselves in critical sectors." He would like to consider a public–private partnership with a corporate and private investor ready to put $500m toward the production and launch of a niche of mini-satellites (up to 400 kg) like the recent Ofeq spacecraft: "... provided they meet the Ministry of Defense's internal security requirements, we can prove there will be return on investment and allow them to create a constellation of satellites based on our already considerable infrastructure and leading technology in this field." He added "for the next generation, if I am working with a strategic investor on the constellation we discussed, I may want to work with an alternative foreign vehicle capable of launching more than one at the same time."[8]

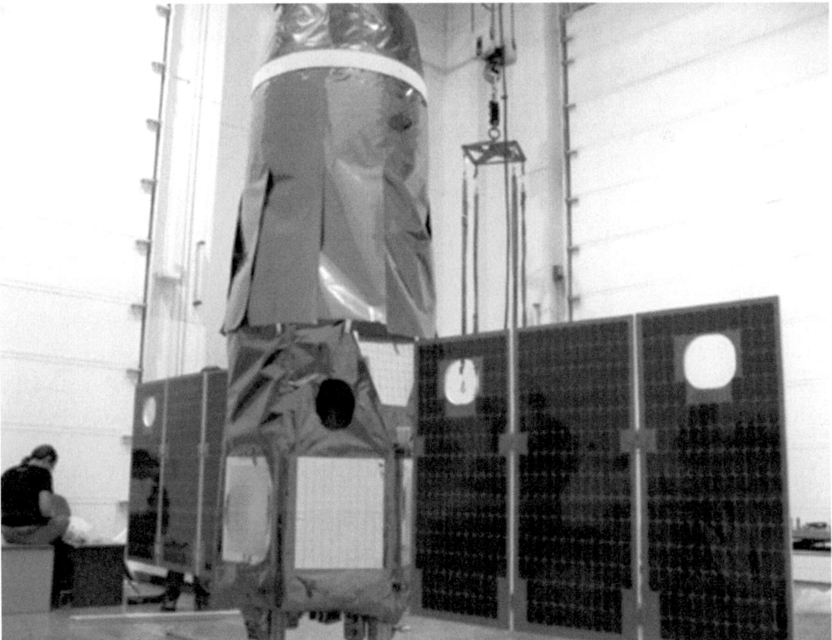

The Optsat 3000 carries a camera with telescope for very high-resolution observations of features on the Earth

At the Paris Air Show 2009 (Le Bourget), IAI exhibited the family of its Optsat/TecSAR spacecraft for optical and radar Earth observations

Professor Isaac Ben Israel, chairman of the ISA (since 2005), when interviewed about Israel's space program for the commemorative publication *Israel at Sixty* (2008), explained that the main difference of space development in Israel, compared to other space programs in the world, was that "we specialize in developing microsatellites weighing 300–400kg. The telescopes should be very light. So should the mirrors, the lens, the electronics. We needed satellites light enough to launch from Israel. Because of the neighbours, we can only launch westwards, over the Mediterranean. Since the Earth rotates in the opposite direction, you lose a lot of energy. So we put a lot of effort into miniaturizing. The launch doesn't cost so much, the subsystems cost less. So we found ourselves with the capability of building sophisticated satellites that were relatively cheap. And the world started to show an interest".

For orbital operations in the future, Israel will continue to improve the Shavit system of indigenous satellite launches from Palmachim base. In May 2004, Colonel Yoram Ilan-Lipovski reported on "micro-satellites and airborne launching new goals in space", in a publication of the Fisher Brothers Institute for Air and Space Strategic Studies.[9] He insisted on the great variety of missions to be achieved by microsatellites (less than 100 kg): electro-optical observations, electro-magnetic surveillance, spectrum analysis and electronic intelligence (Elint) and communications for specific purposes. About airborne launchings, he referred to two major alternatives:

- the heavy airborne launch by using large cargo aircraft would increase Israel's ability to place satellites in orbit from the air – in the eastern or southern direction – mini-satellites of the Ofeq family or groups of microsatellites or nano-satellites;

- the light airborne launch can be carried by fighter aircraft, such as the F-15. Such a launcher has been developed on the basis of the Black Sparrow, a system intended to simulate the Scud missile re-entry process, developed by Rafael Ltd. This system could launch only a single microsatellite or a group of nano-satellites.

The report recommended the initiation of research and development activities in microsatellite technology and in both alternatives of airborne launch, to examine the innovative concept of Launch On Demand (LOD), to integrate innovations into Israel's space program and the possibilities of cooperation with additional countries. In the meantime, Israel decided to go ahead with developing joint activities with India in launch operations and satellite technology.

STRATEGIC PARTNERSHIP WITH DEPARTMENT OF SPACE AND ANTRIX IN INDIA

In mid-2009, the IASF has had in operation up to three "spy satellites" or spacecraft for high-resolution observations and with great autonomy, both based upon the very agile IMPS bus. These are *Ofeq-5* (since 2002) and *Ofeq-6* (since 2007) with optical sensors for less than 1-m resolution images and TecSAR/Polaris with Synthetic Aperture Radar for X-band observations. For this all-weather, day/night observation satellite of 260 kg, a secret launch contract for the IASF (government of Israel) was signed in November 2005 by IAI with Antrix Corp., the business branch of the ISRO (Indian Space Research Organisation). There was no official announcement of this Israeli–Indian space connection. Polaris/TecSAR-1 launch was surrounded by some secrecy, imposed by the requirements of Tsahal. It took place with PSLV C10 from the Indian launch site at SDSC/SHAR (Satish Dhawan Space Center/Sriharikota Range) on 21st January 2008. According to the ISRO announcement, the radar satellite was placed into its planned orbit of 450–580 km, with an inclination of 41°. The launcher was the simplified variant of the Indian PSLV or

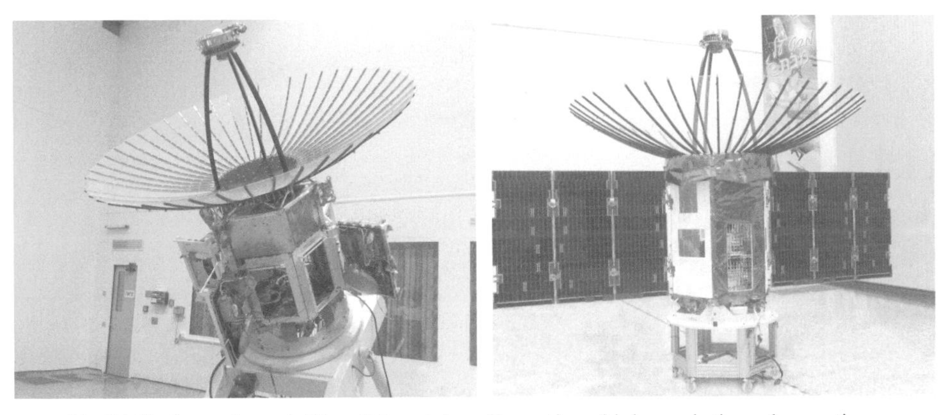

TecSAR, the radar satellite of Israel for all-weather, high-resolution observations

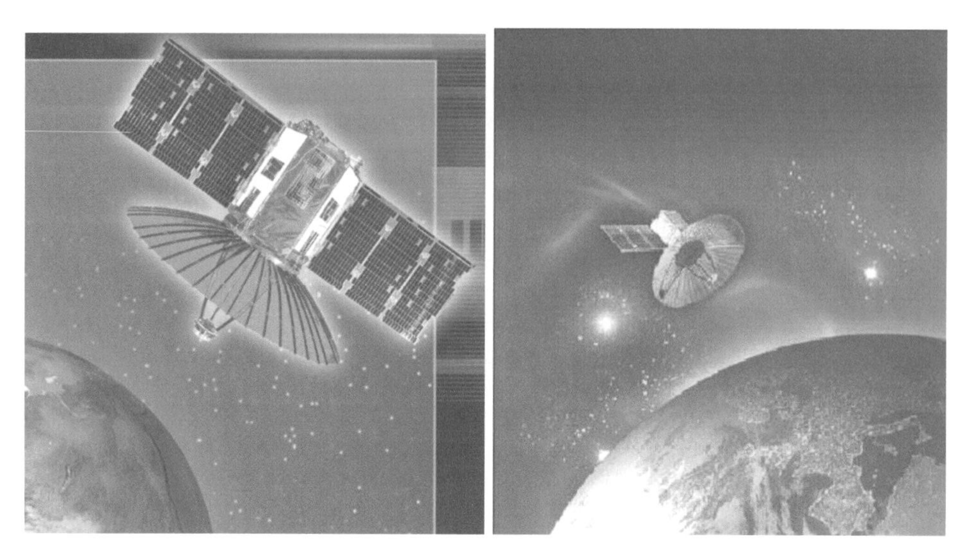

Artist's view of Israeli TecSAR in orbit

Rare view of TecSAR preparation at MBT Space

The spectacular lift-off of the Indian PSLV C10 with the secret TecSAR-1/Polaris radar satellite

PSLV-CA (Core Alone) version, able to launch up to 1.1 tonnes into a 622-km Sun-synchronous orbit.

TecSAR, described as one of the most advanced observation systems, was not really a confidential spacecraft. A mockup of the satellite was exhibited by the ISA and IAI at Bourget Air Show from 2005 and at annual exhibitions of the IAC (International Astronautical Congress). A lot of details about its performances can be found on the Directory of EOportal (http://directory.eoportal.org/). Its design, development and manufacturing activities of the satellite were led by the MBT Space Division of IAI, with the participation of high-tech Israeli industries: ELTA Systems (for the XSAR), Tadiran Spectralink, Rafael and Rokar International. The highly agile bus design – the IMPS II platform – combined with the body-pointing parabolic antenna dish (deployable umbrella of some 3 m in diameter), allows greatly increased viewing capabilities. The spacecraft/antenna may be dynamically redirected into direction of the flight path. A wide FOR (Field of Regard) within the incidence-angle range may be obtained on either side of the ground track for event monitoring coverage. The multi-mode Synthetic Aperture Radar (XSAR) is capable of high-resolution imaging of Spot (1 m), Strip (3 m), Mosaic (4 m) and Wide coverage (8 m).

Fifteen months after TecSAR-1 launch from India, on 20th April 2009, the fourth PSLV-CA was used to launch the unexpected RISAT-2 (Radar Imaging Satellite) spacecraft of 340 kg into 452–456-km orbit, inclined at 41°. It was in a similar orbit to TecSAR-1, so that RISAT-2 observations can be combined with those of the Israeli radar satellite. There was no reference to this mission in the official annual report 2008–2009 of the Department of Space, Government of India, and in the ISRO papers presented at the IAC 2009 in Glasgow. About the RISAT-2 mission, ISRO insisted that the Israeli–Indian spacecraft was not a spy satellite. It looked like the combination of the well proven IRS (Indian Remote Sensing Satellite) platform with the multi-mode X-band SAR system and antenna developed by IAI (ELTA Systems). Due to the aftermath of the 2008 Mumbai attacks against hotels, the

Indian government decided to speed up the acquisition of RISAT-2 with Israeli assistance for 1-m resolution monitoring, in order to meet the federal needs for anti-terrorism security at its borders. RISAT-1, planned for launch in 2010, will be a fully Indian radar satellite with C-band (5,350 GHz) SAR. With a launch mass of 1,780 kg, it will use a 6 by 2-m planar active array antenna, with dual polarization, for ground observations of 2-m resolution (spotlight mode). It will be launched into a 586-km circular orbit with an inclination of 98°.

TECHNION IN SPACE WITH "MADE BY STUDENTS" TECHSAT

Outside the military and commercial programs of Israel in space, there is another program of small satellites in Israel, for technological purposes with educational aspects. It is an initiative of the Faculty of Aerospace Engineering, under the direction of Professor Haïm Eshed, at the Technion Institute of Technology of Haifa. It proposed to develop a constellation of three to four microsatellites for continuous early-warning coverage. Finally, the ISA (Israel Space Agency), the Israeli industry and Amsat–Israel agreed to cooperate with Technion to design, build and test the simplified low-cost, low-power Techsat of some 50 kg, carrying technological experiments. The Techsat program was renamed Gurwin in honor of Joseph and Rosalind Gurwin, whose long-term interest and support for Israel space research enabled Techsat's development.

Techsat, a cubic box-shaped (0.46 m high, 0.43 m across) spacecraft, was covered on four sides by solar cells producing 17 W. Equipped with a three-axis stabilized Earth pointing system, it carried a highly miniaturized payload comprising a radio-amateur digital store-forward transponder, an Elisra CCD camera for Earth observations (ERIP or Earth Remote-sensing Imaging Package for 52-m resolution pictures), a Technion UV spectro-radiometer to monitor Earth's albedo, Technion X-ray detectors using advanced technology and experiments to monitor the space environment. The satellite's total cost, including launch services, was estimated at around $5m, so that the Techsat mission can be considered very cheap. The ground monitoring station is located at the Technion's Asher Space Research Institute. A detailed presentation "Techsat-1 – an Earth-pointing three-axis stabilized microsatellite" was made by Technion during the 45th Congress of the IAF (International Astronautical

The Gurwin/Techsat microsatellite of Technion

Federation) in Jerusalem. Technion considered the development of the multi-purpose Techsat-2 platform for a series of applications in low-orbit: high-resolution remote sensing (Techsat-2A) and digital communications (Techsat-2B). Projects of lunar micro-probe and astrophysics mission were also envisioned.[10]

The flight model of the first Techsat, whose development started in 1991, was completed in October 1994. Its launch with the Russian ICBM Start from Plesetsk cosmodrome failed on 28th March 1995. For the team of professors and students of Technion, this failure represented a great disappointment. It took four years to find the financial and human resources to build and to launch the replacement of the lost spacecraft. Gurwin-2/TecSat-1B of 48 kg was successfully launched as a piggyback payload by the Ukrainian Zenit rocket from Baikonur on 10th July 1998. It was still functioning in February 2009, when Technion opened the new building for the Norman and Helen Asher Space Institute. Professor Haim Eshed attended this event: "... ten years ago, students at the Technion built and launched a much more complicated and complex satellite [than the technological microsatellite which was just launched by Iran]. We predicted that it would remain in space two years and today, on its 10th anniversary, it is still up there, alive and kicking and transmitting." Gurwin-2 is the Israeli member of the Amsat family of radio-amateur satellites (with the name of Gurwin OSCAR-42).

With the arrival of immigrant Jewish scientists from the former Soviet Union, Technion Institute of Technology planned to extend its capabilities into professional satellites. Techsat made Technion one of the few universities and polytechnical institutes to have designed, built and put into service a small satellite. It had a project to develop further microsatellites, even mini-satellites of 250–300 kg. It worked on the creation of a three-axis stabilized multi-purpose platform and looked for international cooperation. However, due to the lack of funding and interest – such an activity was not a priority – from the Israeli government and in spite of the successful Gurwin-2/TecSat-1B mission, Technion had to slow down the Techsat engineering program.

DEVELOPMENT OF STUDENT NANO-SATELLITES FOR AN INTERNATIONAL CONSTELLATION

In the meantime, another project for very small satellites, or nano-satellites, took form at Technion. INSA (Israeli Nano Satellite Association) was founded in June 2006 by engineers of IAI, Rafael, Rokar, AccuBeat, Spacecom and HP-Indigo companies, by professors and students of Technion Israel Institute of Technology and of the Weizmann Institute to start the design and manufacture of a highly miniaturized spacecraft. Derived from Cubesat-type technology, they are planned for deployment in a global constellation (up to 60 nano-satellites). INSA's mission is to promote the development and the use of nano-satellites for academic and commercial purposes. With the technical assistance of the MBT Space Division, it aims at creating in Israel a community that represents the state-of-the-art knowledge of nano-satellite systems, in order to maintain and stimulate the high-tech expertise

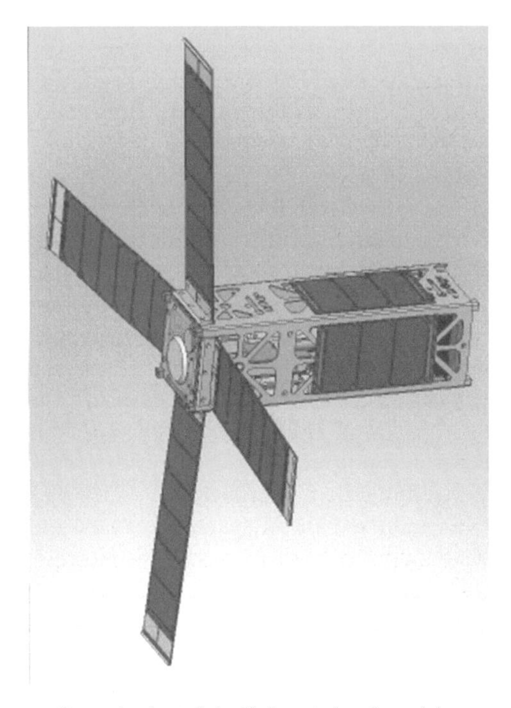

Insat-1, the triple Cubesat developed by the Israeli Nano Satellite Association

of Israel in space. The first of two nano-satellites developed by INSA is InKlajn-1, dedicated to radio-amateur communications around the world. It is planned for launch in 2010 on an Indian PSLV rocket.[11]

Developed with COTS (Commercial of the Shelf) components is the triple-Cubesat developed by Delft University (Faculty of Aerospace Engineering). However, there is no official information about cooperation between INSA and the Dutch University. Its onboard computer employs a new VLSI/ASIC semiconductor developed at Technion with the commercial name of Ramon Chips (in memory of the late Colonel Ilan Ramon, Israel's first astronaut). INSA hopes that Ramon Chips will be integrated in all future satellites of Israel. In order to remember the late Dr Marcel Klajn, a pioneering Israeli space scientist and INSA supporter, the first satellites will be named InKlajn. Another nano-satellite project, named JSat (Jewish satellite), is currently envisioned by INSA.

SCIENTIFIC COOPERATION WITH INDIA (ISRO) AND FRANCE (CNES)

In early 1988, the ISA looked for a National Scientific Satellite (NSS) based upon an Ofeq platform. It called for academic, commercial, research and development groups in Israel to submit proposals for experiments to design a scientific mission. Among the submitted proposals, one was selected with the highest priority: a Tel Aviv University study for a cluster of small, wide-field telescopes for UV (ultraviolet) observations of the sky. The high-tech payload for these astronomical satellites is referred to as TAUVEX (Tel Aviv University UV Explorer). The status of the Israeli space activities in space science can be found in the biannual report to COSPAR.[12]

TAUVEX, developed by El-Op (Elibit Systems Electro-Optics), consists of a three-telescope array mounted on a single bezel and enclosed in a common envelope. Each telescope, with an aperture of 20 cm, uses sophisticated technology with field-flattener corrector lenses. The detectors are photon-counting imaging devices made of a CsTe cathode on the inner surface of the entrance window, a stack of three micro-channel plates and a multi-electrode anode of the wedge-and-stripe type. The ISA was unable to find the funding to develop a fully Israeli spacecraft. It negotiated with the Russian Space Agency to incorporate the TAUVEX experiment with the payload of the astrophysical spacecraft SRG (Spectrum Roentgen Gamma) of Russia, which was originally planned for launch in 1995–1996. The TAUVEX imagers designed a survey of about 10% of the UV sky (1,350–3,500 angstroems) operating on an SRG platform alongside numerous X-ray and gamma-ray experiments developed by Danish, British, Italian, Swiss, French, Canadian and Russian scientists. The specific purpose of TAUVEX is to fill an important gap in the distribution and the nature of celestial UV sources, which are still mostly unknown. The development of the international SRG spacecraft, whose prime contractor is Federal Enterprise Lavochkin Association in Moscow, was affected in the 1990s by the political change of the Soviet Union to Russia and by the decrease in financial resources for Russian scientific missions in space. Instead of using the Spektr bus, it will be based upon the more compact and lighter Navigator platform. The launch of SRG was continuously postponed: the most recent update stated that it could take place in 2010.

But TAUVEX could no longer find a place on the reduced payload of the new SRG model, so the ISA had to find another opportunity to place TAUVEX in orbit. There were then negotiations with the ISRO (Indian Space Research Organisation), which agreed to carry TAUVEX on the technological Gsat-4 spacecraft. On 25th December 2003, an MOU (Memorandum of Understanding) to include TAUVEX on Gsat-4 was signed by Madhavan Nair, Chairman of ISRO, and Aby Har-Even, Director General of the ISA. TAUVEX became an Indo–Israeli mission onboard Gsat-4, which will be put in geosynchronous orbit by the fully Indian GSLV-MkII

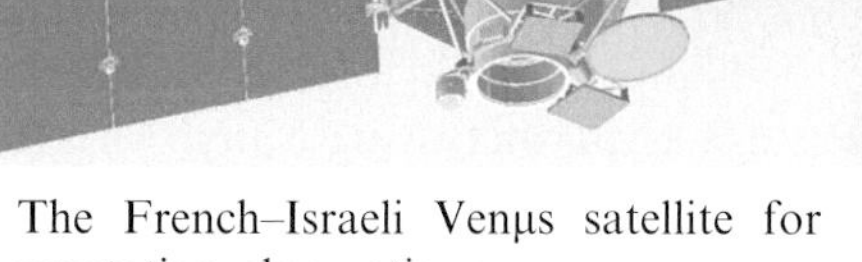

The French–Israeli Venμs satellite for vegetation observations

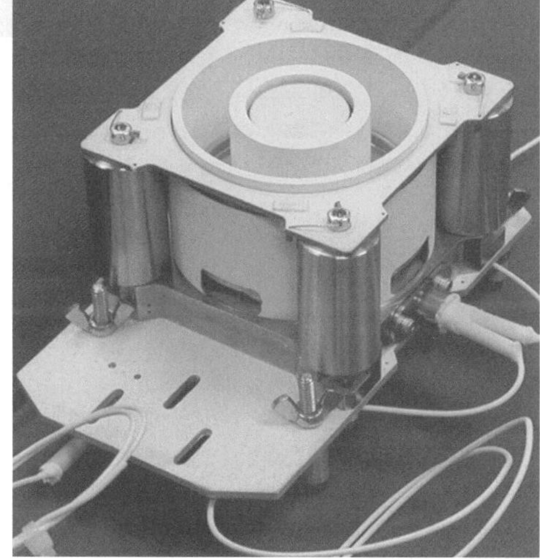

Venμs will test an ion propulsion system for orbital maneuvers

launch vehicle. The maiden flight of this rocket is expected to take place before the end of 2009. TAUVEX will be a collaborative effort between Tel Aviv University and the Indian Institute of Astrophysics, with the scientific data available to all Indian and Israeli scientists.[13]

Venμs (Vegetation and Environment Monitoring on a new microsatellite) is a French–Israeli mission with scientific instruments (Earth images at high resolution and with high repetitivity) and technological payload (low orbit maneuvers with electric propulsion). On 12th April 2005, a MOU was signed between the ISA and CNES (Center National d'Etudes Spatiales) to cooperate on a small and agile satellite for Earth observations in the framework of the European GMES (Global Monitoring for Earth and Security) program. Venμs data will help to prepare the ESA Sentinel-2 mission in 2012–2013. Initially planned for launch before 2010, it is due to fly in mid-2011 for an overall duration of 3.5 years. CNES is responsible for the design of the camera, for launcher interface and science mission center. The ISA is in charge of the spacecraft and of the satellite control center. The launch vehicle has yet to be selected: the Indian PSLV or Russo-Ukrainian Dnepr is the most probable candidate. The allocated budget for the Venμs program is around $7m for the Israeli part. Implemented on a modified IMPS (Improved Multi Purpose Satellite) platform of IAI, the French–Israeli Venμs spacecraft of 265 kg has a double mission:

- The scientific mission is to improve the understanding and the modeling of land surface and vegetation processes to develop new applications such as water balance, crop yield and carbon flux assessments. It will be done for 2.5 years in order to cover two full season cycles in both hemispheres. Every two days, the Venμs Super Spectral Camera (VSCC), specified by CNES and built by El-Op, will acquire images of predefined sites of interest, around the world, with a ground resolution of 5.3 m and with a swath of 27 km.
- The technological mission is aimed at validating the IHET (Israeli Hall Effect Thruster) of 15 mN. Based upon electric propulsion technology, two thrusters of this type will operate on xenon propellant, which is ionized by electrons emitted from the cathode and accelerated as plasma using a high electrical field. They will be used at least 2,500 times, for a cumulative amount of more than 1,000 hr, to demonstrate orbit maintenance, orbit transfer and drag compensation. Venμs will fly at 710-km altitude during the first 2.5 years and then will maneuver with the IHET to achieve a 410-km circular orbit.[14]

GLOBAL COMMERCIAL VENTURES (1): AMOS BY SPACECOM

Because of the limited resources of its satellite program – apart from military needs for strategic observations and communications – Israel has pressed ahead with commercial ventures to enter the global business of space applications. Two operators of satellites have emerged with the financial support of IAI: Spacecom and Imagesat. They are exploiting, as an industrial high-tech return, military needs to use satellites for communications and Earth observations.

The AMOS (Afro-Mediterranean Orbital System) project to develop, build, operate and market a medium-class geosynchronous communication and broadcasting satellite for regional coverage was studied from 1990 by IAI. The development of the satellite, whose name is the acronym of Affordable Modular Optimised Satellite (AMOS), got the full go-ahead in January 1992. IAI, which retains ownership of *Amos-1*, created the commercial body named Spacecom Ltd in 1993 to sell transponder capacity for communications and broadcasts in the Middle East and in Eastern Europe (around Hungary). Spacecom Ltd is a public company controlled by Israeli shareholders: IAI, General Satellite Services Co, Mer Services Group Ltd and Gilat Communications Engineering (owned by Eurocom Holdings since April 2007).

Spacecom is exploiting the 4°W orbital location as a "hot spot" position for communications and broadcasting services. The cost of *Amos-1*'s development, launch and 10-year operations was estimated to be around $250m. Israeli banks provided loans for $100m, guaranteed by the support of government at $15m per year with access to transponders for public and military needs. IAI, the prime contractor, has the ownership of the satellite. *Amos-1*, weighing 961 kg at launch, using a three-axis stabilized bus of 1.21 by 1.61 by 1.93 m, was a small satellite with Ku-band (14/11 GHz) payload consisting of seven active transponders out of nine. Each repeater of 36 W generates up to 55 dBW maximum EIRP (Effective Isotropic Radiated Power), so it is particularly suitable for DTH (Direct To Home) broadcasts. The satellite, designed for a lifetime of 10 years, was developed by IAI MBT as prime contractor with the assistance of European space industries to provide specific systems: French Alcatel (now Thales Alenia Space) for the Ku-band payload, German Dornier and MBB (now Astrium), respectively, for power and thrusters and Teldix for momentum wheels.

During the 45th IAC (International Astronautical Congress) in Jerusalem, on 9th October 1994, IAI invited the media to visit the MBT plant (IAI) in Yehud and to see *Amos-1* in construction. It was an interesting opportunity, with Amitsur Rosenfeld, Amos program manager, to evaluate how significant was the know-how of Israel in the development of space systems. During the presentation, Spacecom insisted on the need for Ku-band communications by satellite in the Middle East, "existing C-band satellites serving the area, have been maturing in orbit, while the need for an advanced Ku-band communication satellite increased rapidly in recent years, as regional economies developed and as social and cultural effects of satellite broadcasting in the region have been felt".[15] *Amos-1* was launched on 16th May 1996 by Ariane 44L (V86) with the Indonesian Palapa-C2 as co-passenger. It is controlled from the MBT Yehud satellite complex, also used for the Ofeq spacecraft. Until mid-2009, it was still operated in inclined orbit at 1.5°W for occasional services. In August 2009, *Amos-1* was sold by IAI, its owner, to Intelsat and renamed Intelsat-24 for operations over the Indian Ocean.

The aim of IAI, as prime contractor of *Amos-1* and shareholder of Spacecom, was to place the three-axis stabilized Amos platform at the right place on the market of small geosynchronous communications and broadcasting satellites of less than 1.5 tonnes at launch. This business was dominated by the successful spin-stabilized

The Israeli Spacecom operator is regularly participating in the annual IBC exhibition and conference in Amsterdam

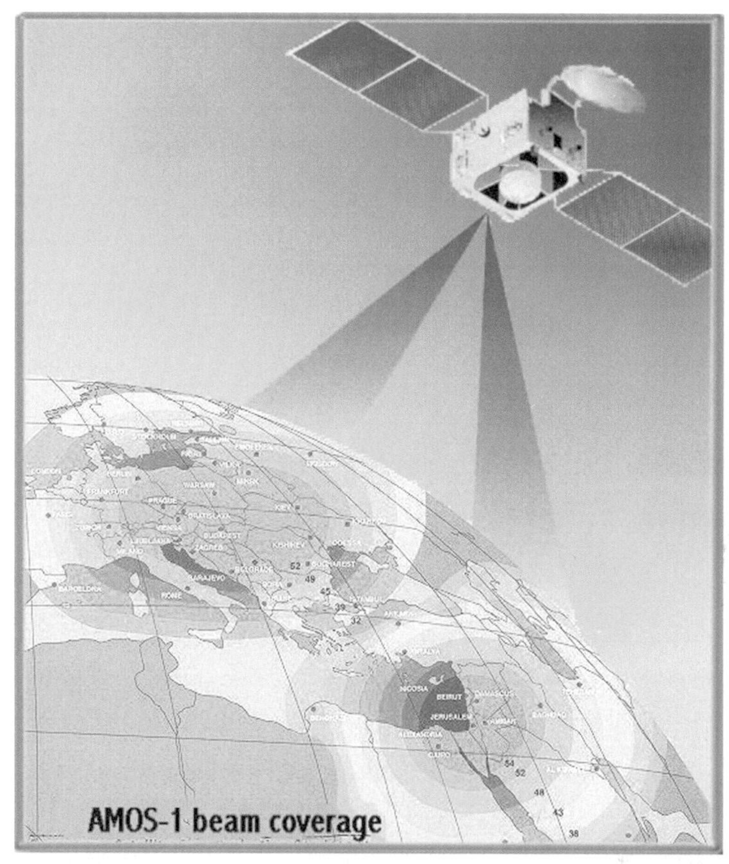
AMOS-1 beam coverage

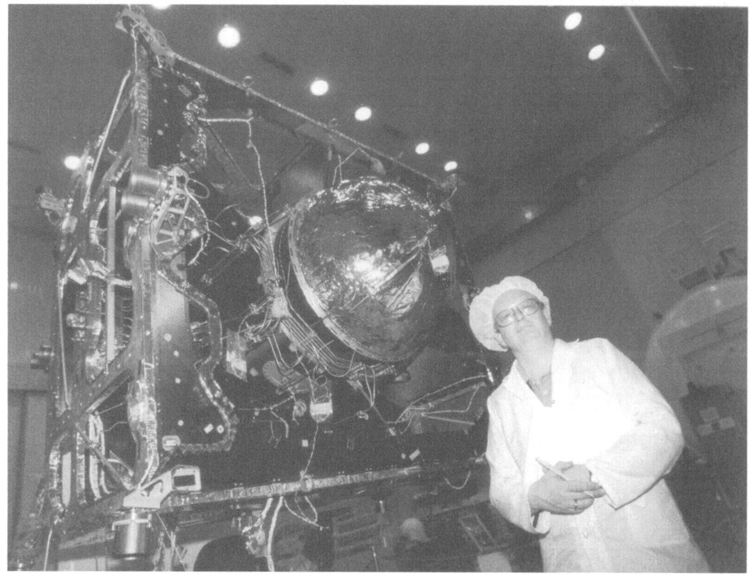

Amos-1, the first "made in Israel" communications and broadcasting satellite, was the first compact and small three-axis stabilized spacecraft for commercial operations in geosynchronous orbit

HS-376/BSS-376 (Hughes, then Boeing) spacecraft. The first customer for the second *Amos-1*-type satellite was Magyarsat Ltd in Hungary – a joint venture 50/50 of IAI and Antenna Hungaria: it was responsible for operating CECS/CERES (Central Europe Communications Satellite/Central European Regional Satellite) to be launched in 1999 and in-orbit backup for *Amos-1*. During the visit of the MBT Space plant, Amritsur Rosenfeld expressed great hope in this venture: "... we envisioned the signature of a contract with Magyarsat Ltd during this summer [1994], but the decision was delayed because of parliamentary elections in Hungary. It will take a maximum of 30 months to build the second Amos satellite and the investment will be around $130m, launch and insurance included."[16] Finally, the Magyarsat project was not considered as viable and failed to attract enough support from the Hungarian authorities. It was revived a few years later under the name of HunStar-1, but without success. IAI had another opportunity to sell an Amos-type spacecraft: Hong Kong Satellite Technology Group (HKSTG) contracted IAI for two HK-Sat based on the improved Amos-HP (High Power) platform (5.5 kW for 20 Ku-band transponders), but it could not meet the financial requirements for its project of a satellite broadcasting system in Eastern Asia.

AMOS by Spacecom is currently expanding, as explained in the documentation released at *Satellite 2009*, Washington, DC: "... the recent surge in activity in the Middle East has significantly increased demand for in-theatre as well as cross-Atlantic connectivity. In a region suffering from poor telecommunications infrastructure, communication satellites are essential in linking the deployed force to the internet and telephony backbone. Exponential growth in bandwidth requirements and the penetration of net-centric operations have made access to satellite capacity a core necessity."[17] It has two operational satellites at 4°W: *Amos-2* and *Amos-3*, both built by IAI with Thales Alenia Space for the payload. They were launched, respectively, by Russian Soyuz–Fregat (Starsem) on 27th December 2003 and by Russo-Ukrainian Zenit 3SLB on 28th April 2008, both from Baikonur cosmodrome. They offer more powerful services in Ku-band, covering Europe, the Middle East and the east coast of the United States (Atlantic Bridge). *Amos-3* has new features, such as Ka-band (28–30/20 GHz) transponders, with steerable beams with global steering range in both Ka-band and Ku-band (possible spotbeams in Europe, Africa and South America).

At IBC 2008, in September, at Amsterdam, AMOS by Spacecom announced an increase in its staff and activities, with three other Amos satellites to be launched between 2010 and 2012. It was looking for international partners to expand into the world as a global operator. The in-orbit delivery of *Amos-5* was ordered on 30th July 2008 to Russian enterprise ISS (Information Satellite Systems) Reshetnev (former NPO PM) with an attractive contract of some $157m or €107m. It is an innovative satellite of 1.6 tonnes (in geosynchronous orbit) and of 5.6 kW, whose advanced bus, the Express 1000N, uses ion propulsion for orbital control. Its payload, developed by Thales Alenia Space, consists of 18 C-band and 18 Ku-band transponders, with three steerable beams. Planned for Proton launch from Baikonur in late 2010, it will be positioned at 17°E to provide a full range of satellite services over Africa.

There is a plan to order *Amos-6* before the end of 2009 for a launch in 2011–2012.

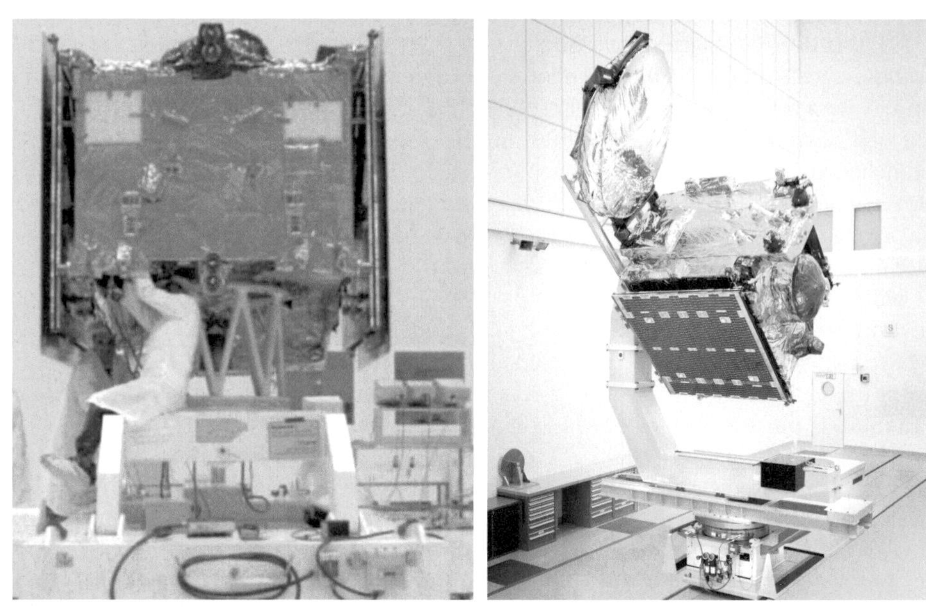

Amos-2 during its final preparation at MBT Space

Amos-2 was the first geosynchronous satellite launched by Starsem Soyuz–Fregat

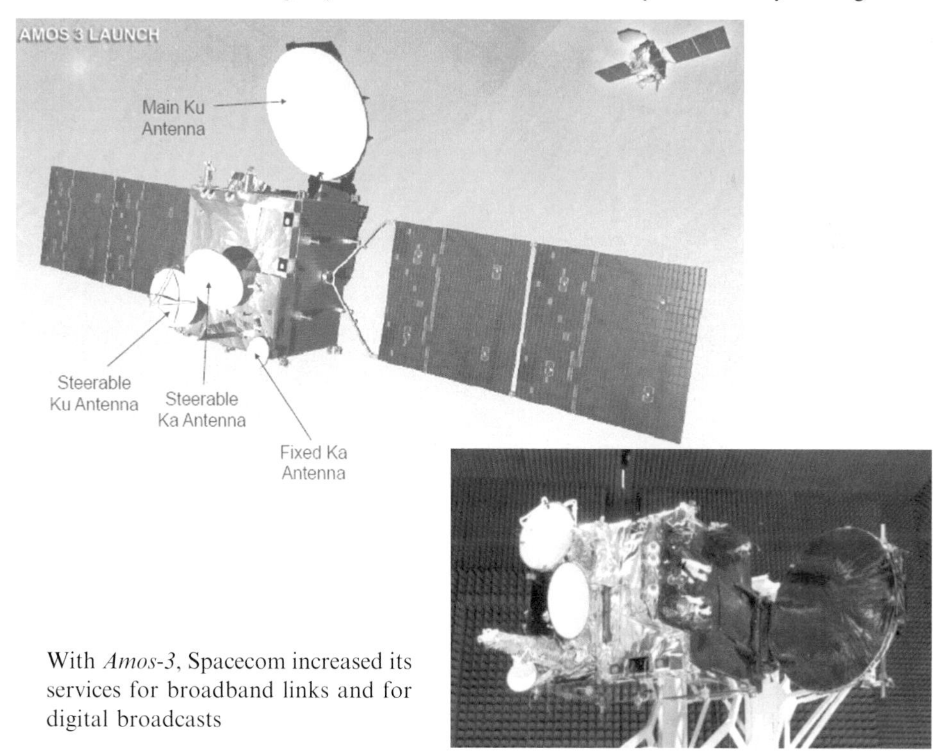

With *Amos-3*, Spacecom increased its services for broadband links and for digital broadcasts

LAND LAUNCH
К Б Ю М

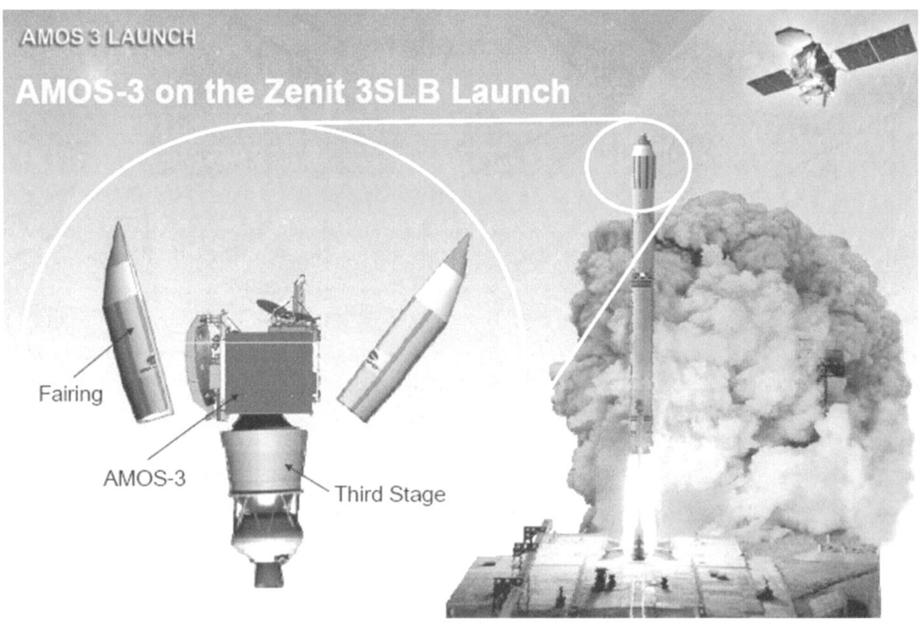
AMOS 3 LAUNCH
AMOS-3 on the Zenit 3SLB Launch
Fairing
AMOS-3
Third Stage

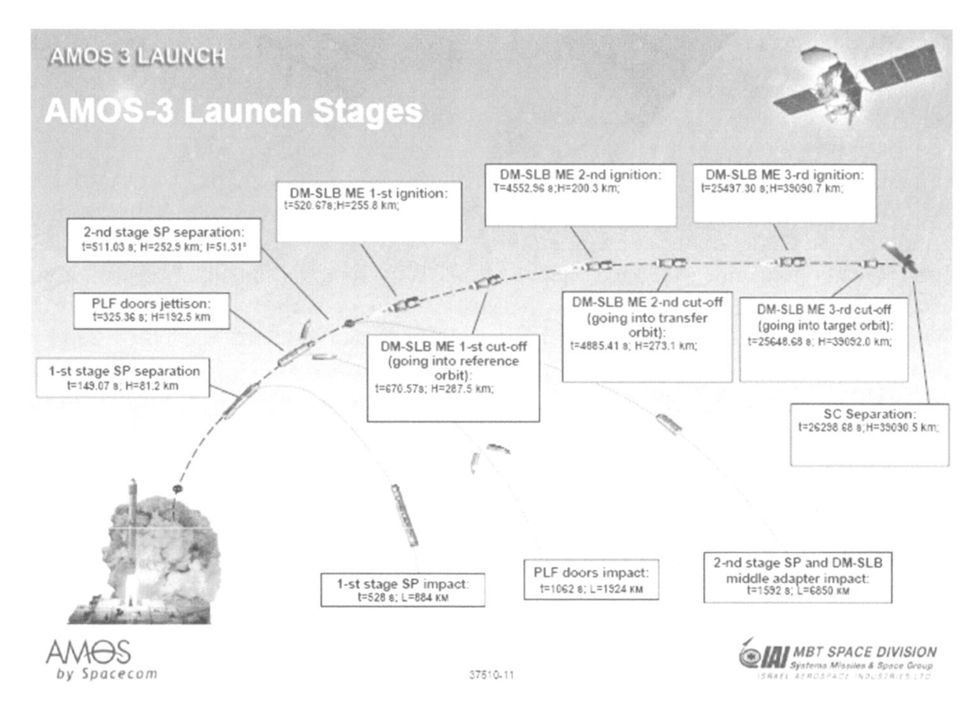

Amos-3 was the first geosynchronous satellite launched by Zenit 3SLB from Baikonour cosmodrome

This satellite will join *Amos-2* and *Amos-3* at 4°W to provide Ku-band and Ka-band coverage over Europe, the Middle East and the East Coast of North America. *Amos-4*, of 3.5 tonnes, based on the more powerful Amos HP bus, is already in construction by MBT Space (IAI), with Ku-band and Ka-band payload provided by Thales Alenia Space: the contract was signed with IAI on 11th July 2007 for the amount of \$365m (some €250m) for the construction and the launch, following a public–private partnership. While Spacecom will pay \$100m, the Israeli government will pay Spacecom \$265m from a pre-launch deal to supply it with services over *Amos-4*'s full lifetime. With a scheduled launch in 2012, it will open a new orbital slot, between 64°E and 76°E – over south-east Asia – in order to add the Asian market to the portfolio of Spacecom customers. Once this satellite is available on the geosynchronous arc, the Israeli Comsat operator will reach close to 80% of the world's population. It already envisions the project of *Amos-7*, whose contract could be finalized in 2010–2011. During an interview at IBC 2009, Spacecom Ltd stated its interest to acquire or lease an in-orbit geosynchronous satellite, by means of a partnership with another satellite operator, in order to start operations in Africa from 17°E.

GLOBAL COMMERCIAL VENTURES (2): IMAGESAT INTERNATIONAL

Imagesat International NV was registered during 1997 in Willemstad, Curaçao (Netherlands Antilles) to provide regional high-resolution satellite imaging capabilities, with complete autonomy, secrecy and flexibility. It has offices in Tel Aviv to supervise the construction of the EROS (Earth Resources Observation Satellites) and the main control station, as well as in Limassol (Cyprus) for commercial operations. Its space segment consists of the Eros constellation with low-cost, remote sensing mini-satellites that use the successful Optsat-2000 platform (*Ofeq-4*). There is no detailed information about the investment needed to deploy this constellation. *Eros-A*, the first satellite of 250 kg, was launched in 480-km Sun-synchronous orbit on 5th December 2000 by a Russian Start-1 missile from Svobodny base in eastern Siberia. With this spacecraft able to make 1.9-m observations with 14-km swath, Imagesat became the second company in the world to successfully deploy a non-governmentally owned satellite for high-resolution imagery (panchromatic mode). *Eros-A*, which has a planned lifetime of 10 years, has been operational for commercial services and products since January 2001.

The second satellite – *Eros-B* of some 350 kg – has superior capabilities because of a larger camera of CCD/TDI type (Charge Coupled Device/Time Delay Integration) to offer panchromatic observations of 0.7-m resolution. Its launch, also with Start-1 from Svobodny, took place on 25th April 2006. Operational in 500-km Sun-synchronous orbit, it allows, like *Eros-A*, a quick revisit of ground features. The main focus in Imagesat International business is the licensing and training of autonomous regional cooperators of Eros remote sensing capabilities following

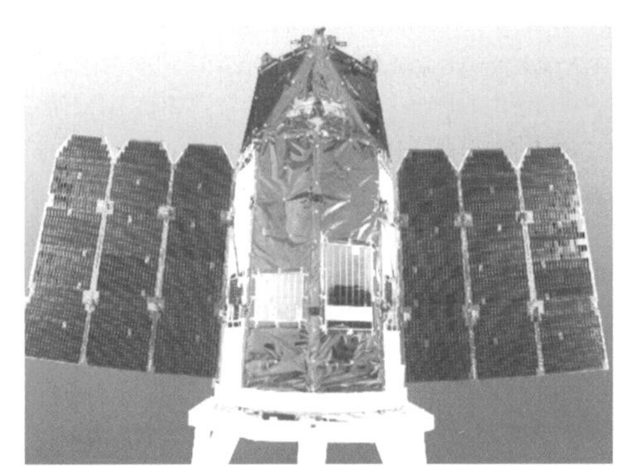

The *Eros-A* satellite of Imagesat International was launched from Eastern Siberia

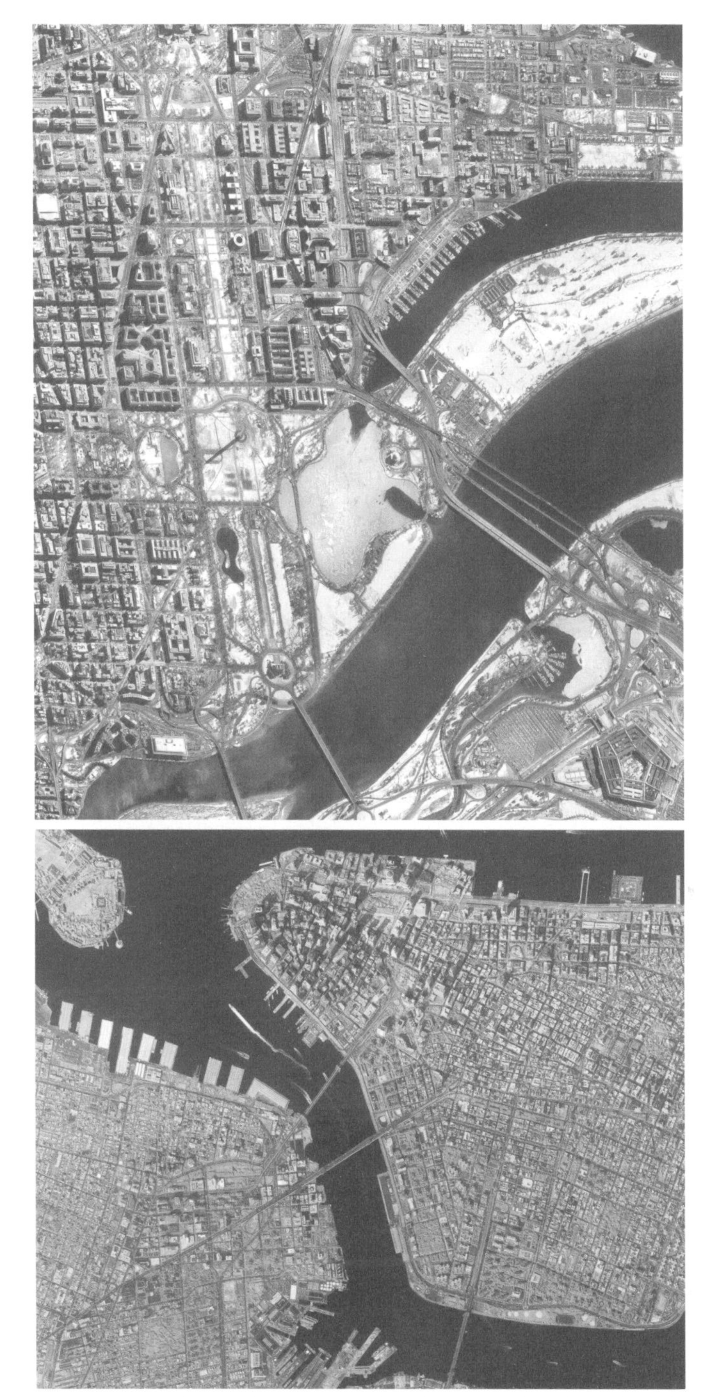

US cities on these panchromatic images taken by *Eros-A* and marketed by Imagesat International, Washington, DC, New York

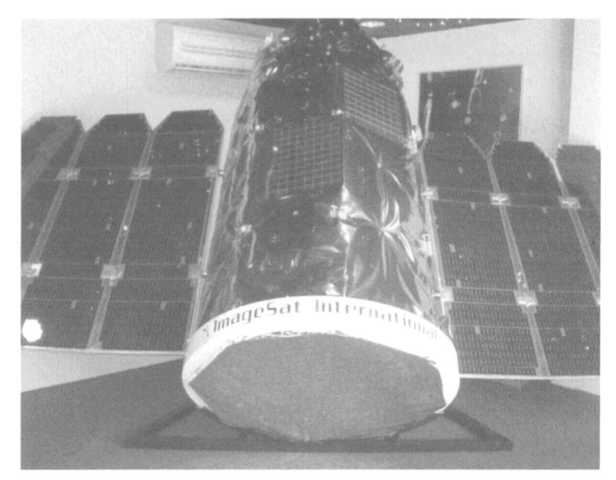

The *Eros-B* offers a higher resolution for its panchromatic Earth observations

specific agreements. These services are called Satellite Operating Partner (SOP) and Exclusive Pass on Demand (EPOD). In the geographic areas in which Imagesat has signed a SOP agreement for access to an Eros satellite camera, imaging services are fully reserved to the customer of the company.

One of the first images transmitted by *Eros-B* on 28th April 2006: it showed the Tabah Dam in Syria

Gibraltar seen from space by *Eros-B*

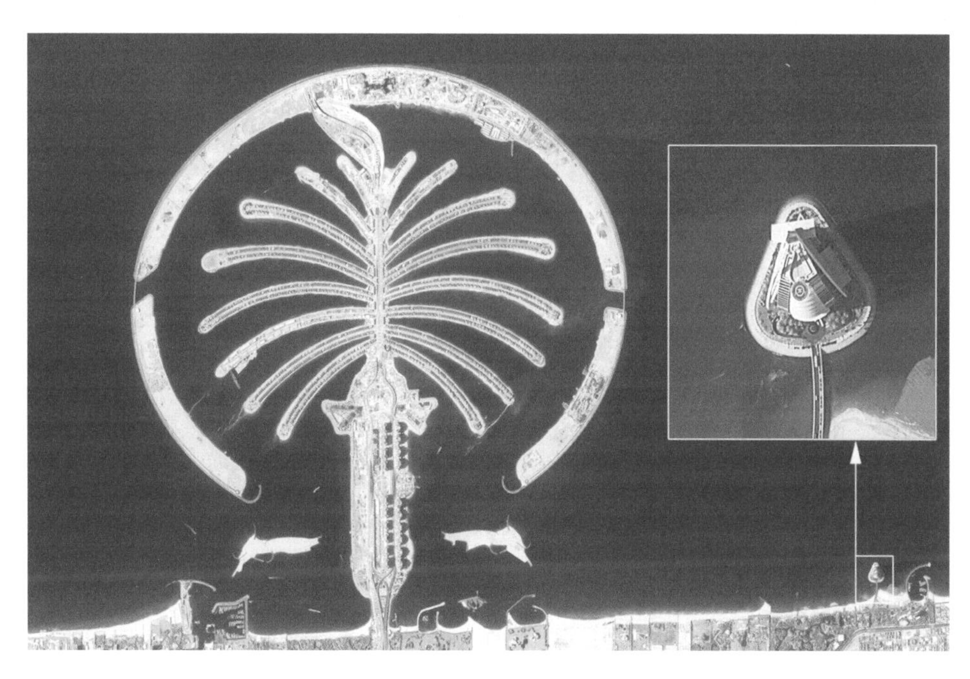

Easily identified from space: the spectacular constructions of Dubai, photographed by *Eros-B*

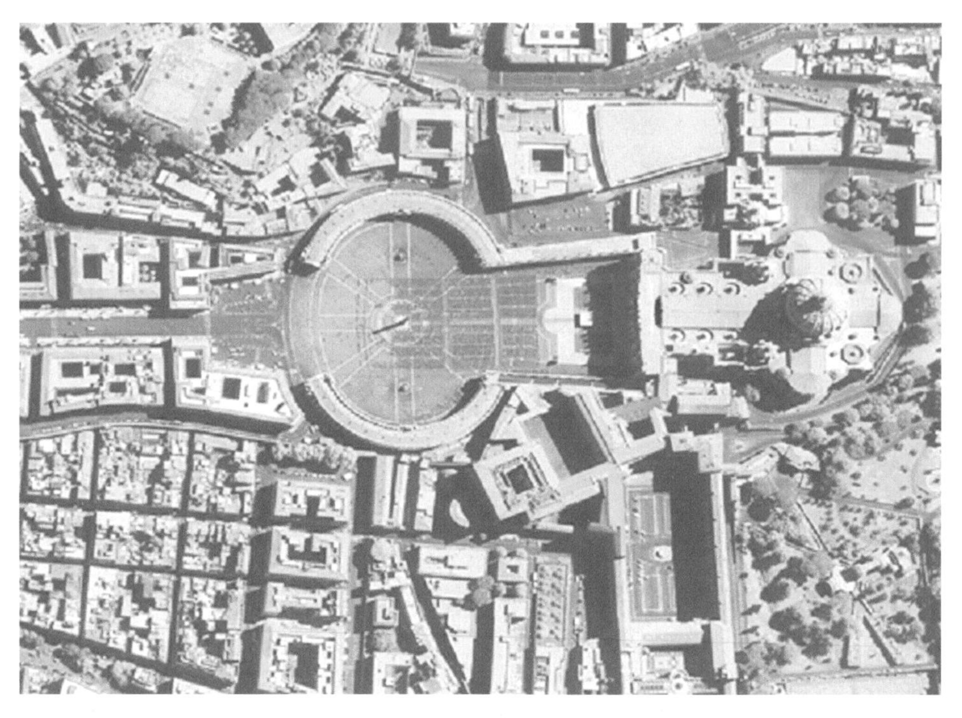

The smallest State of the world: Vatican seen by *Eros-B*

Preparation, observed by *Eros-B*, of the facilities for the Olympic Games 2008 in Beijing

The EROS Satellite Constellation
Sun-Synchronous at ~500 km

Satellite	EROS A	EROS B	EROS C
Weight	250 kg	300 kg	360 kg
Expected Lifespan	2000-2010	2006-2016	2009-2019
Camera	Panchromatic	Panchromatic	Panchromatic / Multispectral
Standard Resolution	1.9 m	0.7 m	0.7 m / 2.8 m
Swath	14 km	7 km	11 km

The planned family of Eros remote sensing satellites. The launch of ERS-C is not yet planned by Imagesat International

What about the future of Imagesat International? The commercial operator initially planned to deploy up to four Eros spacecraft but, because of a troubled situation among the Israeli shareholders – IAI is the most important one – the next step in the development of the system appears unclear. *Eros-C*, of 350 kg, with multispectral capability, was expected to start Earth observations in 2010, but no launch was announced by mid-2009. The website of the operator (www.imagesatintl.com) shows high-quality black and white photographs of various Earth sites, but does not give any indication about the status of the business plan, nor about the concept and performances of the coming spacecraft to be built by IAI.

THE TRAGIC FATE OF THE FIRST ISRAELI ASTRONAUT, COLONEL ILAN RAMON

In 1995, Shimon Peres, who was the Prime Minister of Israel, and Bill Clinton, as President of the United States, signed an agreement for the flight of an Israeli astronaut on board the Space Shuttle. It also was decided that the astronaut would be a payload specialist for an Israeli scientific experiment to be selected by the ISA with the approval of NASA. In the selection process, the ISA had to take into account all the requirements imposed by a manned spaceflight. The MEIDEX (Mediterranean Israeli Dust Experiment) proposed by Tel Aviv University (Department if Geophysics and Planetary Sciences) to study the transport of mineral dust in the atmosphere over the Mediterranean Sea and the tropical Atlantic Ocean was the final choice. The objectives, with the use of a multispectral CCD camera, included measurements of the reflective properties of the Earth's surface and of the visibility in the atmosphere as well as spectral observations of sprites above the tops of thunderclouds. The sea surface albedo was another important area of study.

The official portrait of astronaut Ilan Ramon

Inside *Columbia*

In 1997, the Israeli Space Agency selected Colonel Ilan Ramon, a fighter pilot in the Israeli Air Force. Born on 20th June 1954, he graduated as an electronics and computer engineer from Tel Aviv University. In July 1998, he went to NASA Johnson Space Center in Houston (Texas) for astronaut training. He had to wait until 16th January 2003 to go to space with the Space Shuttle *Columbia*. This STS-107 mission, using the orbiter in solo flight and carrying the Spacehab double research module on its inaugural flight, was dedicated to multidisciplinary microgravity experiments and Earth science research. Colonel Ramon's responsibility, with the MEIDEX instrumentation, was to observe the underlying terrain and identify dust plumes, their location and their extent. He also completed the measurement sequence when flying over dust plumes, using an airplane and ground observation network for *in situ* observations and direct sampling. The video taken to

The crew of STS-107 flight with Colonel Ramon from Israel

study atmospheric dust detected a new atmospheric phenomenon, dubbed TIGER (Transient Ionospheric Glow Emission in Red).[18]

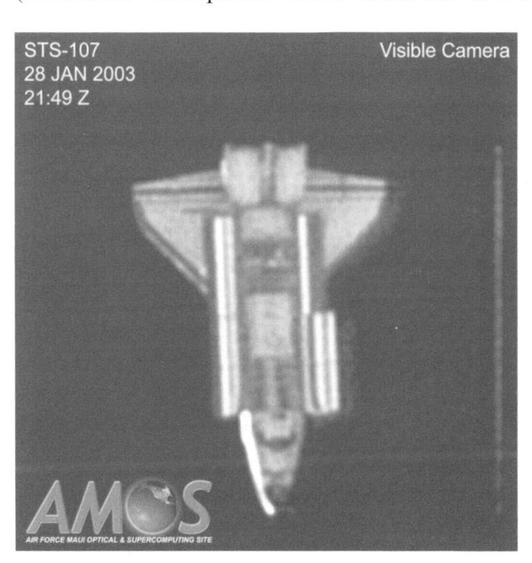

One of the last pictures showing *Columbia* in orbit: it was taken by Amos observatory of the US Department of Defense

The return of STS-107, on 1st February 2003, was a catastrophic failure, with the loss of the seven crew members. *Columbia*, at the end of its 28th mission, disintegrated in the atmosphere over Palestine (Texas) 15 min before planned landing. *Columbia* could not survive during atmospheric re-entry, because the thermal RCC (Reinforced Carbon-Carbon) protection of the *Columbia*'s left wing was struck at high speed and damaged by insulation foam from the external tank during the ascent to orbit. For the first time, a non-US astronaut was killed in a Space Shuttle accident. Israel's only astronaut, who died at the end of his first spaceflight, is considered a national hero.

Activities during STS-107: STS-107 crew in microgravity and at work

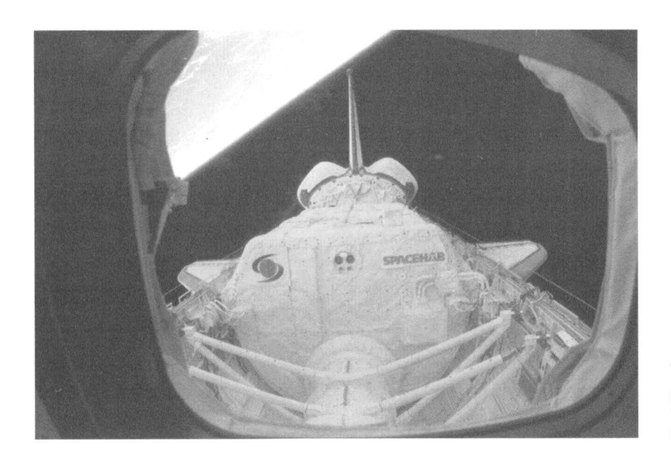

The Spacehab laboratory in the payload bay of *Columbia*

MEIDEX data were not completely lost. A total of more than 8 hr of video obtained during the flight of Colonel Ramon was saved. The results suggest that the occurrence rate of sprites and elves over oceanic and continental storms may be higher than earlier estimates. Strong enhancements of the brightness of the airglow layer above lightning flashes were observed, with lateral dimensions on the order of 400–500 km. There was the "miraculous recovery" of the 37 pages of the diary Colonel Ramon wrote during his spaceflight. His widow brought it to Jerusalem's Israel Museum for restoration. Inscribed in black ink and pencil, the journal covered just the first six days of the 16-day mission. On the last day of the diary, he stated "today was the first day that I felt that I am truly living in space. I have become a man who lives and works in space".

SPACE PROGRAM AT THE TOP OF THE TECHNOLOGY

On 27th July 2005, a guiding Committee defined the vision of the ISA in this statement: "Space research and exploration is an essential instrument of the defense of life on Earth, the lever for technological progress, the key to existing in a modern society, essential for developing an economy based on knowledge, and the central attraction for scientific and qualified human resources." The goals of the ISA are "to preserve and broaden the comparative advantage of Israel and to place it among the group of leading countries in the space research and exploration area". The ISA is mainly involved with the international aspects of the Israeli space program, while IAI, with the MBT subsidiary, is the main contractor of the space systems for military/governmental purposes and for commercial missions (Earth observations with Imagesat, communications and broadcasts with AMOS Spacecom). Israel Defense Forces (IDF/Tsahal) are responsible for launch operations from a simplified facility based upon the mobile platform on the Jericho-2 missile transporter-erector: it is located at Yavne, close to the Palmachim Air Force on the coast south of Tel Aviv.

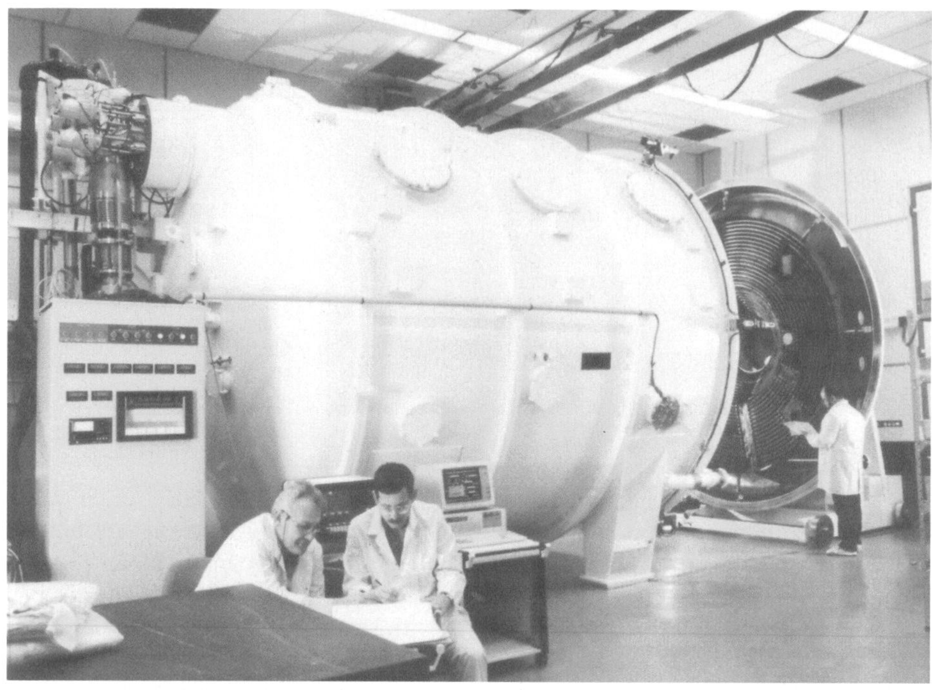

MBT Space is well equipped with satellite test and engineering facilities

Close to MBT Space facilities, Spacecom has its control center to operate the fleet of geosynchronous Amos satellites for communications and broadcasts

The ISA is regularly exhibiting the new spacecraft of Israel at the International Astronautical Congress (IAC), jointly organized by the International Astronautical Federation (IAF) and the International Academy of Astronautics (IAA). During the IAC in Valencia (Spain), in October 2006, Zvi Kaplan, Director General of the ISA, explained some orientations of the civilian space program of Israel: "We have some $3m a year and ISAS has a staff of a few people. We are out-sourcing capabilities to develop experiments and to monitor satellites. Our vision is to create, with a limited amount of resources, a very efficient space program at the top of the technology. We don't want our young people to have to leave Israel for careers in space science, technology and applications. Of course, we see the civilian part of this program, along with the needs of our small country for defense and security."

The ISA is an active member of the International Astronautical Federation. Zvi Kaplan, Director General of the ISA, is seen at the International Astronautical Congress, at Valencia

What about the main priority of the Israeli space program? "Israel keeps today the initial approach of the early 1980s of designing and operating small satellites, ranging from 200 to 350kg, for Earth observations. The trend is to produce advanced small satellites with innovative technologies. It is of economic importance, because of the cost proportionate to the mass. From good panchromatic satellites with high resolution capabilities, we are moving to hyperspectral sensors and synthetic aperture radar systems. Our concept for the future is to have multi-users spacecraft for defense, science, business, society purposes. Two Israeli commercial ventures are using our know-how of space systems: Imagesat International for Earth observations with the Eros spacecraft, Spacecom with the geosynchronous Amos satellites for communications and broadcasts." How can Israel maintain its space activities at high-tech level? "Because of our tight budgets, we have to think international cooperation for their development and operations. We are very proud to cooperate with India for the TAUVEX payload for astrophysics, with France for the Venμs mission of a small satellite to monitor vegetation and water resources. Our project for the next decade could be a remote sensing satellite for hyperspectral observations."

MILITARY SUPERIORITY AND SECURITY AS PRIORITIES FOR SPACE MISSIONS

The Israeli space program remains under the control of the Israel Air Force, which was restructured in 2006 under the name of the Israel Air and Space Force (IASF). The Israeli space budget consists annually of some $70–80m for military activities and some $1–2m for civil programs. Satellite launch operations are interlinked with the Jericho program of Israeli ballistic missiles:

- The single-stage Jericho-1 is a Short-Range Ballistic Missile (SRBM), developed by IAI with the assistance of the French Dassault company in the 1960s. Capable of launching a 450-kg warhead up to a distance of 500 km, it has been replaced by the more powerful Jericho-2.
- The solid two-stage Jericho-2 is a Medium-Range Ballistic Missile (MRBM) operational since 1990: it can carry a 1-tonne nuclear payload to a range of 1,500 km; it forms the basis of the three-stage Shavit launcher and its concept was developed during the 1980s with the cooperation of South Africa using the Overberg Test Range facilities.
- The three-solid-stage Jericho-3, put into service since 2008, is an Inter-continental-range Ballistic Missile (ICBM), estimated to have a range of 5,000–7,000 km with a payload of 750 kg–1 tonne. It is a variant of the Shavit-2 rocket, which was used to launch *Ofeq-6* on 10th June 2007. It remains unclear which will be the next rocketry development in Israel: it is the only country to have an operational anti-ballistic missile defense system, with the very agile Arrow rocket, developed with the technological support of the United States.[19]

There is a close relationship between the Pentagon and the Israeli Ministry of Defense for the development of advanced defense and security systems. However, according to a statement made by IASF commander Major General Eliezer Shkedy in 2006, Israel wants to develop independent operational capabilities in space. An agreement with Washington provides the Mediterranean country with critical early-warning data from US DSP (Defense Support Program) satellites. But, as Israel sees regional air superiority as a priority for its security in the Middle East, the IASF commander proposed to achieve total access to strategic information with the use of space-based systems. In the long term, admitted Major General Shkedy, the IASF must consider assets for defending against and responding to attacks on its spy and relay satellites, as well as capabilities for refueling, repairing and upgrading satellites in space.[20] The IASF has a plan to deploy geosynchronous electronic intelligence (Elint) satellites based upon the *Amos-1* bus of IAI.

TOWARDS AN ISRAELI–AMERICAN CONSTELLATION OF SAR SATELLITES?

Until now, the defense budgets of Israel did not allow for the development of such ambitious military plans in space for the IASF. However, a former Director General of Israel's Ministry of Defense, David Ivry, stated that budgetary resources could be made available, once there is a strong consensus that space is essential for the national security of Israel. IAI has the technological capacity to develop, to operate

At the 56th IAC during October 2005, in Fukuoka (Japan), Israel revealed the TecSAR radar satellite

and to secure defense systems in orbit. Since 2007, Israel established some links with India for joint activities in military space applications. The Indian–Israeli space connection enables the combined use of two SAR satellites as the core of an international constellation.

Israel's Ministry of Defense also has a relationship with the Pentagon to go ahead with the development of SAR mini-satellites.

In April 2007, while it appears that the highly expensive space radar project of the Pentagon – it represented an investment of many billions – would not go ahead, Northrop Grumman Space Technology reached an exclusive teaming arrangement with IAI, the prime contractor of the TecSAR satellite, to provide the TecSAR engineering design to American government customers. This American–Israeli industrial team is proposing the low-cost Trinidad system as an operationally responsive space initiative that can deliver critical new capabilities to Pentagon users about 28 months after authorization to proceed. The initiative is described as a rapid-response, low-risk and affordable space-based radar imaging system that is designed for 24-hr surveillance, day and night, in all weather conditions from a low Earth orbit. Northrop Grumman Corporation (NGC) Space Technology plans to combine its proven capability to integrate intelligence, surveillance and reconnaissance (ISR) mission systems with the TecSAR high-resolution, SAR imaging satellite expertise and operations developed by IAI. Mission assurance capabilities, including secure communications and security requirements, will be incorporated into the spacecraft by Northrop, with final integration and tests at the former TRW facilities in Redondo Beach (California).

An acquisition decision memorandum signed by the Pentagon on 21st July 2008 confirmed the fast-tracked search for small, low-cost radar satellites and supporting ground systems to put them in service as early as 2012. The ultimate goal is to deploy a constellation of 300-kg TecSAR–Trinidad radar satellites for instantaneous observations to face strategic threats on a global scale. High-resolution radar imagery from space can take, with the TecSAR–Trinidad system, an important place in the strategic operations of the Department of Defense to provide global awareness in rapid and cheap mode. A compact, even mobile, ground exploitation station is designed to receive and to process SAR data for defense activities, such as target detection, change monitoring and target cluster indication. These operations must be done in a short time, thanks to quick dissemination of processed data to command posts enabling the commander to rapidly take a decision on military targets.

The TecSAR–Trinidad architecture offers as great advantages the cost and the flexibility of the system, as well as multi-user operations with multiple stations. The initial satellites could be built and launched for less than $200m apiece. In June 2009, the Northrop Grumman–IAI team demonstrated the use of TecSAR data with a receiving and processing station in a van at Join Interagency Task Force (JIATF) headquarters in Key West (Florida). This demonstration was part of the US Southern Command's Project Thunderstorm, an effort funded by the Rapid Reaction Technology Office of the Pentagon to identify and support emerging capabilities against asymmetric threats. In planned future demonstrations, direct tasking of the Israeli satellite will be performed from a Northrop Grumman truck via co-located trailer-mounted antennae. In the meantime, the Pentagon considered the purchase of radar data from Germany (the operational constellation of SAR–Lupe satellites) and from Israel (the TecSAR–Polaris spacecraft).

In addition to the smaller, lighter and less technologically risky satellites envisioned in the new plan, the Pentagon and intelligence community were considering buying radar data from Germany and Israel to supplement American satellites, Hartman said. Israel's TecSAR satellite was launched aboard an Indian rocket on 21st January, while Germany had two radar satellite systems: the civil–commercial TerraSar-X satellite and the military SAR–Lupe constellation.

COOPERATION OF ISRAEL WITH SPACE PROGRAMS OF THE EUROPEAN UNION

From August 1996, Israel has been the only non-European country to be fully associated with the research funding programs of the EU (European Union). It has participated in the research and development activities of the FP4 (fourth research Framework Program), the FP5 and the FP6. In July 2007, the EU and Israel agreed to renew their cooperation in science and technology, giving Israeli researchers, universities and companies full access to the FP7 projects and initiatives. Under this agreement, Israel is set to contribute over €440m to the €50.5bn of the new framework program for 2007–2013. It may take part in the cooperation program, whose collaborative activities and networks concern 10 research areas: health; food, agriculture and biotechnology; information and communication technologies; nanosciences, nanotechnologies, materials and new production technologies; energy; environment (including climate change); transport (including aeronautics and navigation); socio-economic sciences and humanities; space, which especially includes the GMES (Global Monitoring for Environment and Security) program; and security, with the use of remote sensing systems. With the Venμs mission to model land and vegetation processes, the ISA plans to play a very active role in environment monitoring from space.

With the Galileo program, which will represent a public investment of some €5bn until 2013, the EU, with the ESA as prime contractor, is developing its greatest ambition in space applications: to create an operational constellation of 30 medium Earth orbit (MEO) satellites, with the associated ground network, providing highly accurate, global navigation services under civilian control. On 13th July 2004, Israel became the third non-EU country – after the United States and China – to sign a joint agreement with the EU for cooperative activities on satellite navigation, timing science and technology, standardization, frequencies and certification. This agreement allows Israeli industries and system operators to have an active role, along with European partners, in the manufacturing of products and in the marketing of services.

In September 2005, Matimop, the Israeli Industry Center for research and development, became a member and shareholder of the Galileo Joint Undertaking (GJU). Matimop is a non-profit organization under the Ministry of Industry, Trade and Labour, which promotes research and development efforts, cooperation and technology transfer. It has committed to contribute €17.8m for five years to the GJU – this funding has been transferred to the European GNSS Supervisory Authority (GSA) – and to the In-Orbit Validation (IOV) phase of the Galileo program. Israeli industries, universities and institutes are involved with European partners in research and development activities for Galileo products and services. The Rafael company was contracted by Astrium to develop and deliver the attitude control thrusters for the four IOV satellites of the Galileo system.

REFERENCES

1. Johnson, Nicholas L. and Rodvold, David M.: *European & Asia in Space 1993–1994*. Kaman Sciences Corporation/USAF Phillips Laboratory, 1995.
2. Burleson, Daphne: *Space Programs Outside the United States*. McFarland & Company Inc. Publishers, Jefferson, North Carolina, 2005. The review of space activities in Israel (pp. 153–157) summarizes the development of "made in Israel" satellites, space experiments and launchers until 2004.
3. *National Paper: Israel*, Second United Nations Conference on the Exploration and Peaceful Uses of Outer Space, Vienna, 9 August 1982.
4. *Israel at Sixty – From Modest Beginning to a Vibrant State 1948–2008*. Boston Hannah Chicago, 2008, available online at www.israels60th.net/, Chapter 7.6, Out of this world: Israel space program, pp. 99–102. See also Paikowsky, Deganit: *The Israeli Space Effort – Logic and Motivations*. IAC-09.E4.2.1. Tel Aviv University, October 2009.
5. *Israel Missile Update – 2005*, The Risk Report, Volume 11, Number 6, November–December 2005.
6. Zorn, E.L.: *Expanding the Horizon – Israel's Quest for Satellite Intelligence*. CSI (Center for the Study of Intelligence) of the CIA, 2001.
7. Statement made by Mr Aby Har-Even at Unispace III, July 1999. National paper of Israel (184/AB/9), Unispace III.
8. *Defense News*, Chaim Eshed, 22 September 2008.
9. Col. Yoram Ilan-Lipovski: *Micro-Satellites and Airborne Launching New Goals in Space*. The Fishers Brother Institute for Air & Space Strategic Studies (founded by the Israel Air Force Association), paper number 24, May 2004.
10. Professor Shaviv, Giora and Shachar, Moshe: *Techsat-1, an Earth-Pointing Three-Axis Stabilized Microsatellite*. Technion, IAA 94-IAA.11.2.767, October 1994.
11. http://insa.netquire.com/ or www.insasite.com/.
12. The Israeli National Committee for Space Research: *Biannual Report to COSPAR – Space Activities in Israel*. Jerusalem, June 1988.
13. http:/tauvex.tau.ac.il/.
14. http://smsc.cnes.fr/VENUS/, Venμs presentation by the French–Israeli team at the 7th International Symposium of the IAA, Berlin, May 2009.
15. Spacecom press release, *Jerusalem*, 11 October 1994.
16. Personal report of Théo Pirard on the 45th International Astronautical Congress in Jerusalem, October 1994.

17. *AMOS Documentation*, Satellite Week 2009, Washington, DC, March 2009.
18. *STS-107 Providing 24/7 Space Science Research*. Press Kit, NASA, 2002.
19. *Israel – How Far Can Its Missiles Fly?* The Risk Report published by Wisconsin Project on Nuclear Arms Control, June 1995; Abdullah Toukan, *Study on a Possible Israeli Strike on Iran's Nuclear Development Facilities*. CSIS (Center for Strategic & International Studies), March 2009, available online at www.scribd.com/doc/12402056/Israel-Defense-Industries.
20. Opall-Rome, Barbara: Israel puts Air Force in charge of space activities – hotly debated decision renames the service the Israel Air & Space Force. *Space News*, 20 February 2006; Florian Albin Gerl: Outer Space – the emerging market. Diploma thesis, GRIN Dokument V1166794, Munich/Ravensburg, 2008, pp. 78–79.

ANNEXE 1: MAIN PLAYERS IN ISRAEL'S SPACE PROGRAM

Organization	Governmental role	Major activities in the space programme
Israel Air and Space Force (IASF)	Reports to the Ministry of Defence and the Prime Minister	Since 2005, endorsed with the sole responsibility for designing and operating all Israeli military space activities (decision representing a major clarification in the organization of the Israeli space programme)
Israeli Space Agency (ISA)	Part of Ministry of Science and Technology	Coordination of Israel's civil space programme and management of cooperative venture (TAUVEX with India, Venμs with France)
Ministry of Trade and Industry (MTI)	Reports to the Prime Minister	Funding industrial projects, such as those led by IAI, and support to the commercial space enterprises of Israel
Israel Institute of Technology (IIT)	Higher education institute	Consisting of the Asher Space Research Institute, which has a special expertise in small and very small satellites (responsible for the Techsat programme)
Israel Aerospace Industries (IAI)	Government-owned satellite and launcher manufacturer	Main contractor for the design, development and production of Israeli satellites and launch vehicles
Amos–Spacecom	Public–private enterprise to operate the commercial Amos communications and broadcasting satellites on geosynchronous positions	Marketing transponder capacity and comsat services with *Amos-2* and *Amos-3* co-located at 4°W position with coverage of the Middle East, Eastern Europe, African and North American areas
Imagesat International	Israeli–American venture to market images from space	Owner and operator of the commercial EROS satellites (*Eros-A*, *Eros-B*) for high-resolution observations
Matimop	Israeli Industry Centre for R&D	Partner of EU for the research and development framework programmes, member and shareholder of the European GSA (GNSS Supervisory Authority) for the Galileo programme of civilian navigation satellites

Source: Euroconsult 2006–2007.

ANNEXE 2: ISRAELI SPACE LAUNCHERS

Vehicle name	SHAVIT/SHAVIT-1	SHAVIT-2
State	Israel	Israel
Operator	Israeli Military Forces + Israel Space Agency (ISA)	Israeli Military Forces + Israel Space Agency (ISA)
Prime contractor	IAI (Israel Aircraft Industries) + IMI (Israel Military Industries) Rocket Systems Division	IAI (Israel Aircraft Industries) + IMI (Israel Military Industries) Rocket Systems Division
Launch site	Palmachim AFB (mobile platform)	Palmachim AFB (mobile platform)
First flight (planned)	19 September 1988 (Shavit), 5 April 1995 (Shavit-1)	10 June 2007
Reliability rate	4/5	1/2
Performances:		
• LEO mass capacity	160 kg (retrograde orbit)	Up to 400 kg (retrograde orbit)-
• GTO mass capacity	NA	NA
• Fairing height/ diameter	3.3–3.9/1.35–1.5 m?	3.3–3.9/1.35–1.5 m?
• Dual launch system	NA	NA
Basic vehicle:		
• Lift-off mass	23 tonnes	31 tonnes
• Lift-off thrust	413 kN	564 kN
• Height	15.4 m	19.5 m
• Number of stages	3	3
Type of propulsion for each stage (number × engines) [propellants]	1st stage: solid (1 × TAAS ATSM-9 for Shavit, ATSM-13 lengthened about 20% for Shavit-1) [HTPB] 2nd stage: solid (1 × TAAS-II ATSM-9) [HTPB] 3rd stage: solid (1 × spherical ARC-Rafael AUS-51 "Marble") [HTPB]	1st stage: solid (1 × TAAS ATSM-13 lengthened about 20%) [HTPB] 2nd stage: solid (1 × TAAS ATSM-13 lengthened about 20%) [HTPB] 3rd stage: solid (1 × spherical ARC-Rafael AUS-51 "Marble") [HTPB]
Improved versions (description) [performances]	Commercial LeoLink (LK) variants abandoned because of no launch market Study of air-launched Shavit without first stage	Commercial LeoLink (LK) variants abandoned because of no launch market Study of air-launched Shavit without first stage No confirmation about a Shavit-3 version using American ATK Castor 120 as first stage instead of TAAS ATSM-13 motor

Vehicle name	SHAVIT/SHAVIT-1	SHAVIT-2
Technical status (as of 1st July 2009)	Launch to the West over the Mediterranean Sea, in a retrograde orbit	ISA using other launch vehicles (Indian PSLV for TechSAR satellite), because of the low reliability and limited performances for Shavit-2
Commercial status (as of 1st July 2009)	Only used for governmental/ military satellites	Only used for governmental/ military satellites
Estimated launch price	NA	NA

References

Reports of Jonathan McDowell (Jonathan's Space Report).
Reichl, Eugen: *Das Raketentypenbuch*. Motorbuch Verlag, Stuttgart, 2007.
European Space Directory, annually published by ESD Partners, Paris.
Detailed description, with drawings, photos and videos, of Shavit-Next published by Norbert Brügge on Space Launch Vehicles website, www.b14643.de/Spacerockets_1/index.htm.

13

North Korea: The most secret country in space

On 31st August 1998, the Communist regime of Pyongyang achieved its first success in space: but did it really happen? The launch into low orbit of the Kwangmyongsong-1 "musical satellite" using the Taepodong-1 medium-range missile in a three-stage version was officially announced four days later. However, the spacecraft tracking network in US Space Command was unable to confirm the presence of the North Korean microsatellite around the Earth.

At the Unispace Conference in Vienna (July 1999), the North Korean delegation confirmed a national space program for scientific purposes, with many satellites in preparation, but no further space success took place subsequently. For 10 years, some slow and distinct progress was made by the rocketry industry of North Korea, which accepted a self-imposed moratorium of missile launches for the period 2000–2005. A test of the Taepodong-2 intercontinental missile in July 2006 was a failure.

In February 2009, after the successful launch of a small satellite with an Iranian rocket, North Korea reported with many advance announcements the preparation of a satellite launch with the Unha-2 rocket from Musudan-ri center. This launch for the Korean Committee of Space Technology took place on 5th April. Although it appeared to be successful, no payload was found in orbit by the spacecraft-tracking networks of the United States and of Russia. Again, the Democratic People's Republic of Korea (DPRK) did not tell the full story about the exact nature of the Unha-2 test flight, but seemed instead to demonstrate the art of launching "ghost-satellites"!

NORTH KOREA (DPRK), THE MOST SECRET COUNTRY IN SPACE: SPACE BLUFF OR MILITARY CHALLENGE?

The Democratic People's Republic of Korea (DPRK) or North Korea, whose capital is Pyongyang, is mostly a rural country with a population of some 24m inhabitants. The political regime and the economical system are totalitarian Stalinism, with a single party and the de facto dictatorship of Kim Jong-il, son of the eternal President Kim Il-sung (1912–1994). Kim Jong-il is the Supreme Commander of the Korean

People's Army (KPA) and chairman of the National Defense Commission of North Korea. Since the 1950s, North Korea has been at war with the neighboring South Korea. The US Department of State estimates that the DPRK has the fourth largest military in the world, with some 1.21m armed personnel. After failing to reunify Korea by force, Pyongyang tries to destabilize South Korea by demonstrating military superiority. The National Defense Commission is the ultimate command authority for the North Korean missile arsenal.

The economy of North Korea is mainly a military one, giving priority to high-tech development of systems for military purposes. In 1965, USSR rebuffed a request from the DPRK for ballistic missiles. The Hamhung Military Academy was established to train North Korean personnel with techniques of rocket and guidance systems. In October 1966, Kim Il-sung gave instruction for the allocation of military industrial resources for the procurement and production of ballistic weapons. In this strategy, managed by the Second Machine Industry Ministry, the KPA played a key role in technological activities for Weapons of Mass Destruction (WMD). With the technical support of the Soviet Union (Russia, Ukraine) and China, it initiated the development of ballistic missiles for its defense and as export products. During the 1970s, North Korea and China had a cooperative agreement for the joint development of liquid-fuelled ballistic missiles. It took a decade for the North Korean engineers of KPA to design and develop the Nodong missile. It is generally accepted that the first Nodong prototypes were launched from Musudan-ri Missile Test Range in 1990.

Worldwide, it is well known that North Korea has developed its experience in the

A rare view of Taepodong-1/Paektusan-1 rocket at North Korea's missile arsenal (in the factory of Sanum-Dong Design/Research Center)

technology of ballistic missiles by improving the Scud rocket developed and produced by the USSR.[1] The North Korean Scud variants, under the names of Nodong and Hwasong, have customers in Iran, Syria, Egypt and Pakistan.[2] Many American reports, with very detailed information, have been published about the history and dissemination of missile technology from North Korea to foreign countries.[3]

During the early 1990s, just after the testing phase of the Nodong rockets, the DPRK initiated development of two ballistic missile systems or multi-stage rockets that would become known in the West as the Taepodong-1/Paektusan-1 and Taepodong-2/Paektusan-2. We will use the names of Taepodong-1 and Unha-2 to identify the space launch vehicles. Taepodong-1 design objectives were apparently for a system that could deliver a 1,000–1,500-kg warhead to a range of 1,500–2,500 km. Taepodong-2 is intended to carry the same warhead to 4,000–6,000 km.

Launch preparation for the maiden flight of Taepodong-1 began at the Musudan-ri Launch Facility on 7th August 1998. Two weeks later, these preparations were accompanied by movement of Korea People's Navy (KPN) vessels into the East Sea. By 27th August, final preparations for a test were detected by American intelligence, with satellite observations and signal monitoring, and their surveillance systems were moved into position. Lift-off occurred at 12:07 (local time) on 31st August from the Musudan-ri Launch Facility. Without any advance warning, North Korea launched the first medium-range Taepodong-1 ballistic missile from the north-eastern part of North Korea shortly after noon local time (03:07 GMT) on 31st August 1998. The

The North Korean launch of a satellite was a historical event celebrated by the release of postal stamps

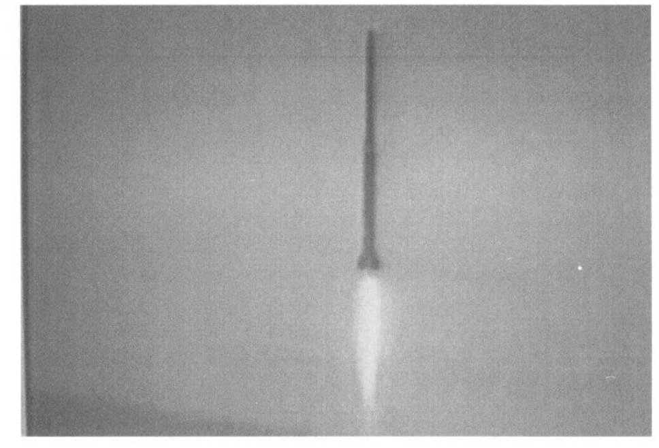

Preparation and launch of Taepodong-1 at the Musudan-ri launch facility

flight was tracked by American intelligence systems in space and on the ground. It was identified as the test of a new long-range missile developed by the DPRK.

Surprisingly, on 4th September 1998, the Korean Central News Agency broadcast a report claiming the successful launch of the first North Korean artificial satellite, Kwangmyongsong-1 ("Bright Lodestar"). The official announcement of the North Korean "first" in space was made four days after the launch of the Taepodong-1 launch vehicle. It gave a lot of details about the flight of the three-stage rocket and the mission of the small satellite, but without indicating its mass. This was followed by a video of the launch, the satellite and an animation of the spacecraft in orbit:

> "The Korean Central News Agency broadcast a report today over the successful launch of the first artificial satellite in the DPRK. The report says: our scientists and technicians have succeeded in launching the first artificial satellite aboard a multi-stage rocket into orbit. The rocket was launched in the direction of 86° at a launching station in Musudan-ri, Hwadae county, North

Hamgyong Province at 1207, 31st August, Juche 87 (1998)* and correctly put the satellite into orbit at 12hr 11min 53sec in 4min 53sec. The rocket is of three stages. The first stage was separated from the rocket 95sec after the launch and fell on the open waters of the East Sea of Korea 253km off the launching station, that is 40° 51′N latitude, 132° 40′E longitude. The second stage opened the capsule in 144sec, separated itself from the rocket in 266sec and fell on the open waters of the Pacific 1,646km from the launching station, that is 40° 13′N latitude, 149° 07′E longitude. The third stage put the satellite into orbit 27sec after the separation of the second stage.

The satellite is running along an oval orbit 218.8km in the nearest distance from the Earth and 6,978km in the farthest distance. Its period is 165min 6sec. The satellite is equipped with necessary sounding instruments. It will contribute to promoting scientific research for peaceful uses of outer space. It is also instrumental in confirming the calculation basis for the launch of practical satellites in the future. The satellite is now transmitting the melody of the immortal revolutionary hymns 'Song of General Kim Il Sung' and 'Song of General Kim Jong Il' and the Morse signals 'Juche Korea' in 27MHz.

The rocket and satellite which our scientists and technicians correctly put into orbit at one launch are a fruition of our wisdom and technology 100%. The successful launch of the first artificial satellite in the DPRK greatly encourages the Korean people in the efforts to build a powerful socialist state under the wise leadership of General Secretary Kim Jong Il."

Korean Central News Agency (KCNA), Pyongyang, 4th September

It is interesting to compare this statement with the official announcement made on 5th April 2009 about the flight of the Taepodong-2 or Unha-2 (see later in this chapter).[4] According to a report published in the 5th September 1998 edition of the *Washington Post*: "Intelligence analysts noticed from the data they intercepted from the missile that it had a flight path that was 'a bit odd'," said the senior official, who, like several other American defense officials, asked not to be named. "It appears that something separated from the second stage and it appears to have some thrust behind it," he said. In trying to explain why American officials did not immediately identify the object, given the intelligence priority the American government says it places on missile proliferation in North Korea, the official said "If there were a device and it was very small, it wouldn't be easily detectable. A small satellite isn't the first thing we would be concerned about."

The fact that the Democratic People's Republic of Korea, a poorly rural but heavily militarized country, was prioritizing space science came as a great surprise and was met with much skepticism. During the following days, some official information, referring to an interview, was published to outline the patriotic efforts

* "Juche" means "independence": adopted in 1955 after the loss of North Korea's Soviet bloc trading partners, it is the North Korean doctrine of self-reliance, establishing the Kim family personality cult. It justifies DPRK military policy reflecting the Kim-Il sung thinking about national development objectives.

of the DPRK to acquire the technology of satellites for scientific purposes. This propaganda statement described how the North Korean scientists were reaching the dimension of space, but did not state which type of activities they were developing with satellites. It was the only source to mention some names and to indicate the purposes of space technology in North Korea:

> "Rodong Sinmun[5] today carries its reporter's interview with scientists about the process of the development of an artificial satellite and its prospect together with a sketch map showing the satellite carrier rocket launching test. Academician, Prof. and Dr. Kwon Tong Hwa, Kim Il-sung Order winner, Labour hero and deputy to the Supreme People's Assembly, who was involved in the launch said that the DPRK has long since developed the science and technology of artificial satellites and laid its solid industrial foundation, not boasting of it. He explained the background of the launch of the satellite: under the wise guidance of General Secretary Kim Jong-il, the DPRK has developed a multi-stage rocket capable of carrying an artificial satellite already in the 1980s and made remarkable successes in researches into satellites as well. When he was alive, the President Kim Il Sung said that it was high time for the DPRK to launch an artificial satellite. All preparations for the launch had been completed in Korea by the beginning of the 1990s.
>
> As instructed by Kim Jong-il, the first artificial satellite was launched to mark the first session of the 10th Supreme People's Assembly and the 50th anniversary of the founding of the DPRK. This was a historic event which made the Korean people and the world know about the might of the local industry and science and technology of satellites the DPRK had independently developed without boasting of it."
>
> KCNA, Pyongyang, 8th September

Candidate academician, Professor and Doctor Han Hae Chol, Kim Haeng Gyong and other scientists, who were involved in the launch of the artificial satellite "Kwangmyongsong No. 1", referred to the purpose of the launch:

> "First, it is to master the technology of putting a satellite correctly into orbit with a multi-stage carrier rocket.
>
> Second, it is to perfect the structural engineering design of a multi-stage carrier rocket and its control technology.
>
> Third, it is to study the circumstances of the space and verify if electronic devices correctly operate in the space.
>
> Fourth, it is to complete the observation system of the carrier rocket and satellite. For this purpose, necessary observation devices were installed at the carrier rocket and satellite.
>
> With the successful launch of the satellite, necessary tests were made in space and a solid foundation was laid to launch a practical satellite. It also provided a turning-point for establishing satellite observation and telecommunications systems in the DPRK. The scientist chose the sky above Tsugaru strait between Hokkaido and Honshu, Japan, as the trajectory of the recent artificial satellite

launch. The carrier rocket was equipped with a device with which to lead the rocket to a safe area and explode it in case the flying rocket deviated from the expected trajectory. Through the recent success the scientists are convinced of the reliability of the carrier rocket before anything else.

The success demonstrated the perfect efficiency of the multistage carrier rocket. A scientist Kim Haeng Gyong, who had participated in the launch, said the carrier rocket correctly flew along the theoretically expected trajectory and that all the apparatuses of the artificial satellite were working properly after it was put into orbit. Scientists said scientific foundations were laid for the immediate launch of practical satellites and the development and use of telecommunications satellites have been made possible. The artificial satellite was moving along its own orbit and sending them survey data, including temperature, pressure and conditions of power source, they said."

The sketch map of the test launch of the carrier rocket of the artificial satellite shows where the carrier rocket was separated stage by stage and where the separated parts of the rocket dropped. It shows that the first stage was separated from the rocket 95 sec after the launch, in the air 35.9 km high and 19.5 km off the launching station, that the second stage separated itself from the rocket in 266 sec, in the air 204 km high and 450.5 km from the launching station and that the artificial satellite was put into orbit in 293 sec, at the speed of 8,980 m/sec, 239.2 km high and 587.91 km from the launching station. It also shows that the first stage separated from the rocket fell 25 km from the launching station, that is 40°51′N latitude, 132°40′E longitude and that the second stage fell 1,646 km from the launching station, that is 40°13′N latitude, 149°07′E longitude.

On the same day, KCNA released this statement about the North Korean satellite in orbit:

"The first artificial satellite of the DPRK launched at Musudan-ri, Hwadae county, North Hamgyong Province, on 31st August is now revolving round the Earth. The multi-stage carrier rocket and satellite launched by Korean scientists and technicians were developed with local strength, wisdom and technology 100%. Today's edition of Rodong Sinmun carries an interview with scientists and technicians involved in the satellite launch and a sketch map of the carrier rocket launching test. The first and second rocket stages are made up of liquid rocket engines and the third stage an engine with highly efficient solid fuel. The control mechanism of the rocket is a product of advanced science and technology, including computers."

KCNA, Pyongyang, 8th September

Why the name of Kwangmyongsong-1?

"The first artificial satellite was launched in socialist Korea. It is wonderful that Korea has the technology of launching a satellite. Korea has become a member of the space state 'club'. This is what foreign mass media reported about the launch of the first artificial satellite in the DPRK. The carrier-rocket and satellite launched in Korea are a home-made product - motors, control

> system, fuel, metal and nonmetal material, launching station equipment and control technology. Putting the artificial satellite into orbit at one time is unthinkable without clairvoyant intelligence and outstanding leadership of General Secretary Kim Jong Il. He has long unfolded a blueprint for space development by ourselves and energetically led the work for its realization. He has also built a strong research group and laid the solid industrial foundations to guarantee the conquest of the cosmos. When scientists were losing heart in the face of difficulties, he encouraged them, saying a good beginning is half done and that he believes they are sure to make a success and that the cause of failure should be found with courage. Under his deep love and wise guidance, the scientists and technicians developed space science by their own efforts and succeeded in launching the first artificial satellite into orbit at one try. They named the satellite "Kwangmyongsong" in praise of Kim Jong Il. "Kwangmyongsong No. 1" and the carrier rocket had been already developed six years ago. Since then, the scientists and technicians have concentrated their efforts on the research into the practical satellite and made a great success, preparing for the satellite test launch. The prospect for space science and rocket technology is promising. Competent scientists and technicians have been built up and the solid industrial foundations laid."
>
> KCNA, Pyongyang, 21st September

The same day, US Space Command published an unequivocal press release about the absence in space of Kwangmyongsong-1:

> "US Space Command has not been able to confirm North Korean assertions that it launched a small satellite on 31st August 1998. US Space Command has not observed any object orbiting the Earth that correlates to the orbital data the North Koreans have provided in their public statements. Additionally, we have not observed any new object orbiting the Earth in any orbital path that could correlate to the North Korean claims. Lastly, no US radio receiver has been able to detect radio transmissions at 27 megahertz corresponding to the North Korean claims. Efforts by US Space Command personnel to locate the alleged North Korean satellite are continuing. The US Space Command has dedicated the resources of its space surveillance network to searching for the satellite the North Koreans claim to have launched. The U.S. Space Surveillance Network consists of ground-based radars and optical sites capable of observing objects orbiting the Earth out to geosynchronous orbit (22,300 miles above the Earth)."
>
> Peterson AFB, Colorado

The American space surveillance capability is impressive and it has been reinforced and upgraded. Its worldwide array of radar and optical sensors can detect a missile launch within seconds. If an object reaches orbit, the US Space Command can detect it, determine its country of origin and compute its orbit very quickly. While the US Space Command was unable to detect a satellite, the conclusion is that the attempt of the DPRK to launch a satellite was a failure. No tracking data indicated that such an object as the one claimed by North Korea ever attained a sustained orbit around the Earth. Circumstantial evidence seems – it is only a hypothetical assessment – to indicate that the satellite achieved orbital velocity before it was destroyed when its solid propellant third stage ruptured catastrophically after 25 sec of its planned 27-sec burn time, with débris impacting some 2,973 km downrange.

However, in December 1998, KCNA published this final statement about the first satellite of the DPRK:

> "100 days have passed since the DPRK's first artificial satellite 'Kwangmyong-song No. 1' was placed into orbit. During the period, the satellite has made over 770 orbits of the earth demonstrating the might of the country. 'Kwangmyongsong No. 1', which has a special position in the world's history of the development of artificial satellites, shows that the DPRK has reached ultra-modern technological level in the domain of the space development. The qualitative level of all devices, elements and materials of the satellite as well as the already-developed multi-stage carrier rocket are very high. Space experts of the world highly estimate the DPRK for successfully solving difficult scientific and technical problems such as development of super heat-resisting material, control of satellite's position and peculiar stage separating method. The observation of the satellite is being made on a scientific basis. It is helpful toward accelerating the space study for peaceful purposes. Technical problems of manufacturing and controlling the multi-stage carrier rocket and separating its engines have already been solved satisfactorily. Complex and difficult technical problems of placing a satellite into orbit and issues of making communications with the satellite, accurately operating electronic devices in the space and establishing a perfect system of observing the satellite in the earth have been solved. Some time ago, scientists, technicians, workers and officials who contributed to the launch of the satellite were awarded state commendations, gifts and state academic degrees and titles."
>
> KCNA, Pyongyang, 9th December

Following a Japanese source in December 1998, the DPRK announced that it would launch Kwangmyongsong-2, but no launch date was fixed. US intelligence source, quoted by the daily *Times of India*, said that North Korea was moving parts of a Taepodong-1 rocket from storage to launch pad and that North Korean Foreign Minister, Kim Kye-Gwan, described the planned launch as part of the DPRK satellite program. This statement was confirmed by a European parliamentary delegation returning from a trip to North Korea.

THE ORIGINS OF A HIDDEN SPACE PROGRAM

The ambition to acquire the access to space through testing powerful launchers received the final impetus during the 1990s. The Workers' Party of Korea decided to include an artificial satellite center as an educational tool at the Three Revolutions Exhibition Hall, which opened in Pyongyang on 12th December 1989 to showcase the achievements of the DPRK. The facility, dedicated to space science and technology in the electronic industry section, became a reality on 9th April 1993, when General Secretary Kim Il-sung approved the development of spacecraft in North Korea. Nowadays, it exhibits a replica of the Kwangmyongsong-1 satellite and explains how the satellite has been put and has worked around the Earth.[6] Other information stated that another satellite displayed at the science museum was the mockup of a communication satellite that was said to be several years from launching (Kwangmyongsong-2?).

North Korean priority in high-technology activities is given to the development of missile systems. The family of the North Korean dictators – Kim Il-sung, then his son, Kim Jong-il – considers satellites as gifts in the sky. Their desire to reach space gives them an opportunity to hide the reality of ballistic rockets, which represent a heavy threat for the neighboring countries. What is the real interest and need of the DPRK for space science and applications? In which fields of scientific activities is North Korea particularly busy? What about the program of the KCST (Korean Committee for Space Technology), under the military control of the KPA (Korean People's Army)?

At a late-1993 or early-1994 meeting of the Korean Workers' Party (KWP) Central Committee, Kim Il-song expressed his desire to place a satellite into orbit. This decision was apparently precipitated by the international recognition Seoul received after its second science microsatellite, KITsat-2/Uribyol-2, in September 1993. Kim's appeal to the Central Committee led to the expansion of the DPRK's nascent space program and the requirement for a satellite launch vehicle. The most likely candidate for use as a launch vehicle was the Taepodong-1. Because the timing of this decision and the start of the Taepodong program were so close, it is possible that there were plans for a satellite launch version from the project's inception. The Sanum-Dong Research and Development Facility, where a white mockup of the Taepodong-1 was detected in June 1994 by American reconnaissance satellites, is responsible for the Taepodong missile program. Located about 14 km north of the center of Pyongyang, it is known as part of the "Number Seven Factory". The Second Natural Science Academy, formerly the Academy of Defense Sciences, is responsible for the design, production and qualification of all weapons systems in North Korea.

The DPRK sent a delegation to Unispace III (Third United Nations Conference on the Exploration and Peaceful Uses of Outer Space) in Vienna. The delegation, made by permanent representatives of the DPRK to the International Organizations in Vienna, had two advisors from Pyongyang: Ju Ryong Hwi and Sim Sun Choi – experts of the KCST. The Democratic People's Republic of Korea (North) reaffirmed the presence in orbit of its first satellite, Kwangmyongsong no. 1 on 31st August 1998. During Unispace III, neither original documentation nor explanation was given about space activities, sounding balloons or rockets.

Jon Yun Hyong, head of the delegation of DPR Korea, stated that "our scientists and technicians developed the multi-phase rocket capable of carrying the artificial satellite in the early 1980's and succeeded in manufacturing the satellite in the early 1990s". He added that the satellite was developed without international assistance "by our own engineers and scientists, with our own materials and technologies", that Kwangmyongsong no. 1 had transmitted Morse signals during nine days, but would circle the Earth for two years. Questioned about the fact that the American tracking services did not detect any Korean satellite in orbit, he laughed, giving two explanations: "First, the level of the technology in USA to track spacecraft is sophisticated but not efficient; second, their claim that our space launch is a missile test had a political purpose." He announced "in a very, very near future, you will be satisfied to hear the news from our country about the launch of a second satellite. While the first satellite was a demonstration one loaded with measuring equipment, the second one will be equipped to test communications". He also stated that the space program of the DPRK was managed by a well developed infrastructure led by the State Academy of Sciences.

During a plenary session of Unispace III, there was a vigorous exchange of views between the two Koreas. The Republic of South Korea criticized the peaceful presentation of the rocket launch made by the DPRK: "This demonstrates capabilities for medium or long-range missiles by North Korea; such a development, if unchecked, would adversely affect the peace and security of the peninsula." The representative of DPR Korea replied "The words of the South Korean delegate simply follow the script written by South Korea's allies. If they were true Koreans, they would be happy to see the first Korean satellite launched into orbit on our first attempt ... Monsieur, ne soyez pas trop excité".[7]

ANALYSIS OF THE LAUNCHER AND ITS MAIDEN FLIGHT

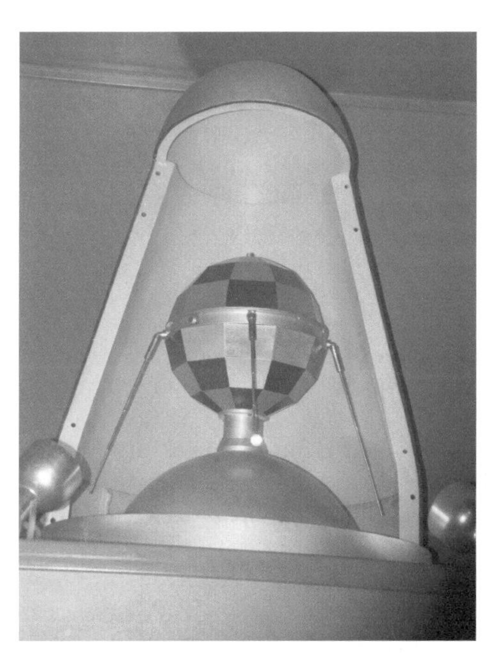

Mockup of the "historical" first satellite developed by the DPRK is part of the Permanent Exhibition at Pyongyang [Philippe Casyn]

Instead of a warhead, the Taepodong-1 missile carried a solid booster as its third stage to which a small satellite, containing a transmitter and batteries, was attached. The mass of the satellite does not exceed 10–20 kg. An expert from the Russian Space Agency reportedly stated that it would be around 13 kg. A poor image and an official video showed a satellite similar to the first Chinese satellite, Dong Fang Hong-1, which looked like a very simplified Telstar-1-type sphere. Later, a drawing on a DPRK postage stamp showed the spherical satellite attached to the solid third stage. According to the North Korean media, Kwangmyonsong-1 completed its 100th orbit on 13th September 1998. Even if it was a failure, even a hoax to reach space, the maiden launch of Taepodong-1 demonstrated the technological progress of North

Mockup of the launch infrastructure, as shown at the Permanent Exhibition at Pyongyang [Philippe Cosyn]

Korea in multi-stage rocketry and guidance systems.

Western analysts intensively used the video footage of the Kwangmyongsong-1 launch from Musudan-ri test site. How was the Taepodong-1/Paektusan-1 launch vehicle designed? What about its characteristics and performance? The military engineers of the DPRK, assisted by Soviet and Chinese rocketry specialists, developed a more powerful booster through the basic concept of the "building blocks". Taepodong-1, as a preliminary step in the development plan of an ICBM (Inter-Continental Ballistic Missile), was a two-stage IRBM (Intermediate-Range Ballistic Missile), which appeared to combine a Nodong derivative as the first stage and a Hwasong 6 as the second stage. It was approximately 25.5 m long and weighed between 18 and 20 tonnes. It could carry an estimated 700–1,000-kg warhead to a distance of 1,500–2,500 km. By adding a solid upper stage (possibly derived from a short-range missile booster), it would be able to put a 20–25-kg payload into low Earth orbit.

The non-profit organization NTI (Nuclear Threat Initiative), on its website, published this interesting description of the three-stage Taepodong-1 launch vehicle, with a table of its estimated performances for access to orbit:[8]

> "The first stage is based on a Nodong-type first stage. However, compared to the basic Nodong missile, a shorter burn time than the standard Nodong was observed in the August 1998 flight. This indicated an increased thrust level since a reduced propellant load must be completely excluded. Similar observations regarding the second stage were made: though this stage is generally thought to comprise a Scud-type system, a significant derivation from

A postal document shows the payload of Taepodong-1: Kwangmyongsong-1 remained attached to the third stage of the launch vehicle. A video release on YouTube gives the same image

the standard burn occurred. Specifically, two thrust levels were noticed with an initial high thrust phase followed by a long low thrust phase. Thus, the second stage was either a significantly modified Scud – for example, through the addition of four vernier engines (possibly with a separate turbo pump) parallel to the main engine – or a different stage was used altogether, possibly a modified SA-5. However, recent North Korean information indicates that the one plus four engine arrangement is more likely. The third stage is an unguided, spinning solid rocket.

The first stage propulsion system is a liquid rocket engine probably using the storable propellant combination inhibited red fuming nitric acrid (IRFNA) and kerosene. Ignition is accomplished by a hypergolic (self-igniting) start fuel designated Tonka – the World War II German designator for this propellant – filled into the fuel line at the main fuel valve. The propellant feed system is a turbo pump driven by a bipropellant gas generator using the main propellants. The start and shut down valves are one shot devices, actuated by pyrotechnic charges. Tank pressurization is performed by air stored in a toroidal high-pressure bottle in front of the missile's guidance section and heated by the turbine exhaust gases.

The second stage is basically of a similar design. The North Korean pre-launch photos seem to show an external propellant line design, similar to that of Scud-A/R-11 or SA-5. Further specific statements are difficult to make. Reconstruction of the launch vehicle's trajectory data indicates that the third stage is a Soviet Tochka/SS-21 rocket motor.

It is estimated that the guidance system basically resembles that of the Nodong arrangement with body-mounted free gyros but with the modification of an additional gyro for accuracy improvement. Prior to launch, the missile is orientated such that the trajectory plane hits the target and the guidance systems keeps the missile in this plane. Two of the three body mounted gyros are used for attitude and the third one lateral acceleration control; the horizon gyro's mounting must be able to move to a horizontal position. A pendulum integration gyro assembly serves for speed measurement. The fins are fixed and thrust vector control accomplished by four jet vanes in the first stage; for the second stage, either four vernier engines are used or thrust vector control is accomplished similar to the first stage by jet vanes. The third stage is spin-stabilized prior to motor ignition.

The third stage failure may be attributed to the cartridge propellant grain design of the Tochka motor. While in the ballistic missile mode the final acceleration is 10G, in the satellite launcher application, the final acceleration reaches nearly 50G, possibly resulting in a cartridge collapse with blocking and subsequent nozzle ejection. Motor rotation and its effect on internal ballistic may also be contributing factors to the failure.

Since the SLV's performance parameters are based on a reconstruction and the assessments by different sources yield a significant variation, a performance chart based on the probable mission objective is given."

Updated May 2003 – from NTI report

From the photograph and video of the Taepodong-1 missile launch released by the KCNA, the American analyst, Charles Vick, of Global Security noted that the payload section appeared to be roughly 4 m long and 1 m in diameter, based on the dimensions of the other stages. This gives a rough estimate of the size of the third stage. In addition, assuming the first two stages were a Nodong and a Scud, from the range at which the second stage is reported to have traveled, one can estimate that the mass of the third stage must be in the range of a tonne.

The reports that débris was carried far into the Pacific Ocean, if true, would imply that the third stage burned for some time after the second stage burned out, which should have been observable to American intelligence. After the North Korean announcement of its launch attempt, an American official was reported as stating that the United States detected something separating from the third stage that appeared to have some thrust behind it. While the satellite may have been small, the third stage may have been 3–4 m long and perhaps 1 m in diameter, as estimated above, which should have been easily observable. The United States presumably got a very good look at the missile launch. American intelligence detected launch preparations two weeks in advance and reportedly monitored the launch with two spy planes and a reconnaissance ship, the *Observation Island*, which had Cobra Judy radars on board. Despite this, the intelligence community appears to have initially missed the third stage with its payload.

The relatively long developmental period for the Taepodong rockets was a result of delays in the Nodong missile program: the technical difficulties of multi-staging, engine clustering, guidance, airframe design and development, combined with the economic turmoil that the DPRK was facing in the 1990s. Many of these technical issues were addressed by the DPRK's employment of rocketry designers and engineers from Russia, Ukraine and from its missile partners such as Pakistan and Iran. DPRK missile designers and engineers continued to travel to China for professional training and possible technology exchanges throughout the 1990s. Initial prototypes for Taepodong missile systems were probably manufactured in 1995 or 1996.

To support the Nodong and Taepodong programs, expansion of the DPRK's ballistic missile infrastructure has continued. Construction of specialized "underground missile bases" or "missile silos" that began during the mid-1980s continued through the 1990s. A small number of these facilities were located throughout the country and are now believed capable of handling both Nodong and Taepodong missiles. Work seemed to begin on a small satellite named Kwangmyongsong-1 in parallel with launch vehicle development. In designing its still hypothetical first satellite, the DPRK was believed to have received considerable assistance from the Chinese Academy of Launch Technology. Kwangmyonsong-1, shown on an official image and on stamps, offers some similarities with the first Chinese Dong Fang Hong satellites.

The launch showed that North Korea had successfully demonstrated staging technology, since the staging of the first two stages appears to have worked successfully. This is not particularly surprising, since staging can be done in ways that are not technically demanding, although it increases the complexity of the missile and may reduce the reliability. The photograph of the launch appeared to show a space between the first and second stages, which suggested that North Korea used a similar technology to Chinese and early Russian missiles in which the upper-stage engine is ignited while the lower stage is still burning.

The third stage apparently suffered a technical failure and failed to insert Kwangmyongsong-1 into orbit. Instead, it continued east, burning up, with a débris trail that apparently extended to approximately 4,000 km. American aircraft and ships tracked the test. Following the test, Japan's Self Defense Forces sent three destroyers and patrol aircraft to search the impact areas in the Pacific for wreckage of the missile and its warhead. These efforts may have been in vain, since the second stage impacted on the edge of the Japan Trench in waters with a depth of some 5,000 m.[9]

The demonstration of staging was important, because it is a necessary step for North Korea to develop long-range missiles. In addition, one clear, verifiable possibility for implementing a limit on North Korea's missile development was to ban the flight testing of multi-stage missiles before North Korea demonstrated its staging capability. The August 1998 rocket test removed that as a possibility. The third stage and satellite capabilities came as a surprise, indicating that the program was further along in developing ICBMs than had previously been estimated. The launch tested a number of critical aspects in ICBM development, such as multi-stage separation, attitude control, guidance techniques, different fuel systems, etc.

MODEST, ISOLATED MUSUDAN-RI LAUNCH COMPLEX

The attempt of the DPRK to launch a satellite was made from the Musudan-ri launch complex or Hwadaegun Missile Test Facility (40°13′N, 149°07′E), in Hwadae County, North Hamgyong province, near the northern tip of the East Korea bay. Since August 1998, great attention was focused on this dispersed site in a hilly rural region, at about 100 km from the city of Chongjin. It is located some distance from major transportation modes, such as harbors or airports. There are no railway connections, or even paved roads connecting the launch complex with the country. While this profound isolation may be only a modest barrier to a test program consisting of a single launch every few years, it is evidently inconsistent with the transportation requirements posed by a serious missile test program requiring rocket launches at frequent rates.

FAS (Federation of American Scientists), using DigitalGlobe and Aegis images, published a well illustrated report, "Musudan-ri Missile Test Facility, North Korea, February 15, 2002–March 26, 2009", available online at www.fas.org/nuke/guide/dprk/facility/musudan-ri.pdf.

Musudan-ri, in spite of a modest and rudimentary infrastructure on an isolated

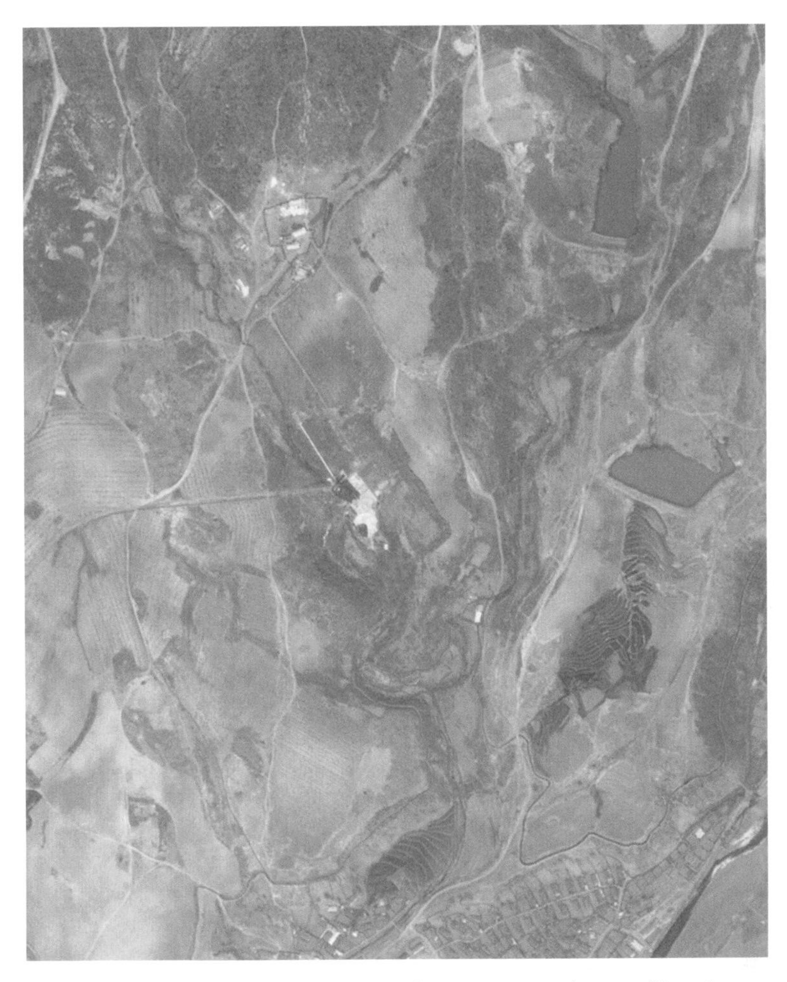

The two American operators of high-resolution remote sensing satellites demonstrated the performances of their respective spacecraft to take detailed images of the facilities and roads at the Musudan-ri launch complex or Tonghae Satellite Launching Ground: DigitalGlobe and GeoEye

portion of North Korea's north-east coast, became one of the most photographed "secret" areas of the world. The Musudan-ri launch facility increasingly mesmerized American security planning during the 1990s, with the various North Korean missiles tested from this site, because it constituted the primary threat that drove American operations and Japanese and South Korean defense programs. Later, the high-resolution of commercial satellite imagery revealed the vaunted Musudan-ri test site as a facility barely worthy of note, consisting of the most minimal imaginable test infrastructure. It is quite evident that this facility was not intended to support, and in many respects is incapable of supporting, the extensive test program that would be needed to fully develop a reliable missile system.

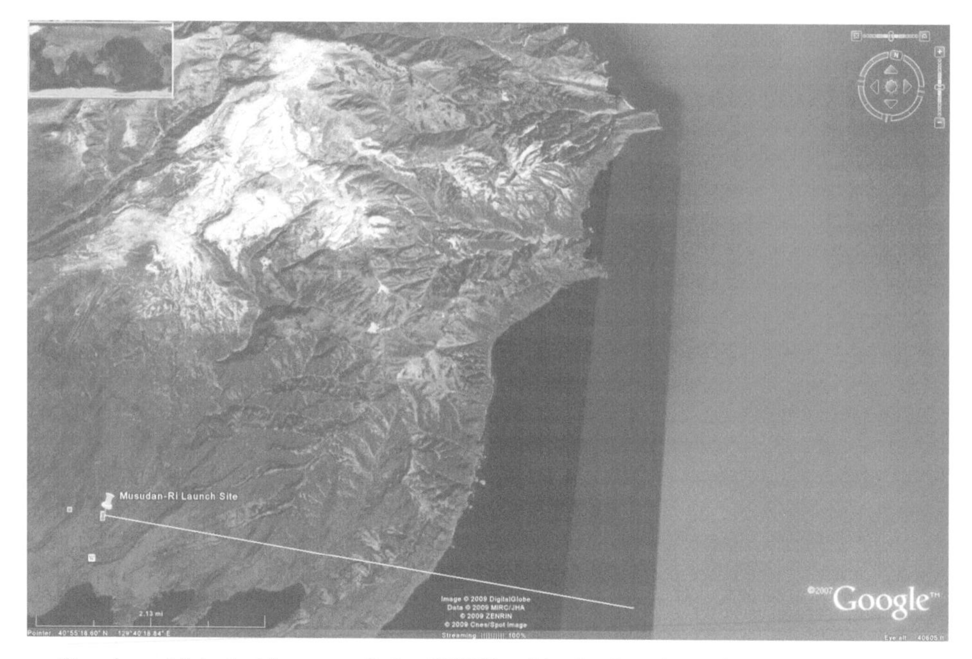

Google published this map of the DPRK with the location of Tonghae Satellite Launching Ground at Musudan-ri

The modest ambitions of the North Korean test program are clearly revealed by the scale and nature of the Nodong test facility, surely the antithesis of Cape Canaveral Air Force Base. The Nodong missile facility betrays no indication of permanent occupancy, but rather gives evidence of a temporary encampment to which launch crews might, from time to time, repair or test their handiwork. There is a complete absence of any manner of industrial support or other test facilities and the bare-bones test infrastructure is connected by no more than a spidery network of unpaved trails. Although the dirt and gravel roads that connect the facilities at the test site may suffice for tests at intervals of years, a serious test program would generate frequent vehicular traffic that would require paved roads. Non-frequent testing can be supported by trucking in precisely that quantity of propellant needed for the test at hand, but missile test facilities normally include separate liquid propellant storage areas that are sufficient to support a number of tests.

The initial construction of the missile base in Musudan-ri was reportedly completed in 1988, following observations made by American reconnaissance satellites that currently monitor the rocketry and nuclear energy activities of the DPRK. Since that time, a total of two missile tests have been conducted from this facility. A prototype of the Nodong-1 missile was detected on a launch pad in May 1990. The single test flight of this missile was conducted in May 1993. There are reports that a February 1994 static engine test was directly related to the Taepodong program. The May 1994 modification of the launch towers at the Musudan-ri

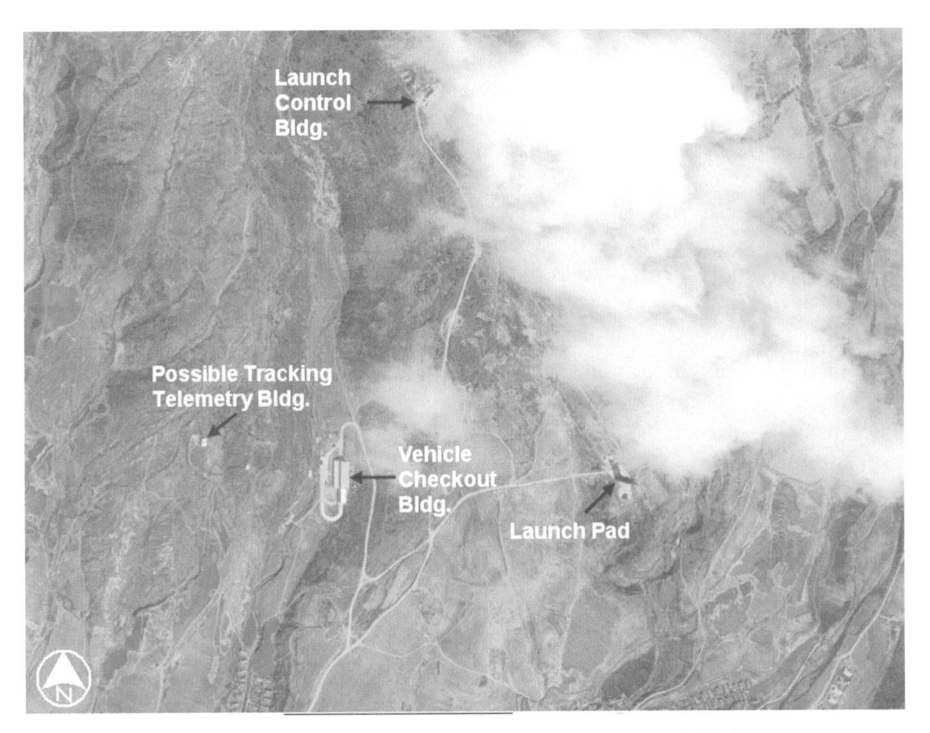

GeoEye and DigitalGlobe took these photographs of Tonghae Satellite Launching Ground facilities on 29th March and on 2nd April 2009

Launch Facility and the erection of a giant shelter pad against propellant jets provided some additional indications of Taepodong's progress.

In August 1998, the multi-stage Taepodong-1 missile, described as a small satellite launcher, made its first flight. During 1999, some preparations were detected for the launch of the larger and more powerful Taepodong-2 missile. The building of a new launch pad since late July 1999 received specific attention. The new complex, with a higher tower, was adapted from the existing one used for Taepodong-1. Compared to the previous pad – with an estimated height of about 22 m – the new pad is 1.5 times taller, standing about 33 m.

As of early August 1999, it was suspected that the Taepodong-2 rocket would now be complete and put in storage near the launch pad. However, no launch attempt was made. It seems that a mockup of Taepodong-2 was used for some infrastructure tests, especially during the night and under cloudy conditions, in order to avoid American observations from space. American intelligence reports that North Korea conducted propulsion tests for the Taepodong-2 missile in April or May 1999 and confirmed that North Korea was refurbishing its missile launch pad and transporting rocket fuel to storage depots. According to some media reports, North Korea has conducted three or four static tests of Taepodong-2 missile engines at Musudan-ri between December 1999 and January 2000.

The Musudan-ri Test Facility is roughly triangular, consisting of a single launch pad, a range control facility located 850 m to the north-west of the launch pad, and a Missile Assembly Building (MAB) located about 500 m directly due west of the launch pad. While the MAB is oriented due north, the remainder of the complex is roughly oriented 35° west of north. These three major elements of the test facility are connected by a network of unpaved roads and trails, some of which are evidently peculiar to the test facility, others of which may be associated with local agricultural activity. The facilities of the North Korean missile test range are interspersed with active agricultural areas, with a number of small agricultural settlements located in close proximity. There is no evident security perimeter separating the missile test facilities from the surrounding agricultural communities. There is a complete absence of residential structures that might be associated with missile test staff, as well as a complete absence of larger structures that might provide "industrial" or other operational support. By comparing the 1999 Ikonos imagery of Space Imaging (now GeoEye) with the 1971 Corona imagery of the Department of Defense spacecraft using recoverable capsules, it is clear that there has been a significant expansion in the number of dwellings and associated structures in these settlements.[10]

MISSILE FLIGHT TEST MORATORIUM 2000–2005

In 2000, the DPRK took the unilateral decision to observe a moratorium in missile flight testing. However, it pursued a steadily expanding ballistic missile development program, in line with its national philosophy of Chu'che (the doctrine of self-reliance). North Korean engineers aggressively worked to improve Taepodong guidance systems (with GPS/star navigation), to design and develop lightweight structures and to conduct static tests of engines (with gimbaled nozzle) at Musudan-ri site. Also in development were re-entry vehicles, warheads and penetration aids in weaponry technology.

Like the Nodong program, it is probable that the Taepodong program benefited from technology exchanges and test flights related to the Pakistani Ghauri and Iranian Shahab-3 programs. It is also believed – but it is not confirmed – that both Iranian and Pakistani observers were present for the Taepodong-1 launch of August 1998. Iran has been involved in the development of the Taepodong family from its inception, including financing and the exchange of information, technology and personnel.

In November–December 2002, an explosion occurred at the North Korean test site, following South Korean sources in April 2003. According to military officials, an American spy satellite detected the explosion in Hwadae-gun, North Hamgyeong province at the time. Débris from the explosion was scattered in the area, and the accident was likely to lead to significant damage in the north's development program for long-range missiles, they said. A 30-m launching pad was restored in late 2003 at the Musudan-ri missile complex. In May 2004, it was reported in Seoul that North Korea had restored facilities for missile engine testing that had been destroyed by the explosion of late 2002 and that it was preparing to test engines for the Taepodong-2 ballistic missile.

Charles P. Vick, who monitors the DPRK missile development and space launch activities as senior fellow of Global Security, gave considerable additional comments on the infrastructure at Musudan-ri for the Taepodong-2 program: a new UDMH propellant farm as well as numerous camera sites, flame bucket concrete slabs and vehicle parking areas and a near pad bunker. Off site, there was an extension for a horizontal assembly building as well as improvement for the range control center and the static test firing facility. Vick stated:

> "The soft nature of the launch site certainly indicates that it is not intended as a strategic facility for strategic ballistic missile operations'.[11] This present Taepodong-2 launch pad with its 2m inside diameter launch pad has been utilized for one static test firing of its first stage. This launch pad indicates that the base diameter of the Taepo-Dong launch vehicle is on the order of 2.2m in diameter. [...] The significant change from the design of the Taepo-Dong 1 gantry to the design of the present Taepo-Dong 2 gantry umbilical tower reflects a very significant rebuild of the facility feature structures. Most significant is the realization that the gantry for Taepo-Dong 2 is far taller than is needed for the existing launch vehicle design. It portends potential growth of future satellite launch vehicles beyond the Taepo-Dong 2. This design change to the tower took place within a year after the Taepo-Dong 1 launch utilizing the same launch facility."

In May and June 2006, DigitalGlobe took these two high-resolution images of the Musudan-ri Launch Complex (Tonghae Satellite Launching Ground). They showed some significant changes in the infrastructure

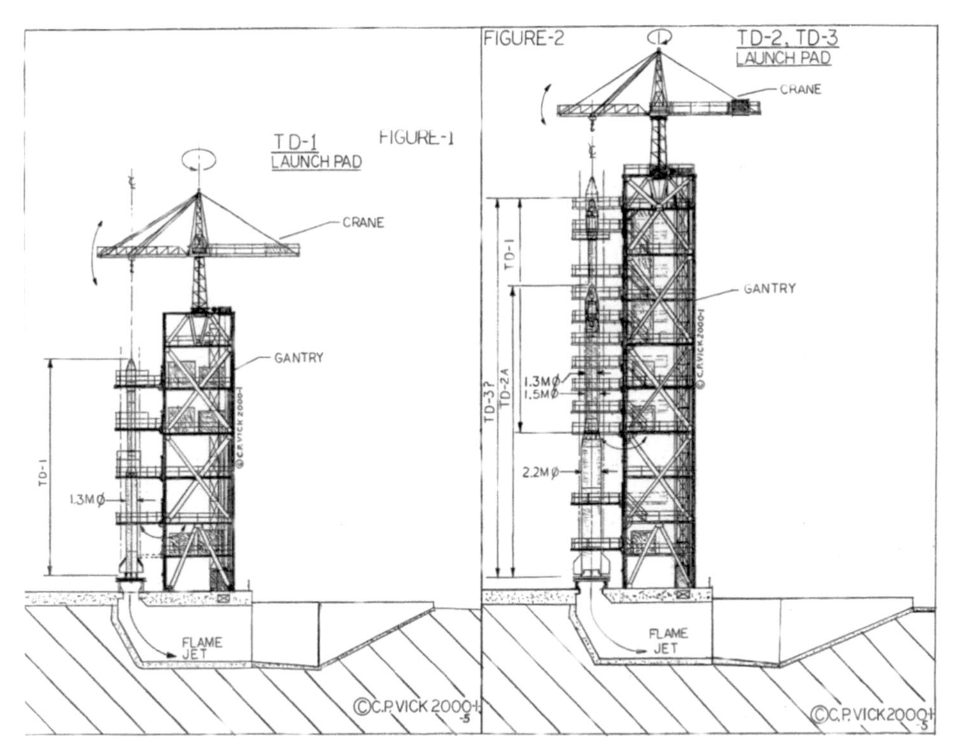

Charles Vick, rocketry specialist of Global Security, made this drawing showing the evolution of the Musudan-ri launch pad from 1998 to 2006

"UNSUCCESSFUL" FIRST FLIGHT OF TAEPODONG-2 IN JULY 2006

On 17th January 2006, the test launch of the No-dong-B missile from an Iranian test site announced the end of the North Korean long-range ballistic missile self-imposed testing moratorium from 2000 until late 2005. This is when North Korea declared it was no longer bound by its self-imposed moratorium. In mid-May, many reports about the DPRK missile activities suggested that North Korea might test the new larger Taepodong-class booster from the upgraded facilities (especially the launch pad) of Musudan-ri. Commercial high-resolution imagery, with American private remote sensing satellites – QuickBird of DigitalGlobe and Ikonos of Space Imaging/ GeoEye – was used to follow the preparation of the long-range ballistic missile at the launch site of North Korea. But the photographs taken from space were kept classified by the US Department of Defense. No image was available for publication in the media.

The successful GeoEye (on 17th and 24th May) and DigitalGlobe (on 9th and 15th June) images are the only ones that have been available to 28th June 2006. Cloud cover prevented successful collection of imagery of the required launch infrastructure points of interest, except for the 9th June image. Charles P. Vick had access to the imagery in order to review the preparation and the flight of Taepodong-2, which took place during 4th July 2006 with five other Nodong-type missile launches.[12] Referring to these observations from space, he speculated that the DPRK was preparing the launch of the Kwangmyongsong-2 communications satellite: "Close examination of the 9th June image of the raised pad under the gantry service levels barely reveals the pad presence and possibly a white booster first stage engine boat tail sitting on the pad, but the resolution is so low as to be irresolvable for the required clarity certainty of this analysis. This may also indicate that the launch vehicle may not be using large base fins but instead is using gimbals mounted vernier thrust chambers for steering."[13]

In the 24th May imagery, a square dolly was observed outside the horizontal assembly facility or Missile Assembly Building (MAB): was it used for the encapsulated satellite payload fairing and third-stage payload, which were hoisted and stacked by the gantry umbilical tower crane? For Charles Vick, it remained unclear whether the satellite payload had been stacked on top of the booster assemblage. Prior to this, the second stage was erected and hoisted for stacking on top of the first stage by the gantry umbilical tower crane. Based on Japanese reports, the encapsulated satellite payload fairing and third-stage payload were apparently stacked some time on 19th June, prior to the satellite imagery spotting technicians in the top of the gantry umbilical tower service levels. The exact configuration of the Taepodong-2 rocket for its maiden flight remained a mystery: two or three active stages or first stage with dummy upper stages?

The 22nd June 2006 Digital Globe imagery showed the on-going fuelling tanker truck operation around the base of the launch pad, with most of the tanker trucks removed from the launch pad area. At least one side of the movable service levels of the gantry umbilical tower around the launch vehicle had been folded back but the view available did not permit a view of the launch vehicle. The missile propellant loading with highly toxic, corrosive, hypergolic UDMH (Unsymmetrical Dimelhyl-Hydrazine) and IRFNA (Inhibited Red Fuming Nitric Acid) would have been completed during the 17–18th June period. On 20th June, North Korea restated its past policy statement that the DPRK, as a sovereign state, has the full autonomy right to develop, deploy, test-fire and export its missiles. The maiden launch of Taepodong-2 was expected to take place a few days later, on 21st June, but bad weather conditions, with heavy rain, kept the rocket fuelled on the pad for two weeks! Exposing the booster tanks, propellant lines and engines to the corrosive propellants is dangerously destructive, unless properly resistant materials and seals are extensively used. It is also suspected that there were some technical problems during the preparation of the rocket systems.

The DPRK designated the impact area as a "no-sail" and "no-fly zone" until 11th July 2006. On 4th July, there was a volley of at least six missile launches from North Korea over a span of four hours, with another seventh at the end of the day. One of those lift-offs was the first Taepodong-2 rocket, perhaps topped by a satellite, according to Charles Vick.[12] The fourth launch, at 4:01 pm, was the Taepodong-2 missile, thought to have a range of around 6,000 km. It failed after between 30 and 60 sec of flight, according to initial reports. Indications, upon further review,

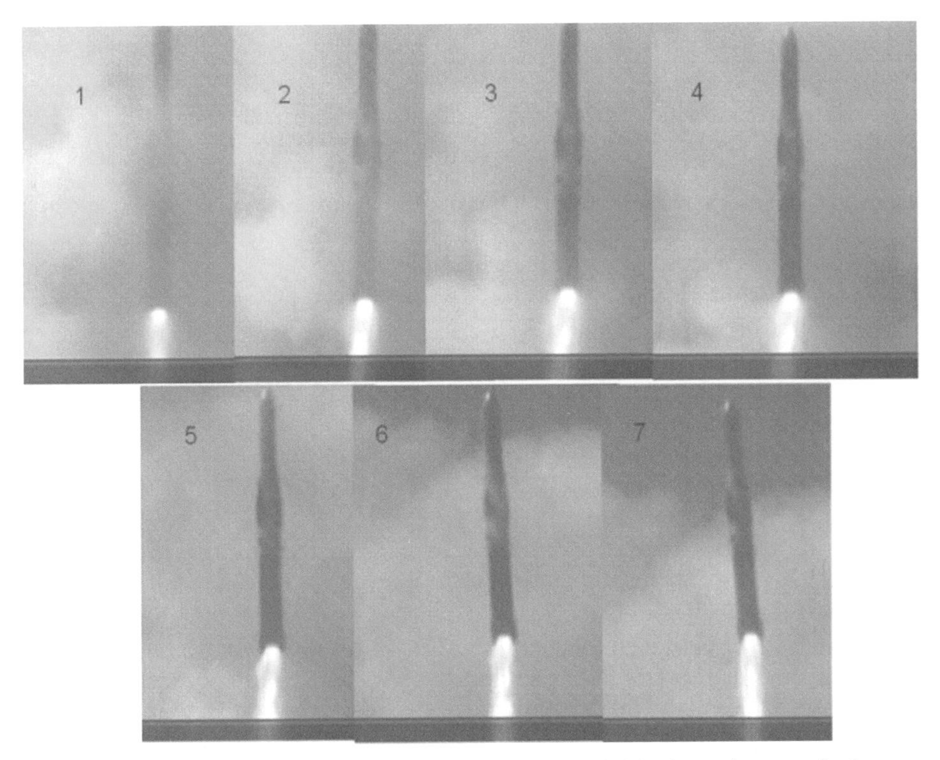

In July 2009, a video report of DPRK TV News showed this short abstract of a long-range missile launch. It is supposed to be the – apparently erratic – lift-off of Taepodong-2 on 4th July 2006

showed that the telemetry was lost after between 50 and 52 sec, perhaps due to its break-up or maybe a sea impact. Department of Defense sources and White House National Security Adviser Stephen Hadley stated that the new long-range missile of North Korea only burned for 42 sec. This launch failure was indicative of a characteristic maximum dynamic pressure failure in mid-air, tilting over, folding and breaking up in several seconds.

The Taepodong-2 rocket was launched on a due east minimum-energy trajectory close to 41° out of the launch site, heading in the general direction of the Pacific Ocean east and south of the Hawaiian island chain. This was entirely typical of the presumed satellite launch attempt trajectory. The satellite, presumably named Kwangmyongsong-2, is assumed to be the payload of this first attempted Taepodong-2 launch. It has also been suggested that a weather satellite may be the payload. The payload mass would be in the 150–250-kg range. Charles Vick stated: "It is unclear whether the Taepodong-2 launch was another attempt to put a satellite into orbit. Details are scarce and partially contradicting, but it seems the rocket broke up 42sec after lift-off, which observers believe to be the time of Max Q (the maximum of the dynamic pressure exerted upon a rocket during acceleration). This may indicate that a structural failure occurred during the first stage burn and that rocket was not blown up intentionally by ground control (for instance because it was off course)."

The failure apparently took place while on its ascending pitch-over program between 35 and 42 sec, causing it to veer off course. It then broke apart shortly afterwards, at 50–52 sec of flight, when the second stage and what was left of the third stage and payload sheared off in their various parts. This is when the telemetry was lost as it was keeling over. The first stage apparently remained "airborne for upwards of two minutes" while still continuing the first-stage burn before it finally tumbled out of control, collapsing into the Sea of Japan. The launch vehicle was not deliberately destroyed, as practiced in the West, because, like the Russian and apparently the Chinese rockets, it carried no in-flight destruct system. According to American and Japanese government officials, the three-stage first Taepodong-2 lost a piece of flight hardware that was sighted falling from the vehicle very early in the launch. Presumably, it was its two-piece encapsulating payload shroud-fairing that broke loose. It probably impacted on the satellite payload and third stage, heavily damaging them.

Following Charles Vick's report, some officials of the Department of Defense indicated that the stable missile boost phase of 42 sec and the subsequent tumbling out of control to impact into the Sea of Japan were only airborne for close to two minutes. The flight path would have taken it east over central northern Japan if the flight had continued. It would imply that the first stage essentially remained intact and functional. Certainly, the North Korean design engineers and scientists learnt a lot from this failed flight test. However, it took some 32 months before going ahead with a new test launch.

The military launch exercise of 4th July showed that the DPRK achieved some technological progress during its self-imposed testing moratorium. In fact, it bought time to design and to develop an improved Taepodong-2 long-range missile or satellite launcher, with no intention of doing otherwise. The redesigned vehicle was static test-fired several times, with at least one failure, at the static test center of Musudan-ri. The damage to the facility was quickly repaired and the testing was quickly renewed successfully. "This moratorium was a success shell game on the world stage with all of its implications," wrote Charles Vick. No information was available about the exact nature of the first Taepodong-2 launch and of its compact payload, be that a small comsat or weather satellite. There was also the question of the role of Iran and China in the development of the Taepodong-2 launch vehicle.

The spectacular test of several missile versions provoked a strong reaction of the international community at the UN (United Nations) Security Council. On 15th July 2006, the Council adopted, with a unanimous vote, Resolution 1695, pressing North Korea to suspend all missile-related activities and requiring all UN Member States to exercise vigilance and prevent the transfer of missile-related materials and technologies to the DPRK. The North Korean Foreign Ministry reacted the next day by refuting the "resolution" of the UN Security Council against the DPRK. He referred to the 5th July launches as "part of the KPA routine military exercises to increase the nation's capacity for self-defense". He emphasized that North Korea is not a member of the MTCR (Missile Technology Control Regime) and was not bound by any international laws or agreements restricting its missile tests.

FURTHER ENGINE TESTS AND NEW LAUNCH COMPLEX

In mid-September 2008, *Jane's Defense Weekly* published a report by Joseph Bermudez and Tim Brown, analysts at Jane's Intelligence Group, about the existence of a new North Korean ballistic missile launch site. With the help of a private satellite imagery analysis company, Talent-keyhole.com, they identified this more sophisticated facility that is being built on the site of a small village called Pongdong-ni. This village was displaced during construction. Bermudez and Brown referred to the launch site as the Pongdong-ni Missile and Space Launch Facility, about 50 km from the Chinese border (39°39′N, 124°42′E) and 12 km north-west of Pyongyang. The new infrastructure presented a challenge to American and foreign reconnaissance, since it was obscured from direct airborne and seaborne observation by nearby hills. Difficulties with airborne reconnaissance were exacerbated by the facility's location at the northern reaches of the Yellow Sea between North Korean and Chinese airspace.

The base had been under construction for eight years and would be capable of launching both the Taepodong-2 ballistic missile and the Taepodong-2 space launch vehicle. The facility also had a rocket engine test stand, capable of supporting test firings of all known North Korean rocket motors. This complex showed that despite continued economic, political and social hardships, North Korea continued to commit precious resources to the development of ballistic missile and space launch capabilities. While the United States was continuing to force an end to North Korea's nuclear program, its ballistic missile and space launch programs appeared to be continuing towards challenging ambitions.[14]

Following the high-resolution images taken by commercial remote sensing satellites, this modern complex could be used to test rocket engines and to launch long-range missiles as well as space launch vehicles. It included a launch pad, a command-and-control tower and a rocket engine test stand. Although the facility was under construction since 2000, it is not fully operationally to proceed with missile flights or space launches and it appears unlikely to do so before 2010. Its utilization would represent a major step for the still hypothetical space program. Following reports in South Korean media quoting American intelligence information, engine tests took place on the new infrastructure. German Norbert Brügge, who published excellent information, with many photos, drawings and tables about space rockets and launch sites in the world (www.b14643.de/Spacerockets_1/index.htm), referred to the launch site of Dongchang-ri.[15]

North Korea has two facilities to perform satellite launches. The older, Musudan-ri ballistic missile and space launch facility, which was built in 1988, is rudimentary compared to the standards of any other space launch complex in a high-tech country. Apparently, there are no facilities for long-term propellant storage, no housing or warehouse space, and no all-weather roads. It has little or no space-

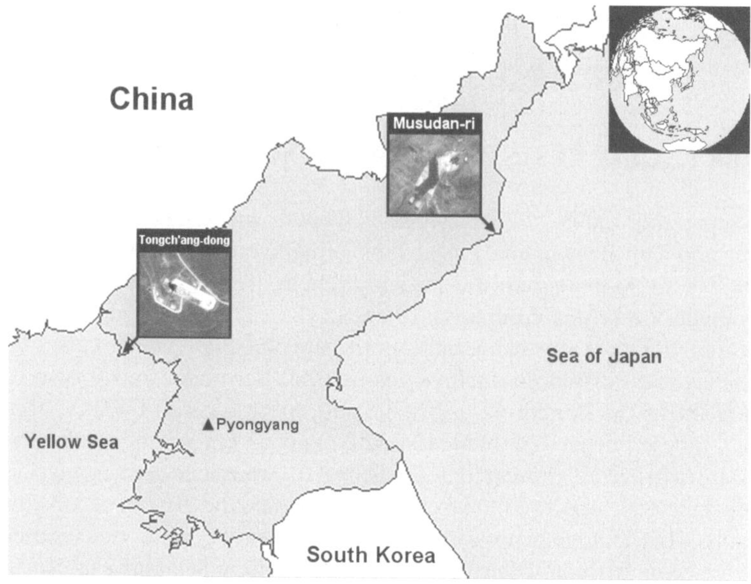

The two launch sites of the DPRK for the Taepodong-2/Unha-2 program

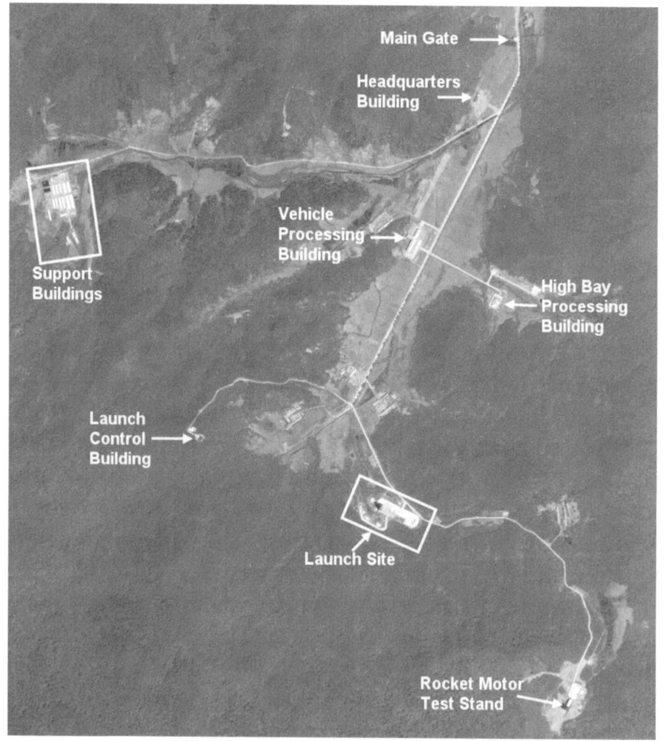

An overview, in June 2009, of the Pongdong-ni/Dongchang-ri launch complex, from space

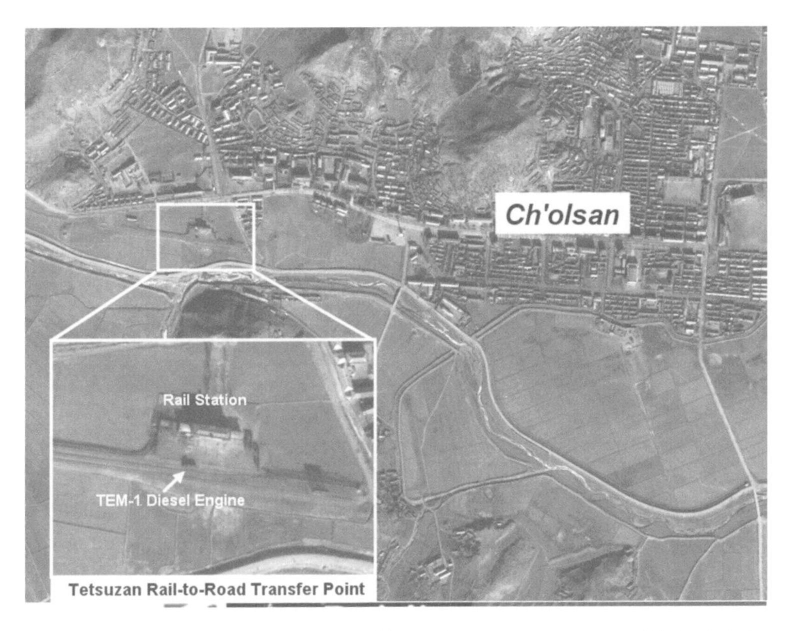

Access by railway to Dongchang-ri launch complex, seen by DigitalGlobe

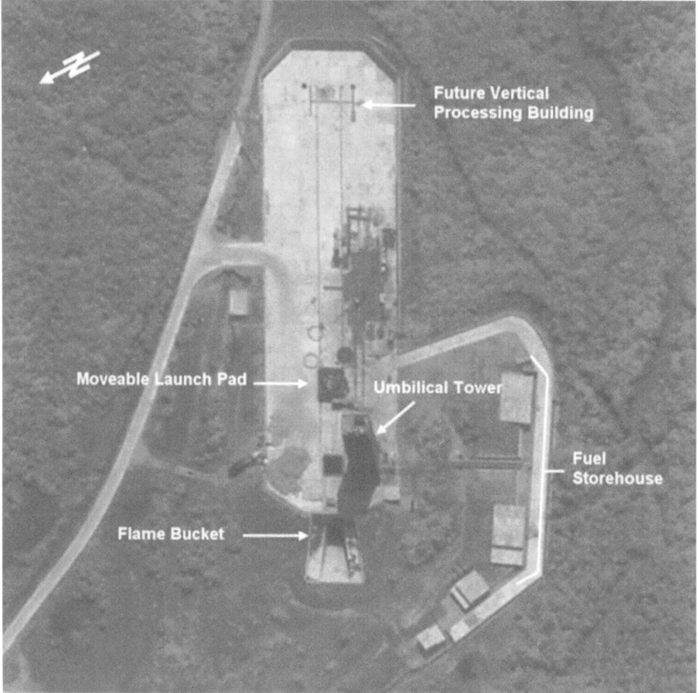

The launch pad for Taepodong-2/Unha-2 at the Pongdong-ni/Dongchang-ri complex

Seen by DigitalGlobe, the construction in progress of rocketry facilities at Pongdong-ni/Dongchang-ri launch complex: engine test stand, high-bay building, assembly hall

tracking capability to determine the success or failure of the launch. The new rocket test and launch facility at Pongdong-ni/Dongchang-ri allows North Korea to demonstrate commitment to its ballistic missile and rocket programs. Yonhap, a South Korean news agency, claims that North Korea tested a missile engine at the Pongdong-ni during May or June 2008. These missile engine tests, along with the construction of the new launch site, demonstrate that North Korea's missile program has decided to move forward towards ambitious objectives. Following an American official, the intelligence community had known about the secret facility since the start of construction eight years ago and was monitoring it closely with air and space systems. It is expected that Dongchang-ri rocket site would become available for missile tests and space launches in 2010.

DPRK PROGRESS TO IMPROVE (SPACE) LAUNCH CAPABILITY

On 2nd February, Iran successfully launched the two-stage Safir rocket with a microsatellite named Omid. Five days later, North Korean daily Rodong Sinmun made this commentary:

"Iran's recent satellite launch has not only demonstrated her national power

> but shown before the world that there can be no longer any monopoly of space development and its use. Iran has shown its will to take a more positive part in the race for advancing into space, defying any accusations of the above-said satellite launch by the United States and other Western countries.
>
> No one should be allowed to take issue with space development and its use or stand in their way as long as they are of peaceful nature and regarded as helpful to promoting the welfare of humankind. Any country which stands for the peaceful use of space is entitled to advance into space any time and use it for peaceful purposes.
>
> The DPRK, a member of the international community, has an option to advance into space and a legitimate right to participate in the space scientific and technological race. It is the DPRK's stand and policy to make a positive use of space resources, wealth common to humankind, in an effort to enable the Korean people to lead a highly civilized and bountiful material and cultural life and bring a bright future to the Korean nation.
>
> The DPRK has already set up institutions specializing in space research and development for peaceful purposes. The DPRK's scientists and technicians are now energetically pushing ahead with the work to make a peaceful use of space as required by the developing reality and the international trend. The DPRK's policy of advance into space and its use for peaceful purposes is just as it meets the requirements of the developing times and no force on earth can block this."

In mid-February 2009, Western sources reported about the preparation of a long-range missile to be launched from Musudan-ri. On 16th February, KCNA confirmed these reports with this announcement: "This is a vicious trick to put a brake on the wheel of not only the DPRK's building of military capability for self-defense but also scientific researches for peaceful purpose under the pretext of missile. One will come to know later what will be launched in the DPRK. Space development is the independent right of the DPRK and the requirement of the developing reality. The DPRK's latest science and technology developing day by day in conformity with the trend of the present times when the world is advancing toward the road of pioneering space have registered big successes in the field of space development, too."

Then, on 24th February, KCNA referred to a spokesman for the KCST (officially mentioned for the first time since Unispace III in July 1999) to the following statement, entitled "Preparations for launch of experimental communications satellite in full gear":

> "Outer space is an asset common to mankind and its use for peaceful purposes has become a global trend. The DPRK has steadily pushed ahead with researches and development for putting satellites into orbit by its own efforts and technology since the 1980s, pursuant to its government's policy for the development of space and its peaceful use. In this course, scientists and technicians of the DPRK registered such great success as putting its first experimental satellite Kwangmyongsong-1 into orbit at one try in August 1998.
>
> Over the past decade since then a dynamic struggle has been waged to put the nation's space science and technology on a higher level, bringing about

> signal progress in the field of satellite launchings. The DPRK envisages launching practical satellites for communications, prospecting of natural resources and weather forecast, etc. essential for the economic development of the country in a few years to come and putting their operation on a normal footing at the first phase of the state long-term plan for space development.
>
> The preparations for launching experimental communications satellite Kwangmyongsong-2 by means of delivery rocket Unha-2 are now making brisk headway at Tonghae Satellite Launching Ground in Hwadae County, North Hamgyong Province.
>
> When this satellite launch proves successful, the nation's space science and technology will make another giant stride forward in building an economic power."

For the first time, Pyongyang announced in advance the campaign of the KCST to launch a rocket (Unha-2) with a satellite (Kwangmyongsong-2). It specified the location of the launch center at Musudan-ri, which got the name of Tonghae Satellite Launching Ground. Such preliminary information was a completely new process for DPRK authorities.[16] The space launch of Taepodong-1 in August 1998 was disclosed only four days after the shot of the missile! The commercial remote sensing satellites, with very high-resolution imagery – WorldView-1 of DigitalGlobe, GeoEye-1 of GeoEye – were used to monitor the activities at the launch site. Global.Security (www.globalsecurity.org) and the Institute for Science and International Security (ISIS, www.isis-online.org) published images of DigitalGlobe and of GeoEye showing the preparation of the launch and the Unha-2 launcher on the launch pad. Pictures were taken on 26th, 27th, 29th March and 2nd and 5th April (the day of the launch). The last one was a surprisingly spectacular one: the satellite photographed the rocket in flight, just a few seconds after lift-off. It was possible to see the missile's exhaust plume at Musudan-ri.

There were also preliminary announcements concerning the trajectory of the North Korean launcher, unlike the previous two attempts to launch a satellite.

The International Civil Aviation Organization (ICAO) and the International Maritime Organization (IMO) got from Pyongyang a letter dated 11th March indicating that the launch would take place between 4th and 8th April, between 2 and 7 am (UTC). This letter, referring to the launch of the Kwangmyongsong-2 experimental communications satellite, identified two potential "danger" areas with coordinates. ICAO produced a map with the North Korean indications.

Following a KCNA statement on 12th March, the DPRK acceded to the Treaty on Principles Governing the Activities of States in the Exploration and Use of Outer Space including the Moon and Celestial Bodies and the Convention on Registration of Objects Launched into Outer Space.

On 20th March, on the website of Union of Concerned Scientists (UCS), David Wright, specialist of DPRK missile and satellite affairs, published "An analysis of North Koreas's Unha-2 Launch Vehicle".[17] He stated: "The Unha-2 launcher is believed to be a version of the Taepodong-2 missile, which North Korea has never flight tested. [It is] much larger than the Taepodong-1 and Iranian Safir-2 launchers.

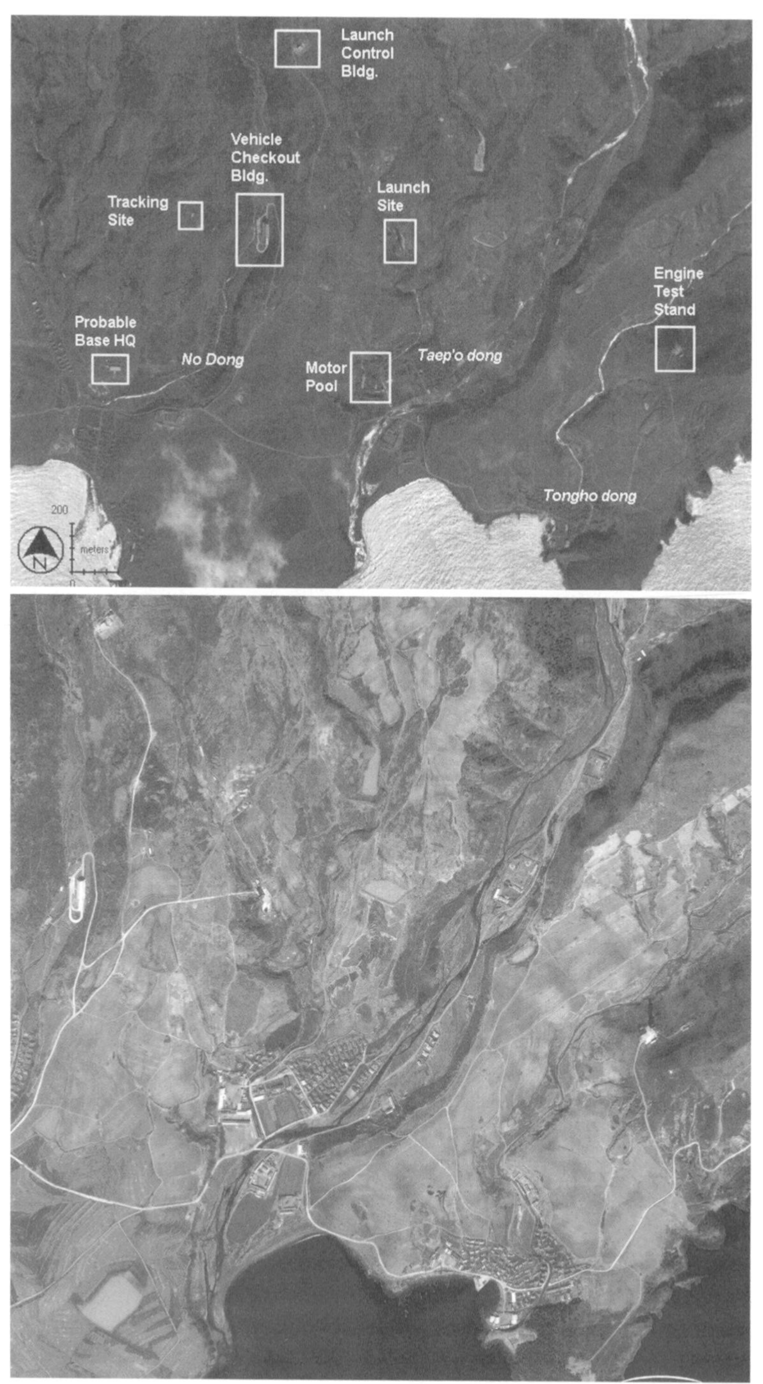

In February, then in April 2009, it was possible to monitor launch activity at Musudan-ri with regular overview of DigitalGlobe satellites

On 29th March, the Quickbird-1 satellite of DigitalGlobe took this view of Musudan-ri launch pad: it is possible to see there the Taepodong-2/Unha-2 rocket

The Safir-2 has a diameter of 1.25m, a length of about 2 m, and a mass of 26 tonnes. In contrast, the Taepodong-2 is believed to have a first stage dimeter of about 2.2 m and a total mass three times larger than the Safir 2 (approximately 80 tonnes)." Commenting the North Korean pre-information about the launch with the zones in which the first two stages are expected to fall into the ocean, he noted: "This announcement indicates that the launcher will have three stages (if the launch is successful, the third stage will remain in orbit with the satellite). Second, it indicates that the launch direction will be due east, similar to the Taepodong-1 launch. Such a direction is what you would expect for a satellite launch since it allows the launcher to gain speed from the Earth's rotation (0.35km/s in this case)." David Wright assumed that the large first stage of the Taepodong-2/Unha-2 launch vehicle would use a cluster of four engines of the same type that the single rocket engine used by the Nodong missile: "By clustering four engines, North Korea could use an existing engine to develop a stage four times the thrust of the Nodong and would be following a development path that was used by other countries in building larger rocket stages." About the second stage, it would be propelled by a Nodong engine modified for use at high altitude: "North Korea might design a stage that uses this engine but has the same diameter as the first stage, which can reduce the mass of the stage." For the American analyst, "the stages of a satellite launcher typically will have longer burn times than a ballistic missile. In addition, an unpowered coast phase is often added before the final stage ignites, especially if the satellite is intended to orbit at altitude above a few hundred kilometers."

Clearly visible from space on images taken in late March and early April 2009 by high-resolution satellites of DigitalGlobe and GeoEye, the white-painted rocket Unha-2 of the DPRK on its launch pad

On 4th April, Musudan-ri was the most monitored launch site by systems in orbit and on the sea. KCNA confirmed that the preparation for launching Kwangmyong-song-2 with the Unha-2 rocket had been completed with information from the Korean Committee of Space Technology: "The satellite will be launched soon. There is no change in the technological indexes necessary for the safe navigation of airliners and ships provided to the international organizations and the countries concerned in advance." The launch had to be delayed because of bad weather conditions in the eastern coast of the DPRK. The willingness of North Korea to launch a satellite can be justified by a "space race" with South Korea, which was preparing the KSLV-1 launcher with a Russian first stage for a first flight in mid-2009. T. Brown, in a report for GlobalSecurity, insisted on the difference in policies between North and South Koreas: "The timing of this latest launch might also be race with South Korea which plans to launch a satellite into orbit in the summer of 2009. South Korea, unlike its neighbor to the North, has a real space program with a real budget, western technical assistance and a commercial partnership with Russia. They even have a press center for the international media to watch the launch of the Korea Space Launch Vehicle this summer at the Naro Space Center located on an island 485km south of Seoul. The South Koreans do not appear to racing with the North, but the North Koreans seem to be racing to launch their satellite into orbit first."[18]

Comparative illustrations of Unha-2 (2009) and Taepodong-1 (1998) on the launch pad of Musudan-ri

The dimensions of Unha-2 estimated from the video of DPRK TV News

Unha-2 detailed by Norbert Brügge, Spacerockets website

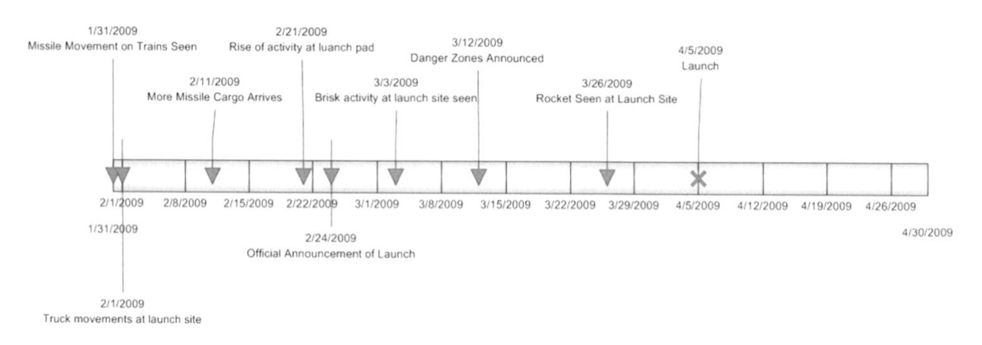

The long duration of the Unha-2 launch campaign (eight weeks)

The impressive Mission Control center for the launch of Unha-2 with a small satellite. The location of this facility is not known

The white Unha-2 rocket lifted off from Tonghae Satellite Launching Ground on 5th April. Its flight was announced by warning systems of the US Department of Defense, of the Japanese Ministry of Defense and carefully monitored by South Korean authorities and media. Initially, it was considered a success. The official announcement of the launch was made by KCNA some six hours after the Unha-2 launch. The United States, Japan and South Korea tracked it with satellites and ships. The information was released without any great triumphalism:

"Scientists and technicians of the DPRK have succeeded in putting the satellite Kwangmyongsong-2, an experimental communications satellite, into orbit by means of carrier rocket Unha-2 under the state long-term plan for the development of outer space. Unha-2, which was launched at the Tonghae Satellite Launching Ground in Hwadae County, North Hamgyong Province at 1120 on 5th April, Juche 98 (2009), accurately put Kwangmyongsong-2 into its orbit at 11:29:02, nine minutes and two seconds after its launch. The satellite is going round the earth along its elliptic orbit at the angle of inclination of 40.6° degrees at 49km perigee and 1,426km apogee. Its cycle is 104min and 12sec. Mounted on the satellite are necessary measuring devices and communications apparatuses.

The satellite is going round on its routine orbit. It is sending to the earth the melodies of the immortal revolutionary paeans 'Song of General Kim Il Sung' and 'Song of General Kim Jong Il' and measured information at 470 MHz. By the use of the satellite, the relay communications is now underway by UHF frequency band. The satellite is of decisive significance in promoting the scientific researches into the peaceful use of outer space and solving scientific and technological problems for the launch of practical satellites in the future.

The carrier rocket Unha-2 has three stages. The carrier rocket and the satellite developed by the indigenous wisdom and technology are the shining results gained in the efforts to develop the nation's space science and technology on a higher level. The successful satellite launch symbolic of the leaping advance made in the nation's space science and technology was conducted against the background of the stirring period when a high-pitched drive for bringing about a fresh great revolutionary surge is under way throughout the country to open the gate to a great prosperous and powerful nation without fail by 2012, the centenary of birth of President Kim Il Sung, under the far-reaching plan of General Secretary Kim Jong Il. This is powerfully encouraging the Korean people all out in the general advance."

This news release has the same tone as the information announcing on September 1998 the successful launch of the Kwangmyongsong-1 satellite. The statement about a success was also wrong: no DPRK object was detected in orbit and no signals were picked up around the Earth! Again, as in 1998, North Korea launched a "ghost satellite"! However, the video of the launch and of a control room (with a large screen) in Pyongyang showed the technological progress the DPRK had achieved since 1998. KCNA also indicated that General Secretary Kim Jong-il visited the General Satellite and Command Center to watch the launch of the experimental communications satellite: "After being briefed on the satellite launch, he observed the whole process of the satellite launch at the center." No photograph confirmed the development and preparation of the Kwangmyongsong-2 satellite, so was the program of the DPRK a domestic reality or an international bluff?

Jonathan McDowell, who regularly updated his space report with the satellites launched everywhere in the world, notes: "The North Korean News Agency KCNA quoted 0220:00 UTC as the launch time. The western press reported the launch time

Many shots of the Unha-2 launch vehicle: images from the official video

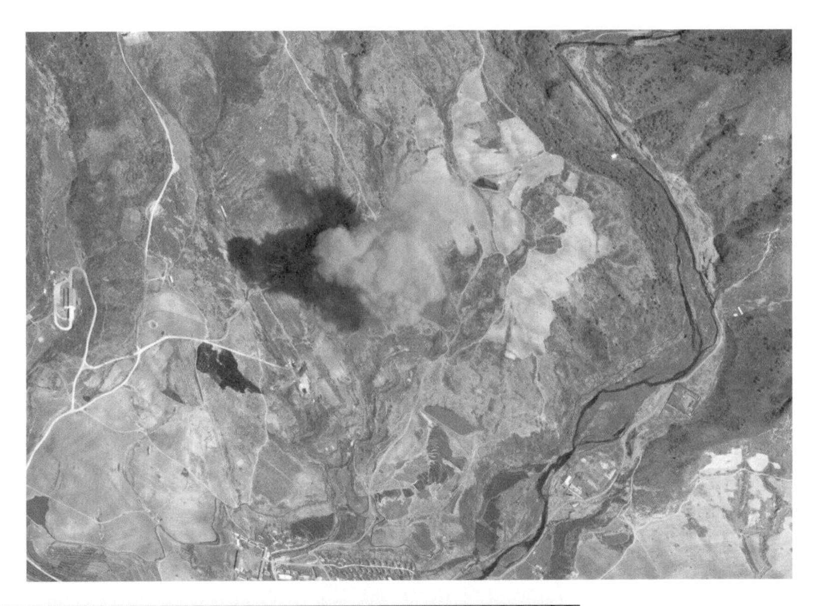

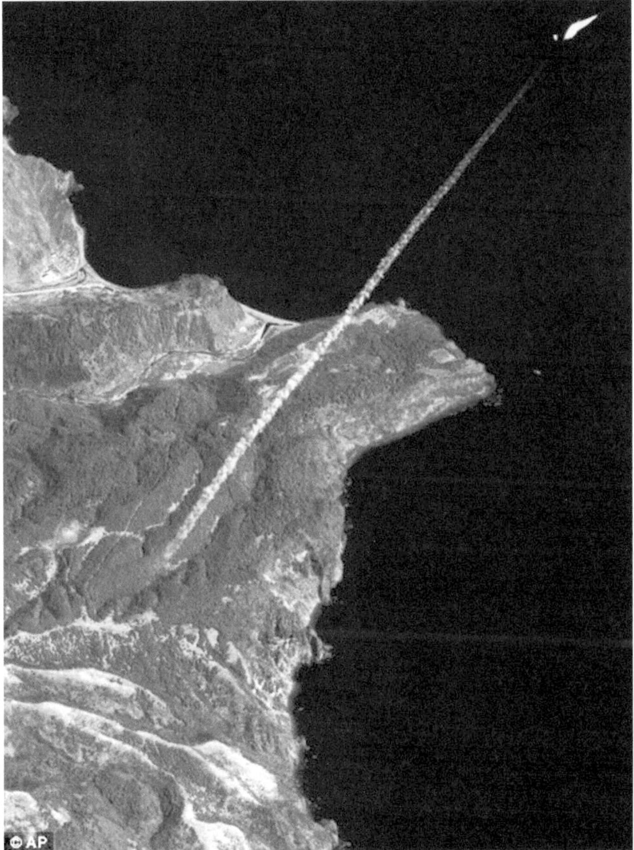

QuickBird-1 of DigitalGlobe was just flying over the Musudan-ri launch complex: it took the picture of Unha-2 just after lift-off. The shadow of the smoke generated by the launch is clearly visible on the ground! The trajectory of the rocket is also revealed on this image taken from space as 0230 UTC, and Russian sources say they tracked the launch at 0232 UTC. The Worldview-1 satellite took an image of the launch contrail from space reportedly at 0231:16 UTC, which must have been within a couple minutes of launch, so I will adopt 0230 UTC as the best current estimate of the launch time."[19]

Western analysts of the North Korean space launch, using the official video released by North Korea, confirm that the Unha-2 rocket of 2009 used a larger-diameter first stage than the Taepodong-1 rocket of the 1998 launch attempt. The pictures are consistent with a rocket about 27 m high and 2.2 m in diameter, including a 14-m long first stage, an 8-m long second stage with a diameter of 1.2–1.4 m and a third stage fairing with a diameter around 1 m. They insisted on the new progress made by the DPRK in the technology of multi-stage rockets. On 13th April, Victoria Samson, for Secure World, reported[20] that "According to publicly available data, the payload failed to separate properly from the launch vehicle and the entire rocket ended up in the Pacific Ocean" and she noted that "From a technical perspective, this test did not prove a good advertisement for North Korea's long-range ballistic missile capabilities, given that all three tests have ended in failure with apparently different root causes. It also raises the question of how North Korea intended to collect critical telemetry data on the rocket performance, since it is unlikely that North Korea had ships collecting data or radar stations tracking it once it went over the horizon from the launch pad. And if North Korea had no way to monitor its missile in-flight, it implies that they are not serious about building an operational long-range ballistic missile weapon system".

Craig Covault, reporting on the Spaceflightnow website,[21] referred to new data of the Japanese Ministry of Defense and the US Department of Defense, to state on 10th April that the North Korean rocket flew further than earlier thought: "The second stage of the rocket performed as planned, rather than failing early in its flight phase. The new information comes from updated radar tracking calculations and possibly also US Air Force Defense Support Program (DSP) missile warning system." Charles P. Vick published his analysis on 17th April:[22] "The launch of Kwangmyongsong-2 satellite was expected to be a technology demonstration flight test. The satellite payload mass was expected to be in the 100–170–250–550 kg range."

He described the launch with many details: "The first stage performed as planned quite cleanly impacting about 280km from Japan. The second stage did separate, completing its full throttled burns. ... The flight was to take 9min 2sec (542sec) to get to orbit with about 4-2sec for start-up. ... In fact the third stage would have had a coast period in its flight plan of perhaps 120sec to the perigee height of the trajectory before firing to complete orbital insertion .... The onboard computer separated the payload and third stage shrouds on time and they impacted down range within the planned Pacific Ocean impact zone ..., the second stage and third stage separated, but the third stage did not fire. ... The payload went down with the third stage. ... The first stage exhibited four thrust chambers, but the steering system is not clearly defined as vanes in the gas jet or vernier or thruster steering."

Norbert Brügge, who published a regularly updated website on Space Launch Vehicles, gave a useful summary, with some rear pictures, about the design and the flight of Unha-2. He described it as an "impressive launch vehicle": "Unha-2 is no great surprise. It was believed for a long time that the first stage was a reproduction of the Chinese DF-3 missile. That is confirmed now. The first stage is derived clearly by the Chinese DF-3. The propulsion is consequently a cluster of four YF-3 engines (Nodong engine). All four fixed engines of the cluster use their own turbopumps, documented by four exhaust exits. In the launch video can be seen the exhaust plumes from the turbopump pulsing. The turbopump exhaust tubes sticking down from the airframe to below the nozzles. Length, diameter and masses of the stage correspond with the DF-3. The stabilizing fins are missing."[23]

THE ART OF LAUNCHING "GHOST SATELLITES"

With the unsuccessful tests of multi-stage Taepodong-1 and Unha-2/Taepodong-2 rockets using obsolete propulsion and control systems, the DPRK demonstrates the media art, through spectacular videos, of launching "ghost-satellites". Was it behind the false revelations of the scientific activities developed by the KCST? Has North Korea the technological capability to design, build, launch and operate even modest satellites?

Access to space for peaceful purposes is used by the Pyongyang government – under the dictatorship of Kim Jong-il – to hide the military achievements of the

American spacecraft revealed the facilities and the activities at the secret Tonghae Satellite Launching Ground, Musudan-ri

Korean People's Army (KPA). It demonstrates how the technology in North Korea is progressing, well but slowly, in rocketry and defense systems. It creates an image of efficiency for the Stalinist–Communist regime to rival the liberal-capitalist South Korea (involved in space activities for science and applications) as well as to market its production of new missiles for exportation contracts. Both "space shots" of the DPRK could be considered as subtle tests of atmospheric re-entry devices that are crucial for the final trajectory of ballistic missiles carrying mass-destruction payloads. They can be considered as failures.

While Kwangmyongsong-1, in a rudimentary picture, looked like the first Chinese satellite, no photograph of Kwangmyongsong-2 has been published. The flight of Taepodong-1 was not announced in advance: the black-painted rocket, later described as the carrier of the first DPRK satellite, could be seen as a ballistic missile. The preparation of Unha-2, also unofficially named Taepodong-2, was described in advance by the official KCNA. Indication was given about the schedule and time (from 4th to 8th April) and the impact location of the first two stages. The white-painted rocket that could be observed from space on the launch pad by commercial remote sensing satellites appeared more like a satellite launch vehicle than a missile.

It will be interesting to know what, after the unsuccessful attempt of 5th April 2009, will be the next steps of the DPRK in rocketry and space development. Can we expect a "space race" between North Korea and South Korea to place a satellite around the Earth? The DPRK launched a satellite, but the spacecraft could not reach space! The construction at Pongdong-ni – another more modernized facility with an important test site – was identified from imagery of commercial remote sensing. This site could be the launch center for the next attempt of the DPRK satellite launch. Great secrecy surrounds North Korean activities in rocket engines and space systems. Until now, the official media of the DPRK did not release any photographs of rocket engine, of rocketry preparation, of satellite development, spacecraft integration: in fact, the DPRK is the most secret state among the emerging space powers. What about the real purposes behind its orbital presence that represents a challenging achievement for this economically poor country isolated by a Stalinist–Communist regime?

REFERENCES

1. From 2000, GlobalSecurity (www.globalsecurity.org), in Alexandria (Virginia), as a public policy organization in the United States, has reviewed the activities of WMDs in the world. It has a special focus on North Korea, with Charles Vick as specialist, whose intelligence assessments are considered reliable by the American Central Intelligence Agency (CIA) and the Defense Intelligence Agency (DIA). The FAS (Federation of American Scientists – www.fas.org) also publishes updated information about North Korea's Taepodong and Unha missiles. See also IAC-09.E4.3.5, Cosyn, Philippe: The DPRK's Road to Space: A brief history. IAA History of Astronautics Symposium, Daejeon, October 2009.

2. Burleson, Daphne: *Space Programs Outside the United States*. McFarland & Company Inc. Publishers, Jefferson, North Carolina, 2005. The review of space activities in North Korea (pp. 206–209) describes the progress from Nodong-Scud to the multi-stage Taepodong-1 rocket.
3. Bermudez, Joseph S., Jr: *A History of Ballistic Missile Development in the DPRK*. Center for Nonproliferation Studies, Monterey Institute of International Studies, Monterey, California, 1999. This covers intensively the developments from 1960 and 1999 of the ballistic missile capability in North Korea. Pinkston, Daniel A.: *The North Korean Ballistic Missile Program*. Strategic Studies Institute, Carlisle, Pennsylvania, February 2008. This gives a more recent review of DPRK efforts in missile technology. These two American reports are reference sources to understand how North Korea became a major player in rocketry development, missile production and business.
4. NK News (www.nk-news.net) reviews all the press releases of KCNA (Korean Central News Agency).
5. *Rodong Sinmun* (newspaper of the workers) is the official daily newspaper of the Central Committee of the Workers' Parti of Korea (WPK). It is regarded as a source of official viewpoints on many issues in the DPRK.
6. Information of Korean Central Broadcasting, 24 July 2002, published by Space Daily website.
7. Report and interview of Théo Pirard, who attended as a journalist Unispace III (July 1999).
8. *Paektusan 1 Space Launch Vehicle: Technical Assessment*. Nuclear Threat Initiative website, updated May 2003, www.nti.org/e_research/profiles/NK/Missile/1709_1713.html.
9. Clark, Phillip: Fact and fiction: North Korea's satellite launch. *Spacelaunch* 4, January–February 1999.
10. Vick, Charles P.: *Musudan-ri/Musudan-ni Missile Test Facility*. GlobalSecurity.org, with updated information and drawings: 2000, 2005, 2009.
11. Vick, Charles P.: *Musudan-ri 2005 update*. GlobalSecurity.org.
12. Vick, Charles P.: May, June, July 2006 buildup to the Taepo-Dong 2C/3 satellite launch attempt, 10 July 2006. GlobalSecurity.org. Reichl, Eugen: *Das Raketentypenbuch*. Motor Buch Verlag, Stuttgart, 2007. Reviewing all types of space launch vehicles and sites, this German book gives an interesting presentation (with map and table) of the North Korean efforts to reach the space dimension with the Taepodong-2 rocket.
13. To the report of Charles P. Vick about the satellite launch attempt, Leonard David referred with his article, North Korea's rocket failure – intelligence windfall, posted on 9th July 2006 on the Space News website.
14. *Jane's Defense Weekly*, London, 11 September 2008.
15. *BBC News*, North Korea builds new missile site, 11 September 2008. Sang-Hun, Choe: North Korea said to have tested missile engine. *New York Times*, 16 September, 2008. Brügge, Norbert: Space Launch Vehicles website (www.b14643.de/Spacerockets_1/index.htm), has excellent information, with many photos, drawings and tables, about space rockets and launch sites in the world.

16. Comments about the North Korean space launch: Jeff Kueter and James Mazol, George C. Marshall Institute, Washington, DC, March 2009.
17. Wright, David: *An Analysis of North Korea Unha 2 Launch Vehicle*. Union of Concerned Scientists, Cambridge, Maryland, 20 March 2009.
18. Brown, Tim: *The Meaning of North Korean Missile Launch*. GlobalSecurity.org, published 5 April 2009.
19. Jonathan's Space Report no. 610, Sommerville, Maryland, April 2009 (http://planet4589.org/space/jsr/).
20. Samson, Victoria: *North Korea's Launch*. Secure World Foundation, 13 April 2009.
21. Covault, Craig: North Korean rocket flew further than earlier thought. Spaceflightnow, 10 April 2009.
22. Vick, Charles P.: Kwangmyongsong-2 satellite payload. www.globalsecurity.org/, 17 April 2009.
23. A detailed description, with drawings, photos and videos, of Unha-2 is published by Norbert Brügge on the Space Launch Vehicles website (www.b14643.de/Spacerockets_1/index.htm).

ANNEXE: NORTH KOREAN LAUNCH VEHICLES

Vehicle name	Taepodong-1/Paektusan-1	Taepodong-2/Unha-2
Operator	KCST (Korean Committee of Space Technology)	KCST (Korean Committee of Space Technology)
Prime contractor	Sanum Dong MDC (Missile Design Center)/Factory 125	Sanum Dong MDC (Missile Design Center)/Factory 125
Launch site	Musudan-ri/Tonghae Satellite Launching Ground, Hwadae country	Musudan-ri/Tonghae Satellite Launching Ground, Hwadae country
First flight	31 August 1998 (mid-failure)	5 July 2006? 5 April 2009 (mid-failure)
Reliability rate	0/1	0/2?
Performances:		
• LEO mass capacity	Up to 25kg? (attached to third stage)	Up to 0.5 tonnes
• GTO mass capacity	NA	NA
• Fairing height/diameter	2 m/0.8 m?	3 m/1.2 m?
Basic vehicle:		
• Lift-off mass	From 21 to 31 tonnes for Taepodong-1	From 68 to 79 tonnes for Taepodong-2
• Lift-off thrust	300–400 kN?	1,120 kN?

Vehicle name	Taepodong-1/Paektusan-1	Taepodong-2/Unha-2
• Height	27 m?	30 m?
• Number of stages	3	3
Type of propulsion for each stage (number × engines) [propellants]	1st stage: liquid (1 × Nodong YF-2 engine) [IRFNA/UDMH] 2nd stage: liquid (1 × improved Nodong YF-3 engine/Hwangsong 6 missile adapted for propulsion in vacuum) [N204?/UDMH] 3rd stage: solid	1st stage: liquid (4 × Nodong YF-2 engines) [IRFNA/UDMH] 2nd stage: liquid (1 × YF-3 improved engine/Hwangsong 6 missile adapted for propulsion in vacuum) [N2O4/UDMH] 3rd stage: solid
Improvements	Cooperation with Pakistan and Iran for joint development of Taepodong derivatives: the Pakistani Shaheen/Ghaznavi and the Iranian Shahab launch vehicles	Development of the more powerful first stage, ground-tested since 2004 Cooperation with Iran for joint development of the Iranian Safir-2/IRILV launch vehicles
Technical status	Pakistani and Iranian engineers contributing to development of upgraded rocketry engines Taepodong tests stopped from 2000 until 2006 because of political and economic pressures from Japan and United States	First flight failure because of the first-stage collision with upper structure (fairing?) Construction of a second launch site at Pongdong-ni, about 50 km from the Chinese border Problems with the solid third stage (ignition, stabilization?)
Economic status	Described in UN meetings (Unispace III, Vienna, July 1999) as the "access to space" for developing countries Replaced by Unha-2 booster	Part of the "Chuche" (self-reliance) policy of the DPRK Political competiton with the KSLV-1 jointly developed by South Korea and Russia

References

Reports of Jonathan McDowell (Jonathan's Space Report).
David Wright (Union of Concerned Scientists), Victoria Samson (Secure World Foundation), Charles P. Vick (GlobalSecurity.org).
Reichl, Eugen: *Das Raketentypenbuch*. Motorbuch Verlag, Stuttgart, 2007.
Detailed description, with drawings, photos and videos, of Unha-2 published by Norbert Brügge on Space Launch Vehicles website (www.b14643.de/Spacerockets_1/index.htm).

14

South Korea: New entrant for space systems

The authorities of Seoul, for technological purposes, began a space program through the microsatellites of KAIST (Korea Advanced Institute of Science & Technology) and with some foreign assistance (SSTL in Britain). Earth observations and scientific measurements from space became the priority of the KARI (Korea Aerospace Research Institute) with the KOMPsat (Korean Multi-Purpose Satellite) program, cooperating with American, French and German manufacturers of space systems. At the same time, sounding rockets were developed as demonstrators of propulsion systems, while a national system of communications and broadcasting satellites was put in place under the name of Koreasat.

In 1999, South Korea defined a 15-year space program to establish a domestic industry of launchers and satellites. A growing family of KSLV launchers (Korea Space Launch Vehicle) was developed with Russian assistance (Khrunichev Space Center). The first launch took place from a new complex, Naro Space Center, on 14th August 2009. The South Korean satellite industry worked mainly with European partners on advanced remote sensing satellites for multispectral and radar observations and on miniaturized technological satellites. The COMS (Communication, Ocean and Meteorological Satellite) program will create a South Korean multipurpose platform in geosynchronous orbit. The aim of South Korea is to become a major player in the business of space systems and services as well as in manned ventures at the international level.

SOUTH KOREA: RECENT ENTRANT FOR NEW SPACE SYSTEMS AND MICRO-ELECTRONICS LEADERSHIP

The Republic of Korea (South) is a young player in space technology, with a modest secondary role until the 1980s.[1] It began a program of space activities in the late 1980s, somewhat later compared to other developed countries. Space in Korea was driven by the view of national needs and a technology development strategy. The Ministry of Science and Technology had the leading role in space development, which became a reality in 1989, 20 years after the first steps of American astronauts on the Moon and 10 years after the flight of the first European Ariane launcher.

The Korean upper part of the KSLV-1 rocket: the second stage with solid engine and the 100-kg satellite for science and technology

A spectacular view of KSLV-1 preparation: the ground test vehicle is moved on the road to the launch pad

The South Korean move from space dreams to ambitions in astronautics was made by the establishing in August 1989 of the Satellite Technology Research Center (SaTReC) inside KAIST. It was a university-based research facility to promote the education and training of satellite engineers through research programs in satellite engineering, space science and remote sensing. With SaTReC, which was selected in 1990 as an Engineering Research Center by the Korea Science and Engineering Foundation, South Korea broke into spacecraft technology. It had to find a fast and reliable way for the acquisition of expertise to develop low-cost small satellites. It initiated a collaborative partnership with the University of Surrey, in Guildford in Britain, where Professor Martin Sweeting and academic research staff had developed since 1981 the Surrey Space Center for design, manufacture and operations of micro-satellites. In 1985, the team of Professor Sweeting created the company SSTL (Surrey Satellite Technology Ltd) to build the UoSat family. SaTReC/KAIST was one of the first customers of SSTL to train engineers and students with the construction and use of the KITsat (Korean Institute of Technology Satellite) spacecraft.

Less than three months after the creation of SaTReC, on 10th October 1989, the Korean Aerospace Research Institute, known as KARI, was established as the key institute to promote advanced aerospace research activities, in order to create a competitive air and space industry in Korea. It was a subsidiary of the Korea Institute of Machinery & Metals (KIMM), but, in October 1996, it became an independent aerospace agency (as KARI Incorporated Foundation), responsible for the long-term program of Korean missions in space. While SaTReC developed the Korean capability in the niche of experimental microsatellites using miniaturized components, KARI worked specifically on space applications with the Korea Multi-Purpose Satellite (KOMPsat) program and on rocketry propulsion with the Korean Sounding Rocket (KSR) program. International cooperation represented the main card to play for South Korea in order to accelerate its satellite industry and space business. The industrial contract for KOMPsat-1, an engineering satellite with payloads for science, Earth observations and communications, was won by TRW Space & Electronics (now part of Northrop Grumman Corp.).

A third actor in the space development of South Korea was, since the early 1990s, Korea Telecom (KT) with the Koreasat system of geosynchronous communications and broadcasting satellites for domestic and regional purposes, achieved through contracts with foreign industries that will compete following an international RFP (Request For Proposals). For the Koreasat-1 contract, American industry was selected with Martin Marietta (now Lockheed Martin) for the satellites and with McDonnell Douglas Aerospace (now Boeing) for their launches from Cape Canaveral. A fourth player for the use of Earth observations from space was the Korea Meteorological Agency (KMA), which will operate its own geosynchronous observatory. Its satellite office grows up to receive and to process the data from geosynchronous weather satellites of the United States, Japan, then later, of China.[2]

During two decades of activities in satellite technology and space applications, South Korea established cooperative links with a great number of space agencies and industries around the world. It established partnerships or signed contracts with the United Kingdom (SSTL for KITsat), South Africa (University of Stellenbosch),

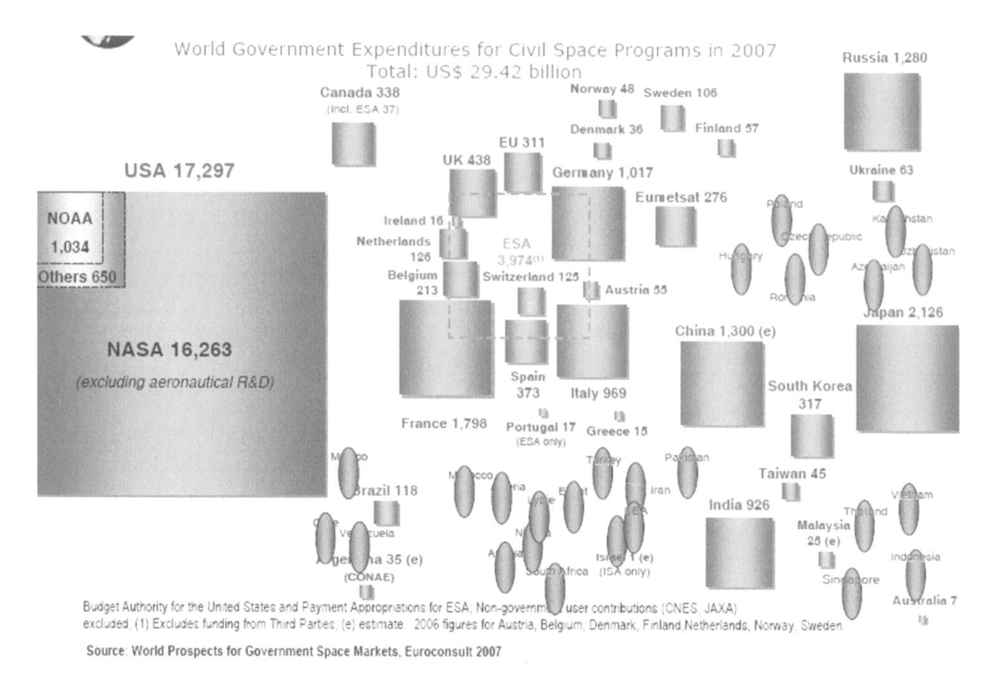

The Korean space effort compared to the budgets of other countries in the world

Israel (El-Op), United States (Lockheed Martin, Boeing), France (CNES, Thales Alenia Space, Astrium, SPOT Image and Infoterra), Italy (Thales Alenia Space), Russia (Roscosmos, Kosmotras, Eurockot) and Japan (JAXA, Mitsubishi Heavy Industries). It had commercial contracts through the SaTReC Initiative (SI) in Malaysia (ASTB for Razaksat), Dubai (EIAST for Dubaisat), Turkey (Tübitak for Rasat) and Singapore (CREST/Nanyang Technological Institute for Xsat).[3]

What about the ambitions of Seoul in space exploration, applications and transport? From 1996 to 2015, over 20 years, South Korea plans to invest up to €2bn to develop its indigenous technology in space and rocket systems. The Korean Ministry of Science & Technology (MOST) published in 2005[4] a table that summarizes its efforts (Table 14.1).

In the early 1990s, South Korea had an annual budget of less than €40m (some $50m). Since then, this budget has doubled, even tripled. In order to meet new

Table 14.1. South Korea's investments in space technology and rocket systems

Program 1996–2015	Budget € ($)
Satellites	€1,046m ($1,486m)
Launchers	€619m ($878m)
Naro Space Center	€252m ($358m)
Total	€1,917m ($2,721m)

ambitions in space from 2015 to 2025 – with the powerful KSLV-2 rocket and with the development of lunar probes – and to affirm its space expertise for the international community, South Korea would have to increase its budgetary efforts by another €2.5bn during the next decade.

In the middle of the 2010s, for only five years, the investments shown in Table 14.2 are planned.

Table 14.2. South Korea's planned investments

Program 2012–2016	Budget € ($)
Satellites	€516.8m ($734m)
Launchers/Naro Space	€684.9m ($971m)
Basic research	€59.2m ($84m)
Total	€1,260.9m ($1,789m)

FIRST STEP (1989–1999): THE KITSAT/URIBYOL MICROSATELLITES

KITsat (Korea Institute of Technology Satellite) is a science satellite program run by KAIST, which had the privilege of operating the first spacecraft for South Korea: KITsat-1 was made by SSTL (Surrey Satellite Technology Ltd), with the training of South Korean engineers and through a technology transfer agreement, while KITsat-2 – a copy of the first one – was achieved at SaTReC. Both satellites of around 50 kg, like the first UOSat microsatellites, were stabilized by gravity-gradient boom and magnetorquers.

KITsat-1/Uribyol-1, the first satellite of South Korea, during its preparation at the British University of Surrey

On 10th August 1992, KITsat-1, weighing 48.6 kg (once in orbit renamed Uribyol-1/OSCAR-23) and South Korea's first satellite, was launched as a piggyback payload on Ariane 4 (flight V52) with the French–American TOPEX/Poseidon spacecraft for oceanography measurements. KITsat-1 was a research project encompassing the installation of a ground station in South Korea. Surrey Satellite Technology Ltd (SSTL) designed the satellite, which was built by students of

KAIST under a technology transfer program at the University of Surrey's Spacecraft Engineering Research Group. Korean engineers were involved in the UoSAT-5 mission, through technology transfer and training of students on the MSc courses at Surrey.

The miniaturized engineering payload of KITsat-1, based upon the use of the modular microsatellite bus proven on UoSAT-3, -4 and -5, included:

- an Earth Imaging System (EIS) consisting of two CCD (Charge Coupled Device) imagers, two lenses and a Transputer Image Processor;
- a Cosmic Ray Experiment (CRE) to monitor radiation environment and its effects;
- a Digital Signal Processing Experiment (DPSE) comprising two Texas Instruments, DSP, which are used to relay compressed speech in real time and implement advanced data link modulation techniques;
- the Pacsat Communications System (PCS), which provides digital communications for stations in the amateur satellite (Amsat) service, as part of the OSCAR (Orbiting Satellite Carrying Amateur Radio) family.

KITsat-2, alias Uribyol-2/OSCAR-25, of 47.5 kg was nearly identical to the first one. It was launched on 26th September 1993 by Ariane 4 (flight V59, with French SPOT-3). The main challenge, for South Korea, was to develop and build its own spacecraft from start to finish. It also offered the opportunity to verify and to enhance the technology used in the KITsat-1 mission, to promote the Korean microelectronics industry by using domestic components, to qualify the SaTReC team for the manufacturing and to test a microsatellite in Korean facilities. The payload consisted of KAIST Satellite Computer (KAScom), the CCD Earth Imaging System and a Digital Signal Processing Experiment with some new instrumentation: a Low-Energy Electron Detector (LEED), Infrared Sensor Experiment (IREX) and Digital Store and Forward Communication Experiment (DSFCE).

It took more time – over five years – to get KITsat-3 ready for launch: it used a three-axis stabilized bus for a low-cost remote sensing satellite system. The spacecraft was heavier (with a total mass of 110 kg) and more powerful (with a maximum power of 180 W). Its main payload was the Multispectral Earth Imaging System (MEIS) for 13.5-m resolution images, which made it the pioneering precursor of small satellites for Earth observations. MEIS, a linear pushbroom CCD camera, was developed in cooperation with the South African University of Stellenbosch, in exchange for the assistance of SaTReC for the compact 15-m resolution imager of the 63-kg SUNsat-1 (Stellenbosch University satellite): the first microsatellite of South Africa was launched on 23rd February 1999.

Another package of instruments was the Space Environment Scientific Experiment (SENSE) to make *in situ* plasma research by studying high-energy particles, radiation effects on micro-electronics, electron temperature and the magnetosphere. KITsat-3 was launched on 26th May 1999 by an Indian PSLV rocket into a Sun-synchronous orbit with an altitude of 720 km. While SaTReC was responsible as the prime contractor for the development of the KITsat satellites, KARI (Korean Aerospace Institute), with its test facilities, contributed to the environmental tests.

Kuwait seen by KITsat-3 in February 2000

The capital of Egypt, close to the desert, seen by KITsat-3

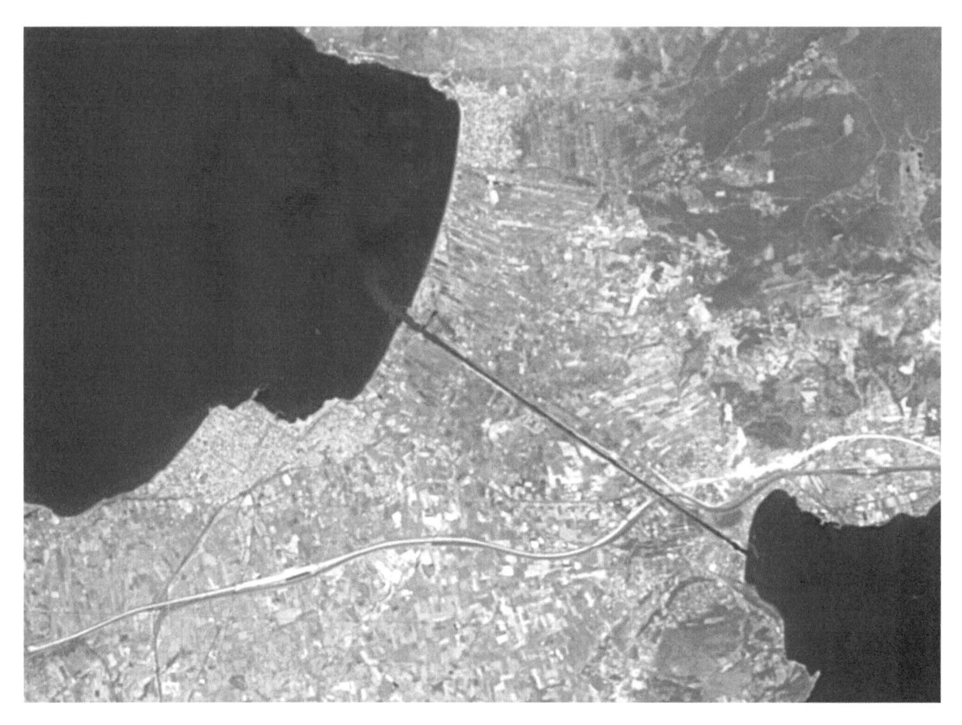

The channel of Corinth in Greece, clearly visible on this KITsat-3 image

With the KITsat-3 mission, SaTReC demonstrated its technological know-how to design, develop and operate microsatellites for high-quality Earth observations. Its indigenous achievements of the KITsat program marked the beginning of commercial space activities in South Korea. In January 2000, the SaTReC Initiative (SI) was established as a space technology spin-off of KAIST by the team of experienced engineers who carried out the successful development of the first three microsatellites in South Korea. SaTReC became the technological unit for the space systems of the KARI with the STsat (Science and Technology Satellite) program, while the SaTReC Initiative is the commercial arm of the Korean satellite industry through microsatellite platforms and payloads.[5]

The STsat-1, also named KITsat-4 or KAISTsat-4, was developed and operated by SaTReC. Launched on 27th September 2003 by the Russian Cosmos 3M rocket from Plesetsk, it was put into Sun-synchronous orbit at a 685-km altitude as the first satellite fully dedicated to the national space science program of South Korea. Using the SI-100 bus, whose performances had been tested by KITsat-3, it was an observatory for astronomy and astrophysics. Among the instruments were: a Far-ultraviolet Imaging Spectrograph (FIMS), a Solid State Telescope (SST), a Narrow Angle Star Sensor (NAST) and a payload for Spectroscopy of Plasma Evolution from Astrophysical Radiation (SPEAR).

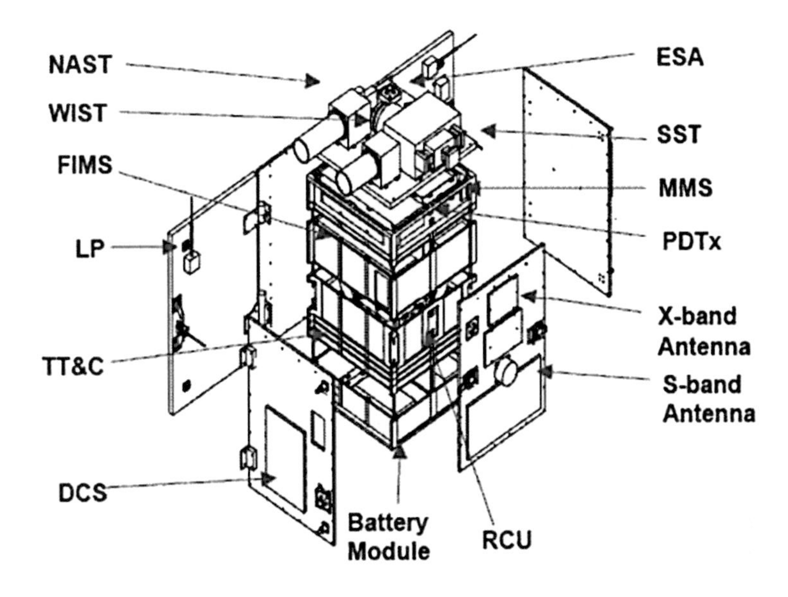

The very compact and modular STsat-1 microsatellite, using the SI-100 platform

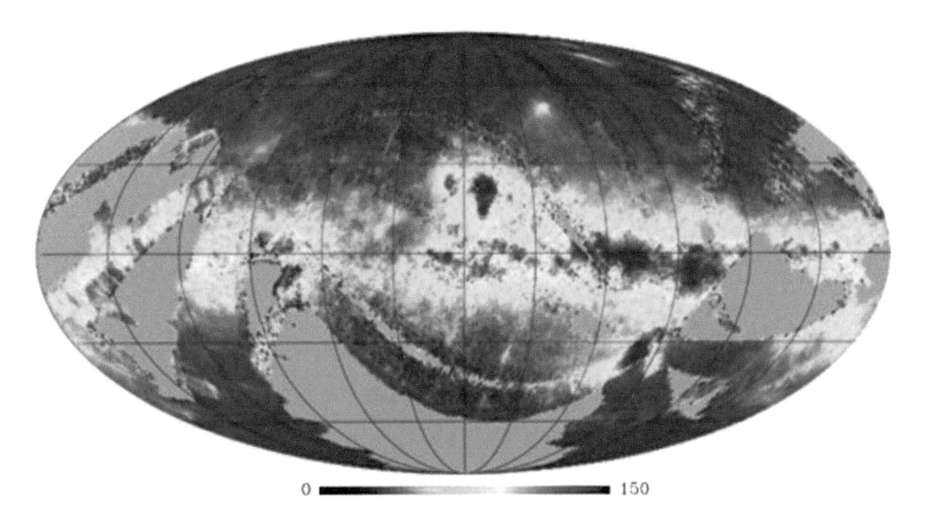

The center of the Milky Way, our galaxy, observed by the STsat-1 astrophysics payload

The mission of FIMS was to observe the universe and aurora. A simple and reliable strategy was adopted in STsat-1 to synchronize time between On-board Computer (OBC) and FIMS. This strategy was devised to maintain the reliability of the satellite system and to reduce implementation cost by using minimized electronic circuits. The Far-ultraviolet IMaging Spectrograph was to determine which of the highly contentious global models best described the behavior of the hot, highly ionized interstellar medium in the Milky Way by creating the first ever maps of the far-UV sky, including CIV and OVI emission. FIMS was used to evaluate the impact of high-mass stars on the surrounding, star-forming ISM, known as "feedback". FIMS also mapped super-bubble emissions, constraining hot gas breakout models, which determined feedback effects.

The SaTReC and SaTReC Initiative are the strategic providers of KARI in the framework of the national space plan for specific systems in large and heavy spacecraft and for small satellites missions in science and technology (with the STsat program, with microsatellites designed for the Korean satellite launcher). They demonstrated Korean independence in the business of miniaturized spacecraft for advanced engineering and for remote sensing missions.

SaTReC Initiative/KAIST developed the first microsatellite to be launched by KSLV-1: STsat-2

Twenty years after the establishment of SaTReC inside KAIST, the company SaTReC Initiative is marketing two types of three-axis stabilized platforms: the SI-100 (100–150 kg) and SI-200 (300 kg). It won many contracts for the complete or partial development of agile satellites for medium-resolution Earth observations:

- In November 2001, SI formed a joint engineering team with Astronautic Technology Sdn. Bhd. (ASTB) of Malaysia to design and to develop the MACsat (Medium-sized Aperture Camera satellite) of 190 kg (330 W of power), using the SI-200 bus from the KITsat/STsat heritage of South Korea. The spacecraft, renamed Razaksat in 2003, is able to take and record images of 2.5 m (panchromatic mode) and of 5 m (four spectral bands).
- Since 2001, SI has had a technological partnership with the CREST (Center for Research in Satellite Technologies) and Nayang Technological University of Singapore for the Xsat microsatellite, which is planned for launch in late 2009. The first spacecraft of Singapore, derived from the SI-100 bus, will carry the multispectral Iris camera, an upgraded version of the MEIS imager developed for KITsat-3.

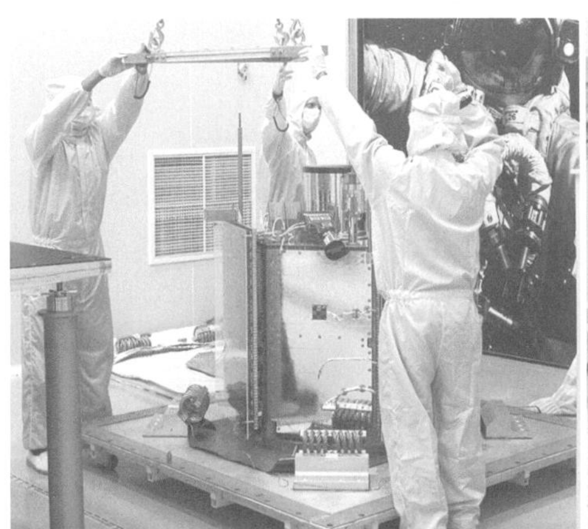

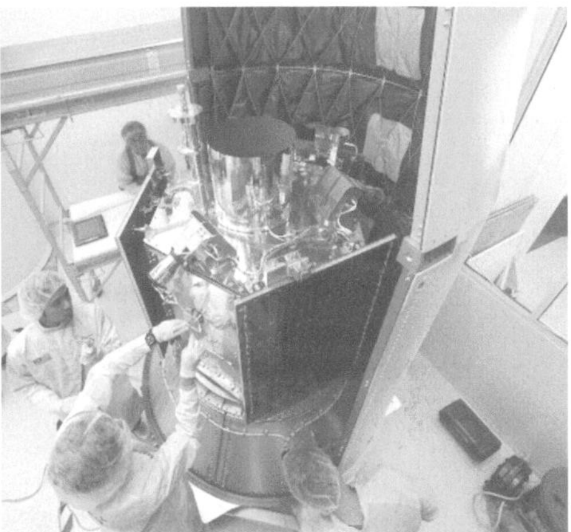

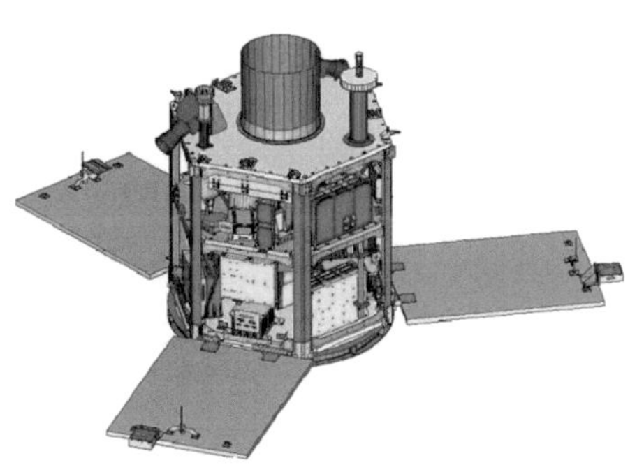

The "made in Korea" Razaksat is being prepared for launch at the SpaceX launch complex on Omelek Island, Kwajalein Atoll, Pacific Ocean

The private launch, with the SpaceX Falcon-1 rocket, of Razaksat for ASTB in Malaysia

- Since 2004, SI has the same type of cooperation with Turkish Tübitak-Uzay for the development of Rasat, whose launch is expected to take place in 2010. SI is responsible for its payload, which will consist of OIS (Optical Imaging System) to take panchromatic and multispectral images of 7.5 and 15-m resolution, respectively.
- In May 2006, EIAST (Emirates Institution for Advanced Science & Technology) of Dubai in UAE (United Arab Emirates) awarded a contract to SI for the Dubaisat-1 mission around a technology transfer program. Dubaisat-1 of around 200 kg (330 W) is a SI-200 bus equipped with DMAC (Dubaisat-1 Medium Aperture Camera): it offers the same performances as Razaksat.

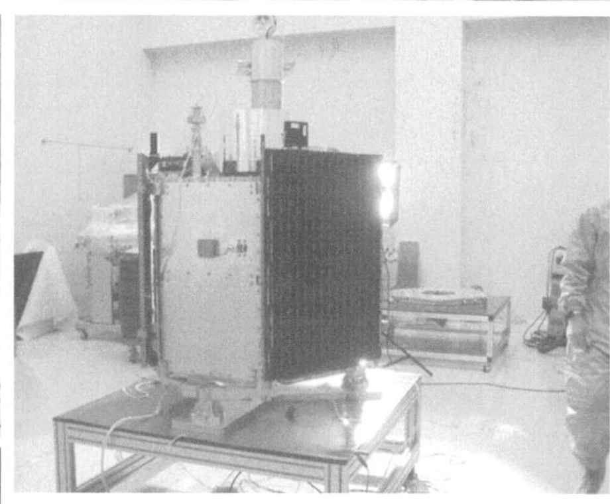

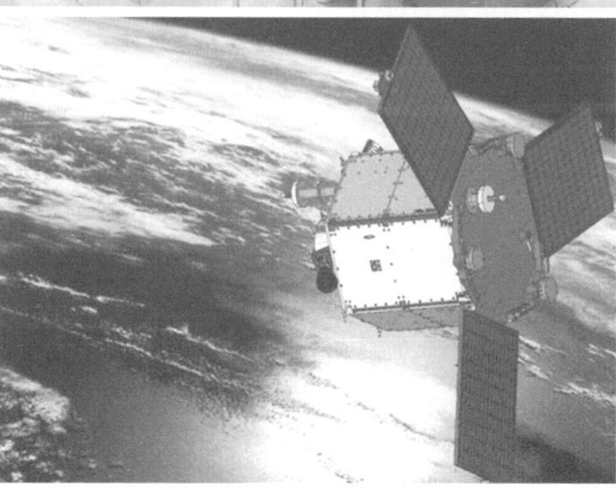

SI (SaTReC Initiative) engineers preparing the Dubaisat-1 remote sensing satellite in the clean room facilities of Dnepr launch complex, Baikonur

One of the first images taken of Dubaisat-1 after its successful launch from Baikonur

During July 2009, the satellite teams of SI were busy in the simultaneous management of two launch campaigns:

- on 14th July, Razaksat was launched by the Space X *Falcon 1* rocket (two liquid stages), from the Omelek Island of Kwajalein Atoll, Pacific Ocean;
- on 29th July, Dubaisat-1 was sent to orbit by the Kosmotras Dnepr rocket (three liquid stages) from the cosmodrome of Baikonour in central Asia.

Currently, the main commercial product of SaTReC Initiative is its high-resolution Earth observation satellite currently under development for Dubaï (Dubaïsat-2). This 250-kg remote sensing satellite, with 400 W of power, has a challenging electro-optical payload able to take images of 1.5-m resolution in panchromatic mode, 6-m in multispectral bands (visible and near infrared), with a swath width of 18 km. The three-axis stabilized platform has a high agility for better revisit viewing.[6]

SECOND STEP (1999–2009): APPLICATIONS SATELLITES AND ACCESS TO SPACE

In parallel with the microsatellite activities of SaTReC/KAIST, KARI (Korea Aerospace Research Institute) was created, along with the Korean Research Institute of Ships and Ocean Engineering (KRISCO), as part of the national development plan for establishing a strong technology basis for the Republic of Korea. Both research institutes were created in October 1989 as new subsidiaries to that of the

Korean Institute of Machinery and Metals (KIMM), which was established in December 1976. KARI was established as a leading Korean research institute to promote advanced research activities in aeronautics and astronautics. KARI had a specific responsibility of upgrading the level of Korean aerospace research to world class and helping to develop competitive aerospace know-how in Korea.

Also, in October 1999, after many years of sensitive discussions, Daewoo Heavy Industries, Samsung Aerospace and Hyundai Space & Aircraft pooled their aerospace operations into a new, independent company: Korean Aerospace Industries Ltd (KAI). The three firms hold equal shares in KAI for combined annual sales of approximately $700m, with a total employment of 3,200 people. The merger was sought to bring Korean aerospace industry in line with the current trend of consolidation. Now, as a single operation, any overlap among the three will be eliminated. The goal is to build KAI into a total integrator for the development and production of aircraft and space systems by the merger of three leading aerospace companies.

KARI, as the central institution for the space program of Korea, has the main objectives of pushing ahead indigenous satellite technology in space applications – especially for Earth observations – and autonomous access to space with sounding rockets and satellite launch vehicles (through the development of solid and liquid propulsion systems). It took a decade for KARI to have its first KOMPsat (Korea Multi-Purpose Satellite), which was developed with foreign assistance – American contractors – for dual-purpose applications: observations and communications. Nowadays, KARI and the Korean space industry have built the KOMPsat-1 and

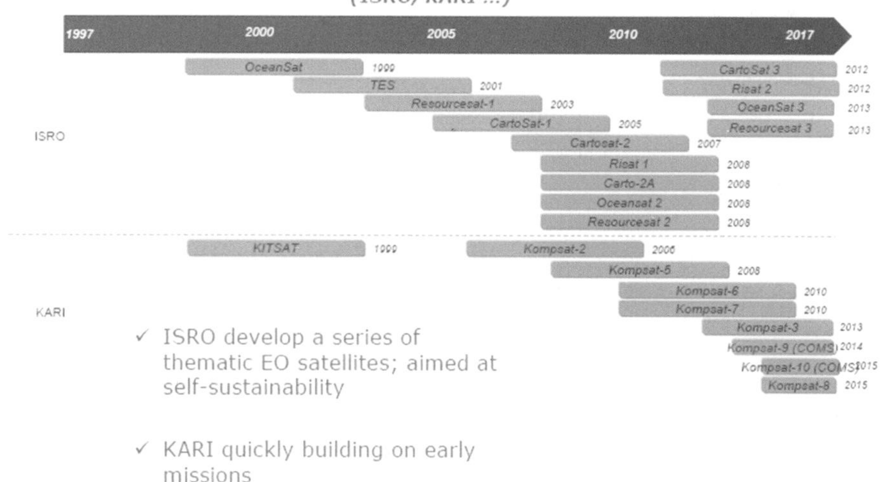

Comparison of the Korean and Indian strategies for Earth observation satellites (environment and security)

KOMPsat-2 spacecraft for Sun-synchronous orbit. In its plan for a further 10 satellites in the pipeline for the coming decade, it is pursuing an ambitious spacecraft program of remote sensing missions: KOMPsat-5 radar satellite in 2010 and KOMPsat-3 optical satellite in 2011 for high-resolution imagery. It is also involved with COMS-1, a communications, oceanography and meteorology satellite in geosynchronous orbit. It plans to have in Sun-synchronous orbit KOMPsat-3A (optical) in 2012, KOMPsat-6 (radar) in 2015 and KOMPsat-7 (optical) in 2017. This program refers to the international cooperation with space industries in the United States, France, Italy, Germany, Israel and Japan. A partnership with China has been envisioned during preliminary contacts but could not be achieved.[7]

The peninsula of Korea viewed from space, by using a montage of satellite imagery

The KOMPsat program was initiated in 1995 as a major space investment in Korea. Its main objective is to develop and operate a national space segment for Earth observations. An efficient infrastructure was installed to provide valuable services to remote sensing users in various fields of applications, specifically to provide Earth resources research and surveys of natural resources, to conduct surveillance of large-scale disasters and mitigation efforts, to collect high-resolution images for South Korea GIS (Geographical Information System) and to contribute to the publication of printed and digitized maps. For that purpose, KARI put in place at Daejeon a Satellite Integration and Test Center (SITC) with 11,408 m of floor area for KOMPsat-1, extended to 15,000 m for KOMPsat-2. In February 2008, SITC had an extra building to accommodate geostationary satellites with their payload. The KOMPsat Ground Station facilities, located at KARI headquarters in Daejeon, include S-band and X-band antennae, data-storage and -processing equipment, satellite operation software, mission analysis and planning software and a satellite simulator.

SPACE SYSTEMS FOR A KEY PROGRAM OF EARTH OBSERVATIONS

On 21st December 1999, KOMPsat-1, of 510 kg at an altitude of 685 km, represented the first step of KARI in space: launched by a solid Taurus launcher of Orbital Sciences, it was renamed Arirang-1 in orbit. This first spacecraft of KARI was built by TRW (now part of Northrop Grumman Corp.) on its Eagle class of stable, lightweight, modular platforms for whom it was the first export contract. It has three-axis stabilization with dual solar rays with three payloads: the Electro-Optical Camera (EOC, a 6.6-m resolution CCD imaging system) for land use mapping and environment monitoring, the Ocean Scanning Multispectral Imager (OSMI) for 1-km resolution ocean surveys in six spectral bands and a Space Physics Sensor (SPS) for ionosphere measurements and high-energy particle detection. In order to acquire satellite design technologies, 25 technical staff of KARI have joined

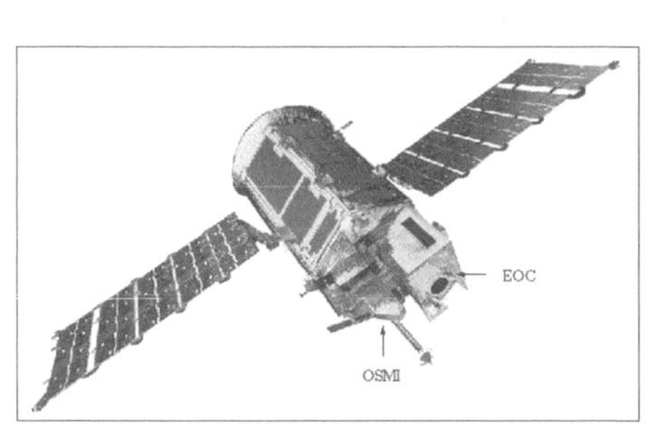

KOMPsat-1/Arirang-1, first Korean satellite for land mapping and ocean monitoring

In the sky of California, the successful flight of Taurus carrying in orbit KOMPsat-1/Arirang-1

the some 125-member TRW design team. Seven Korean industrial enterprises have also dispatched some 30 engineers for the same program. The participating Korean industries were responsible for the "koreanization" of satellite components, while KARI mobilized some 50 researchers to study satellite design data at TRW and to learn about satellite systems and components.

During the development of KOMPsat-1, there were discussions by KARI with German OHB-System and DLR and with Israeli Electro-Optic Industries Ltd (El-Op) to develop the EKOsat-IR (ELOP/KARI/OHB satellite-InfraRed) spacecraft. The KOMPsat-1 flight spare model available was proposed to carry two instruments: the Multi-Spectral Remote sensing System (MSRS) for 5-m resolution images in 12 spectral bands and the Hot-Spot Recognition System (HSRS) for thermal infrared observations with 260-m resolution. This project, initiated in 2002 for a possible launch in 2005, was to continue the mission of the German BIRD (Bi-spectral InfraRed Detection) microsatellite. It could not be finalized into a concrete mission, despite real efforts: funding from South Korea, Germany and Israel was the main problem.

The KOMPsat-2/Arirang-2 remote sensing satellite is heavier (800 kg) and more powerful (955 W), built by KARI with the engineering assistance of EADS Astrium in France. Launched on 28th July 2006 by Rockot launcher from the Russian cosmodrome of Plesetsk, it was injected into 685-km Sun-synchronous circular orbit at an inclination of 98.14° (passing the ascending node at 10:50 am). Its remote sensing payload consists of a Muti-Spectral Camera, jointly developed by KARI with El-Op (Elbit Systems Electro-Optics Industries Ltd) of Rehovot (Israel) and OHB-System of Bremen (Germany): this linear pushbroom instrument is capable of taking images of 1-m panchromatic resolution and 4-m multispectral resolution,

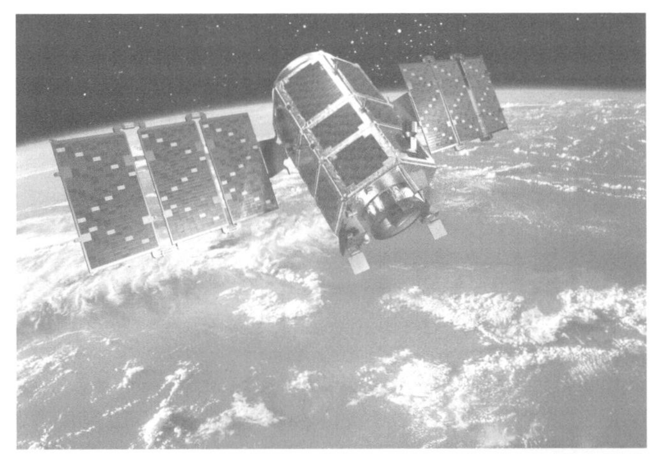

Artist's view of the KOMPsat-2/Arirang-2 high-resolution imaging satellite, jointly developed by Astrium, KARI, El-Op and OHB

The harbor facilities of Nagoya are clearly identified on this multispectral image of KOMPsat-2

with a swath width of 15 km. The South Korean daily, *Korea Times*, had these comments about the Arirang-2 launch: "Korea became the world's n°6 satellite power. It had joined the ranks of a handful of nations with the capability of taking satellite pictures with a 1-m resolution."[8]

The Sydney Olympic Park, seen in detail (1-m resolution) by KOMPsat-2 in August 2006

North Korean features under the high-performance eye of KOMPsat-2: the Baekdu Mountain

The optical imaging satellite has the capability of acquiring up to 7,500 images, with a ground footprint of 15 by 15 km every day, equivalent to 1.7m km a day. Such imagery performance, ideal for detecting and identifying ground features, makes Arirang-2 a key asset for mapping at scales of 1:5,000 to 1:2,000 for urban planning and hazard management. In October 2005, the French SPOT Image company (part of the Infoterra group) signed with KARI an exclusive agreement for marketing and distribution outside South Korea, the United States and the Middle East the images taken by Arirang-2. SPOT Image is marketing KOMPsat as "the alternative metric solution". High-resolution images taken by Arirang-2 are available through an internet search system. The images can be purchased directly online. For private use and other uses, KAI (Korea Aerospace Industries), acting as the sales agency of KOMPsat images in Korea, will distribute them at a commercial price. In order to accelerate the use of satellite images and stimulate related industries in Korea, the commercial distribution price will be set at one-fifth of the price of foreign satellite images and, in certain cases, the final price will be set at one-twentieth.

KARI decided to make KOMPsat the key program of remote sensing missions from space, not only to meet the needs of South Korea in geo-management and environment control, but to conduct permanent observations of the neighboring North Korea and to play an active role in the international Sentinel Asia initiative. For the 2010s, up to five spacecraft are scheduled in the Korean space program:

- KOMPsat-3, with a high-resolution optical sensor, had an initially scheduled launch date of 2008, which later slipped to 2011. While the objective is to ensure the continuity in Earth observations from the successful KOMPsat-1 and KOMPsat-2 system, KOMPsat-3 is capable of higher image resolution and can provide high-resolution electro-optical (EO) images – resolution of 0.7 m in panchromatic mode, 3.2 m for multispectral observations – required for geographical information systems (GIS) and other environmental, agricultural and oceanographic monitoring applications. This spacecraft, of 800 kg, 2 m in diameter and 3.5 m in height, is developed in a joint effort by Astrium Satellites and KARI.

  For the KOMPsat-3 launch, Mitsubishi Heavy Industries, Ltd (MHI) received an order from KARI. The transaction, in which Mr Hideaki Omiya, President of MHI, signed the agreement in Korea in January 2009, represents the first satellite launch services order placed to MHI by an overseas customer. The launch is slated to take place in the fiscal year ending 31st March 2012. The satellite will be transported by ship from Korea to Tanegashima Island in Japan (Kagoshima Prefecture) for launching by means of MHI's H-IIA at the Tanegashima Space Center of the Japan Aerospace Exploration Agency (JAXA). Simultaneously launched by the same launch vehicle will be JAXA's GCOM-W (Global Change Observation Mission-Water), which is to undertake water-related observations from the same Sun-synchronous orbit.
- There is no KOMPsat-4 mission, because of the fact that the Sino–Korean

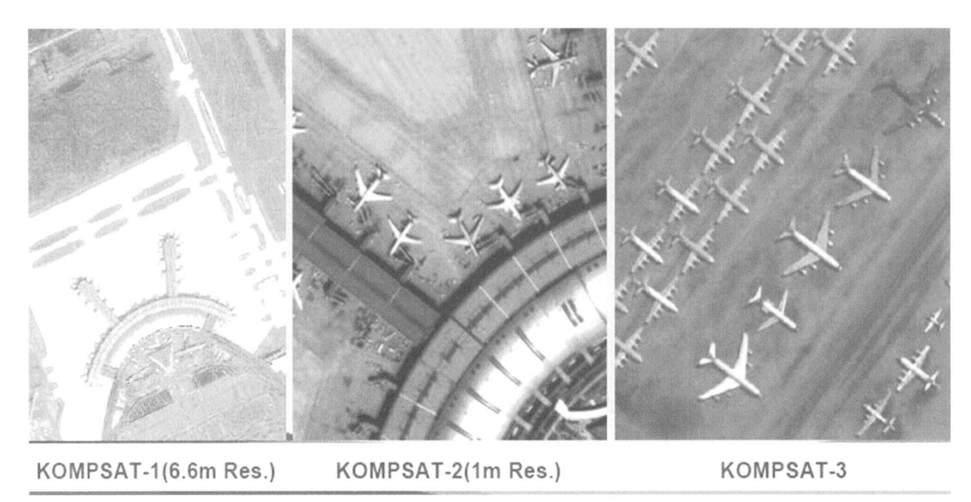

Comparison between resolutions of the Earth observations made by KOMPsat-1 (retired), KOMPsat-2 (operational) and KOMPsat-3 (simulation)

The KOMPsat-3 platform taking form in the integration facility of KARI

word for the number four, "sa", is a homonym of the Chinese character for death. There is the KOMPsat-3A mission, with an optical satellite for high-resolution images including infrared observations, planned for launch in 2012. The spacecraft will be equipped with the KOMPsat-3A Infrared Sensor System (KISS) in addition to KOMPsat-3 sensors. In February 2008, KARI commissioned the German company AIM Infrarot-Module GmbH for the development and manufacture of the KISS module, including the Integrated Detector Dewar Cooler Assembly (IDDCA).

- KOMPsat-5, whose payload consists of the X-band Synthetic Aperture Radar (SAR) as active sensor, will operate in Sun-synchronous orbit between 500 and 600 km. With its capacity to achieve all-weather, day–night observations of 1-m resolution, it represents a major progress for space technology in South Korea. Its launch is scheduled for 2019, with the Russo-

KOMPsat-5 looks like radar satellites developed by Thales Alenia Space in Italy

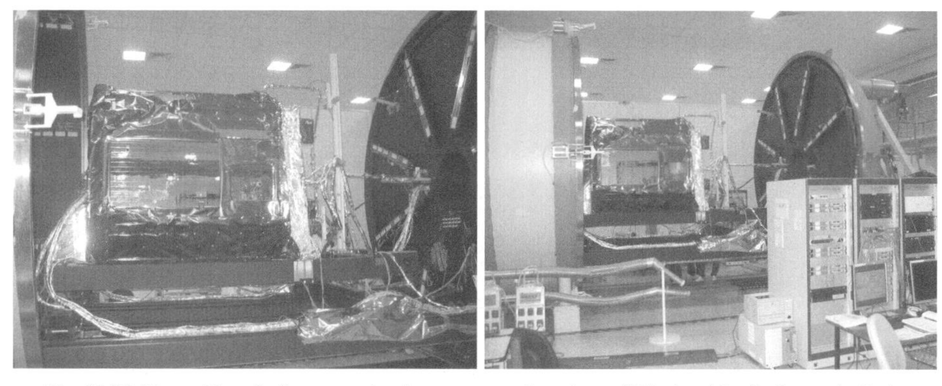

The KOMPsat-5 bus before tests in the vacuum chamber of Thales Alenia Space in Italy

-Ukrainian Dnepr rocket. In March 2006, Alcatel Alenia Space (now Thales Alenia Space Italy) signed the contract with KARI to provide the SAR payload system, which is similar to the Italian Cosmo-SkyMed and ESA Sentinel-1 radars.

NATIONAL LAUNCH CAPABILITIES WITH SOUNDING ROCKETS

KARI was in charge of achieving the Long-Term Plan for National Space Development Promotion. The first basic version was published in 1997. A 15-year space program was defined in 1999 to establish a domestic industry of launchers and satellites. It is revised every two years by the National Science & Technology

Committee. The last revision of the roadmap for space development in Korea dates from 2007: it determines missions until 2025, with ambitious priorities for the powerful KSLV-2 rocket, as well as space exploration with probes around the Moon and on the lunar surface. A Space Development Institute is currently taking form to support and strengthen the government's space policy office and to manage the national program of space activities by establishing a close network among research and development institutes, universities and industries. In parallel to the acquisition of satellite technology, KARI has another priority: the development of rocket systems for indigenous access to space, with the KSR (Korea Sounding Rocket) program as the first step to a national satellite launch vehicle.

As a part of the Korea National Space Plan, the government of Seoul led the Korea Sounding Rocket (KSR) program during 1993–2002. KARI provided enhanced capabilities across a spectrum of Earth sciences and/or astrophysics-related disciplines. This program includes KSR-I, KSR-II and KSR-III, with a total of five launches carried out. Two flew in the KSR-I series, two in KSR-II and one in KSR-III. The main mission objective was to provide the opportunity for scientists to research Earth sciences and/or astrophysics and to improve Korean rocket development technologies, including payload instruments, flight systems, ground systems, operational techniques and interfaces between subsystems. As demand increased for launching of low-orbit Earth satellites internally and externally, the Korean government gave the green light to the Korea Space Launch Vehicle (KSLV) program in 2002.[9]

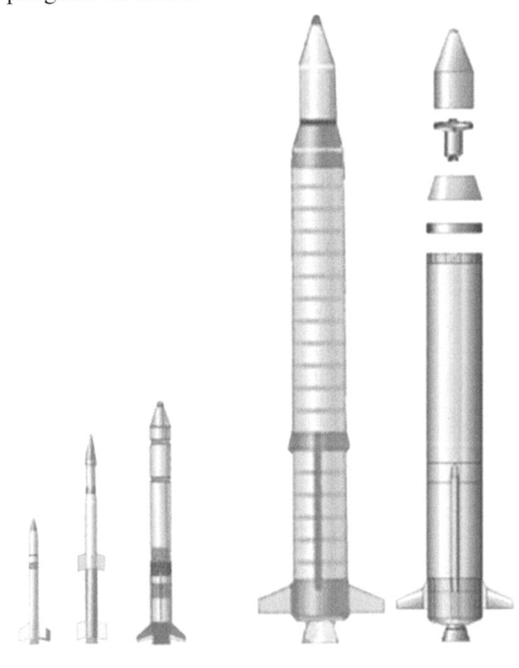

From KSR-I to KSLV-1: South Korea shows its ambitions to have its own access to space

In 1990, just after its establishment, the Korea Aerospace Research Institute (KARI) embarked on a research and development program for the scientific rocket, KSR-I (Korea Sounding Rocket-I), also named KSR-420S (because of booster diameter), which was the first domestic single-stage unguided solid-propellant scientific rocket, with a length of 6.7 m, a diameter of 0.42 m and lift-off weight of 1.3 tonnes. On 4th June, South Korea successfully launched the KSR-I, its first ever scientific research rocket, from Anheung launching site in the central province of Chungcheong. KSR-I, with payload capacity of up to 50 kg, could reach the altitude of 75 km. The second one was successfully launched on 1st September 1993. Two stratospheric

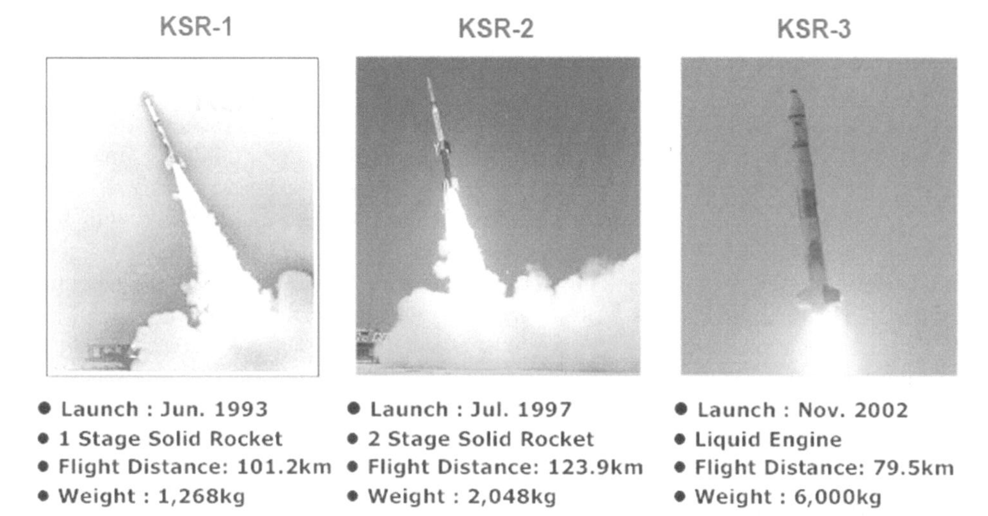

The first steps of KARI in the development of sounding rockets for space experiments

ozone profiles have been obtained using rocket-based solar absorption UV radiometry. The ozone sensor consists of four radiometers, measuring the attenuation of solar ultraviolet radiation as a function of altitude during ascent. According to Jane's Strategic Weapons Systems, it was possible for South Korea to modify the KSR-1 into a ballistic missile to carry a 200-kg payload with a range of 150 km.

KSR-II was a two-stage solid-propellant scientific rocket developed for scientific experiments in the upper atmosphere. Based on the experience acquired through the development and launch of single-stage rockets, KARI was able to build the KSR-II and was powerful enough to reach an altitude of maximum 150 km and beyond. The vehicle has a length of 11.04 m, a total weight of about 2 tonnes and a diameter of 0.42 m. On 10th July 1997, Korea launched its first independently developed two-stage booster, the KSR-II, again to measure the vertical distribution of ozone by using an ultraviolet radiometer. On its initial test flight, the rocket carried a 150-kg scientific observation unit to an altitude of 151.5 km. On 11th June 1998, the second KSR-II was launched with, as payload, an X-ray observation system developed at the Korea Astronomy Observatory during the period 1995–1997. This instrument was designed for the observations of compact X-ray sources.

The final purpose of Korean rocketry activities was to develop not a ballistic missile, but a small launcher with the aim, within a decade, of placing in orbit a multi-purpose satellite of several hundred kilograms. Some sources in the late 1990s suggested that the KSR-II program was a step towards a South Korean space launch program. According to an article published by *Jane's International Defense Review* in February 1997, South Korea was developing a three-stage version of the rocket that may be completed by 1999, but the official announcement of KARI in 2002 was only to develop a small satellite launch vehicle from the technology of the KSR-III rocket that tested the first liquid-propellant engine of South Korea. The original concept for the KSLV-1 was to cluster indigenous KSR-III rockets in order to form a multistage launcher.

For the KSR-III, KARI embarked on its first liquid-fuel rocket project in December 1997. It started a lot of contacts with space rocket manufacturers in order to evaluate the technical challenges for such a development, which was a new venture in South Korea. A 13-tonne-thrust liquid propellant engine was designed with the sounding rocket KSR-III. It was mainly used by KARI to acquire technological expertise in liquid rocketry and to incorporate core technologies for the satellite launch vehicle in areas such as propulsion, guidance/control and mission design. The KSR-III used a pressure-fed-type liquid rocket engine with gaseous helium as a pressurizer, with LOX (liquid oxygen) and kerosene as propellants. No complicated turbopump machinery was used, so that the first liquid rocket of South Korea had limited power and flexibility. KSR-III had a feeding system with both a pressure regulator and a venturi as a flow control device. The purpose of the propulsion feeding system was to feed a certain amount of propellant from the propellant tank to the combustion chamber manifold inlet during combustion.

On 28th November 2002, the first test flight of KSR-III was made by the KARI rocketry team from Anheung Proving Ground, 160 km south-west of Seoul. It reached an altitude of 42.7 km and flew over 84 km. It was described as successful, developing a feeding system with a simple configuration and a small number of operational parts. There was only one KSR-III launch. At this time, Jane's reported that a three-stage KSR-3 rocket capable of reaching an altitude of 350 km was planned for development. In the meantime, a significant event occurred in Seoul: in 2001, Korea became a member of the Missile Technology Control Regime (MTCR) association, sharing the goals of non-proliferation of unmanned delivery systems capable of delivering weapons of mass destruction. It put itself in a strategic position to purchase technologies from other countries for non-military rockets. After many visits of rocketry industries in Russia, after more than two years of technical negotiations, KARI decided to

The unique flight of KSR-III demonstrated to KARI how difficult it is to develop powerful liquid rocket engines

adopt the cooperative solution of Russian liquid first stage for the KSLV-1 concept. In 2004, Russia agreed with South Korea to jointly develop a liquid-propellant rocket engine with turbopumps, the key component for the KSLV-1. The maiden flight of this Russo–Korean launcher, in order to put into low orbit a 100-kg engineering microsatellite, was announced for late 2007.

MUGUNGHWA: COMMERCIAL SATELLITES FOR TELECOMMUNICATIONS AND BROADCASTS

South Korea was among the pioneering nations to acquire and to operate, for regional purposes, a domestic system of geosynchronous satellites for communications and broadcasts. In March 1991, Korea Telecom (now KT Corporation), property of the South Korean government, asked satellite suppliers to supply a turnkey satellite communications system in Ku-band frequencies through a total investment of $260m. Four industrial teams submitted proposals: GE Astro (now Lockheed Martin), Space Systems/Loral and Hughes (Boeing) in the United States, and British Aerospace (Astrium Satellites) in the United Kingdom. In December 1991, KT signed a contract of $145m with GE Astro/Lockheed Martin (with Matra Marconi Space/Astrium as payload provider) for two AS-3000 series satellites and a ground segment to be deployed during 1995. Goldstar Information & Communications, Korean Air and Hughes Network Systems were to provide a VSAT network of 70–80-cm dishes.[10]

Finally, Lockheed Martin manufactured Koreasat-1, -2 and -3 – renamed in orbit Mugunghwa-1, -2 and -3 – with subcontractors in South Korea. The first two geosynchronous satellites of the Koreasat system had a mass of about 1,460 kg at launch and 830 kg on station. Their payload consisted of 15 Ku-band transponders, of which three had high-power repeaters (120 W) for direct broadcasts. Both spacecraft were launched by Delta 7925 rockets from Cape Canaveral and positioned at 116°E for a planned design lifetime of 10 years. The primary purpose of the communication satellite project was to go ahead proactively with delivering advanced telecommunication services: broadband, high-speed voice and images to the general public. Koreasat had to establish the groundwork for South Korea to enter the worldwide business of space communications and to push the country into the rank of technologically advanced nations in geosynchronous orbit.

The launch of Koreasat-1/Mugungwha-1 on 5th August 1995 was a partial failure. The shortfall in thrust for injection in geostationary transfer orbit reduced the time of stable operations to some five years. The satellite had to use a part of its maneuvering and attitude control propellant to reach its final position over the Solomon Islands. Still very useful in its orbit, it was sold by KT in mid-2000 to the Europe Star company (owned by Alcatel, before becoming in 2005 part of the Panamsat/Intelsat system) in order to develop Ku-band services in Eastern Europe.

Koreasat-2/Mugungwha-2, whose payload consisted of 12 FSS (Fixed Satellite Service) transponders for general communications and three DBS (Direct Broadcasting System) repeaters, was launched successfully on 14th January 1996. Its program involved upgrading existing systems and capabilities. It allowed South Korea to become the second country in the world to start digital broadcasts by satellite in July 1996. In July 2009, Koreasat-2 was sold to the Asia Broadcast Satellite company of Hong Kong, which operated it for occasional services under the name of ABS-1A.

Koreasat-3/Mugungwha-3, launched on 4th September 1999 by an Ariane 42 from French Guyana, was also configured to provide both fixed and direct broadcast services. The spacecraft, weighing 2,790 kg at launch, was fitted with a steerable antenna to improve coverage capability on the Asian continent. The payload, jointly developed by Lockheed-Martin and SPAR Aerospace of Canada, had 30 Ku-band and three Ka-band transponders. The satellite, at 116°E, is to be replaced in 2010 by Koreasat-6.

Koreasat-4 will not exist, because the Sino–Korean word for the number four, "sa", is a homonym of the Chinese character for death and means "misfortune" (like number 13 in Western countries). Koreasat-5 is made by Alcatel Space (now Thales Alenia Space France) following a contract of €148m signed in June 2003 with the KT Corporation and the Korean Agency for Defense Development (ADD). The Spacebus-4000 C1 spacecraft, of 4,448 kg (mass in geostationary transfer orbit), features a state-of-the-art broadband payload with a total of 36 transponders: 24 Ku-band, eight SHF (for military use) and four Ka-band (for broadband links). On 21st August 2006, it was lofted successfully by the Russo–Ukrainian Zenit 3SL from the Sea Launch platform in the Pacific Ocean. Positioned at 113°E for a 15-year operational lifetime, it provides both commercial and military communications services.

The next Koreasat-6 developed by Thales Alenia Space with Orbital bus will be a compact communications and broadcasting satellite offering high performance in Ku-band frequencies

Koreasat-6, of 2,750 kg (with 3.4 kW payload power), is in construction for Arianespace launch in late 2010 (with Ariane 5-ECA or Soyuz-2 from French Guyana). On 14th May 2008, Thales Alenia Space France announced a contract with KT to provide its first telecommunications satellite of the 2010s. For this contract, Thales Alenia Space teamed with American manufacturer Orbital Sciences Corporation (Orbital), who will provide the satellite platform STAR-2. As prime contractor, Thales Alenia Space has overall responsibility for design, manufacturing and test of the Koreasat-6 satellite and ground segment, as well as Launch and Early Operations Phase (LEOP), In-Orbit Test (IOT) support and associated services. Designed for a 15-year lifetime, Koreasat-6 will have 30 active Ku-band transponders for services at 116°E.

To expand C-band services, KT leases capacity from SES Americom-New Skies on the geostationary NSS-703 and NSS-12 satellites at 57°E. It is keenly interested in technological Ka-band experiments made by KARI with its geostationary multipurpose COMS-1 (Communication, Ocean and Meteorological Satellite) at 116° or 138°E.

SK Telecom, the other operator of wireless communications in South Korea – especially for mobile services – is involved, along with Japanese partners (industries, operators and investors), in the development and business of MBsat-1 services. The Mobile Broadcasting Corporation (MBCO) was established by some 55 shareholders in Japan and by Korean TU Media Corporation (subsidiary is SK Telecom with some 150 investor companies in South Korea) to offer S-band direct broadcasts (audio, video, multimedia data services) by satellite to mobile terminals, personal digital assistants, cellular phones and home portables.

The MBsat system, for a total investment of €500m, was authorized and registered by Japan with the ITU (International Telecommunications Union). The first high-power satellite, equipped with a 12-m unfurlable reflector, was made by Space Systems/Loral and launched by Atlas-III rocket from Cape Canaveral on 13th March 2004. Five years later, the Japanese–Korean MBCO venture does not look so promising because of competition with terrestrial transmitters for digital services, especially on the peninsula of South Korea. The MBsat system is the first operational experience in S-band mobile services for video broadcasts. TU Media recorded some 1.85m subscribers for its personal broadcasts by satellite.

With MBsat, South Korea and Japan are the first countries to receive digital TV on portable receivers from a geostationary broadcasting satellite

KOREA METEOROLOGICAL ADMINISTRATION (KMA) AND COMS-1 SERVICES

On 15th December 1998, the Korea Meteorological Administration (KMA) headquarters moved to a new 10-storey building located about 10 km south-west of central Seoul. It contains the Satellite Office equipped with a new dual-satellite data receiving/analyzing system and with the receiving system for geosynchronous and polar weather satellites. For the general public, satellite imagery has been serviced through the KMA home page since 1997. Currently, infra-red, visible, fog and yellow sand images for the latest 24 hours are available. In the case of the numerical weather prediction model, the global data assimilation and prediction system (GDAPS) uses satellite data for data assimilation.

In the long-term plan of the National Space Program, established since 1996, the KMA has to accommodate the public and civilian demand for satellite utilization and maintain the continuity of satellite services. For this purpose, it cooperates with the Korean Aerospace Research Institute (KARI) for the development of the Communication, Ocean and Meteorological Satellite (COMS).

The COMS program concerns the only geostationary satellite with a meteorological mission in Korea. Four ministries in the government of Seoul are involved in the COMS development and each ministry is responsible for individual requirements and development objectives: the Ministry of Science and Technology (MOST) for satellite system and bus development, the Ministry of Information and Communication (MIC) for communications payload and satellite control system development, the Ministry of Maritime Affairs and Fisheries (MOMAF) for ocean color sensor development, and the KMA for meteorological payload development. There are two geostationary satellites planned in the National Space Development Plan: the first in 2009 and the second in 2014. The feasibility study for COMS – by establishing user needs, mission concept, operational concept and performance requirements as well as cost, schedule and risk estimation – was initiated in 2001. Preliminary analysis of system specifications and cooperation plans with foreign satellite manufacturers began one year later.[11]

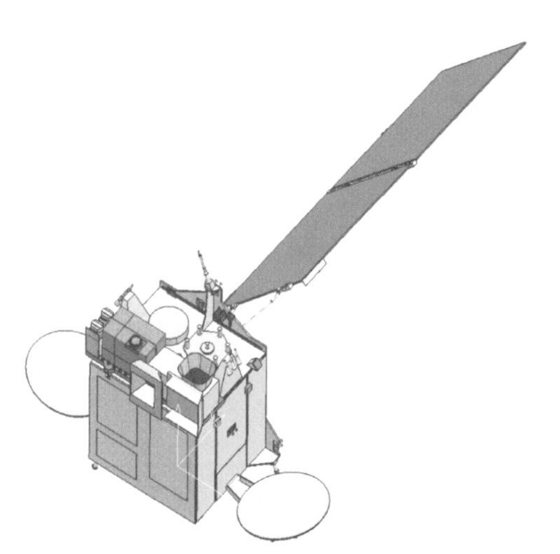

The multipurpose COMS-1 is developed by Astrium Satellites for KARI and KMA

The main development period started – after an international competition – on 17th May 2005 with the selection of the EADS Astrium proposal, to last until the launch of COMS-1 in 2009–2010 by means of an Ariane 5-ECA rocket. According to the development schedule, the Preliminary Design Review

(PDR) for the satellite, jointly built by Astrium Satellites (with the Eurostar 3000 platform) and KARI, was completed in January 2006. COMS-1, of 2.5 tonnes (3 kW of power), is planned to begin operating in early 2010 at one of the orbital positions, 116.2°E or 128.2°E, for a designed lifetime of seven years. It has three missions: one for meteorology, another for ocean observation and the other for experimental communications in Ka-band frequencies. Each of these missions is implemented with a dedicated payload:

- the Meteorological Imager (MI), procured by ITT (United States) and referred to as CAGI (Commercial Advanced Geo-Imager) for real-time observation and early detection of severe weather phenomena; imagery and radiometry information will be provided every 30 min, following three different modes (global, regional, local), in one visible channel (with 1-km resolution) and four infrared channels (with 4-km resolution);
- the Meteorology Data Dissemination will receive data in S-band and relay them in L-band;
- the Geostationary Ocean Colour Imager (GOCI), developed by Astrium Satellites, will monitor the marine ecosystem with data over eight imaging bands in the visible and near-infrared spectrum (0.5-m resolution over Korea);
- the Ka-band communication payload – an in-house development of Electronics and Telecommunication Research Institute (ETRI) with SaTReC Initiative – will offer broadband multimedia services through three spotbeams (South Korea, North Korea and Northeast China); it will also function for communication services for natural disaster coverage within the governmental network of south Korea.

SATELLITE LAUNCHES WITH THE RUSSO–KOREAN KSLV PROGRAM

The main priority of South Korea is in the worldwide business of space applications and satellite launches, as defined since 1996 in the National Space Plan. In November 2002, KARI tested the KSR-III, its first sounding rocket with a liquid engine. It planned to have an operational launch vehicle for small satellites by 2005, with the hope that such a development will encourage its high-technology industries. The project was to design a small three-stage launcher by combining liquid and solid KSR boosters: this KSLV-1 concept would use the liquid KSR-III as the core vehicle, with a pair of strap-on boosters derived from the KSR-III and with solid KSR-I as an upper stage. It was abandoned in favor of using foreign rocketry to accelerate the KSLV-1 development.[12]

In 2004, Russia agreed with South Korea to jointly develop a liquid-propellant rocket engine for the KSLV-1 project. On 26th October 2004, the State Research & Production Space Center of Khrunichev signed an initial contract of some $200m for the design and development of the South Korean launcher as well as for the construction of the space launch center. It took two years to define the Russo–

우주개발 중장기 계획(1996~2015)

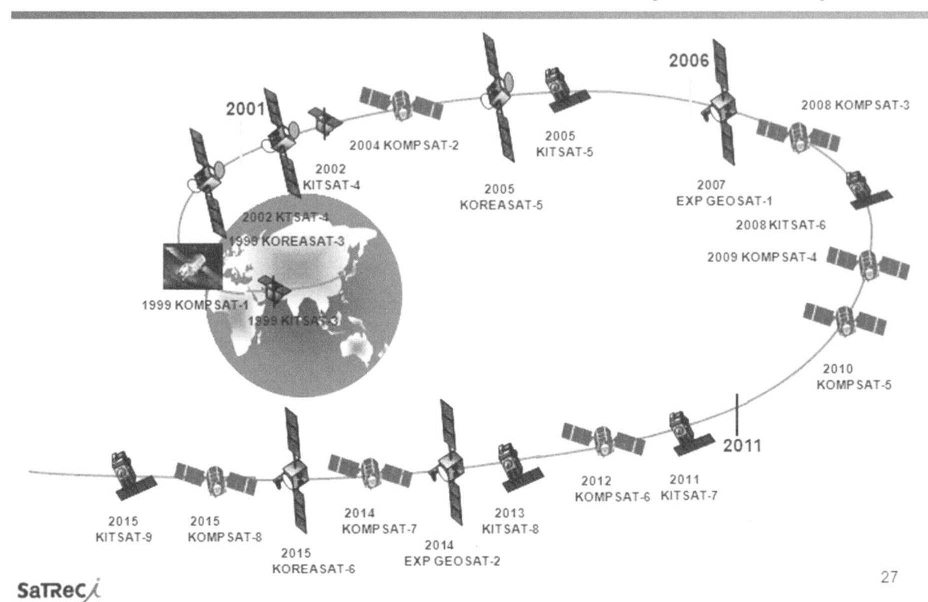

National Space Program in Korea

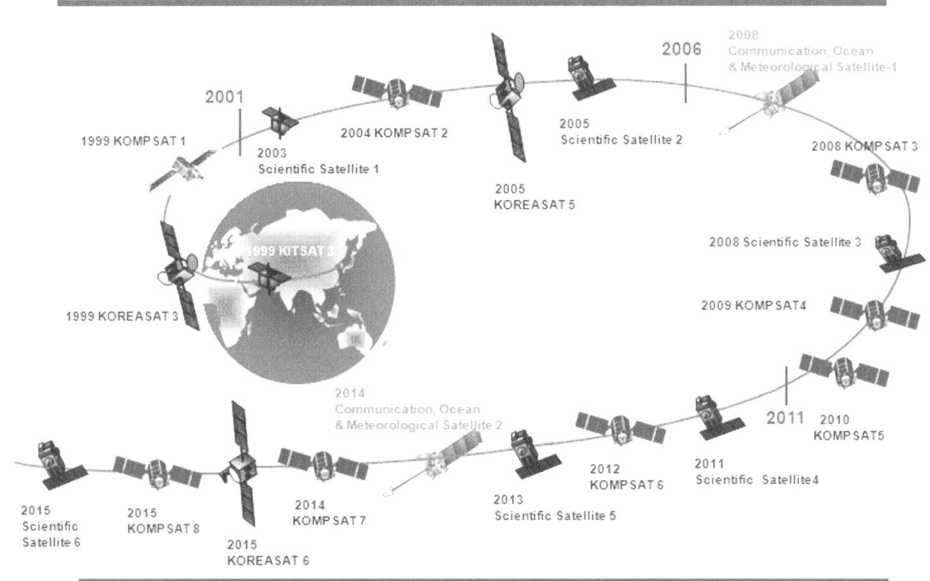

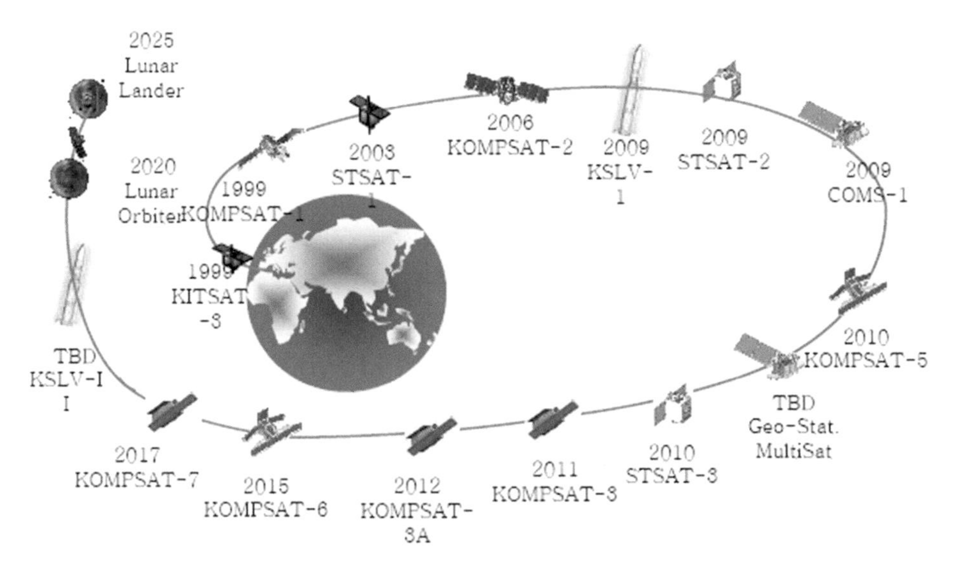

By comparing the National Space Plans of 1996, of 2004 and of 2008, it is possible to see how recent is the priority given by South Korea to indigenous launch capacity

Korean partnership for the KSLV-1 development. The final agreement was signed on 23rd October 2006 by Professor Kim Woo-sik, Korean Deputy Prime Minister and Minister of Science and Technology and Russian Roscosmos (Federal Space Agency) head, Anatoly Perminov. Vladimir Nesterov, manager of the Khrunichev Space Center, made this statement: "We have obtained a contract to develop a first stage propulsion system for a launch vehicle capable of launching a 100kg spacecraft. We will supply components and materials needed for this development without transferring technology." Nine months later, on 7th June 2007, the Russian parliament ratified the technology cooperation pact allowing the transfer and protection of sensitive rocket technology and parts. Finally, the partnership agreement was signed by Russian Vladimir Putin on 28th June and by Korean President Roh Moo-hyun (1946–2009).

The Russo-Korean KSLV-1 program is the most ambitious activity in space for South Korea. It concerns the creation and the operations of the Space Rocket Complex (SRC), which consists of a two-stage launcher for small satellites as well as a range of facilities for its testing, processing, launch and control (with power supply, office areas, living quarters and educational center). Following July 2009 data published by the Korean media, it is a total investment of some \$635m (€449m). The country spent about \$391m (€277m) on developing the KSLV-1, of which about \$198m (€140m) went to the Russians, who are contracted for at least two launches. Some \$243m (€172m) were required to complete, with Russian assistance, the new and modern launch site, called Naro Space Center, which makes Korea the 13th nation in the world to have its own spaceport.[13]

KSLV-1, renamed Naro-ho (Naro-I), is a major step for South Korea, compared to the liquid sounding rocket KSR-III of 2002. The two-stage launcher is 33 m high

For the 2010s, South Korea, from its state-of-the-art Naro Space Center, will be able to launch microsatellites with high performance

The Russo-Korean KSLV-1 (ground test vehicle) is seen on its erector at the launch pad of the Naro Space Center

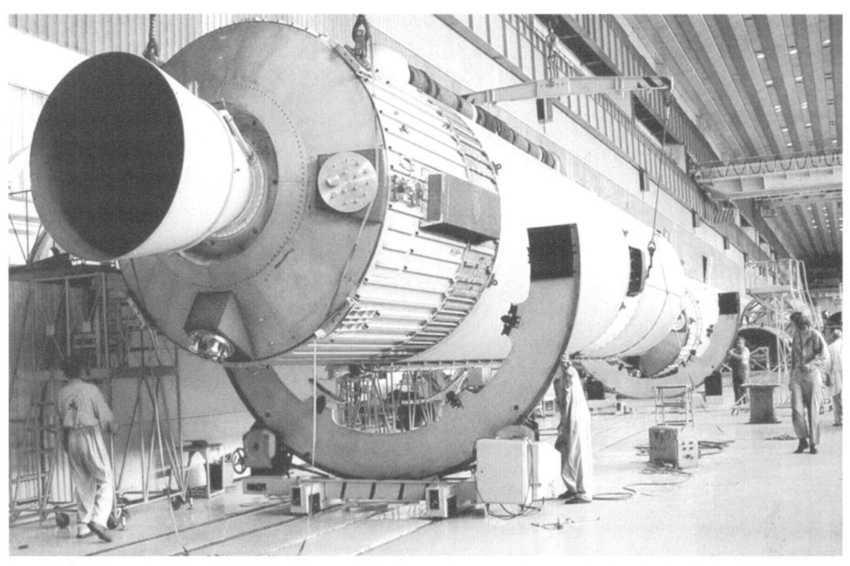

In the factory of Khrunichev Space Center, the mockup of the Angara URM booster offers a great similarity with the first stage of KSLV-1

with a maximum diameter of 2.9 m. It has a mass of 140 tonnes (with 130 tonnes of liquid and solid propellant) at lift-off. It uses environmentally clean propellants (kerosene and liquid oxygen) on the first "made in Russia" stage. The Khrunichev Space Center, which provides the rocketry technology for the Korean SRC, has developed the 25.8-m-long lower assembly. This first stage is derived from the Russian URM-1 (Universal Rocket Module) developed by Khrunichev for the Angara family of modular launch vehicles, to be operated by the Russian Space Forces from the Northern cosmodrome of Plesetsk. Some 80% are systems in development for the Angara program in Russia.[14]

NITs RKP is firing an Energomash RD-151 or RD-191 engine in its testing center located near the town of Peresvet, region of Moscow

KARI is testing the solid booster that will propel the second stage of KSLV-1

Identical to the basic concept of the URM-1 booster, which is a key element of the Russian Angara launchers, the first stage of KSLV-1 presents a great difference in the propulsion system. While the rocket module of Angara is propelled by the NPO Energomash RD-191 engine, which develops a thrust of 1,800–1,900 kN with a specific impulse of 310–337 sec, it is propelled by its weaker variant: the RD-151, which has a reduced thrust of 1,600–1,700 kN, according to statements in the South Korean media. This single-combustion chamber engine is derived from the technology of the RD-170, which was originally used in the 1980s to power the liquid-propellant boosters of the giant Energia rocket. It works in an extremely efficient, high-pressure combustion cycle. The KSLV-1 launch from the Naro Space Center in South Korea is an important event for both Russian Khrunichev and Energomash public enterprises. It is the maiden flight of the RD-191 engine, as well as the URM-1/Angara model. The first stage of KSLV-1 has to work for nearly four minutes (some 237 sec).

The Korean Aerospace Research Institute (KARI) is responsible for the design, development, test and production of the solid Kick Motor (KM) that forms the second stage and upper part of the rocket. The payload to be placed in low orbit is installed over the KM and the KSLV-1 avionics. The "made in Korea" KM is an enlarged by-product of the solid KSR-I/420S booster. Developing a thrust of 42 kN for 66 sec, it must function at 300-km altitude in a vacuum environment. The second stage is made up of core parts entirely developed in Korea, which include, in addition to the motor, the Inertial Navigation System (INS), the power, control and flight safety systems, plus the nose fairing. Designing, manufacturing, test/evaluation and assembly of these core components of the upper stage have been carried out

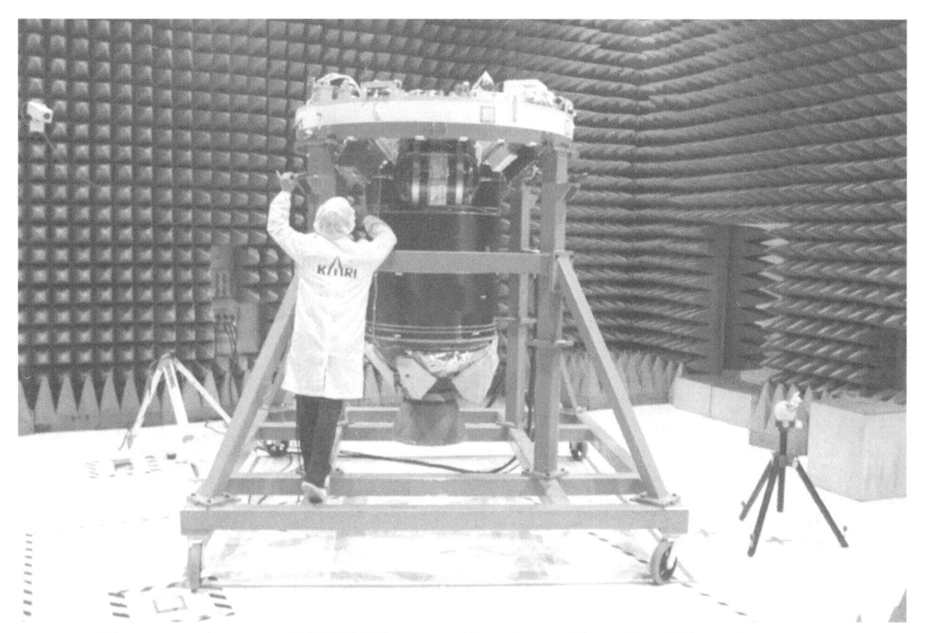

The second stage of KSLV-1 uses a "made in Korea" solid rocket motor

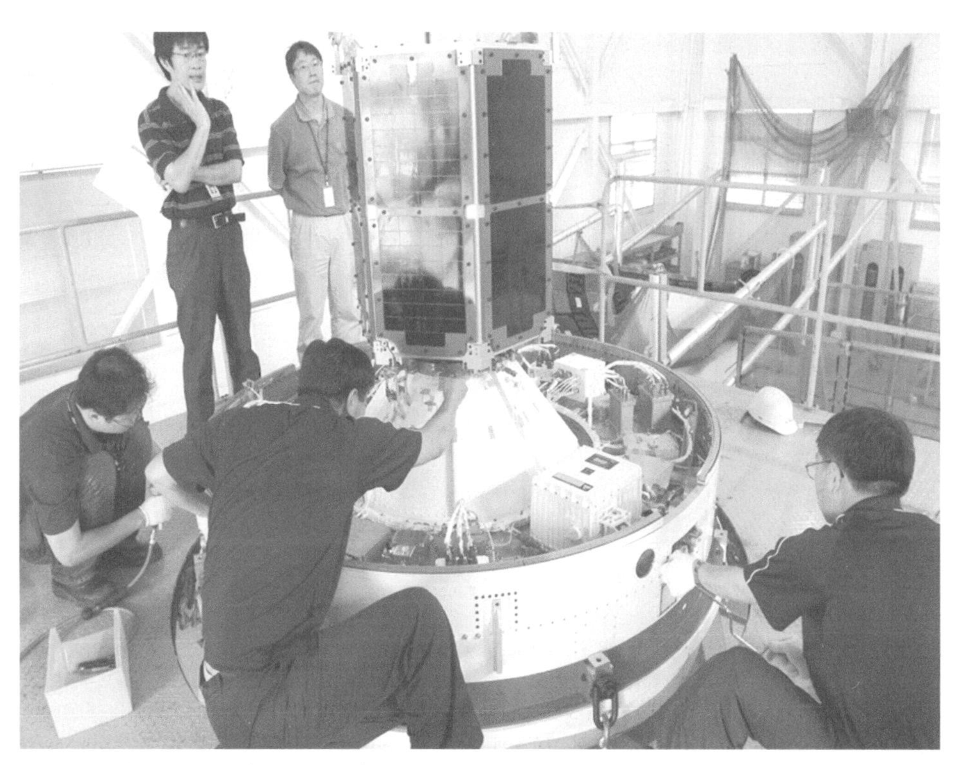

At the top of the second stage, the avionics and the payload of KSLV-1

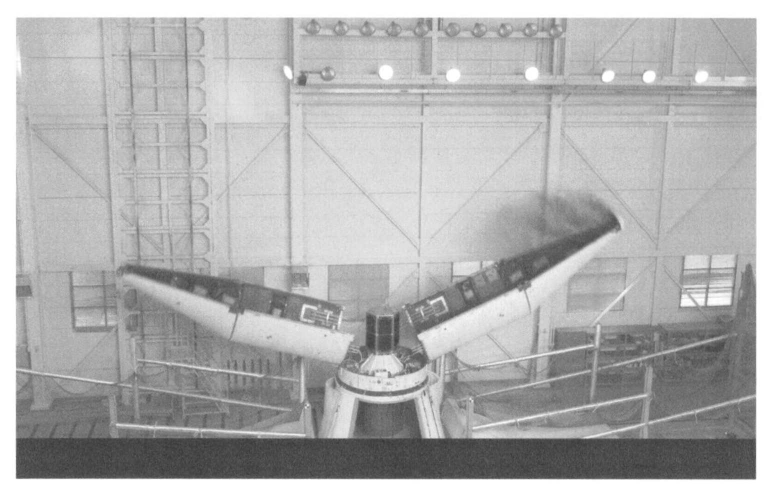

KARI, in a special facility, has tested the opening of the KSLV-1 fairing

Preparation of the fairing that protects the payload of KSLV-1

domestically. In doing so, Korea has secured core technologies required to develop the future fully "made in Korea" Korean Space Launch Vehicle or KSLV-2. KARI built a Ground Test Vehicle with the second stage and payload shroud. The testing of the preparation/integration facilities and the launch pad infrastructure was completed in April 2009.

The KSLV-1 telemetry system is composed of two data streams: a lower (first) stage telemetry stream and an upper (second) stage telemetry stream. Different from the previously developed telemetry system by KARI for the KSR (Korea Sounding Rocket) series, the new telemetry system has been required to meet enhanced requirements: in summer 2005, ABSL Power Solutions (United Kingdom) was awarded the contract by KARI to provide the batteries for the KSLV-1 launch vehicle. Under this contract, ABSL Space Products was to provide four types of batteries, each crucial for successful flight operations.

Cooperation between Russia and South Korea for the KSLV-1 program was seen as a "win–win" partnership for Khrunichev, prime contractor of the modular Angara family and KARI, main contractor of the Korean Space Rocket Complex. This bilateral partnership was not an easy one. Because of the "double hat" position of the Khrunichev enterprise in Moscow, the development of the first stage was affected by delays. The Russian partner had to face budgetary hurdles on the part of the Russian government for the Angara program. It was tempted to rely on the money invested by Seoul in order to progress with the Angara system. Instead of having the maiden KSLV-1 flight in late 2007 as planned, Khrunichev postponed it – up to six times – from October 2006 . The delays for the maiden flight of KSLV-1, due to problems with the "made in Russia" first stage, were commented on critically by the press in South Korea. These critics raised suspicions that the Russians were using Korean money to get precious expertise for their own rockets, named Angara.[15]

During his visit to South Korea in October–November 2007, Roscosmos (Russian Space Agency) head Anatoly Perminov accepted that Seoul was keeping up the pressure on Moscow for the manufacture of KSLV-1 elements. The first launch was scheduled for October 2008. Taking into account actual progress, this deadline was highly doubtful. In April 2008, Professor Hong-yul Paik, president of KARI, announced that the upper stage of the KSLV-1 had been completed, confirming that KSLV-1 was scheduled for launch from the Naro Space Center located in Goheung, Southwestern Korea, in December 2008. Starting on 3rd April 2008, the final comprehensive operation tests took place to check overall operation and function along every step of the flight sequence after the launch, especially the set of events to ensure a successful flight of the "made in Korea" upper part of KSLV-1: fairing separation, ignition of the KM, attitude control, satellite separation and flight completion. Specifically, the test checked whether the rocket shroud of STsat-2 (Science and Technology Satellite-2) would work at 166-km altitude and if the second stage kick motor would be ignited at 300-km altitude in order to put the microsatellite into its assigned orbit (300–1,500 km).

The Ministry of Education, Science and Technology held a KSLV-1 launch review committee meeting on 31st July 2008. After reviewing the status of the launch preparations, the second half of 2009 was decided as the possible launch timeframe for KSLV-1, renamed Naro-ho before its maiden flight in summer 2009. During the meeting, experts assessed that the launch schedule would have to be adjusted because of a delay in the installation of the launch complex system due to the recent earthquake in Sichuan, China, which resulted in the late delivery of parts

Some scenes showing the preparation of the KSLV-1, from the assembly building to the launch pad at the Naro Space Center

manufactured in that region, as well as the addition of extra performance tests to guarantee the success of Korea's first launch. Meanwhile, the Ministry stated that it would continue to thoroughly monitor the progress of the launch preparations for the successful launch and safety of Korea's first space launch vehicle as top priority.

It took not a few months, but more than one further year to get KSLV-1 ready for its maiden launch, which is described by the South Korean media as one of the most important moments in Korean science history. This first launch was announced for 30th July, then postponed to 11th, then 19th August 2009. Finally, it was made on 25th August. The flight followed a nominal path, but the orbital velocity could not be achieved because of the abnormal separation of one of the fairing halves (due to a software glitch?). Another KSLV-1 could be ready for launch nine months later for a new test flight in May–June 2010. The Russians agreed to participate in a third launch, if the first two attempts failed. Some unofficial statements reported that Khrunichev Space Center anticipated producing up to 10 KSLV-1 first stages based on the Angara technology. However, KARI reported that only two KSLV-1/Naro-ho would be built, just to demonstrate the capability of South Korea to have access to orbit, with the technological purpose of delivering in low orbit Korean engineering microsatellites of 100 kg. The second phase in the rocketry program of South Korea, planned to start in 2010, aimed at developing the KSLV-2, a three-stage rocket (two liquid plus a solid third stage), which would be capable of injecting, into orbit around the Earth, spacecraft of up to 1.5 tonnes in 2018 and some 550 kg in translunar orbit during 2020.

The role of the Russian rocket industry in the development of the next KSLV is not yet well established. Its cooperation with KARI has come under scrutiny, as it was reported in this severe editorial published by the South Korea daily *Korea Times*: "South Korean scientists and engineers have come to realize how difficult it is to join the ranks of the world's space powers without having sufficient technology for a rocket. ... Russia is under attack for its unwillingness to transfer its rocket technology to Korea. The Russian space center [Khrunichev] has provided an RD-151 rocket engine to Korea, but it has conducted combustion tests for the KSLV by using an RD-191 engine. The RD-151 engine is inferior to the RD-191 designed for Russia's next generation space rocket Angara scheduled to be launched in 2011. This may back up the allegations that Russia has been exploiting Korea to finance its own space program. The Korean government and its space agency [KARI] should give a clear explanation about such allegations. It is hard for them to avoid criticism that they were pushing the space rocket project too quickly and recklessly without securing enough technologies and finding a faithful partner for the development of the KSLV-1."[16]

THE NARO SPACE CENTER: FROM HILLY ISLAND TO SPACE

South Korea made significant efforts to create and operate its new and impressive Naro Space Center on Oenaro Island at the south-western part of the Korean peninsula, about 48 km south of Seoul . It is built with technical assistance of experts from the Russian KBTM enterprise, in the hilly coast of Goheung County, in South Jeolla Province. This center, covering almost 5 km^2 of land, plays a central role in Korea's space development. Its construction began in December 2000. The Naro Space Center is equipped with state-of-the-art facilities: a launch complex including storage and supply facilities for liquid propellants, assembly building, rocket tracking station and Mission Control center. A visitor facility, with a museum on

On 25th August 2009, the Russo-South Korean Naro-ho (KSLV-1) made its maiden flight, with a successful lift-off. It failed to reach the orbital speed because of a technical glitch that prevented half of the payload fairing from separating correctly from the second stage [KARI]

space technology, welcomes thousands of guests, especially young people, to tell the story of the challenges of South Korea in the odyssey of space.

The Russian first stage of KSLV-1/Naro-ho was delivered from Khrunichev Space Center (Moscow) by Antonov-124 aircraft to the Busan Gimhae International Airport. The metropolitan city of Busan (also known as Pusan), the second in importance after Seoul, has the largest harbor in South Korea. From there, the rocket was moved by ship to Goheung, then by road to the integration building of the Naro Space Center. The launch infrastructure consists of a launch pad and erector, a launch control center, a ground-based observation facility, a tracking radar, an electro-optical monitoring system and the flight termination system (in case of wrong trajectory or malfunctioning). Korean Air, the country's largest airline, is responsible for the handling assembly of the rocket, while Hyundai Heavy Industries, working on a blueprint provided by the Khrunichev Space Center (with KBTM assistance), recently completed the construction of the facilities. The government of Seoul envisions expanding and upgrading the center for the KSLV-2 program, with the construction of a new launch pad. The objective is to make the Naro Space Center an international spaceport: "Our launch system passed 358 categories on the functionality test with flying colours, and we think that our experience of building this facility will provide the base to design another spaceport with our own technology," said Muin Kyung-ju, the director of the Naro Space Center.

Once the integration of first and second stages of the satellite and of the fairing is completed by engineers and technicians of Khrunichev and KARI, the fully assembled rocket is moved by road to the launch pad. On the morning of the launch, the rocket is installed on the stationary erector, which will then rotate 90° and reach vertical position. After the final checkup by engineers, the lower assembly or first stage of the rocket is fuelled with liquid oxygen and kerosene. The launch will be

General view of the facilities at the Naro Space Center, on the hilly coast of Oenaro Island (Gohueng)

The Naro Control Center, heart of South Korean launch facilities

guided by the launch control and observation center that is about 2 km away from the launch pad.

After launch, the rocket ascends vertically for 25 sec, enough time to reach 900 m height, then rotates 10°E and heads towards Okinawa. The rocket is flying 290 km over the surface when it passes the Japanese island chain, just to avoid Japan's airspace. Then, 225 sec from the launch, the spacecraft's nose fairing is jettisoned from the upper part of the rocket. The liquid propellants of the first stage will be consumed after 238 sec. The lower assembly is then separated from the upper part of the rocket and falls to sea, while the Kick Motor of the second stage positions itself and the satellite in the planned direction. The solid motor of the upper part is exhausted after 540 sec (nine minutes) and separates from the satellite, which is about 306 km above the ground at that point. The 100-kg microsatellite is injected into low Earth orbit. The maiden flight of KSLV-1, on 25th August 2009, could not successfully put into low orbit STsat-2A, made by SaTReC. During the 60th IAC (International Astronautical Congress) at Daejeon, it was possible to visit SaTReC/KAIST: STsat-2B was finalizing its tests for a possible launch in May–June 2010.

THE STSAT-2 (SCIENCE & TECHNOLOGY SATELLITE-2) SERIES OF MICROSATELLITES

For its maiden flight, the KSLV-1 rocket carries a real spacecraft with scientific payload and technological experiments. On its ellipsoidal orbit of 300–1,500 km, the first STsat-2 of some 100 kg (160 W of power), which is three-axis-stabilized and has a two-year lifespan, will use its microwave radiometer to scan the intensity of the Earth's radiant energy and it will also be used to measure satellite orbits. The Science & Technology Satellite-2 (STsat-2) program has been developed since October 2002 as a sequel mission to KAISTsat-4/STsat-1 in orbit since September 2003. The objectives of STsat-2 are for three missions to demonstrate the domestic achievement of 100-kg satellites for KSLV-1 launches, the progress of advanced technology for small spacecraft and the development and operation of world-class space science payloads.

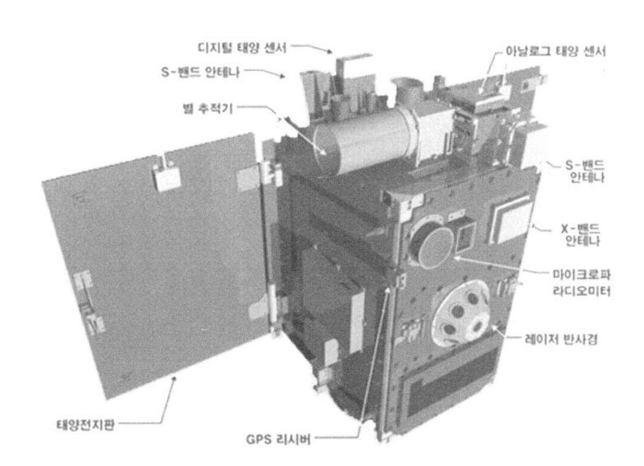

STsat-2, the three-axis-stabilized microsatellite to be launched by the first KSLV-1

Jointly designed by KAIST and the Gwangju Institute of Science and Technology (GIST), the first STsat-2 (STsat-2A) consists of two payloads for space research. The main payload is DREAM (Dual-channel Radiometer for Earth and Atmosphere Monitoring): the goal of this remote sensing mission is to acquire brightness temperature of the Earth at 23.8 and 37 GHz as well as physical parameters such as cold liquid water and water vapor through data processing. The secondary payload is LRA (Laser Retroreflector Array) for the Satellite Laser Ranging (SLR) mission to determine the more precise orbit of STsat-2 than is possible with S-band tracking data alone, to calibrate the main payload (DREAM), and finally to support the science research such as Earth science and geodynamics.

For spacecraft technology experiments, STsat-2 has the following instrumentation on board: a Pulsed Plasma Thruster (PPT), Dual-Head Star Tracker (DHST), Fine Digital Sun Sensor (FDSS) and a compact onboard computer with high-speed data transmission (up to 10 Mbps). The technological mission aims at developing a thermally, mechanically, electrically stable and radial-resistant spacecraft system having high-precision attitude determination and control capability in the

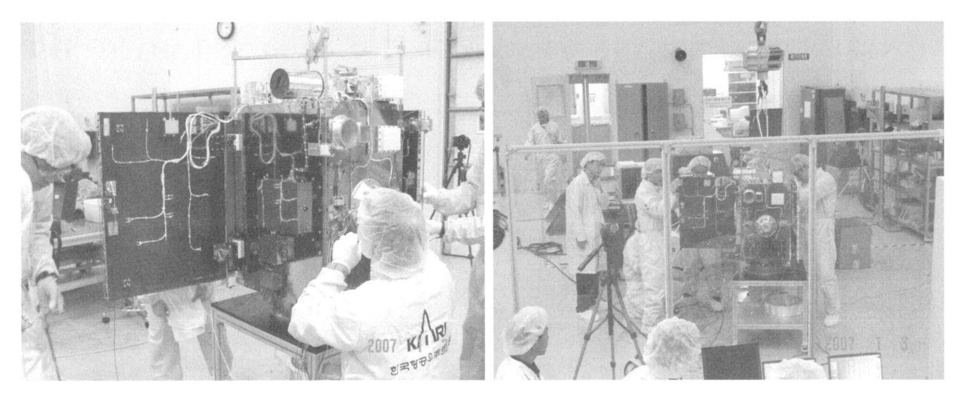

Preparation of the STsat-2 microsatellite, which contains many scientific and technological experiments

The STsat-2 microsatellite at the top of KSLV-1, waiting for the fairing

environment of space along its ellipsoidal orbit. Additionally, the Precise Orbit Determination (POD) of STsat-2 can be used to evaluate the performance of the first Korean launcher (KSLV-1).[17]

The STsat-3 is an engineering mini-satellite of about 150 kg, developed by SaTReC. It will be launched in 2011 from abroad in order to test new technologies (such as an electric thruster, Li-ion battery, small solar power regulator, etc.) to take hyperspectral images of the Earth (with COMIS/Compact Imaging Spectrometer) and to make infrared observations of our planet and our galaxy (with the MIRIS/ Multi Infrared Imaging System).

MANNED SPACEFLIGHT IN 2008: ODYSSEY OF A YOUNG LADY TO THE ISS

Russian–Korean cooperation in space, apart from the joint KSLV-1 program of access to space, has another spectacular aspect: the spaceflight of a South Korean citizen in the International Space Station (ISS). It took place from 8th to 19th April 2008 using Soyuz TMA-12 for departure and Soyuz TMA-11 for return. The government of Seoul paid Roscosmos the ticket for this 10-day mission: $27m (€17m).

The selection of the two Korean candidates for spaceflight training in Zvezdny Gorodok (Star City) near Moscow was a great popular success throughout South Korea during 2005. There were some 36,206 applicants to become the first Korean in space. Two researchers, Mr Ko San and Ms Yi So-yeon, were the finalists chosen on 25th December 2006 for the Korean astronaut program. They started their training

Scenes of training of South Korean astronauts in Russia

in Moscow for the Soyuz TMA spaceship and in Houston for conditions in the ISS. On 5th September 2007, the Korean Ministry of Science and Technology announced Ko San as primary astronaut and Yi So-yeon as backup, following performance in tests during training in Russia. However, on 10th March 2008, less than one month before the flight, this decision was reversed, after Roscosmos, the Federal Space Agency, asked for a replacement. Ko twice violated the regulations of the training protocol in Russia by reading – without authorization – some sensitive documents and by mailing one of them to Korea! It was the first time that a candidate for spaceflight had to be replaced for a breach of training regulations.

A woman replaced the male candidate as the first Korean in space. Yi So-yeon, born on 2nd June 1978, became a member of the Soyuz TMA-12 crew with Russian Sergei Volkov, commander, and Oleg Kononenko, flight engineer. Her lift-off from the Baikonour cosmodrome took place on 8th April, drawing much public interest and support in the Korean peninsula. As a scientist graduated in mechanics, then obtaining a doctorate in biotechnology from KAIST, where SaTReC Initiative, the manufacturer of small satellites, is located, she became the first South Korean astronaut. She became the second Asian woman to fly in space, after Japanese Chiaki Mukai. Soyuz TMA-12 docked with the ISS on 10th April. The crew was welcomed by the inhabitants of the station, American astronauts Peggy Whitson and Garrett Reisman, with Russian cosmonaut Yuri Malenchenko.

The Korean astronaut spent about nine days aboard the ISS and was busy carrying out 18 scientific experiments for KARI, South Korean laboratories and institutes. She took with her 1,000 fruit flies in a special air-conditioned container box for Konkuk University: she monitored how changes in gravity and space environment had an impact on the behavior of the flies or their genome. Other experiments concerned the growth of plants in space, the study of her body (especially her heart) and the effects of gravity change on the pressure on her eyes and shape of her face. By means of a 3D camera especially designed by Samsung, Yi took shots of her face every day to see how it swelled in microgravity. She also made Earth observations to monitor the movement of dust storms from China to Korea. She measured the noise level inside the ISS. She took time to promote Korean

Yi So-yeon, the first South Korean astronaut, ready to go for her mission to the ISS

culture by having a Korean-style dinner and using the national flag Taegeukgi. South Korean scientists created a special low-calorie and vitamin-rich version of kimchi for the orbital mission of Yi.

On 19th April, Yi So-yeon, with Peggy Whitson and Yuri Malenchenko, said goodbye to the crew remaining in the station. Her return in the Soyuz TMA-11 capsule, after the jettisoning of the orbital module (with power and propulsion systems), was a dramatic event, because the spacecraft followed a ballistic re-entry that subjects the crew to extreme force (up to 10 G). As a result of this unusual

Lift-off of Soyuz TMA-12 with Russo-Korean crew, on 8th April 2008

Two images of Yi So-yeon on board the ISS

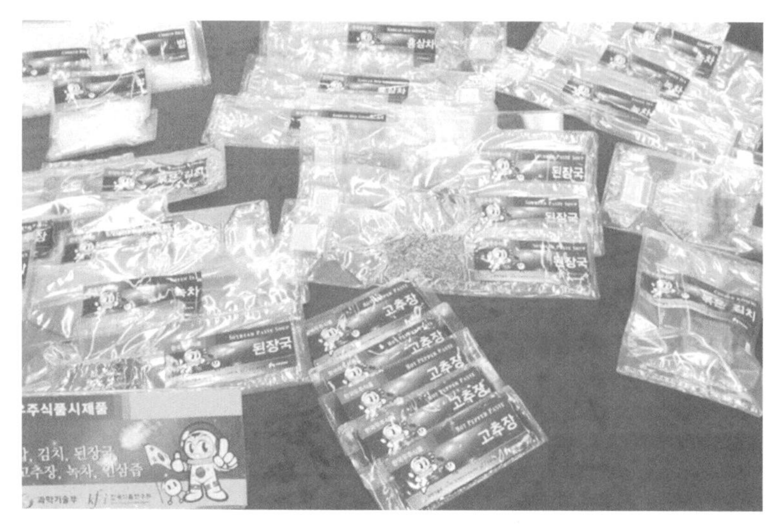

The inhabitants of the ISS experienced some specimens of Korean space food

Russian helicopters on the landing site of the Soyuz TMA-11 capsule after a hard re-entry in the atmosphere

trajectory, Soyuz TMA-11 landed 418 km off-course from its target for touchdown in the steppe of Kazakhstan. The crew required special inspection by medical personnel of the Gagarin training center. Yi, who suffered from the extreme deceleration, had to be hospitalized after her return, due to back pain caused by the tough return trip.

After her flight, Yi So-yeon worked as a researcher at KARI and as an ambassador for Korea in space. She attended as an honored guest the 59th IAC

(International Astronautical Congress) in Glasgow (Scotland). In a TV interview, she commented her experience – the first one for a lady! – of the shock of the return in the atmosphere: "I had a little back pain. It was not serious. But, by chance, I've never had a car accident, so I have never experienced a small shock at all. ... At the time, I thought I was so unlucky. Now I think it was chance. There have been about 400 astronauts and cosmonauts, but only a few have experience a ballistic re-entry. So that was a very unique experience."[18]

THE FUTURE (2009–2019): SATELLITE LAUNCHES AND SPACE EXPLORATION

Since 2005, South Korea has kept stable the annual budget for its space projects and missions, considered as a high priority for the development of advanced systems. For 2008, the budget of Korea in space – mainly to support the programs at KARI and KAIST – was around $338m (€237m). The aim of Seoul is to make South Korea a major player in the business of satellite systems and services as well as in the international ventures of manned spaceflight and space exploration at an international level. South Korea is a young entrant in the activities of astronautics and still has a long way to go before becoming a real space power.

On 20th November 2007, KARI announced an ambitious plan to join Asia's space race by launching a lunar orbiter by 2020 and sending a probe to the Moon five years later. The Korean Minister of Science & Technology unveiled the project one month after China launched its first lunar orbiter and two months after Japan.

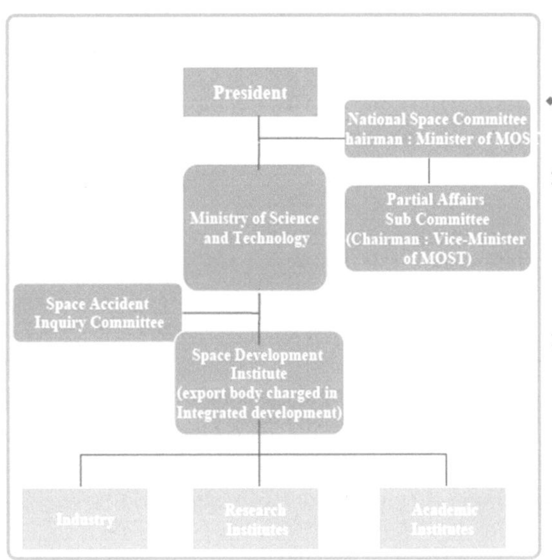

The national space program of South Korea, managed by the Ministry of Science & Technology

➢ **16 satellites are developed, developing, or planned**

	Developed	Developing	Planned
Science Satellite	KITSAT1,2,3 STSAT1	STSAT2,3	low budget small satellite
Multi Purpose Satellite	KOMPSAT1,2	KOMPSAT3, 3A, 5	KOMPSAT6,7
Geostationary Satellite		COMS	Geostationary multi purpose satellite

Summary of 25 years of South Korean presence in space: spacecraft from 1992 to 2017

According to the road map for national implementation of space development, which includes the development of a 300-tonne launcher or KSLV-2 and of robots for the exploration of the Moon, the estimated investment would be $3.9bn (€2.74bn) during the 2010s. Korea does not exclude participation in an innovating project to build an automated observatory at the lunar southern pole, in the framework of the International Lunar Network, which involves eight countries.

The next decade of South Korea in space will be marked by these planned events:

- deployment of a constellation of Earth observation satellites (optical sensors, radar) through the KOMPsat and COMS programs to meet the needs of the Sentinel Asia initiative led by the APRSAF (Asia–Pacific Regional Space Agency Forum);
- industrial production of geosynchronous space systems for meteorology, communications, broadband links and digital broadcasts;
- indigenous development of sophisticated robotics, with nano-miniaturized intelligent systems, for the exploration of the Moon and the solar system;
- improvement of national access to space – up to the Moon – from the Naro Space Center, through the KSLV-2 program of a more powerful launcher, whose first flight is planned for 2018.

Rocketry represents a great challenge for the future of South Korea in space. Its rocket technology is still at the initial stage. It is imperative for the nation to step up cooperation between the government, research institutes and corporations in developing up-to-date technology for rockets. KARI hopes that the experience acquired with the Russo–Korean KSLV-1 will accelerate its efforts to develop a more advanced KSLV-2 with its own technology. "Seoul has decided to go solo on the next rocket based on know-how acquired in the building of the KSLV-1 with Russia", according to a statement of Professor Lee Sang-mok, Deputy Minister of Education, Science & Technology.[19] The brochure of KARI (2007 edition) described the domestic efforts to progress the technology of liquid propulsion systems: "To develop high-performance engines, we ascertain the design requirements of the subcomponents, such as combustion chamber, gas-generator, turbopump unit and propellant fed system."[20] For the KSLV-2 program, KARI is pushing ahead research on structures, onboard electronics, launcher control system, thermal environment, aerodynamics, mission design and engine testing.

In an interview during the preparation of the Naro-ho maiden flight on 25th August, Lee Joo-jin, president of KARI, gave these details of the future of space rocketry in South Korea: "We are in the advanced stages of development of a 30 tonne [270kN] thrust rocket that includes a combustion chamber for a liquid fuel rocket, turbopumps that mix fuel with oxidation agents and an engineering model for the gas generator." He added that a full-size model of the 270-kN engine, propelled by a liquid oxygen and kerosene mixture, had been built with numerous tests conducted to check its capabilities. "Based on the experience gained, work has started on a more powerful 75 tonne [670-kN] thrust rocket."

The design of KSLV-2 is not yet fully finalized. In a brochure in 2007, KARI published a drawing of a first stage with four liquid engines. It seems that the second stage will be propelled by a single 670-kN engine. South Korea has still a long way to go before having a fully home-made launcher of satellites. It must master its own

South Korea, as a young space power, achieves commercial success by selling mini-satellites for Earth observations. SaTReC Initiative is recognized as a very active actor in space business around the world

technology of liquid rocket propulsion. In this field, North Korea is largely ahead by upgrading old rocket engines (with toxic propellants) that have been developed for liquid missiles by the former Soviet Union.

The government of Seoul recognizes space technology as a national strategic stepping stone. Compared to the Space Development Plan when it was established in 1996, the new version of 2007 increases five-fold the budget for the Korean space program. In the meantime, as manpower for space activities has also steadily expanded, 1,700 people (2006 data) are now working in space sectors such as satellite, launcher and space applications. In the last 10 years, the gap between South Korea and the space powers has been greatly reduced. Because the needs of satellite application services currently are increasing, space development will continue to grow. With further investment in space research and the expansion of manpower through the Space Fundamental Core Technology project from 2008, advancements are expected to be steady. Government efforts have popular support in the Korean population.

REFERENCES

1. *World-Wide Space Activities*, report of Committee on Science & Technology, US House of Representatives, published in September 1977, stated that South Korea is "a member of Intelsat and the WMO (World Meteorological Organization) and participates with the United States in the analysis of data from Landsat and from lunar samples".
2. *Unispace III – National Paper of the Republic of Korea*. Vienna, June 1999.
3. Johnson, Nicholas L. and Rodvold, David M.: *European & Asia in Space 1993–1994*. Kaman Sciences Corporation/USAF Phillips Laboratory, 1995; Burleson, Daphne: *Space Programs Outside the United States*. McFarland & Company Inc. Publishers, Jefferson, North Carolina, 2005, pp. 270–271. *L'industrie spatiale en Corée du Sud*, Etude (68 pages) de la Mission économique de Séoul pour Spheris, Paris, juin 2009. It published a list of the international partners in the Korean space program.
4. UBIFrance: *L'industrie spatiale coréenne*. Fact sheet (four pages) released 28 July 2008.
5. www.skyrocket.de/space/; http://satrec.kaist.ac.kr/english/SaTReC.html.
6. SaTReC Initiative, documentation about products, 2008, www.satreci.com.
7. www.globalsecurity.org/space/world/rok/kompsat.htm; www.kari.re.kr/.
8. *Korea Times*: Korea joins world's satellite powers, 22 August 2006 (AsiaMedia).
9. www.globalsecurity.org/space/world/rok/ksr.htm.
10. *Jane's Space Directory 2006–2007*, edited by Bill Sweetman, Coulsdon (Surrey), 2006.
11. COMS-1, http://directory.eoportal.org/.
12. UNOOSA (United Nations Office for Outer space Affairs)/COPUOS (Committee on the Peaceful Uses of Outer Space): *National Report of Korea*, A/AC.105/907/Add.1, 2007.

13. *www.globalsecurity.org/space/world/rok/kslv.htm.*
14. *European Space Directory 2009*, 24th edition. ESD, Paris, 2009. Report about South Korea and Table *Access to Space: Launch Vehicles.*
15. Korean reports of *Yonhap News* and daily *Korea Times* about the maiden flight of KSLV-1, July–August 2009.
16. *Korea Times*: A set of liftoff delays. 12 August 2009.
17. STsat-2, http://directory.eoportal.org.
18. Rincon, Paul: Korean astronaut's bumpy return. BBC News, 6 October 2008.
19. Press information of RIA Novosti, 10 June 2009.
20. KARI publication, 2007, www.kari.re.kr.

ANNEXE

Vehicle name	KSLV-1/NARO-HO
State	South Korea
Operator	KARI (Korea Aerospace Research Institute)
Prime contractor	KARI (Korea Aerospace Research Institute) + Khrunichev Space Center + Shinyoung Heavy Industries
Launch site	Gohung Island (at the end of the Korean penisula)
First flight [planned]	25 August 2009 (failure), [May–June 2010]
Reliability rate	NA
Performances:	
• LEO mass capacity	100 kg
• GTO mass capacity	NA
• Fairing height/diameter	5 m/2 m
• Dual launch system	NA
Basic vehicle:	
• Lift-off mass	139 tonnes
• Lift-off thrust	1.667–1.800 kN
• Height	33 m
• Number of stages	2
Type of propulsion for each stage (number x engines) [propellants]	1st stage: 1 x RD-191 Glushko engine, developed by NPO Energomash (RP-1/LO2) for the Angara URM (Universal Rocket Module) booster 2nd stage for KSLV-1: solid KSR-I motor of 86 kN, developed by KARI
Improved versions (description) [performances]	KSLV-2 in 2018 [1.5 tonnes in LEO], but design not yet finalized

Vehicle name	KSLV-1/NARO-HO
Technical status (as of 1st July 2009)	Test flight, on 28th November 2002, of first liquid sounding rocket KSR-III (three stages, 14 m long, 6 tonnes at lift-off) with an indigenously developed engine using pressurized propellants Agreement signed on 26th October 2004 with Khrunichev Space Center, NPO Energomash (engine) and KBTM (ground infrastructure) to develop a family of launch vehicles based on the Angara booster (1st stage) Delays of the ground testing for the launch vehicle and the launch infrastructure
Commercial status (as of 1st July 2009)	Mainly used to launch "made in Korea" small satellites
Estimated launch price	NA

References

Reports of Jonathan McDowell (Jonathan's Space Report).

Reichl, Eugen: *Das Raketentypenbuch*. Motorbuch Verlag, Stuttgart, 2007.

European Space Directory, annually published by ESD Partners, Paris.

Detailed description, with drawings and photos of KSLV-1, published by Norbert Brügge on Space Launch Vehicles website, www.b14643.de/Spacerockets_1/index.htm.

15

Contrasts and comparisons

Finally, *Emerging Space Powers* sets these space programs in a global context, commenting on their distinctive features.

Looking back over the period of the space age, the "space powers" may be divided into three categories: the original space super-powers, Russia and the United States; a larger group of middle-range space powers; and the very new entrants. The space super-powers first launched satellites in 1957–1958; the middle-range group followed in the 1960s to 1980; and the new entrants followed thereafter. This historical evolution is illustrated in Table 15.1.

Table 15.1. First satellite launches by the space powers

Super-powers	
Russia/Soviet Union	1957
United States	1958
Middle-range	
France	1965
Japan	1970
China	1970
Britain	1971
India	1980
New entrants	
Israel	1988
(Brazil	1997 (first attempt))
(North Korea	1998 (first attempt (?))
Iran	2009
(South Korea	2009 (first attempt, unsuccessful) – new attempt in 2010)

To this day, Russia and the United States remain the dominant super-powers, the United States having a larger space budget than all the other nations put together and Russia launching more satellites annually than any other nation. The middle-range powers have grown up into the significant, mature space programs that we now know as China, Japan and India, while the French achievements formed the basis for the European space program (the British launcher was cancelled). The sizes of the respective space programs over time are reflected in Table 15.2.

Table 15.2. Total space launches, 1957–2008

USSR/Russia/Ukraine*	2,850
United States**	1,352
Europe***	186
China	115
Japan	71
India	22
Israel	5

* Includes Sea Launch. ** Includes American launches from Woomera, Australia; San Marco platform, Kenya; and from Canary Islands, Spain. *** Includes four French launches from Hammaguir, Algeria and one British launch from Woomera, Australia. Source: Théo Pirard: *European Space Directory*, 2009.

From 2009, we can add Iran to the list and may soon be able to add South Korea. Had launch attempts been more fortunate, we should have been able to add Brazil, North Korea and South Korea. Japan is therefore the fifth spacefaring group or nation, followed by India at sixth and Israel seventh. The emerging space powers described in this book account for only 2% of total launches. This table has a strong historical bias, for many of the launches in it came from the 1970s and 1980s when Russia was launching 100 satellites a year and before the Indian and Japanese space programs were as developed. If we look at more recent years, we may get a more helpful, current picture of the level of space activities (Table 15.3).

Table 15.3. Space launches, 2006–2008

Russia/Ukraine	86
United States	40
China	26
Europe	17
Japan	9
India	6
Israel	1

Russia remains the leading spacefaring nation in terms of launch rate, followed by the United States. In the past number of years, China's rapidly expanding program has moved it into third place and ahead of Europe – a situation unlikely to change. Japan and India remain in fifth and sixth place, respectively, but Japan and India comprise a higher overall proportion of launches: 11 out of 169, 6.5%.

Space spending is another way of looking at the role of the emerging space powers in the space programs of the world. Here, we get the picture shown in Table 15.4.

Table 15.4. Space spending, estimated 2008, in €m

United States	30,221
Europe	6,862
Russia	3,000
China	3,000
Japan	1,714
India	540

Source: Théo Pirard: *European Space Directory*, 2009.

The dominance of the American space program is very evident here, the Americans spending almost three times more than the rest of the world put together. Here, Europe is the second largest spending group, ahead of China and Russia. The Chinese figure reflects relatively low wages there and understates the high level of modern infrastructure, while the Russian figure understates its substantial, old but very usable infrastructure. Then come our emerging space powers of Japan and India, Japan's spending being three times that of India, reflecting its much broader technological base.

DEVELOPMENT AND FIELDS OF WORK COMPARED

Despite differences of scale and, as we have seen, budget, the Japanese and Indian space programs had many similarities in their development. Both started from small beginnings, using primitive sounding rockets, led by charismatic father figures, Hideo Itokawa and Vikram Sarabhai. Both began with the placing of small satellites in Earth orbit using solid-fuelled rockets. Both pursued the path of indigenization – learning how to borrow technology from abroad and rebuild it at home – an avenue pursued by the Indians with more conviction. The Indian program has been tightly focused on applications (communications, Earth observations and meteorology) and may be one of the most concentrated in the world. India remained faithful to the vision and route mapped out by the founder of its space program, Vikram Sarabhai. The success of India in indigenization and quality observation work seems to have been vindicated, for India is now the world leader in the use of satellite technology and remote sensing as a force for economic and social development. The benefits that rural India has gained from the satellite program have repaid many times over the original investment.

Following a common start, the Japanese and Indian programs diverged to some degree. For the 1980s and 1990s, India concentrated entirely on Earth observations, communications and meteorological satellites and these remain the main lines of development. Apart from the *Rohini* series in the 1980s, the main purpose of which was to develop a launcher, and Chadrayan very recently, India launched no dedicated scientific satellites, though some satellites carried scientific instruments as secondary payloads. By contrast, with its greater economic resources, Japan's program covered a much broader range of activities, with a significant investment in engineering test satellites, scientific missions and manned spaceflight in partnership with the United States. Japan's deep space program is especially notable and at one stage in the 1990s, it was the only country in the world to have such a program, apart from the United States (Russia's had temporarily disappeared, due to lack of funding). Especially impressive has been the investment in engineering test satellites, particularly to develop more sophisticated means of communications. The continuous investment in testing cutting-edge technologies shows how Japan has used its program to successfully develop and maintain its lead in the worldwide development of advanced communications and industrial engineering. No less impressive is the imaginative Japanese scientific program, not least because of its low cost, the use of micro-technologies and, in the case of deep space missions, the employment of relatively small rockets using idiosyncratic trajectories to fly to the Moon and Mars, demonstrating how much can be achieved with little.

The new century saw India begin to close the gap with Japan. The latter space program was a casualty of Japanese economic retrenchment in the late 1990s, with missions deferred, slowed or even cancelled. The program was slimmed down and Japanese ambitions for manned spaceflight were shelved. With the retirement of the *Mu* launcher, Japan was left with only one operational rocket, the H-IIA, while India had two, the PSLV and the GSLV. The Indian space budget rose significantly in the decade of the 2000s, to the point that India could contemplate its own manned spaceflight by 2014.

Turning to our new entrants, it is difficult to see their space programs reaching the same scale or level of operations as Japan or India. Israel's political and military situation means that its program is highly focused on intelligence and security – something that is unlikely to change, but with secondary lines of development in Earth resources and telecommunications. Both South Korea and Brazil are demonstrations, following the Indian model, of how spaceflight can serve as a tool for development in such fields as communications, Earth resources and the environment. The space programs of North Korea and Iran are more ambiguous, for their foreign critics are likely to see them in terms of national prestige, missile development and the legitimization of the régime. Iran is clearly much more than this, for Chapters 7 and 8 illustrate clearly that investment in the space program is part of a sustained investment in applications, communications, engineering and technology. *Emerging Space Powers* indicates that, as the story of the 21st century continues to unfold, more countries may come to see spaceflight as a vital tool in economic and social development.

Annexes

ANNEXE 1: LIST OF LAUNCHES

Japan

11 Feb. 1970	*Ohsumi*	*Lambda-4S*	Uchinoura
16 Feb. 1971	*Tansei*	*Mu-4S*	Uchinoura
28 Sep. 1971	*Shinsei*	*Mu-4S*	Uchinoura
19 Aug. 1972	*Denpa*	*Mu-4S*	Uchinoura
16 Feb. 1974	*Tansei 2*	*Mu-3C*	Uchinoura
24 Feb. 1975	*Taiyo*	*Mu-3C*	Uchinoura
9 Sep. 1975	ETS I (*Kiku 1*)	N-I	Tanegashima
29 Feb. 1976	*Ume 1*	N-I	Tanegashima
19 Feb. 1977	*Tansei 3*	*Mu-3H*	Uchinoura
23 Feb. 1977	ETS II/*Kiku 2*	N-I	Tanegashima
4 Feb. 1978	Exos A/*Kyokko*	*Mu-3H*	Uchinoura
16 Feb. 1978	*Ume 2*	N-I	Tanegashima
16 Sep. 1978	Exos B/*Jikken*	*Mu-3H*	Uchinoura
6 Feb. 1979	ECS 1/*Ayame 1*	N-I	Tanegashima
21 Feb. 1979	CORSA/*Hakucho*	*Mu-3C*	Uchinoura
17 Feb. 1980	ECS 2/*Ayame 2*	N-I	Tanegashima
11 Feb. 1981	ETS IV/*Kiku 3*	N-II	Tanegashima
21 Feb. 1981	*Hinotori*	*Mu-3S*	Uchinoura
11 Aug. 1981	*Himawari 2*	N-II	Tanegashima
3 Sep. 1982	ETS III/*Kiku 4*	N-II	Tanegashima
4 Feb. 1983	*Sakura 2A*	N-II	Tanegashima
20 Feb. 1983	Astro B/*Tenma*	*Mu-3S*	Uchinoura
5 Aug. 1983	*Sakura 2B*	N-II	Tanegashima
23 Jan. 1984	BS 2A/*Yuri 2A*	N-II	Tanegashima
14 Feb. 1984	Exo C/*Ohzora*	*Mu-3S*	Uchinoura
3 Aug. 1984	*Himawari 3*	N-II	Tanegashima
8 Jan. 1985	MS-T5/*Sakigake*	*Mu-3S*	Uchinoura
19 Aug. 1985	Planet A/*Suisei*	*Mu-3S*	Uchinoura
12 Feb. 1986	BS 2B/*Yuri 2B*	N-II	Tanegashima
13 Aug. 1986	*Ajisei* *Fuji*	H-I	Tanegashima
5 Feb. 1987	*Ginja*	*Mu-3S*	Uchinoura
19 Feb. 1987	MOS 1A/*Momo 1A*	N-II	Tanegashima
27 Aug. 1987	ETS V/*Kiku 5*	H-I	Tanegashima
19 Feb. 1988	*Sakura 3A*	H-I	Tanegashima
16 Sep. 1988	*Sakura 3B*	H-I	Tanegashima
21 Feb. 1989	Exos D/*Akebono*	*Mu-2SII*	Uchinoura
5 Sep. 1989	*Himawari 4*	H-I	Tanegashima
24 Jan. 1990	Muses A: *Hiten* *Hagoromo*	*Mu-3SII*	Uchinoura
7 Feb. 1990	MOS 1B/*Momo 1B* *Orizuru* *Fuji 2*	H-I	Tanegashima

25 Aug. 1990	BS 3A/*Yuri 3A*	H-I	Tanegashima
25 Aug. 1991	BS 3B/*Yuri 3B*	H-I	Tanegsahima
30 Aug. 1991	Solar A/*Yohkoh*	*Mu-3SII*	Uchinoura
11 Feb. 1992	JERS/*Fuyo*	H-I	Tanegashima
20 Feb. 1993	Astro D/*Asuka*	*Mu-3SII*	Uchinoura
4 Feb. 1994	OREX/*Ryusei*	H-II	Tanegashima
	EP/*Myojo*		
28 Aug. 1994	ETS VI/*Kiku 6*	H-II	Tanegashima
15 Feb. 1995	*Express*	*Mu-3SII*	Uchinoura
18 Mar. 1995	SFU	H-II	Tanegashima
	Himawari 5		
12 Feb. 1996	HYFLEX	J-1	Tanegashima (sub-orbit)
17 Aug. 1996	ADEOS/*Midori 1*	H-II	Tanegashima
	Fuji 3		
12 Feb. 1997	Muses B/*Haruka*	*Mu-5*	Uchinoura
28 Nov. 1997	ETS VII/*Kiku 7*	H-II	Tanegashima
	TRMM		
21 Feb. 1998	COMETS/*Kakehashi*	H-II	Tanegashima
4 Jul. 1998	Planet B/*Nozomi*	*Mu-5*	Uchinoura
29 Aug. 2001	Laser test payload	H-IIA	Tanegashima
4 Feb. 2002	*Tsubasa*	H-IIA	Tanegashima
10 Sep. 2002	DRTS/*Kodama*	H-IIA	Tanegashima
	USERS		
14 Dec. 2002	ADEOS 2/*Midori 2*	H-IIA	Tanegashima
28 Mar. 2003	Optical 1, Radar	H-IIA	Tanegashima
9 May 2003	Muses C/*Hayabusa*	*Mu-5*	Uchinoura
26 Feb. 2005	MTSAT 1R/*Himarawi 6*	H-IIA	Tanegashima
10 Jul. 2005	Astro E-2/*Suzaki*	*Mu-5*	Uchinoura
24 Jan. 2006	ALOS/*Daichi*	H-IIA	Tanegashima
18 Feb. 2006	MTSAT 2	H-IIA	Tanegashima
21 Feb. 2006	Astro F/*Akari*	*Mu-5*	Uchinoura
	Cute 1.7		
11 Sep. 2006	Optical 2	H-IIa	Tanegashima
21 Sep. 2006	Solar B/*Hinode*	*Mu-5*	Uchinoura
	Hitsat		
18 Dec. 2006	ETS VIII/*Kiku 8*	H-IIA	Tanegashima
24 Feb. 2007	Optical 3	H-IIA	Tanegashima
	Radar 2		
14 Sep. 2007	Selene/*Kaguya*	H-IIA	Tanegashima
23 Feb. 2008	WINDS/*Kizuna*	H-IIA	Tanegashima

Japan: failures

26 Sep. 1966	Test satellite	*Lambda*	Uchinoura
20 Dec. 1966	Test satellite	*Lambda*	Uchinoura
13 Apr. 1967	Test satellite	*Lambda*	Uchinoura
22 Sep. 1969	Test satellite	*Lambda*	Uchinoura
25 Sep. 1970	Test satellite	*Mu-4S*	Uchinoura
15 Nov. 1999	MTS	H-II	Tanegashima
10 Feb. 2000	Astro E	*Mu-5*	Uchinoura
29 Nov. 2003	Optical 2, Radar 2	H-IIA	Tanegashima

Japan's launches by other countries

14 Jul. 1977	GMS 1/*Himawari 1*	Delta	Cape Canaveral
15 Dec. 1977	*Sakura 1*	Delta	Cape Canaveral
7 Apr. 1978	BS 1/*Yuri 1*	Delta	Cape Canaveral
6 Mar. 1989	JCSat 1	Ariane	Kourou
5 Jun. 1989	Superbird A	Ariane	Kourou
1 Jan. 1990	JCSat 2	Titan 3	Cape Canaveral
26 Feb. 1992	Superbird B	Ariane	Kourou
24 Jul. 1992	Geotail	Delta	Cape Canaveral
1 Dec. 1992	Superbird A-1	Ariane	Kourou
9 Jul. 1994	BS 3N	Ariane	Kourou
29 Aug. 1995	JCSat 3	Atlas 2AS	Cape Canaveral
29 Aug. 1995	N-Star A	Ariane	Kourou
5 Feb. 1996	N Star B	Ariane	Kourou
17 Apr. 1997	BS 4	Ariane	Kourou
27 Jul. 1997	Superbird C	Atlas 2AS	Cape Canaveral
3 Dec. 1997	JCSat 4	Ariane	Kourou
29 Apr. 1998	Bsat 1B	Ariane	Kourou
16 Feb. 1999	JCSat 6	Atlas 2AS	Cape Canaveral
8 Mar. 2001	Bsat 2a	Ariane 5	Kourou
12 Jul. 2001	Artemis Bsat 2b	Ariane 5	Kourou
5 Jul. 2002	N-Star C	Ariane 5	Kourou
11 Jun. 2003	Bsat 2c	Ariane 5	Kourou
24 Aug. 2005	OICETS/*Kirari* INDEX/*Rimei*	Dnepr	Baikonour
12 Apr. 2006	JCSat 9	Zenit 3SL	Odyssey platform
11 Aug. 2006	JCSat 10	Ariane 5	Kourou
13 Oct. 2006	LDREX	Ariane 5	Kourou
14 Aug. 2006	BSat 3a	Ariane 5	Kourou
5 Sep. 2007	JCSat 11	Proton M	Baikonour (fail)
18 Dec. 2007	New Horizons	Ariane 5	Europe

India

18 Jul. 1980	*Rohini*	SLV	Sriharikota
31 May 1981	*Rohini* 2	SLV	Sriharikota
17 Apr. 1983	*Rohini* 3	SLV	Sriharikota
20 May 1992	SROSS C	ASLV	Sriharikota
4 May 1992	SROSS C2	ASLV	Sriharikota
15 Oct. 1994	IRS P2	PSLV	Sriharikota
21 Mar. 1996	IRS P3	PSLV	Sriharikota
28 Sep. 1997	IRS 1D	PSLV	Sriharikota
25 May 1999	IRS P4 Oceansat	PSLV	Sriharikota
	KITSAT (Rep Korea)		
	TUBSAT (Germany)		
18 Apr. 2001	GSAT-1	GSLV	Sriharikota
22 Oct. 2001	TES	PSLV	Sriharikota
	PROBA (Belgium)		
	Bird (Germany)		
12 Sep. 2002	Metsat 1 *Kalpana*	PSLV	Sriharikota
8 May 2003	GSAT-2	GSLV	Sriharikota
17 Oct. 2003	Resourcesat 1	PSLV	Sriharikota
25 Sep. 2004	Edusat	GSLV	Sriharikota
5 May 2005	Cartosat 1	PSLV	Sriharikota
	Hamsat		
10 Jan. 2007	SRE-1	PSLV	Sriharikota
	Lapan (Indonenisa)		
	Pehvensat (Argentina)		
23 Apr. 2007	AGILE (Italy)	PSLV	Sriharikota
2 Sep. 2007	INSAT 4CR	GSLV	Sriharikota
21 Jan. 2008	Polaris (Israel)	PSLV	Sirharikota
28 May 2008	Cartosat 2	PSLV	Sriharikota
	IMS 1		
	AAUSAT II (Denmark)		
	Cute 1 (Japan)		
	Can X-2 (Canada)		
	Can X-6 (Canada)		
	Compass 1 (Germany)		
	Delfi C3 (Netherlands)		
	SEEDS 2 (Japan)		
	Rubin 8 (Germany)		
22 Oct. 2008	Chandrayan	PSLV	Sriharikota
20 Apr. 2009	Risat 2	PSLV	Sriharikota

India: failures

10 Aug. 1979	*Rohini*	SLV	Sriharikota
24 Mar. 1987	SROSS A	ASLV	Sriharikota
13 Jul. 1988	SROSS B	ASLV	Sriharikota
20 Sep. 1993		PSLV	Sriharikota
10 Jul. 2006	Insat 4C	GSLV	Sriharikota

India's launches by other countries

19 Apr. 1975	*Aryabhata*	Cosmos 3M	Kapustin Yar
7 Jun. 1979	*Bhashkara 1*	Cosmos 3M	Kapustin Yar
19 Jun. 1981	APPLE	Ariane 1	Kourou
20 Nov. 1981	*Bhashkara 2*	Cosmos 3M	Kapustin Yar
10 Apr. 1982	INSAT 1A	Delta	Cape Canaveral
30 Aug. 1983	INSAT 1B	Delta	Cape Canaveral
17 Mar. 1988	IRS 1A	Vostok	Baikonour
21 Jul. 1988	INSAT 1C	Ariane	Kourou
12 Jun. 1990	INSAT 1D	Delta	Cape Canaveral
29 Aug. 1991	IRS 1B	Vostok	Baikonour
10 Jul. 1992	INSAT 2A	Ariane	Kourou
23 Jul. 1993	INSAT 2B	Ariane	Kourou
7 Dec. 1995	INSAT 2C	Ariane	Kourou
28 Dec. 1995	IRS 1C	Molniya M	Baikonour
4 Jun. 1997	INSAT 2D	Ariane	Kourou
2 Apr. 1999	INSAT 2E	Ariane 42P	Kourou
21 Mar. 2000	INSAT 3B	Ariane 5	Kourou
10 Apr. 2003	INSAT 3A	Ariane 5	Kourou
28 Sep. 2003	INSAT 3E	Ariane 5	Kourou
21 Dec. 2005	INSAT 4A	Ariane 5	Kourou
11 Mar. 2007	INSAT 4B	Ariane 5	Kourou
20 Dec. 2008	W2M	Ariane 5	Kourou

Brazil

<table>
<tr><th>Launch Date|UTC</th><th>COSPAR Designator</th><th>US Cat Log or Flight Number</th><th>Name or Mission</th><th>Launch Vehicle</th><th>Launch Site</th><th>Information</th></tr>
<tr><td>19650401|</td><td>–</td><td>1</td><td>Test</td><td>SONDA-I</td><td>CLBI</td><td>GETEPE Sounding operations</td></tr>
<tr><td colspan="7">SONDA-I sounding rocket first launch. Decommissioned in 1966. Two stage vehicle. Development: IAE. Launches: 9.
First Launch Date: 19650401. Last Launch Date: 19661112.</td></tr>
<tr><td>19661105|13.55</td><td>–</td><td>4C</td><td>Ionosphere</td><td>SONDA-ID</td><td>CBR</td><td>GETEPE Sounding operations</td></tr>
<tr><td colspan="7">Altitude: 65 km.</td></tr>
<tr><td>19661112|13.38</td><td>–</td><td>3C</td><td>Eclipse</td><td>SONDA-ID</td><td>CBR</td><td>GETEPE</td></tr>
<tr><td colspan="7">Altitude: 65 km. November 12, 1966, hundreds of scientist, researchers and technicians from many different countries, participated in cooperative programs for atmospheric research during the total eclipse of the sun. This day, seven SONDA-1D rockets were launched to support the Brazilian eclipse mission.</td></tr>
<tr><td>19661112|14.08</td><td>–</td><td>10C</td><td>Eclipse</td><td>SONDA-ID</td><td>CBR</td><td>GETEPE</td></tr>
<tr><td colspan="7">Altitude: 65 km. [see 1966112|13.38]</td></tr>
<tr><td>19661112|14.50</td><td>–</td><td>15C</td><td>Eclipse</td><td>SONDA-ID</td><td>CBR</td><td>GETEPE</td></tr>
<tr><td colspan="7">Altitude: 65 km. [see 1966112|13.38]</td></tr>
<tr><td>19661112|14.55</td><td>–</td><td>4</td><td>Eclipse</td><td>SONDA-ID</td><td>CBR</td><td>GETEPE</td></tr>
<tr><td colspan="7">Altitude: 65 km. [see 1966112|13.38]</td></tr>
<tr><td>19661112|15.08</td><td>–</td><td>11</td><td>Eclipse</td><td>SONDA-ID</td><td>CBR</td><td>GETEPE</td></tr>
<tr><td colspan="7">Altitude: 65 km. [see 1966112|13.38]</td></tr>
<tr><td>19661112|15.22</td><td>–</td><td>13</td><td>Eclipse</td><td>SONDA-ID</td><td>CBR</td><td>GETEPE</td></tr>
<tr><td colspan="7">Altitude: 65 km. [see 1966112|13.38]</td></tr>
<tr><td>19661112|15.35</td><td>–</td><td>16C</td><td>Eclipse</td><td>SONDA-ID</td><td>CBR</td><td>GETEPE</td></tr>
<tr><td colspan="7">Altitude: 65 km. [see 1966112|13.38]</td></tr>
</table>

Launch Date|UTC	COSPAR Designator	US Cat Log or Flight Number	**Name or Mission**	Launch Vehicle	Launch Site	Information
19760226|	–	XV01	**Serido test**	SONDA-III	CLBI	IAE Sounding operations
SONDA-III sounding rocket first launch. The program is active. Two stage vehicle. Development: IAE. Launches: 28 [20020512]. First Launch Date: 19760226. Last Launch Date: TBD. Altitude: 47 km. This mission is considered a failure.						
19760921|	–	XV02	**Test**	SONDA-III	CLBI	IAE Sounding operations
Sounding rocket launch. Two stage vehicle. Development: IAE. Altitude: 543 km.						
19770331|	–	XV03	**Ionosphere**	SONDA-III	CLBI	IAE Sounding operations
Sounding rocket launch. Two stage vehicle. Development: IAE. Altitude: 546 km.						
19771019|	–	XV04	**Ionosphere**	SONDA-III	CLBI	IAE Sounding operations
Sounding rocket launch. Two stage vehicle. Development: IAE. Altitude: 590 km.						
19780426|	–	XV05	**Ionosphere**	SONDA-III	CLBI	IAE Sounding operations
Sounding rocket launch. Two stage vehicle. Development: IAE. Altitude: 615 km.						
19780927|	–	XV06	**Ionosphere**	SONDA-III M1	CLBI	IAE Sounding operations
Sounding rocket launch. Two stage vehicle. Development: IAE. Altitude: 230 km.						
19790823|	–	XV07	**Ionosphere**	SONDA-III M2	CLBI	IAE Sounding operations
Sounding rocket launch. Two stage vehicle. Development: IAE. Altitude: 248 km.						
19800827|	–	XV08	**Ionosphere**	SONDA-III M1	CLBI	IAE Sounding operations
Sounding rocket launch. Two stage vehicle. Development: IAE. Altitude: 245 km.						
19801222|	–	XV09	**Ionosphere**	SONDA-III	CLBI	IAE Sounding operations
Sounding rocket launch. Two stage vehicle. Development: IAE. Altitude: 599 km.						
19811210|	–	XV10	**Ionosphere**	SONDA-III M1	CLBI	IAE Sounding operations
Sounding rocket launch. Two stage vehicle. Development: IAE. Altitude: 239 km.						

19820630| – XV11 **BIME test** SONDA-III CLBI IAE Sounding operations
Sounding rocket launch. Two stage vehicle. Development: IAE. Altitude: 549 km.

19820908|21.52 – XV12 **BIME 1 Ionosphere** SONDA-III CLBI IAE Sounding operations
Sounding rocket launch. Two stage vehicle. Development: IAE. Altitude: 525 km.

19820913|21.05 – XV13 **BIME 2 Ionosphere** SONDA-III CLBI IAE Sounding operations
Sounding rocket launch. Two stage vehicle. Development: IAE. Altitude: 506 km.

19820917|20.56 – XV14 **Colored Bubbles Ionosphere** SONDA-III CLBI IAE/DLR/US Sounding operations
Sounding rocket launch. Two stage vehicle. Development: IAE. Altitude: 335 km.

19820918|20.45 – XV15 **Colored Bubbles Ionosphere** SONDA-III CLBI IAE/DLR/US Sounding operations
Sounding rocket launch. Two stage vehicle. Development: IAE. Altitude: 335 km.

19830601| – XV16 **Manicore Ionosphere** SONDA-III CLBI IAE Sounding operations
Sounding rocket launch. Two stage vehicle. Development: IAE. Altitude: 599 km.

19831031|22.35 Failure XV17 **Ionosphere** SONDA-III Wallops Island IAE/USAF Sounding operations
Sounding rocket launch. Two stage vehicle. Development: IAE. Altitude: 81 km.

19831114|10.20 Failure XV18 **Ionosphere** SONDA-III Wallops Island IAE/USAF Sounding operations
Sounding rocket launch. Two stage vehicle. Development: IAE. Altitude: 55 km.

19840726|18.05 – XV19 **Ionosphere** SONDA-III CLBI IAE Sounding operations
Sounding rocket launch. Two stage vehicle. Development: IAE. Altitude: 566 km.

19841121| – - **Test** SONDA-IV CLBI IAE Sounding operations
SONDA-IV sounding rocket first launch. Retired 1989. Two stage vehicle. Development: IAE. Launches: 4. First Launch Date: 19841121. Last Launch Date: 19890328.
Altitude: 616 km.

Launch Date|UTC	COSPAR Designator	US Cat Log or Flight Number	**Name or Mission**	Launch Vehicle	Launch Site	Information
19850208|23.22	1985-015B	15561	**Brasilsat A1/ SBTS A1**	Ariane 3	Kourou SLC	Communications
Commercial geostationary (65°W) communications satellite (670 kg) for Telebras based on Hughes HS-376 space bus, with 24 C-band transponders providing telecommunications for Brazil. The satellite was built by Spar Aerospace (Canada) under license from Hughes Space and Communications. Brasilsat A1 and A2 were the initial segments of the national Brazilian network Sistema Barasilero de Telecommunicacoes por Satelite (SBTS).						
19851119|13.00	–	-	**HARP test**	SONDA-IV	CLBI	IAE Sounding operations
Sounding rocket launch. Two stage vehicle. Development: IAE. Altitude: 667 km.						
19851201|	Failure	XV01	**Test**	VLS-R1	CLBI	IAE
VLS-R1 all solid propellant test vehicle. Development: IAE. Launches: 2. First Launch Date: 19851201. Last Launch Date: 19890518. Altitude: 10 km.						
19851211|23.30	–	XV20	**Aeronomy**	SONDA-III	CLBI	IAE Sounding operations
Sounding rocket launch. Two stage vehicle. Development: IAE. Altitude: 516 km.						
19860328|23.30	1986-026B	16650	**Brasilsat A2/ SBTS A2**	Ariane 3	Kourou SLC	Communications
Commercial geostationary (105°W) communications satellite (670 kg) for Telebras based on Hughes HS-376 space bus, with 24 C-band transponders. The satellite was built by Spar Aerospace (Canada) under license from Hughes Space and Communications. Brasilsat A1 and A2 were the initial segments of the national Brazilian network Sistema Barasilero de Telecommunicacoes por Satelite (SBTS). Replaced by **Brasilsat B4** (2000-046A).						
19861101|01.59	–	XV21	**Aeronomy**	SONDA-III	CLBI	IAE Sounding operations
Sounding rocket launch. Two stage vehicle. Development: IAE. Altitude: 444 km.						
19871108|	–	-	**Test**	SONDA-IV	CLBI	IAE Operations
Sounding rocket launch. Two stage vehicle. Development: IAE. Altitude: 130 km.						

19890328|	–	-	**Test**	SONDA-IV	CLBI	IAE Operations
Sounding rocket launch. Two stage vehicle. Development: IAE. Altitude: 821 km.						
19890518|	–	XV02	**Test**	VLS-R1	CLBI	IAE
VLS-R1 all solid propellant test vehicle. Development: IAE. Altitude: 50 km.						
19900122|01.35	1990-005E	20440	**Dove OSC-17**	Ariane 40	Kourou SLC	Communications
Radio amateur Communications microsatellite (12 kg), aka **Oscar 17**.						
19900221|	–	XV53	**Alcantara Ionosphere**	SONDA-II	CLA	IAE/INPE Sounding operations
SONDA-II sounding rocket first launch. Retired in 1996. Single stage vehicle. Development: IAE. Launches: 7. First Launch Date: 19900221. Last Launch Date: 19960828. Altitude: 100 km.						
19901126|	–	XV54	**Manival Ionosphere**	SONDA-II	CLA	IAE/INPE Sounding operations
Sounding rocket launch. Single stage vehicle. Development: IAE. Altitude: 91 km.						
19901130|	–	XV22	**Ionosphere**	SONDA-III	CLBI	IAE Sounding operations
Sounding rocket launch. Two stage vehicle. Development: IAE. Altitude: 500 km.						
19910429|	–	XV23	**Ionosphere**	SONDA-III	CLBI	IAE Sounding operations
Sounding rocket launch. Two stage vehicle. Development: IAE. Altitude: 440 km.						
19911209|21.05	–	XV55	**Aguas Belas Aeronomy**	SONDA-II	CLA	IAE/INPE Sounding operations
Sounding rocket launch. Single stage vehicle. Development: IAE. Altitude: 88 km.						
19920601|02.52	–	XV24	**Aeronomy**	SONDA-III	CLA	IAE Sounding operations
Sounding rocket launch. Two stage vehicle. Development: IAE. Altitude: 282 km.						
19921031|	–	XV56	**Ponta de Areia Ionosphere**	SONDA-II	CLA	IAE/INPE Sounding operations
Sounding rocket launch. Single stage vehicle. Development: IAE. Altitude: 32 km.						
19930209|14.30	1993-009B	22490	**SCD-1**	Pegasus	B-52 a/c	Data Collection/Relay
Satelite de Coleta de Dados (Data Collection Satellite). Environmental data relay satellite. The SCD-1 is a satellite (mass˜ 110 kg) designed for the collection of meteorological data relayed by data collection platforms scattered over Brazilian territory.						

Launch Date|UTC	COSPAR Designator	US Cat Log or Flight Number	**Name or Mission**	Launch Vehicle	Launch Site	Information
19930322|	–	XV57	**Maruda Ionosphere**	SONDA-II	CLA	IAE/INPE Sounding operations
Sounding rocket launch. Single stage vehicle. Development: IAE. Altitude: 102 km.						
19930402|10.34	–	XV01	**Operation Santa Maria/PT-01**	VS-40	CLA	INPE
VS-40 all solid propellant, two stage, test vehicle. Launches: 2. First Launch Date: 19930402. Last Launch Date: 19980321. Experimental vehicle composed of the last two stages of the VLS-1. Altitude: 1250 km. Also reportedly: 950 km.						
19931117|	–	XV25	**Ionosphere**	SONDA-III	CLBI	IAE Sounding operations
Sounding rocket launch. Two stage vehicle. Development: IAE. Altitude: 555 km.						
19940622|	–	XV58	**Oituia Ionosphere**	SONDA-II	CLBI	IAE/INPE Sounding operations
Sounding rocket launch. Single stage vehicle. Development: IAE. Altitude: 93 km.						
19940810|23.05	1994-049A	23199	**Brasilsat B1/ SBTS B1**	Ariane 44LP	Kourou SLC	Communications
Brazil Star One (SES Global) Commercial/Military geostationary (70°W) communications satellite (1765 kg) for Embratel (Tele Mex). HS 376W spacecraft with 28 C-band transponders and one military X-band channel. The satellite was built by Hughes Space and Communications, now Boeing Satellite Systems USA. Design lifetime 12 years. Since about September 2001 drifting westwards.						
19950328|23.14	1995-016A	23536	**Brasilsat B2/ SBTS B2**	Ariane 44L	Kourou SLC	Communications
Brazil Star One (SES Global) Commercial/Military geostationary (65.1°W) communications satellite (1780 kg) for Embratel (Telebras). HS 376W spacecraft with 24 C-band (telephone, television, and data transmission services to Brazil and its southern neighbors) and several X-band (military) transponders for domestic use. The satellite was built by Hughes Space and Communications, now Boeing Satellite Systems USA. Design lifetime 12 years. Since about September 2001 drifting westwards. Replaced by **Star One C1** (2007-056A).						
19951219|	–	XV26	**Ionosphere**	SONDA-III	CLA	IAE Sounding operations
Sounding rocket launch. Two stage vehicle. Development: IAE. Altitude: 406 km.						

19960828| – XV59 **Ionosphere** SONDA-II CLBI IAE/INPE Sounding operations

Sounding rocket launch. Single stage vehicle. Development: IAE. Altitude: 87 km.

19970428|16.42 – XV01 **Operation Santana** VS-30 CLA INPE/DLR

VS-30 suborbital, single stage, launch vehicle first launch. DLR AL-VS30-223 test. Development: IAE. Launches: 7 [20071216].First Launch Date: 19970428. Last Launch Date: TBD. Altitude: 128 km.

19971012|01.19 – XV02 **RONALD** VS-30 Andøya LC INPE/DLR

Suborbital, single stage vehicle. DLR AL-VS30-226 test. Development: IAE. Altitude:˜120 km. Rocketborne Optical Neutral Gas Analyzer with Laser Diodes (RONALD)

19971102|12.25 1997-Failure XV03 **SCD-2A/ Operation Brasil** VLS-1 CLA Data Collection/Relay. INPE

Satelite de Coleta de Dados (Data Collection Satellite). Environmental data relay. The SCD-2A, build by INPE, is a satellite (mass˜ 110 kg) designed for the collection, from 750 km polar orbit, of meteorological (environmental and agricultural) data relayed by data collection platforms scattered over Brazilian territory. The satellite was lost 65s (altitude 300 km) after launch.
VLS-1 all solid propellant orbital launch vehicle, first launch. Development: IAE. Launches: 2 [19991211]. First Launch Date: 19971102. Last Launch Date: TBD.

19980131|23.43 – XV03 **RONALD 2** VS-30 Andøya LC INPE/DLR

Suborbital, single stage vehicle. DLR AL-VS30-229 test. Development: IAE. Altitude:˜120 km. Rocketborne Optical Neutral Gas Analyzer with Laser Diodes (RONALD)

19980204|23.29 98-006A 25152 **Brasilsat B3/ SBTS B3** Ariane 44LP Kourou SLC Communications

Brazil Star One (SES Global) Commercial/Military geostationary (84°W) communications satellite (1780 kg) for Embratel (Telebras). HS 376W spacecraft with 24 C-band (telephone, television, and data transmission services to Brazil and its southern neighbors) and several X-band (military) transponders for domestic use. The satellite was built by Hughes Space and Communications, now Boeing Satellite Systems USA. Design lifetime 12 years. Since about September 2001 drifting westwards.

<table>
<tr><th>Launch Date|UTC</th><th>COSPAR Designator</th><th>US Cat Log or Flight Number</th><th>Name or Mission</th><th>Launch Vehicle</th><th>Launch Site</th><th>Information</th></tr>
<tr><td>19980321|</td><td>–</td><td>XV02</td><td>Operation Livramento/PT-02</td><td>VS-40</td><td>CLA</td><td>IAE/INPE</td></tr>
<tr><td colspan="7">VLS(-1) test in sounding configuration. Development: IAE. Reportedly reached some 900 km altitude.</td></tr>
<tr><td>19981023|00.02</td><td>1998-060A</td><td>25504</td><td>SCD-2</td><td>Pegasus</td><td>L-1011 a/c</td><td>Data Collection/Relay</td></tr>
<tr><td colspan="7">Satelite de Coleta de Dados (Data Collection Satellite). Environmental data relay. The SCD-2, build by INPE, is a satellite (mass˜ 115 kg) designed for the collection, from 750 km, 25° inclination orbit, of meteorological (environmental and agricultural) data relayed by data collection platforms scattered over Brazilian territory.</td></tr>
<tr><td>19990315|09.52</td><td>–</td><td>XV04</td><td>Operation San Marcos</td><td>VS-30</td><td>CLA</td><td>AEB</td></tr>
<tr><td colspan="7">Suborbital single stage vehicle. Development: IAE. Altitude: 128 km. Aka Operacao San Marcos ìgravity Mission.</td></tr>
<tr><td>19991014|03.16</td><td>1999-057A</td><td>25940</td><td>CBERS-1/ Zi Yuan-1 1</td><td>CZ-4B</td><td>Taiyuan SLC</td><td>Remote sensing</td></tr>
<tr><td colspan="7">Aka ZY-1; China Brazil Earth Resources Satellite; Design lifetime 2 years. In general, CBERS satellites orbit at about 780 km altitude (sun-synchronous). They perform about 14 revolutions a day and obtain a complete coverage of the Earth in 26 days (repeating ground track pattern). The 1450 kg satellite is box-shaped with a 985W capacity single solar wing consisting of three panels. The Earth observation payload includes three primary sensors: a five band CCD Camera, a four band IR Multi-Spectral Scanner and a two band Wide-Field Imager. CBERS-1 also carries a Data Collection System and a Space Environment Monitor.</td></tr>
<tr><td>19991014|03.16</td><td>1999-057B</td><td>25941</td><td>SACI-1</td><td>CZ-4B</td><td>Taiyuan SLC</td><td>Science/Technology</td></tr>
<tr><td colspan="7">Satélite Avançado de Comunicações Interdisciplinares (Advanced Interdisciplinary Communications Satellite) from INPE, piggybacked on 1999-057A. Experimental scientific/technologic microsatellite (60 kg) with a magnetometer, particle detectors and an atmospheric experiment. INPE lost contact with the satellite almost immediately. [733*744 km|98.6°].</td></tr>
<tr><td>19991211|18.25</td><td>1999-Failure</td><td>XV04</td><td>SACI-2</td><td>VLS-1</td><td>CLA</td><td>Science/Technology. INPE</td></tr>
<tr><td colspan="7">Satélite Avançado de Comunicações Interdisciplinares (Advanced Interdisciplinary Communications Satellite). Second VLS-1 attempt to launch a microsatellite (85 kg). The VLS rocket flew for about 60 s, burning the first stage boosters. Ignition of the second stage at approximately 32 km altitude led to a high disturbance of the vehicle so that it was destroyed by safety officers. Brazil cancelled the SACI satellite program after this launch failure.</td></tr>
</table>

20000206|16.37 – XV05 **Lençóis Maranhenses ìgravity** VS-30 CLA AEB

Suborbital single stage vehicle. Development: IAE. Altitude: 148 km. Aka Lençóis Maranhenses ìgravity Mission.

20000817|23.16 2000-046A 26469 **Brasilsat B4/ SBTS B4** Ariane 44LP Kourou SLC Communications

Brazil Star One (SES Global) commercial HS 376W type geostationary (75°W) communications satellite (1760 kg) for Embratel with 28 C-band transponders for voice and video communications for the entire South American continent. Design lifetime 12 years. The satellite was built by Hughes Space and Communications, now Boeing Satellite Systems USA. Replaces Brasilsat A2 (1986-026B). Since about September 2001 drifting westwards.

20000821|20.14 – XV01 **Baronesa ìgravity** VS-30/Orion CLA AEB/DLR

VS-30/Orion suborbital launch vehicle first launch. Two stage vehicle. Development: IAE. Launches: 4 [20080131]. First Launch Date: 20000821.

Last Launch Date: TBD. Altitude: 330 km. Payload: 160 kg. 420 s of ìgravity operations.

20001209|12.20 – XV30 **Alecrim ìgravity** SONDA-III CLBI IAE/CTA

Two stage vehicle. Development: IAE. Altitude: 430 km. PSO (ìgravity operations) platform developed by INPE.

CANCELLED – - **SSR** – - Remote sensing

Two small earth observation satellites, SSR-1 and SSR-2, to provide synoptic images of the entire Amazon region, several times a day. The SSR will be a 400 kg satellite using the Brazilian multi mission platform for small satellites, which is being developed under INPE's coordination and manufactured by the Brazilian industry. The SSR imaging system consists of a VIS/NIR sensor and a MIR sensor. The satellites are projected to operate in equatorial circular orbit at an altitude of 900 km and were scheduled for launch in the years 2000 and 2003. *This program has been cancelled. Missions have been taken over by other PMM programs such as Amazônia.*

20020512|04.47 – XV31 **Parnamirim F2 Ionosphere** SONDA-III CLBI AEB Sounding operations

Sounding rocket launch. Two stage vehicle. Development: IAE. Altitude: 593 km.

20021123|22.18 – XV02 **Pirapema Ionosphere** VS-30/Orion CLA AEB

Suborbital, two stage, launch vehicle. Development: IAE. Altitude: 434 km.

20021201|12.33 Failure XV06 **Cumã ìgravity** VS-30 CLA AEB

Suborbital single stage vehicle. Development: IAE. Altitude: 145 km.

Launch Date|UTC	COSPAR Designator	US Cat Log or Flight Number	**Name or Mission**	Launch Vehicle	Launch Site	Information
2002 +	–	-	–	VLM	TBD	AEB
VLM Launch Vehicle (Veiculo Lancador de Microsatelites). Satellite launcher using all solid propellant core of VLS only. Planned for launch of microsatellites. First Launch Date: TBD (2002 or later). Development ended 1997.						
20030822|16.30	2003-Failure-01	–	**SATEC**	VLS-1	CLA	Science/INPE
VLS third qualification flight. Launch vehicle exploded on the launch pad during the launch campaign for a launch on August 25, 2003. The launch had already been delayed from October 2002, May 7 and June 20, 2003. A strap-on booster ignited prematurely and destroyed the rocket and the launch pad. 21 people were killed, none injured. The two satellites and the launch pad were destroyed. SATEC's mass was 57 kg.						
20030822|16.30	2003-Failure-02	–	**UNOSAT**	VLS-1	CLA	
Microsatellite (9 kg) from Universidade Norte de Parana. Launch vehicle exploded on the launch pad during the launch campaign. [see 20030822|16.30|SATEC]						
20031021|03.16	2003-049A	28057	**CBERS-2A/ Zi Yuan-1 2A**	CZ-4B	Taiyuan SLC	Remote sensing
Aka **ZY-1.** Electro-optical photoreconnaissance satellite. Design lifetime > 2 years; Planned for 18 months Chinese control, then Brazilian control. CBERS satellites orbit at about 780 km altitude (sun-synchronous). They perform about 14 revolutions a day and obtain a complete coverage of the Earth in 26 days (repeating ground track pattern). The 1450 kg satellite is box-shaped with a 985W capacity single solar wing consisting of three panels. The Earth observation payload includes three primary sensors: a five band CCD Camera, a four band IR Multi-Spectral Scanner and a two band Wide-Field Imager.						
20040111|04.13	2004-001A	28137	**Estrela do Sul**	Zenit 3SL	Odyssey	Communications
Aka **Telstar-14** and **Skynet Brazil**. Commercial FS 1300 type geostationary (63.0°W) communications satellite (4700 kg) for Loral Skynet do Brasil. Equipped with 41 Ku-band transponders for video and Internet services to Brazil and North America. Originally planned for launch on an American Delta SLV. Reportedly, it encountered problems in power management (solar panels deployment). Since about May 2007 drifting westwards.						

20040804|22.32 2004-031A 28393 **Amazonas** Proton M Baikonur SLC Communications

Commercial Eurostar E3000 type (EADS Astrium) geostationary (61.0°W) communications satellite (4,545 kg) for Hispamar, Brazil. Equipped with 36 Ku-band and 27 C-band transponders for broadcasting and Internet services to South and North America and Europe. Expected life time: 15 years. [35,775*35,798|0.01°]

20041023|16.51 – XV01 **Cajuana Test** VSB-30 CLA AEB

VSB-30 suborbital launch vehicle first launch. Unguided test launch. Two stage vehicle. Development: IAE. Launches 4 [20070719] First Launch Date: 20041023. Last Launch Date: TBD. Payload, reportedly some 400 kg, was carried to an altitude of about 260 km. 7 minutes of ìgravity.

20051027|13.45 – - **SHEFEX Re-entry test** VS-30/Orion Andøya LC AEB/DLR

Suborbital, two stage, launch vehicle. Development: IAE. Altitude: 212 km.

20051201|09.04 – - **DLR TEXUS 42** VSB-30 Esrange SC AEB/DLR

Suborbital, two stage, launch vehicle. Development: IAE. Altitude: 263 km. EML-1 mission (ìgravity operations).

20060510|08.12 – - **DLR TEXUS 43** VSB-30 Esrange SC AEB/DLR

Suborbital, two stage, launch vehicle. Development: IAE. Altitude: 237 km. 6 minutes of ìgravity operations with a 407 kg payload.

20070719|15.13 – XV02 **Cumã II ìgravity** VSB-30 CLA AEB

Suborbital, two stage, launch vehicle. Development: IAE. Altitude:280 km. 7 minutes of ìgravity.

20070919|03.26 2007-042A 32062 **CBERS-2B/ Zi Yuan-1 2** CZ-4B Taiyuan SLC Remote sensing

CBERS satellites orbit at about 780 km altitude (sun-synchronous). They perform about 14 revolutions a day and obtain a complete coverage of the Earth in 26 days (repeating ground track pattern). The 1450 kg satellite is box-shaped with a 985W capacity single solar wing consisting of three panels. The Earth observation payload includes three primary sensors: a five band CCD Camera, a four band IR Multi-Spectral Scanner and a two band Wide-Field Imager.

Launch Date|UTC	COSPAR Designator	US Cat Log or Flight Number	**Name or Mission**	Launch Vehicle	Launch Site	Information
20071114|22.06	2007-056A	32293	**Star One C1**	Ariane 5ECA	Kourou SLC	Communications
Aka **Brasilsat C1 / SBTS C1**. Commercial Spacebus 3000B3 (Thales Alenia Space) type geostationary (61.0°W) communications satellite (4,100 kg) for a joint venture between several South American countries, under Embratel leadership, named Star One, with 28 C-band, 16 Ku-band and 1 X-band transponders for voice and video communications for the entire South American continent. Estimated life tima: 15 years. Replaces Brasilsat B2. [35,779*35,795|0.04°]						
20071216|09.15	–	XV07	**Angicos GPS test**	VS-30	CLBI	AEB
Suborbital single stage vehicle. Development: IAE. Altitude: 120 km.						
20080131|13.00	–	-	**HotPay-2**	VS-30/Orion	Andøya LC	International/ARR
Suborbital, two stage, launch vehicle. Development: IAE. Altitude: 380 km. On board were 9 instruments from 8 countries. (ARR = Andøya Rocket Range Payload Services)						
20080207|11.30	–	-	**DLR TEXUS 44**	VSB-30	Esrange SC	AEB/DLR/SSC
Suborbital, two stage, launch vehicle. Development: IAE. Altitude: 264 km. EML-2 mission (igravity operations).						
20080221|06.15	–	-	**DLR TEXUS 45**	VSB-30	Esrange SC	AEB/DLR/SSC
Suborbital, two stage, launch vehicle. Development: IAE. Altitude: 273 km. (igravity operations)						
20080418|22.17	2008-018B	32768	**Star One C2**	Ariane 5ECA	Kourou SLC	Communications
Aka **Brasilsat C2 / SBTS C2**. Commercial Spacebus 3000B3 (Thales Alenia Space) type geostationary (70.0°W) communications satellite (4,100 kg) for a joint venture between several South American countries, under Embratel leadership, named Star One, with 28 C-band, 16 Ku-band and 1 X-band transponders for voice and video communications for the entire South American continent. Estimated life span: 15 years. [35,777*35,796|0.02°]						

20081205|10.35	–	-	**ICI-2**	VSB-30/Orion	SvalRak Norway	International/ARR
Suborbital, two stage, launch vehicle. Development: IAE. Altitude: 330 km. Auroral research. 10 minutes flight time. Research successful.						
2009	–	-	**SGB-1**	–	-	Communications
Brazilian Geostationary Satellite (CTA/Air Force) for national civil and military air traffic services. Communications, Navigation & Surveillance (CSN) Air Traffic Management (ATM).						
2009	–	-	**SGB-2**	–	-	Communications
Brazilian Geostationary Satellite (CTA/Air Force) for national civil and military air traffic services. Communications, Navigation & Surveillance (CSN) Air Traffic Management (ATM).						
2009	–	-	**SGB-3**	–	-	Communications
Brazilian Geostationary Satellite (CTA/Air Force) for national civil and military air traffic services. Communications, Navigation & Surveillance (CSN) Air Traffic Management (ATM).						
2010	–	-	**SHEFEX 2**	VS-40	CLA	AEB/DLR
VS-40 all solid propellant, two stage, test vehicle. Objective is to test parts and components of the VLS-1 third stage. Aka: EN-03						
2010	–	-	**SARA Suborbital 1**	VS-40	CLA	AEB
VS-40 all solid propellant, two stage, test vehicle. SARA (Satélite de Reentrada Atmosférica) suborbital mission to flight-test the SARA recovery system.						
2010	–	-	**CBERS-3**	–	-	Remote sensing
Electro-optical photoreconnaissance satellite.						
2011	–	-	**Amazônia-1**	–	-	Remote sensing
INPE/Multi Mission Platform (PMM) will be used to carry a UK made high resolution camera and two Brazilian cameras to monitor deforestation and urban expansion from a polar, 780 km, circular orbit.						
2013	–	-	**Lattes-1**	–	-	Science
INPE/Multi Mission Platform (PMM) will be used for space weather monitoring and to carry an X-ray astrophysics observatory. (EQUARS multi experiment).						

Launch Date|UTC	COSPAR Designator	US Cat Log or Flight Number	**Name or Mission**	Launch Vehicle	Launch Site	Information
2013	–	-	**CBERS-4**	–	-	Remote sensing
Electro-optical photoreconnaissance satellite.						
2014	–	-	**GPM-Br**	–	-	AEB/EMBRAPA
Global Precipitation Mission. Satellite for precipitation measurement in tropical orbit using passive microwave imager.						
2014	–	-	**SABIA-1**	–	-	Science
[Under consideration]						
2015	–	-	**MAPSAR-1**	–	-	Remote sensing
DLR. L-band SAR. 3-20 m resolution. [Under consideration]						
2015	–	-	**CBSAR**	–	-	Remote sensing
Chinese-Brasil SAR satellite. [Under consideration]						
2016	–	-	**Amazônia-2**	–	-	Remote sensing
[Under consideration]						
2016	–	-	**CBERS-5**	–	-	Remote sensing
Electro-optical photoreconnaissance satellite.						
2016	–	-	**GEO Met Br**	–	-	Meteorology
[Under consideration]						
2017	–	-	**Lattes-2**	–	-	Science
[Under consideration]						
2018	–	-	**SABIA-2**	–	-	Science
[Under consideration]						
2020	–	-	**CBERS-6**	–	-	Remote sensing
Electro-optical photoreconnaissance satellite.						

Israel

Date	Launcher/site (orbit)	Payload (mass)	Results of the flight (mid-2009)
19 Sep. 1988	Shavit/Palmachim (250–1,580 km)	Ofeq-1 (156 kg)	First satellite launch achieved with success and demonstration of Israeli technology in launch and satellite operations
3 Apr. 1990	Shavit/Palmachim (150–251 km)	Ofeq-2 (160 kg)	Second demonstration satellite to test communications and sensors
28 Mar. 1995	*Start-1/Plesetsk (NA)*	*Gurwin-1/ Techsat-1 (50 kg)*	*Launch failure of the technology satellite made by Technion, carrying CCD cameras, radiation and ozone detectors, amateur radio transmitters*
5 Apr. 1995	Shavit-1/Palmachim (368–729 km)	Ofeq-3 (189 kg)	Imaging surveillance satellite demonstrating great agility and providing visible and ultraviolet images of about 1-m resolution
16 May 1996	*Ariane 44L/Kourou (geosynchronous, at 4° W)*	*Amos-1 (996 kg)*	*Commercial Ku-band satellite for communications and broadcasts, owned by IAI and operated by Spacecom. Still op erational in inclined orbit. Sold in August 2009 by IAI to Intelsat to become Intelsat-24 over the Indian Ocean*
22 Jan. 1998	Shavit-1/Palmachim (NA)	Ofeq-4 (190 kg?)	Planned replacement of Ofeq-3, but launch failure due to the second stage
10 Jul. 1998	*Zenit-2/Baykonur (817–818 km)*	*Gurwin-2/Techsat-1B (50 kg)*	*Technology satellite made by Technion, carrying CCD cameras, radiation and ozone detectors, amateur radio transmitters. Still functioning in orbit*
5 Dec. 2000	*Start-1/Svobodny (488–503 km)*	*Eros-A1 (240 kg)*	*Commercial remote sensing satellite for Imagesat International, based on Optsat-2000 bus, with CCD camera to deliver panchromatic images of 1.8-m resolution*
28 May 2002	Shavit-1/Palmachim (370–757 km)	Ofeq-5 (300 kg)	Military observation satellite, very agile, to take images of less than 1-m resolution
27 Dec. 2003	*Soyuz-Fregat/ Baykonur (geo-synchronous at 4° W)*	*Amos-2 (1.370 kg)*	*Commercial Ku-band satellite (improved version) for communications and broadcasts, operated by Spacecom*

Date	Launcher/site (orbit)	Payload (mass)	Results of the flight (mid-2009)
6 Sep. 2004	Shavit-1/Palmachim (NA)	Ofeq 6 (300 kg)	Launch failure due to the third stage and loss of military observation satellite
25 Apr. 2006	*Start-1/Svobodny (505–514 km)*	*Eros-B (350 kg)*	*Commercial remote sensing satellite for Imagesat International, with improved camera to provide panchromatic images of 0.7-m resolution*
10 Jun. 2007	Shavit-2/Palmachim (340–576 km)	Ofeq-7 (300 kg)	Optical reconnaissance satellite for very high-resolution imagery
21 Jan. 2008	*PSLV-CA/Sriharikota (405–580 km)*	*TecSAR/Polaris (260 kg)*	*Military radar satellite for all-weather, day/night observations of high resolution*
28 Apr. 2008	*Zenit 3SLB/Baykonur (geosynchronous at 4°W)*	*Amos-3 (1.250 kg)*	*Commercial Ku-band and Ka-band satellite with improved payload for communications and broadcasts, operated by Spacecom*

In italics: launches from outside Israel.

South Korean satellites

Date (planned)	Launcher/site (orbit)	Payload (mass)	Results of the flight (mid-2009)
10 Aug. 1992	*Ariane 4/Kourou (1.301–1.402 km)*	*KITsat-1/Uribyol-1 (50 kg)*	*First step of South Korea in space, with engineering micro-satellite developed by British SSTL(Surrey Satellite Technology Ltd) for SaTReC*
26 Sep. 1993	Ariane 4/Kourou (795–805 km)	KITsat-2/Uribyol-2 (45 kg)	Engineering microsatellite for Earth observations and space environment measurements, developed by SaTReC with the assistance of SSTL
5 Aug. 1995	*Delta 7925/Cape Canaveral (geo-synchronous at 116°E)*	*Koreasat-1 /Mugungwha-1 (1.459 kg)*	*First communications and broadcasting satellite (Ku-band) of Lockheed Martin for Korea Telecom (KT). Lifetime limited because of launch problem. Sold in inclined orbit to Europe*Star to start operations in Central Europe under the name of EuropeStarB*

Date (planned)	Launcher/site (orbit)	Payload (mass)	Results of the flight (mid-2009)
14 Jan. 1996	*Delta/Cape Canaveral*	*Koreasat-2 / Mugungwha-2 (1.464 kg)*	*Second satellite, similar to Koreasat-1, of KT Koreasat system, also built by Lockheed Martin. Used in inclined orbit and sold in July 2009 to Asia Broadcast Satellite, Hong Kong, to become ABS-1A*
26 May 1999	*PSLV/Sriharikota (circular at 720 km)*	*KITsat-3 (110 kg)*	*Third microsatellite of South Korea, fully built by SaTReC as test satellite for the STsat bus, used for Earth observations with 13.5-m resolution*
4 Sep. 1999	*Ariane 42/Kourou (geosynchronous at 116°E)*	*Koreasat-/ Mugungwha-3 (2.790 kg)*	*Third comsat (Ku-band and Ka-band) by Lockheed-Martin with the participation of South Korean industries*
21 Dec. 1999	*Taurus/Vandenberg AFB (694–729 km)*	*KOMPsat-1 /Arirang-1 (469 kg)*	*First KARI satellite for multi-purpose mission, mainly for cartography and ocean remote sensing observations, jointly developed by KARI and TRW*
27 Sep. 2003	*Cosmos-3M/Plesetsk*	*STsat-1/ KAISTsat-4 (106 kg)*	*In-orbit demonstration of the three-axis stabilized bus for commercial purposes. Science and technology payload consisting of experiments in astronomy and astrophysics, developed and operated by KAIST/SaTReC Initiative and KARI*
26 Jul. 2006	*Dnepr/Baïkonur (launch failure)*	*HAUsat-1 (1 kg)*	*First technological cubesat/ pico-satellite made by professors and students of Hankuk Aviation University (SSRL/Space System Research Laboratory), lost in the failure of Dnepr 1st stage*
28 Jul. 2006	*Rokot/Plesetsk (circular at 685 km)*	*KOMPsat-2/ Arirang-2 (800 kg)*	*Earth observation satellite jointly developed by KARI and EADS Astrium for high-resolution images with panchromatic (1-m resolution) and multispectral (4-m resolution) camera (provided by Israeli El-Op)*

Date (planned)	Launcher/site (orbit)	Payload (mass)	Results of the flight (mid-2009)
22 Aug. 2006	*Zenit 3SL/Pacific Ocean (geo-synchronous at 116°E)*	*Koreasat-5/ Mugungwha-5 (4.465 kg)*	*Heavy communications and broadcasting satellite (C-band, Ku-band, SHF band), purchased from Alcatel Alenia Space, used by KT Corporation and Korean Agency for Defense Development (ADD)*
25 Aug. 2009 (failure)	*KSLV-1/Naro-1/ Naro space center, Gohung*	*STsat-2 (100 kg)*	*First microsatellite to be launched by South Korean rocket. Payload consisting of a telescope for solar observations. Orbit not achieved because of partial fairing separation*
[Early 2010]	*Ariane 5/Kourou (geosynchronous at 138°E)*	*COMS-1 (2.4 kg)*	*Communications, Ocean & Meteorological Satellite with meteo and ocean imagers, purchased from EADS Astrium and developed with KARI participation for the Ka-band communications payload*
[2010]	*Ariane 5 or Soyuz-2-Fregat/Kourou (geosynchronous at 116°E)*	*Koreasat-6 (2.622 kg)*	*Medium-class communications and broadcasting satellite (Ku-band), built by Thales Alenia Space with the STAR-2 bus of Orbital Sciences*
[2010]	*Dnepr/Baikour (500–600 km)*	*KOMPsat-5 (1.280 kg)*	*First radar satellite of South Korea, with X-band SAR for 3-m resolution observations, purchased from Thales Alenia Space (Italy), based on Cosmo-SkyMed/Sentinel-1 technology, built with the participation of KARI and Korean industries*
[2011]	*H2A/Tanegashima (circular at 685 km)*	*KOMPsat-3 (900 kg)*	*High-resolution (submetric resolution) and multispectral (3.2-m resolution) remote sensing satellite, jointly built by Astrium and KARI to continue the KOMPsat-2 mission of Earth observations*
[2011]	*H2A/Tanegashima? (650–800 km)*	*HAUsat-2 (24 kg)*	*Compact three-axis stabilized nano/microsatellite developed by Hankuk Aviation University (SSRL/Space System Research Laboratory) for educational purposes, to track animals and to study the space environment*

In italics: built outside South Korea.

ANNEXE 2: SPACE INSTITUTES IN IRAN

Space education

Iran has a large network of private, public and state-affiliated universities offering degrees in higher education. State-run (technical) universities of Iran are under the direct supervision of Iran's Ministry of Science, Research and Technology. Below are the Iranian universities that are related to space education.

Sharif University of Technology (SUT)
Department of Aerospace Engineering
Address: Azadi Ave, Tehran, Iran
PO Box: 11365-8639, Tehran, Iran
Tel: +98(21)66164911
Fax: +98(21)66022715 & +98(21)66022731
URL: http://www.sharif.ir/en
E-mail: info@sharif.ir

Formerly Aryamehr University of Technology. The Aerospace Engineering Department at the Sharif University of Technology is located in Tehran, Iran. It was established in 1987 and offers undergraduate, bachelor and graduate master programs and recently Ph.D. programs in the field of aerospace engineering and aeronautical engineering.

AmirKabir University of Technology (AUT)
Aerospace Engineering Department
Address: 424 Hafez Ave, Tehran, Iran
Tel: +98(21)64540
Fax: +98(21)66413969
URL: http://www.aut.ac.ir
E-mail: info@aut.ac.ir

Formerly known as Tehran Polytechnic. The AmirKabir University of Technology, located in Tehran, Iran, has been designated as a center of excellence by the Ministry of Science, Research and Technology in Iran. The Aerospace Engineering Department at AUT offers undergraduate and graduate programs in the field of aerospace engineering. The Departments of Aerospace Engineering and Physics of AUT plan in the near future to offer jointly a program in the field of cosmology and astrophysics.

K.N. Toosi University of Technology (KNTU)
Faculty of Aerospace Engineering
Communication Department
Geodesy and Geomatics Faculty
Address: 322 Mirdamad Ave West, 19697, Postal Code: 19697 64499, Tehran, Iran

PO Box: 15875-4416, Tehran, Iran
Tel: +98(21)8888 2991-3; +98(21)88770218; +98(21)88786212; +98(21)88779473
Fax: +98(21)8879 7469; +98(21)88786213
URL: http://www.kntu.ac.ir
E-mail: info@kntu.ac.ir

The K.N. Toosi University of Technology is located in Tehran, Iran. The Faculty of Aerospace engineering of KNTU offers undergraduate, graduate and postgraduate programs in the field of (aero) space engineering. The Communication Department of this faculty also offers bachelor and master programs in satellite communications. At the Geodesy and Geomatics Faculty, master and Ph.D. degree programs are offered in remote sensing and GIS (http://www.kntu.ac.ir/geodesy/Departments/tabid/409/ctl/Login/Default.aspx).

Malek Ashtar University
Also known as Malek Ashtar University of Technology (MUT), Malek-Ashtar Industrial University and Malek Ashtar University of Defense Technology. The university has branches in Tehran and Isfahan. About the Tehran branch, no address or other information is available via internet search. The URL www.mut.ac.ir is blocked from access (31st March 2009). The Malek Ashtar University is reportedly involved in (military) aerospace research and education.

Malek Ashtar University of Technology (MUT), Isfahan
Department of Mechanical and Aerospace Engineering
Address: Malek Ashtar University of Technology, Ferdowsi Ave, Sahin-shahr, Isfahan, Iran
PO Box: 83145-115, Isfahan, Iran
Tel: +98(312)5226001
Fax: +98(312)
URL: http://www.mut-es.ac.ir/
E-mail: it@mut-es.ac.ir

The Department of Mechanical and Aerospace Engineering of MUT was established in 1987 in Isfahan. The activities of the department include research and education in mechanics, aerospace, aerospace motor engines and avionics. It offers bachelor and graduate master programs in the field of aerospace engineering and mechanics.

Imam Hossein University (IHU)
Department of Aerospace Engineering
Address: Imam Hossein University, Tehran Pars, Shahid Babaei Highway, Tehran, Iran
PO Box: 16535-415, Tehran, Iran
Tel: +98(21)
Fax: +98(21)
URL: http://www.ihu.ac.ir

The Imam Hossein University, affiliated with the Islamic Revolution Guards Corps (Pasdaran), is a public university of engineering, science and military in Iran. The university was established in 1986. The Department of Aerospace Engineering at the Faculty of Engineering conducts both undergraduate and graduate programs in aerospace engineering.

Islamic Azad University, Science and Research Branch (SRBIAU)
Department of Aerospace Engineering
Address: Azad University Campus (Hesarak), Pounak Square, Ashrafi-Isfahani Highway, Postal Code: 1477893855, Tehran, Iran
PO Box: 14515-775, Tehran, Iran
Tel: +98(21) 44865100-3
Fax: +98(21)-
URL: http://www.srbiau.ac.ir; http://www.azad.ac.ir
E-mail: info@srbiau.ac.ir

Headquartered in Tehran, the Islamic Azad University was founded in 1982 and is a private chain of universities in Iran. It currently has a total enrollment of some 1.3m students. The Aerospace Research Institute and the Department of Aerospace Engineering of SRBIAU are involved in aerospace research and offer aerospace engineering courses in bachelor and master degrees.

Aerospace Research Institute (ARI)
Address: 15th Street, Mahestan Street, Iranzamin Ave, Shahrak-e Ghods, Tehran, Iran
PO Box: 14665-834, Tehran, Iran
Tel: +98(21) 88366030
Fax: +98(21) 88362011
URL: http://www.ari.ac.ir
E-mail: info@ari.ac.ir

The Aerospace Research Institute, affiliated with the Ministry of Science, Research and Technology of Iran, is involved in research and education of different aspects of air and space science and technology. Established in 1997, ARI provides the possibility for a number of researchers and master and Ph.D. degree students to benefit from the institute's available achievements, sources and expertise for their educational and research means.

University of Tehran
Solar Physics and Astronomy Section
Remote Sensing Division
Address: University of Tehran, 16 Azar Street, Enghelab Ave, Tehran, Iran
Tel: +98(21)61113358
Fax: +98(21)66409348
URL: http://www.ut.ac.ir
E-mail: info@ut.ac.ir

Known as Tehran University, the University of Tehran is the oldest (1934) and largest scientific, educational and research center of Iran and is called the "Mother University" or the "Symbol of Higher Education of the Country". The university is considered as one of the pioneers of the society in important scientific, cultural, political and social affairs. The Solar Physics and Astronomy Section was established at the Institute of Geophysics of the University of Tehran in 1963. It has gradually developed into a center for research in the different fields of astronomy, solar and solar–Earth studies, and the magnetic storms on the Earth caused by the Sun. A solar observatory was established in the section that provides short-term training programs and facilities for visitors to observe the Sun in the daytime. The section supports extensively the research and education programs in master and Ph.D. levels in astronomy and astrophysics. Established in 1998, the Remote Sensing Division at the Department of Surveying and Geomatics Engineering in the Engineering Faculty at the University of Tehran is involved with research and education in space technology applications, including remote sensing and photogrammetry in master and Ph.D. levels. The Division's training and research activities have been continued by establishing the Remote Sensing Laboratory that is the core of all research activities in the Division of Remote Sensing at the department. The laboratory consists of facilities for both master and Ph.D. students in the division. As a center for remote sensing activities, it also includes an archive for satellite images.

Iran University of Science & Technology (IUST)
Department of Electrical Engineering
Address: Narmak, 16846, Tehran, Iran
Postal Code: 16846-13114Tel: +98-21-77240492Fax: +98-21-77240490
URL: http://www.iust.ac.ir
E-mail: info@iust.ac.ir

Known as Elm-o Sanat University of Iran, Iran University of Science and Technology is one of the major technical universities in Iran, composed of 14 schools and departments. The university was established in 1929. The main campus is located in Tehran; two branches are in Arak and Behshahr. The Department of Electrical Engineering offers research and educational bachelor, master and Ph.D. degree programs in communications, including satellite communications. The Power Electronic, Electrical and Magnetic Fields Research Laboratory (PEEMFRL) is involved in the study and research programs on electromagnetic fields and propagation in space. The laboratory benefits from the potentials and abilities of the researchers and students studying in the Department.

Civil Aviation Technology College (CATC)
Aviation Communication Engineering Department
Address: Mehrabad International Airport, Meraj St, Postal Code: 1387835791, Tehran, Iran
PO Box: 13445-418, Tehran, Iran

Tel: +98(21) 66025120; +98(21) 61022852
Fax: +98(21) 66019426; +98(21) 66009765
URL: http://www.catc.ac.ir/
E-mail: http://mail.catc.ac.ir/

Located in Tehran, CATC trains in the ability to freely exchange messages and navigation data through switching systems, teleprinter, microwave circuits and satellites and to be able to operate and maintain transmission and communication systems. It has been working since 1949 and offers bachelor graduate programs in the fields of atmospheric and aviation communications using satellites.

Shiraz University of Technology
Faculty of Mechanical Engineering and Aerospace
Address: Shiraz University of Technology, Modarres Blvd, Shiraz, Fars, Iran
PO Box: 313-71555
Tel: +98(711) 7262102
Fax: +98(711) 7262102
URL: http://www.sutech.ac.ir/
E-mail: yousefi@sutech.ac.ir

Shiraz University of Technology (SUTECH) was one of the Shiraz University main colleges (Electrical Engineering College). In 2004, the university begins its independent activity as a public university under the name of "Shiraz University of Technology". SUTECH is the second state institute in the Fars Province in higher technological education, basic and applied research. Since 2006, the Faculty of Mechanical Engineering and Aerospace offers research and educational programs in master and Ph.D. levels in Aerospace and Mechanics, focusing on energy transfer.

Shiraz University
Department of Physics
Address: Shiraz University Main Building, Jam-e Jam Blvd, Shiraz, Fars, Iran
Tel: +98(711)6286418
Fax: +98(711) 6286446
URL: http://shirazu.ac.ir
E-mail: shupubr@shirazu.ac.ir

The Shiraz University is one of the high-ranking universities of Iran. The initial nucleus of Shiraz University was formed in 1946. The Physics Department of the Faculty of Science of the university offers master and Ph.D. degree programs in Astrophysics.

Tabriz University
Faculty of Human and Social Sciences
Faculty of Electrical Engineering
Faculty of Physics
Research Institute Applied Physics and Astronomy (RIAPA)

Address: Tabriz University, Imam Khomeini Street, 29th Bahman Blvd, Postal Code: 5166616471, Tabriz, Iran
Tel: +98(411) 3341300; +98(411) 3393599
Fax: +98 (411) 3344013
URL: http://www.tabrizu.ac.ir/
E-mail: http://www.tabrizu.ac.ir/show.asp?id=73

Formerly Azarabadegan University, Tabriz University was established in June 1946. The Faculty of Human and Social Sciences of Tabriz University offers master degree programs in Remote Sensing and GIS. The Faculty of Electrical Engineering conducts master and Ph.D. programs in communications using space. The Faculty of Physics offers master and Ph.D. research and education programs in astrophysics. The Research Institute for Applied Physics and Astronomy (RIAPA) was established at the University of Tabriz in 1972 in order to develop research and promote the institute in terms of both quantity and quality in various fields of applied physics. The Department of Astronomy of RIAPA conducts research work in astronomy and astrophysics, which involves students working for relevant master and Ph.D. degrees.
Further information on RAIPA is available at http://www.tabrizu.ac.ir/riapa/index.htm.

Research Institute for Astronomy and Astrophysics of Maragheh (RIAAM)
Address: Maragheh, Eastern Azerbaijan, Iran
PO Box: 55134-441, Maragheh, Iran
Tel: +98(421)325 55 91
Fax: +98(421)325 55 92
URL: http://www.riaam.ac.ir
E-mail: info@riaam.ac.ir

The Research Institute for Astronomy and Astrophysics of Maragheh was established in 2002. It aims at revival and introducing the scientific and research mission of the Maragheh Observatory, which was built by the famous scientist and astronomer of the Middle Age, Khajeh Nasiruddin Toosi, in 1259 and was active for about 55 years. RIAAM contributes to producing and introducing up-to-date science in astronomy and cosmology and doing research projects at international level. It offers education and research programs in postgraduate (Ph.D. and post-doctoral) levels in astronomy and astrophysics.

International Center for Science and High Technology and Environmental Science (ICSHTES)
Research Institute of Environmental Sciences
Address: ICSHTES, the Road between Mahan and Kerman, Kerman, IranTel: +98(342)6226611-13
Fax: +98(342)6226617URL: http://newsite.icst.ac.ir/default.aspx
E-mail: info@icst.ac.ir

Established in 1996, ICSHTES focuses on the use of new and advanced technologies for studying and monitoring the environment. The Research Institute of Environmental Sciences uses remote sensing and GIS technologies in environmental studies and modeling. The Institute offers master degree programs in environmental studies.

Institute for Advanced Studies in Basic Science (IASBS)
Department of Physics
Address: Gava Zang, Postal Code: 45195, Zanjan, Iran
PO Box: 45195-1159, Zanjan, IranTel: +98(241)415 3133; +98(241)415 2106
Fax: +98(241)421 4949; +98(241)415 2104URL: www.iasbs.ac.ir
E-mail: physweb@iasbs.ac.ir

This Institute for Advanced Studies in Basic Science (IASBS) is based in Zanjan and focuses on the research and education in basic sciences at postgraduate level. The Physics Department started its activity in 1992 as the first department of IASBS. Its education and research programs include theoretical condensed matter physics, astrophysics and mathematical physics. In 2000, a seven-year program leading to a Ph.D. degree was established. Students in this program, after completing their undergraduate and graduate courses, directly proceed to their Ph.D. thesis, thereby skipping the master thesis. At present, the Physics Department is the only place in Iran offering this program. The Department of Physics of IASBS offers master and Ph.D. programs in Astrophysics.

University of Zanjan
Physics Department
Address: University of Zanjan, Tabriz Road Km 5, University Boulevard, Postal Code: 45371-38111, Zanjan, Iran
PO Box: 313, Zanjan, IranTel: +98(241) 5151; +98(241)5152355
Fax: +98(241) 2283203URL: http://www.znu.ac.ir/
E-mail: info@znu.ac.ir

Also known as Zanjan University. The Physics Department of Zanjan University began its activity in 1991. Educating astrophysics is the main part of the activity of the department. It offers research and master and Ph.D. degree programs in astrophysics and astronomy.

Imam Khomeini International University (IKIU)
Astronomy and Astrophysics Department
Address: Qazvin, Postal Code: 34149-16818, Iran
Tel: +98(281)3780021; +98(281)8371592
Fax: +98(281)3780041
URL: http://www.ikiu.ac.ir/
E-mail: http://mail.ikiu.ac.ir/

Imam Khomeini International University (IKIU) was established in 1992 following the merger of Iran International Islamic University with Dehkhoda Higher

Education Institute. The Astronomy and Astrophysics Department of the Faculty of Basic Sciences offers master degree programs in astronomy and astrophysics.

Sh. Chamran University of Ahvaz
Physics Department/Remote Sensing Department
Address: Central University Campus, Golestan Blvd, Ahvaz, Khuzestan, Iran
Tel: +98(611)3360018; +98(611)3337009
Fax: +98(611)3332041
URL: http://portal.scu.ac.ir/
E-mail: comm_center@scu.ac.ir

The Physics Department of the Faculty of Science of Sh. Chamran University of Ahvaz, which was established in 1970, offers research as well as bachelor, master and Ph.D. degree programs in Astrophysics and Radio Astronomy. The Remote Sensing Department of the Faculty of Science of Sh. Chamran University of Ahvaz was established in 2001 and offers bachelor and master degree programs in remote sensing technology.

University of Isfahan (UI)
Surveying Department
Address: The University of Isfahan, Hezar Jerib Street, Isfahan, Iran
Postal Code: 81746-73441
Tel: + 98(311)7932128; + 98(311)7932038
Fax: + 98(311)7932037
URL: http://www.ui.ac.ir/
E-mail: webmaster@ui.ac.ir

Established in 1950, the University of Isfahan, with more than 50 years' experience, is one of the major universities in the fields of science, human science and engineering. The Surveying Department of the Technical and Engineering Faculty that was established in 1988 offers master degree programs in remote sensing.

University of Kashan
Department of Astronomy
Address: University of Kashan, Km 6 Ravand Road, Postal Code: 87317-51167, Kashan, Iran
PO Box: 87317-5116
Tel: +98(361)5555 333
Fax: +98(361) 5552 930
URL: http://www.kashanu.ac.ir
E-mail: webmaster@kashanu.ac.ir

The University of Kashan was founded in 1974. At the time of its establishment, only undergraduate courses in physics and mathematics were offered. The current activities of the university are classified into the four sections of education, research, development and side activities. The Observatory of the University of Kashan, which was established in 2000, is actively involved with the education of astronomy and astrophysics at bachelor level under the Department of Astronomy. The observatory, which is considered as one of the important observatories throughout the country, has seven telescopes with the variety of apertures.

Tarbiat Modares University (TMU)
Department of Geography
Address: Tarbiat Modares University, Nasr Bridge, Jalal Al Ahmad Highway, Tehran, Iran
PO Box: 14115-111
Tel: +98(21)88009720; +98(21)88009735
Fax: +98(21)88007598
URL: http://www.modares.ac.ir
E-mail: edu@modares.ac.ir

Tarbiat Modares University (TMU) was founded in 1982 after the Islamic Revolution in Iran. The main goal of this university is training academic staff of universities as well as researchers required for universities and higher-education centers. TMU endeavors to provide an environment to promote a high level of professional preparation and performance for students from various backgrounds and areas. The Geography Department of the Faculty of Humanities offers master degree programs in Remote Sensing and GIS, focusing on technology applications.

Alzahra University
Physics Department
Address: Alzahra University, Vanak, Tehran, Postal Code: 1993891176, Iran
Tel: +98(21)88044040
Fax: +98(21)88035187
URL: http://www.alzahra.ac.ir
E-mail: webmaster@alzahra.ac.ir

The formerly Farah Diba University, exclusively for women, was founded in 1964. The university began giving its services as the "Higher Educational Institute for Girls" at first. After the advent of the revolution in Iran, the name of the university changed to Alzahra University. The Physics Department of the Faculty of Science at the Alzahra University, which was established in 1977, offers master and Ph.D. degree programs in astronomy and astrophysics.

Shahid Beheshti University (SBU)
Department of Remote Sensing and GIS
Address: Shahid Beheshti University, Daneshjoo Blvd, Chamran Highway, Tehran, Iran
PO Box: 1983963113
Tel: +98(21)29902232; +98(21)29902233
Fax: +98(21)22431919

URL: http://www.sbu.ac.ir
E-mail: info@sbu.ac.ir

Formerly the National University of Iran, Shahid Beheshti University (SBU) was established in 1959 as the National University of Iran and started its academic activity in 1960. The Department of Remote Sensing and GIS of the Faculty of Earth Sciences offers research and educational programs in master degree level. In 1999, the Remote Sensing and GIS Research Center was established in the Earth Sciences Faculty to improve the expertise of human resources in the application of geomantic technologies. It has three subsections: Remote Sensing, Geographical Information System (GIS), and Global Positioning System (GPS).

Islamic Azad University of Maragheh
Department of Astronomy Physics
Address: Azad University Campus, Sh. Derakhshi Blvd, Maragheh, Eastern Azerbaijan, Iran
Tel: +98(421)3254506-9
Fax: +98(421)-
URL: http://www.iau-maragheh.ac.ir/
E-mail: info@iau-maragheh.ac.ir

Established in 1985, the Islamic Azad University of Maragheh offers a variety of educational programs in science and engineering. The Department of Astronomy Physics of the university offers programs for bachelor degree in astronomy and astrophysics. Its educational programs will develop to higher degrees in the near future. It is the only unit amongst all of the Islamic Azad Universities throughout Iran that offers the educational program in astrophysics and astronomy.

Industries

Satellite and space launch vehicle development and production need, apart from basic skills and technologies, high-tech industries. The following industries are assessed, in one way or the other, to be related to the space endeavors of Iran. As other (military) developments and production are in need of virtually the same high-tech industries, information about those industries is hard to obtain. Web pages often do not open properly and when opened, mostly are not of usable value (probably for security reasons). If the respective address details are left blank, those details could not be acquired by the author.

Aerospace Industries Organization (AIO) [Sazemane Sanaye Hava-Faza (SSH)]
Address: 28 Shian 5, Lavizan, Tehran, Iran
Tel: +98(21)2949508
Fax: +98(21)2948301
URL: http://www.aio.ir/latin (English home page is virtually empty)
E-mail: info@aio.ir

It is the leading high-tech industrial and military subsidiary of the Sanam Industrial Group, also known as Department 140 of the Defense Industries Organization of the Ministry of Defense and the Ministry of the Armed Forces Logistics of Iran. AIO is the manufacturer, amongst others, of Shahab ballistic missiles, launchers, rocket/booster propellants and components; it also supplies non-military items and services such as fuel pumps, technical and engineering services, and research and development. AIO is the obvious organization to lead the development and production of the space assets of Iran. AIO manages a number of factories and research centers, including: the Missile Center of Saltanat-Abad, the Vanak Missile Center, the Parchin Missile Industries factories, the Baqeri base factories Numbers 1–3, the Tabriz Bakeri base factory, the Bakeri Missile Industries factory, the Hemmat Missile Industries factory, the Bagh Shian (Almehdi) Missile Industries, the Shah-Abadi Industrial Complex, the Khojir Complex, the Baqerololum Missile Research Center, the Mostafa Khomeini base factory, and the Quadiri Base factory. The AIO is also named the Air and Space Organization (ASO).

Shahid Hemmat Industrial Group (SHIG)
Address: Damavand Tehran Highway 2, Abali Road, Tehran, Iran
Tel: +98(
Fax: +98(
URL:
E-mail:

Shahid Hemmat Industrial Group is subordinated to the Aerospace Industries Organization and comprises several divisions that are related to building and operating launch vehicles, such as, Kalhor Industry [launchers]; Karimi Industry [parts that transfer propellants to the engine and other parts of the launch vehicle]; Cheraghi Industry [production of propellants]; Rastegar Industry [launch vehicle engine production]; Varamini Industry [launch vehicle guidance and control systems]; and Movahed Industry [manufacturing and assembly of launch vehicles].

Iran Electronics Industries (IEI) [Sanaye Eletronic-e Iran (SEI); SAIran]
Address (1): Shahid Langari St, Nobonyad Sq., Tehran, Iran
PO Box: 19575-365, Tehran, Iran
Address (2): Hossain Abad, Ardakan Rd, Shiraz, Iran
PO Box: 71365-1174, Shiraz, Iran
Tel: +98(21)22827321 (Tehran)
Fax: +98(21)22827322 (Tehran)
Tel: +98(71)641000 (Shiraz)
Fax: +98(71)641061 (Shiraz)
URL: http://www.ieicorp.com; http://ieimil.com
E-mail: info@ieicorp.com

Iran Electronics Industries was established in 1972 and presently is the major producer of electronic systems and products in Iran. Subsidiaries are: Shiraz Electronics Industries (SEI) [Electronic Technology]; Iran Communication Industries (ICI) [Communications Technology]; Information Systems of Iran (ISIRAN) [Information Technology]; Electronic Components Industries (ECI) [Microelectronic Technology] in Tehran and Shiraz; Isfahan Optics Industries (IOI) [Optics Technology]; Iran Electronics Research Center (IERC) [R&D]; Iran Space Industries Group (ISIG) [Satellites]; Security of Telecommunication and Information Technology (STI) [Communications security].

Iran Space Industries Group (ISIG)
Address:
Tel: +98(
Fax: +98(
URL:
E-mail:

The founding of the Iran Space Industries Group was announced on the occasion of the launch on 4th February 2008 of the Kavoshgar-1 rocket. ISIG is subordinated to IEI.

Shiraz Electronics Industries (SEI)
Address: Mirzaie Shirazi Shiraz Fars 71365-1589, Iran
PO Box: 71365-1589, Shiraz, Iran
Tel: +98(71)16246253
Fax: +98(71)16242000
URL: http://www.sashirazco.com (not accessible)
E-mail: info@sashirazco.com (not accessible)

Since 1973, Shiraz Electronics Industries has been professionally engaged in electronic products and projects. It employs highly skilled personnel who have combined their experiences with advanced equipment, abundant know-how and motivation to form a well organized and powerful technological industrial group. SEI is currently involved in electronic warfare, control and automation, radar and microwaves, weapon electronics, avionics, computers and electro-optics.

Iran Communication Industries (ICI) [Sanaye Mokhaberat Iran (SMI)]
Address (1): Pasdaran Avenue, Tehran, Iran
PO Box: 19295-4731, Tehran, Iran
Address (2): Apadana Avenue, Tehran, Iran
PO Box: 19575-131, Tehran, Iran
Address (3): 34 Khorramshar Street, Tehran, Iran
PO Box: 15875-4337, Tehran, Iran
Tel: +98(21)22827481
Fax: +98(21) 22827480
URL: http://www:.
E-mail:

The Iran Communication Industries is Iran's leading manufacturer of military and civil communication equipment and systems. More than 75 products in the field of tactical communications and encryption systems meet a wide range of military requirements.

Information Systems of Iran (ISIRAN)
Address: Shahid Sarafroz St 30, Tehran, Iran
PO Box: 15875-4337, Tehran, Iran
Tel: +98(21)8738811
Fax: +98(21)8739197
URL: http://www.isiran.ir (Farsi)
E-mail: info@isiran.ir

Information Systems of Iran is a state-owned company founded in 1971. It is one of the largest and most experienced information companies in Iran. This company claims to be the number one (IT) company in terms of revenue, market share, variety and quality of products and services and a leading company in modern solutions. ISIRAN assists and provides its clients with supreme councils on information systems.

Electronic Components Industries (ECI)
Address: Ghasroddasht Ave, Hossainabad, Shiraz, Iran
PO Box: 71955-887, Shiraz, Iran
Tel: +98(71)644714
Fax: +98(71)644711
URL: http://www.eci.company.ir
E-mail: info@eci.company.ir

Electronic Components Industries was founded in 1976 and comprises two complexes, one in Shiraz and the other in Tehran. Its activities include, among others, designing and manufacturing of semiconductor devices, quartz crystal, multilayer PCB and thick film hybrid, infantry field wire, optical cable and access systems.

Isfahan Optics Industries (IOI)
Address: Kaveh Rd, Isfahan, Iran
PO Box: 81465-1117, Isfahan, Iran
Tel: 98(311)4511740
Fax: 98(311)4518085
URL: http://www.ioico.ir; http://www.ioicivil.ir
E-mail: info@ioico.ir

Isfahan Optics Industries was founded in 1987 with the aim of setting up a vigorous and modern optics industry. Reportedly, the employment of highly qualified engineers and state-of-the art equipment has made IOI one of the most capable industries in Iran. The design and manufacture of complex lenses and prisms, multilayer coatings, a wide range of daylight sights and various types of aircraft windshields are carried out at IOI.

Iran Electronics Research Center (IERC)
Address:
Tel: +98(
Fax: +98(
URL:
E-mail:

Iran Electronics Research Center, founded in 1997, is a scientific, educational and research institute, with active research teams in the fields of electronics, communications, microprocessors, microelectronics, optics, electro optics and radars. The research groups at the IERC include electronics, communications, microprocessors, microelectronics, optics and electro-optics.

Security of Telecommunication and Information Technology (STI)
Address:
Tel: +98(
Fax: +98(
URL:
E-mail:

Shahid Bagheri Industrial Group (SBIG)
Also known as the Iran Technical Organization (IRTO)
Address: Km 9, Old Karaj Road, Tehran, Iran
PO Box: 15745-699, Tehran, Iran
Tel: +98(21)6027002
Fax: +98(21)6027507
URL:
E-mail:

Shahid Bagheri Industrial Group is part of the Defense Industries Organization (DIO), and reportedly cooperates with Russia's Baltic State Technical University and the Sanam Industries Group to create the Persepolis (Takht-I-Jamshid) joint missile education center in Iran to transfer missile technology from Russia to Iran.

Iran Telecommunication Manufacturing Company (ITMC)
Address: Central Office, Opposite Shariati Park after Sayed Khandan Bridge Dr Ali Shariati St, Tehran, Iran
Shiraz Office, ITMC, Km 2, Modaress Blvd, Shiraz, Iran
PO Box: 311-71555, Shiraz, Iran
Tel: +98(0711)7268091; +98 (021)2848016 2845093
Fax: +98(0711)7268094; +98 (021)2844057
URL: http://irantelecom.ir/eng.asp?page=7&code=1&sm=8#top
E-mail:

ITMC, with the goal of producing systems of high-capacity telecommunication centers and on-the-table telephone, was established in 1967 and taken into operation in 1969.

Telecommunication Company of Iran (TCI)
Address: TCI, Shahid Ghodousi Cross, Shariati Street, Tehran, Iran
Tel: +98(21)88113938
Fax: +98(21)88113938
URL: http://www.irantelecom.ir/eng.asp
E-mail: info@irantelecom.ir

TCI is a widespread company involved in telecommunications, subordinated to the Ministry of Communications and Information Technology. It has branches in almost all of the provinces throughout Iran and is responsible for development and management of the country's communication infrastructure, particularly satellite- and ground-based telecommunications.

National Cartographic Center (NCC)
Address: Me-raj Street, Azadi Square, Tehran, Iran
PO Box: 13185-1684, Tehran, Iran
Tel: +98(21)66071001-9
Fax: +98(21)66071000
URL: http://www.ncc.org.ir/
E-mail: info@ncc.org.ir

Established in 1953, NCC is an organization subordinated to the Management and Planning Organization of Iran and is involved in setting up the country's geo-information and map data. NCC exercises a leading role in preparing reference map information of Iran. It is also involved in promoting and education of mapping and geo-information technologies. NCC conducts a variety of relevant projects using space technologies such as remote sensing and GIS.

Iranian organizations related to space policy, development and applications

Supreme Space Council (SSC)
Address: No. 25, East Armaghan Street, Africa Street, Postal Code: 1915633711, Tehran, Iran
Tel: +98(21)22019991
Fax: +98(21)22035623
URL: [As of ISA]
E-mail: [As of ISA]

Iran's Supreme Space Council (SSC) held its first session on 20th July 2005, with President Mohammad Khatami, to which the council is directly responsible, attending the debate. The council's main goals include policy making for the application of space technologies, manufacturing, launching and use of national research satellites, approving space-related state and private sector programs, promoting the partnership of the private and cooperative sectors in efficient uses of space, and identifying guidelines concerning the regional and international cooperation in space issues (see also the box article on the Supreme Space Council in Chapter 7).

Ministry of Communications and Information Technology (Ministry of CIT)
Address: Main Building of MCIT, Sh. Ghodousi Cross, Shariati Street, Tehran, Iran
Tel: +98(21)88114315; +98(21)88114325
Fax: +98(21)88467210
URL: http://www.ict.gov.ir
E-mail: info@ict.gov.ir

The Ministry of Communications and Information Technology (Ministry of CIT) is responsible for exercising frequency spectrum management and protecting the national radio rights at the regional and international levels; centralization of policy making; formulation of regulations and standards and supervision of their implementation in different areas of post, communications and telecommunications such as common and new services in post, telecommunications, space communications, radio communications, data transmission, sound and picture transmission, remote sensing and computer communications; development of a conducive environment for communications, testing, information processing and remote-sensing methods and supporting them; and also policy making for the development of the aforementioned communications facilities and services, in line with the state-of-the-art scientific, experimental and information technology. The Ministry of CIT was formerly active as the Ministry of Post, Telegraph and Telephone (PTT) but in 2003, began its activity as the Ministry of Communications and Information Technology.

Iranian Space Agency (ISA)
Address: No 57/2, Sayeh Street, Vali-e-Asr Street, Tehran, Iran
Tel: +98(21)22029100
Fax: +98(21)22029100
URL: http://www.isa.ir
E-mail: info@isa.ir

The Iranian Space Agency (ISA) is the national (governmental) space agency of Iran. Its president is deputy minister of the Ministry of CIT and secretary for the SSC. The ISA is responsible for the execution of the space policy of the SSC. Practically, the ISA conducts research in the fields of space technology, such as remote sensing and communications. Former presidents of the ISA were Messrs Hassan Shafti and Ahmad Talebzadeh. The current (March 2009) president is Mr Reza Taghipour, the former Deputy Director General of SAIran/IEI.

Ministry of Science, Research and Technology (MSRT)
Address: Corner of the South Piroozan Street, Hormozan Street, Khovardin Street, Sanat Square, Shahrak-e Ghods, Tehran, Iran
Tel: +98(21) 82231000
Fax: +98(21) 82234006

URL: http://www.msrt.ir/default.aspx
E-mail: http://mail.msrt.ir/

MSRT is responsible for the higher education, research and technology promotion in Iran. It conducts its activities mainly through seven deputyships, including education, planning and development, technology, student affairs, research, culture and social issues, and legal affairs. Furthermore, the Supreme Council of Cultural Revolution, the Research Institute for Education Planning, the Iranian Research Organization for Science and Technology (IROST), the Education Assessment Organization of Iran, the Student Welfare Fund, the Central Board for Selection of Educators, Students and Officials, Universities, Science and Technology Parks, and the Institutes of Higher Education and Technology are the institutions and bodies active in the realm of MSRT.

Iranian Research Organization for Science and Technology (IROST)
Address: No. 71 Sh. Mousavi St, Enghelab Ave, Tehran, Iran
PO Box: 15819-3538, Tehran, Iran
Tel: +98(21)88828051-7
Fax: +98(21)88838341
URL: http://www.irost.org
E-mail: admin@irost.org

The Iranian Research Organization for Science and Technology (IROST) was approved and ratified by the Revolutionary Council of the Islamic Republic of Iran in 1980. It is a comprehensive science policy research center directly attached to the Ministry of Science, Research and Technology. IROST is engaged in development of strategies, policies, R&D systems, management, foresight and evaluation of related S&T development and economic progress.

Electrical and Computer Science Engineering Department (ECEDEP)
Address: No. 71 Shahid Mousavi St, Enghelab Ave, Tehran, Iran
PO Box: 15815-3538 Tehran, Iran
Tel: +98(21)66281001-8
Fax: +98(21)66281011
URL: http://www.irost.org
E-mail: ecedep@irost.org

The Electrical and Computer Science Engineering Department (ECEDEP) was established in 1980 as an IROST subdivision (with the goal of supporting researchers and talented people). The objectives of the department are, amongst others, accomplishment of research, applicable semi-industrial projects, compiling technical knowledge and transferring this knowledge to the industry. ECEDEP has a Space Technology Group that emphasizes on satellite payloads, ground stations and space applications. The Space Technology Center of ECEDEP uses the following technology laboratories: Satellite Signal Processing & Data Center; Space Battery Lab; Space Simulator; Solar Cell Test Bed; Space Quality Assurance; TMTC

Laboratory; Space Software Test-bed; Telemedicine Laboratory; Space Sensor, Monitoring and Control Laboratory; and EMC and EGSE Laboratories.

Islamic Republic of Iran Broadcasting Organization (IRIB)
Address: IRIB Main Building, Vali-e-Asr Street, Tehran, Iran
Tel: +98(21) 2291095
Fax: +98(21) 2204 6964
URL: http://www.irib.ir/English/
E-mail: webmaster@irib.ir

The Islamic Republic of Iran Broadcasting (IRIB) organization is a state-run organization that belongs to the so-called cultural institutions and, as such, is subordinated to the Secretariat of Supreme Council of Cultural Revolution.

Applied Science and Research Association (ASRA)
Address: Now-Bonyad Sq., Tehran, Iran
Tel: +98(21) 2814323
Fax: +98(21) 2814323
URL:
E-mail:

ASRA functions as the Iranian member of the Inter-Islamic Network on Space Sciences & Technology (ISNET). It is subordinated to the Department of Mechanical Engineering of the K.N. Toosi University of Technology in Tehran.

Iran Telecommunication Research Center (ITRC)
Address: End of North Karegar, Post Code: 1439955471, Tehran, Iran
PO Box: 3961-14155, Tehran, Iran
Tel: +98(21)8630360
Fax: +98(21)8027762
URL: http://www.itrc.ac.ir
E-mail: info@itrc.ac.ir

ITRC is subordinated to the Ministry of CIT. It is an experienced research entity in the fields of information and communication technology. ITRC runs advanced research facilities and laboratories that enable research teams to conduct studies and carry out experiments.

Islamic Republic of Iran Meteorological Organization (IRIMO)
Address: Meraj St, Azadi Sq., Tehran, Iran
PO Box: 13185-461, Tehran, Iran
Tel: +98(21)66070017-20
Fax: +98(21)664690440
URL: http://www.irimo.ir; http://www.weather.ir
E-mail: pririmo@irimo.ir

IRIMO is responsible for all meteorological information and weather forecasting in

Iran. Data from meteorological satellites are used by the IRIMO forecasting center not only for weather-forecasting purposes, but also for atmospheric disaster mitigation objectives.

Iranian National Center for Oceanography (INCO)
Address: No. 9 Etemadzadeh Ave, West Fatemi St, Post Code: 1411813389, Tehran, Iran
PO Box: 14155-4781, Tehran, Iran
Tel: +98(21)6944873-6
Fax: +98(21)66944869; +98(21)66944872
URL: http://www.inco.ac.ir
E-mail: info@inco.ac.ir

This center operates under the auspices of the Ministry of Science, Research and Technology with the aim of research in all fields related to marine science and also presenting a proposal for better usage of marine resources, promotion of commercial utilization of marine activities and also to facilitate defining the marine strategies of the country within the framework of the government's activities and also improving the level of knowledge, research and marine technology.

National Committee on Natural Disaster Reduction (NCNDR)
Address: NCNDR Bureau at MOI, Fatemi Street, Tehran, Iran
Tel: +98(21)658 024; +98(21)657247
Fax: +98(21)613 12166
URL: http://www.moi.ir
E-mail: info@moi.ir

The Ministry of Interior (MOI) of Iran is responsible for the disaster management mechanism at the national level. The responsibilities and functions related to disasters were formally assigned to MOI in 1991. To manage the assigned disaster management functions, MOI has formed the Bureau for Research and Coordination of Safety and Reconstruction Affairs (BRCSR) and a National Disaster Task Force (NDTF) as well. The NDTF is a coordinating inter-organizational body whose activities vary during different phases of disasters. It is headquartered at the MOI in Tehran and relies for its activities on the BRCSR, whose director is also the manager of NDTF. A total of 4,550 staff, mostly dealing with administrative and logistic support services, perform their duties at national, provincial and local levels. In line with the International Decade For Natural Disaster Reduction (IDNDR), the Islamic Consultative Assembly approved the formation of the National Committee for Natural Disaster Reduction (NCNDR) in 1991 headed by the Ministers of Energy, Agriculture, Health, Commerce, Jahad of Agriculture, Roads, and Transportation and Housing and Urban Development. The Directors of the Planning and Budget Organization, Environment Protection Organization, Meteorology Organization, Forestry and Rangeland Organization, Institute of Geophysics and the Red Crescent Society of Iran are also included. Army and Disciplinary Forces and any other organizations that the Chair of the Committee deems appropriate are also allowed to participate in the Committee. The Committee was designed as a policy-making body to provide for the exchange of information and to allow the government to have the authority to support and follow up the related activities. The National Committee has set up nine specialized sub-committees presided by deputy ministers, 30 provincial Committees presided by General Governors and also a coordination committee presided by the Minister of Interior.

Geological Survey of Iran (GSI)
Address: Geological Survey of Iran, Azadi Sq., Meraj-street, Tehran, Iran
PO Box: 13185-1494, Tehran, Iran
Tel: +98(21)66070532
Fax: +98(21) -
URL: http://www.gsi.ir/
E-mail: info@gsi.ir

The Geological Survey of Iran (GSI) was established in 1962, through a special fund project of the United Nations. The GSI is authorized to carry out geological and mineral investigations throughout the country, to collect results of activities performed in this respect, to establish interrelationship and coordination between such activities and to prepare, complete and publish geological maps of Iran. In 1999, the exploration duties of the Ministry of Mines and Metals were totally assigned to the GSI. The GSI is now responsible for the geological study of the country, and the exploration–evaluation of the mineral resources (except hydrocarbons). This commitment is accomplished by the activities of different groups of the GSI, such as Stratigraphy, Petrology, Sedimentology, Marine Geology, Paleontology, Tectonics, Seismotectonics, Exploration, Geophysics, Geochemistry, Geomatics and different labs, according to the general directions laid down by the Ministry of Mines and Metals, in the framework of the recently approved "Mining Law". Presently, about 700 people are working at the GSI, where their skills along with laboratory and computer facilities are used to carry out research and exploration projects of high quality. The central headquarters of the GSI is in Tehran, and there are five other headquarters in the north-west (Tabriz), north-east (Mashhad), south (Shiraz), south-west (Ahwaz) and south-east (Kerman) of the country. These headquarters carry out local functions of the GSI. The GSI also cooperates with other organizations in Iran and abroad through bilateral cooperation or joint research programs. In order to transfer the geological knowledge and publish new scientific findings, the GSI started publishing the *Geosciences Scientific Quarterly Journal* in autumn 1992.

Iranian Remote Sensing Center (IRSC)/Remote Sensing Administration of ISA
Address: No. 22, 14th Street, Saadat Abad Avenue, 1997994313, Tehran, Iran
Tel: +98(21)2064469-73
Fax: +98(21)2064474
URL: http://www.iran-irsc.com
E-mail: admin@iran-irsc.com

The official tasks of the former IRSC are currently executed by the Remote Sensing Administration of the ISA. There is an office for remote sensing located at the ISA headquarters, but remote sensing activities are mostly conducted at the Mahdasht Space Center. (The contact information may be not valid any more.)

Mahdasht Space Center (MSC)
Address: Km 5 of the Ismail Abab to Shour Ghaleh Road, Next to Payam Airport, Karaj to Mahdasht Road, Mehrshar, Postal Code: 3187713111, Karaj, Iran
PO Box: 31895-135, Mahdasht, Iran
Tel: +98(21)3269005-9
Fax: +98(21)3269031
URL: [As of ISA]
E-mail: [As of ISA]

The Mahdasht Space Center, affiliated to the ISA, is reconstructed and developed on the former site of the Mahdasht Satellite Receiving Station located approximately 65 km west of Tehran whose mission was receiving data from Landsat three decades earlier. The station was established in 1972. The site will comprise the most comprehensive and multi-task ground space complexes as well as work, living and leisure facilities for the Iran's space science and technology specialists, scientists and officials. At present, the main activity of ISA's Remote Sensing Administration, which carries out the functions and duties of the former Iranian Remote Sensing Center, is concentrated at Mahdasht Space Center.

Soil Conservation and Watershed Management Research Center (SCWMRI)
Address: Sh. Shafiee Street, Sh. Asheri Street, Km 9 of Tehran to Karaj Road, Tehran, Iran
PO Box: 13445-1136, Tehran, Iran
Tel: +98(21) 44901214-18; +98(21) 44901240-47
Fax: +98(21) 44905709
URL: http://www.scwmri.ac.ir/main.aspx
E-mail: info@scwmri.ac.ir

The Soil Conservation and Watershed Management Research Institute of the Agricultural Research and Education Organization (AREO) is the focal point for Soil Conservation, Watershed Management, Flood Management and exploitation, River Engineering and training, Coastal Protection, Hydrology and Water Resources Development in Ministry of Jihad-e-Agriculture, in Iran. SCWMRI is located in a campus in Tehran and focuses on research topics on the mentioned areas.

ANNEXE 3: SPACE INSTITUTES IN BRAZIL

Brazil gradually developed organization and infrastructure required for space operations. Following is a list of bodies that are important for Brazilian space policy. This list has been composed from internet data.

AEB – Brazilian Space Agency
Address: SPO Sul Área 5 Quadra 3 Bloco A, CEP 70610-200 Brasília, DF, Brazil
Tel: +55(61)4115500/4115568
Fax: +55(61)4115523
URL: http://www.aeb.gov.br
E-mail: ccs@aeb.gov.br

The Agência Espacial Brasileira (AEB) is the civilian authority in Brazil and acts as the National Space Agency of Brazil in charge of the Brazilian space program. Founded on 10th February 1994, it is subordinated to the Ministry of Science and Technology and replaced COBAE, the Commission for Space Activities. It is in charge of overseeing Brazilian space activities and cooperates with national and international entities to fulfill Brazil's goals in space.

ANATEL – National Telecommunications Agency
Address: SAUS Quadra 06 Blocos C, E, F, H, CEP 70070-940 Brasília, DF, Brazil
Tel: +55(61)23122000
Fax: +55(61)23122002
URL: http://www.anatel.gov.br
E-mail: imprensa@anatel.gov.br

The Agência Nacional de Telecomunicações (ANATEL) functions as Brazil's telecommunications regulator that decides, monitors and controls the use of the Brazilian orbital positions for commercial telecom services. It handles licensing and assigns bandwidth.

CLA – Alcantara Launch Center/CEA – Alcantara Space Center
Address: ROD. MA-106 – Km 7, CEP 65250-000 Alcantara, Maranhão, Brazil
Tel: +55(98)33119000/32291720/32169335
Fax: +55(98)33119325/32291721
URL: http://www.cla.aer.mil.br/
E-mail:

Location Alcantara Ground Station: 2°20'S/44°24'W. The names Alcantara Space Center (CEA) and Alcantara Launch Center (CLA) are used interchangeably, but CEA in fact encompasses the CLA. Centro Espacial de Alcantara (CEA) is subordinated to the Brazilian Air Force (Comando da Aeronáutica) and the Brazilian Space Agency (AEB). The area of some 1,500 km^2 includes general infrastructure and launch complexes. Amongst others, CEA has an active military airfield with a 2,600-m runway, since 1989, a command and control area equipped with a TT&C station, communication, meteorology, radar and electricity supply systems, and an administrative center with hotels, a hospital and administrative buildings. Centro de Lançamento de Alcantara (CLA) is the launch site from which space launches are conducted. It is close to the city of Alcantara, located on Brazil's northern Atlantic coast (2°17′S/44°23′W). It is operated by the Brazilian Air Force. CLA is the space launch site closest to the equator in the world. The construction of the base began in 1982 and the first launch took place on 21st February 1990, when the Sonda II sounding rocket XV-53 was launched.

CLBI – Barreira do Inferno Launch Center
Address: Rodovia RN 063 – Km 11, CEP 59140-970 Caixa Postal 54, Parnamirim, RN, Brazil
Tel: +55(84)32161400
Fax: +55(84)32161421
URL: http://www.clbi.cta.br
E-mail: scs@clbi.cta.br

Location: 5°55′S/35°9′W. The Centro de Lançamento da Barreira do Inferno ("Barrier of Hell" launch center) is a small launch site near Natal, established by the Minister of Aeronautics in 1964 for the Sonda I sounding rocket and has been used until the end of the 1980s for the other Sonda sounding rocket launches. It also has been used for international launch vehicles such as the Skylark. All launches have now been transferred to the Alcantara launch site.

CNAE – National Council for Space Activities
In 1961, a Presidential decree created the GOCNAE (Grupo de Organização da Comissão Nacional de Atividades Espaciais), the Group for the Organization of the National Space Activities Commission. In 1963, GOCNAE became the CNAE (Comissão Nacional de Atividades Espaciais), the National Commission of Space Activities, and existed until 1971. In that year, a decree ended GOCNAE and created the National Institute of Space Research (INPE).

CNPq – National Council for Scientific and Technological Development
The Conselho Nacional de Desenvolvimento Científico e Tecnológico (CNPq) is an organization of the federal government subordinated to the Ministry of Science of Technology, devoted to funding of science and technology in Brazil. CNPq runs an extensive program of fellowships and research grants and has several research institutes of its own. These institutes vary in quality and size, and many of them have their own graduate education programs.

COBAE – Brazilian Commission for Space Activities
The Comissão Brasileira de Atividades Espaciais (COBAE) was established in 1971 as a joint civilian–military committee and placed under the National Security Council. It was chaired by the head of the Armed Forces General Staff and since 1981 in charge of the Missão Espacial Completa Brasileira (MECB), the "Complete

Brazilian Space Mission" aimed at attaining self-sufficiency in space technology. The MECB was created to coordinate launch vehicles, launch sites and the manufacturing of satellites. In February 1994, the Brazilian government created the Brazilian Space Agency to take over the space program from COBAE.

COMAer – Air Force Command
The Brazilian Air Force is the aerospace branch of the Brazilian armed forces and is managed by the Comando da Aeronáutica (COMAer). COMAer was created in 1999 and replaced the Ministry of Aeronautics. COMAer is one of the three armed forces assigned to the Ministry of Defense and comprises four General Commands and two Departments. One of the Commands is the Comando-Geral de Tecnologia Aeroespacial (see CTA).

CRC – Satellite Tracking and Command Center
Address: Av. dos Astronautas, 1758, CEP 12227-590 São José dos Campos, SP, Brazil
Tel: +55(12) 39456380
Fax: +55(12) 39411873
URL: http://www.ccs.inpe.br
E-mail:

Location: 23°11′S/45°53′W (São José dos Campos). The Centro de Rastreio e Controle de Satélites (CRC) is responsible for the planning, management and execution of all INPE satellite operations activities. It is a group of facilities that allows INPE to support launch operations, to control satellites in orbit, and to receive data and retransmit this data to processing centers. The CRC comprises the Satellite Control Center (Centro de Controle de Satélites – CCS), located in São José dos Campos (23°17′S/45°51′W), and two ground stations, Cuiabá (15°33′S/56°4′W) and Alcantara (2°20′S/44°24′W), which are linked by a data communication network called RECDAS. A further telemetry receiving station is located at Fortaleza.

CTA – General-Command for Aerospace Technology (Technical Aerospace Center)
Address: Avenida Brigadeiro Faria Lima, 1941 Parque Martim Cererê, CEP 12227-000 São José dos Campos, SP, Brazil
Tel: +55(12)39476654
Fax: +55(12)39413700
URL: http://www.cta.br
E-mail:

Location: 23°12′S/45°52′W (São José dos Campos). The Comando-Geral de Tecnologia Aeroespacial (CTA), formerly the Centro Tecnico Aeroespacial, is one of the four technical divisions of the COMAer and subordinated to the Brazilian Air Force. It currently employs several thousand civilian and military personnel and is, amongst others, responsible for operating the Alcantara Launch Center. CTA's Aeronautics and Space Institute, the Instituto de Aeronaútica e Espaço (IAE), develops projects in aeronautical, airspace and defense sectors, and is co-responsible for the execution of the Brazilian space program.

DGI – Image Generation Division
Address: Rod. Dutra, Km 40, CEP 12630-000 Cachoeira Paulista, SP, Brazil
Tel: +55(12)
Fax: +55(12)
URL: http://www.dgi.impe.br
E-mail:

The Divisão de Geração de Imagens (DGI), subordinated to INPE, is in charge of the reception, processing and distribution of remote sensing satellite (LANDSAT, SPOT, ERS, RADARSAT, CBERS)-derived images. Although the image processing laboratory is located at Cachoeira Paulista, the satellite data receiving station is located at Cuiabá Ground Station. A high-speed internet connection is used to download raw data to the Data Center in Cachoeira Paulista. A second receiving station, with a new to be developed system, is planned for the city of Boa Vista.

Embratel – Brazilian Telecommunications Company
Address: Avenida Presidente Vargas 1012, CEP 20177-700 Rio de Janeiro, Brazil
Tel: +55(21)2534438
Fax: +55(21)2336446
URL: http://www.embratel.com.br
E-mail:

The Empresa Brasileira de Telecomunicacões (Embratel) was created in 1965 as a state-controlled agency and is responsible for the domestic Brazilian communications network and links through the Intelsat and Inmarsat systems. It operates the national Brasilsat satellite system and works through the Intelsat system via its Tangua ground station for international links. It is also Brazil's national Inmarsat signatory. In 1999, Embratel started to operate the Brazilian Satellite Telecommunications System (SBTS). The satellites are operated from a Satellite System Operational Center located in Guaratiba near Rio de Janeiro. This center comprises a Spacecraft Control Center (SCC) and a Communications Operations Control Center (COCC). The SCC provides TT&C through 14.2 and 6.0-m dish antennae; the COCC primary traffic routing is through a 16.2-m dish antenna.

ETC – Cuiabá Ground Station
Address: Rua Hélio Ponce de Arruda, s/n Bosque da Saúde, CEP 78050-007 Caixa Postal 714, Cuiabá, MT, Brazil
Tel: +55(65)21234111
Fax: +55(65)21234140
URL: http://www.cba.inpe.br
E-mail:

Location: 15°33′S/56°04′W. The Estação Terrena de Cuiabá (ETC) is located in Cuiabá, the capital of the western state of Mato Grosso and the geographic center of South America. The use of satellite images for the assessment of Brazilian natural resources began in the 1970s, when INPE and the Ministry of Science and

Technology established a receiving station in Cuiabá for the ERTS-1 (Landsat) satellite. ETC also houses a TT&C station.

IAE – Institute of Aeronautics and Space
Address: Praça Mal. Eduardo Gomes, 50 Vila das Acácias, CEP 12228-904 São José dos Campos, SP, Brazil
Tel: +55(12)39474889/39475130
Fax: +55(12)39412333
URL: http://www.iae.cta.br
E-mail: comunicaiae@iae.cta.br

The Instituto de Aeronautica e Espaço (IAE) is responsible for developing the Sonda sounding rockets and the VLS orbital launcher and its guidance equipment. IAE's goal is "to enlarge its knowledge and to develop scientific-technological solutions to strengthen Brazilian power in aeronautical, space and defense systems".

INMET – National Institute of Meteorology
Address: Eixo Monumental Via S1, CEP 70680-900 Brasília, DF, Brazil
Tel: +55(61)33443333
Fax: +55(61)33440700
URL: http://www.inmet.gov.br
E-mail: diretor.inmet@inmet.gov.br

The Instituto Nacional de Meteorologia (INMET) is Brazil's national meteorological organization with its headquarters in Brasília. Related meteorology receiving sites in Brazil are: Fortaleze (Funceme) Ground Station; Natal (Navy/CMH) Centro de Hidrografia da Marinha; Cachoeira Paulista (INPE) Center; Brasília INMET Center; Cuiabá (INPE) Ground Station; Manaus (SIVAM) Ground Station.

INPE – National Institute for Space Research
Address: Av. dos Astronautas, 1758, CEP 12227-010 São José dos Campos, SP, Brazil
Tel: +55(12)39456000
Fax: +55(12)39229285
URL: http://www.inpe.br
E-mail:

The Instituto Nacional de Pesquisas Espaciais (INPE) has its headquarters in São José dos Campos, São Paulo, and was founded in 1971, subordinated to the Ministry of Aeronautics. It succeeded the National Space Activities Commission and became an organization within the newly created Ministry of Science and Technology in 1985. INPE's mission is "to promote and carry on studies, scientific research, technological development, and human resources capacity, in the fields of space and atmosphere sciences, space applications, meteorology, and space engineering and technology, as well as in related domains, in accordance with the policies and guidelines set forth by the Science and Technology Ministry". In this framework,

INPE is responsible for the development of the ground and space segments of Brazil's satellite programs.

INPE de Cachoeira Paulista
Address: Rodovia Presidente Dutra, Km 40 SP/RJ, CEP 12630-970 Cachoeira Paulista, SP, BrazilTel: +55(12)31869200
Fax: +55(12)31012324
URL: http://www.inpe.br
E-mail:

Location: 22°40′S/44°59′W. The INPE unit at Cachoeira Paulista was established in 1970. The idea was to gradually transfer there all the technical and scientific activities taking place in São José dos Campos, especially taking into account that those installations only were temporarily ceded by the CTA. However, the initial plan of moving gradually from São José dos Campos was abandoned and Cachoeira Paulista became destined for more operational activities. In 1973, the Image Production Laboratory (currently DGI), designed to process data obtained by satellites, was established. Ten years later, all of the equipment and relative routine activities, such as the reception, processing and distribution of data of meteorological satellites, were transferred to Cachoeira Paulista. Other space-related entities at Cachoeira Paulista are the Laboratory of Combustion and Propulsion, a GPS station, and the Center of Forecast of Time and Climatic Studies (CPTEC) (http://www.cptec.inpe.br). CPTEC is part of the research network of the Ministry of Science and Technology, and has developed, produced and disseminated weather forecasts as well as seasonal climate forecasts since early 1995.

ITA – Technical Institute of Aeronautics
Address: São José dos Campos, SP, Brazil
Tel: +55(12)39475835
Fax: +55(12)39413500
URL: http://www.ita.br
E-mail: reitor@ita.br

The Instituto Tecnológico de Aeronáutica (ITA) is an institution of higher education and advanced research, with emphasis in aerospace science and technology, maintained by the Brazilian federal government. The institute is located in São José dos Campos. ITA is rated as one of the top and most prestigious engineering schools in Brazil. It is one of four institutes of the CTA, having its facilities, along with its laboratories and R&D centers, inside the campus of the CTA. ITA is often mistakenly considered to be a military institution because of its location, but it is civilian in nature, being composed of a vast majority of civilian teachers, directors and students. However, the Air Force does help to maintain the institution and, in exchange, uses it for the training of its engineers. The institution was created in 1950, being responsible and contributing to a great extent to the research and development of the aerospace and defense sectors in Brazil, including INPE, Embraer and Avibrás.

LCP – Combustion and Propulsion Laboratory
Address: Rodovia Pres. Dutra, Km 40, CEP 12630-000 Cachoeira Paulista, SP, Brazil
Tel: +55(12)
Fax: +55(12)
URL: http://www.lcp.inpe.br
E-mail:

Location: 22°40′S/44°59′W. The Laboratório de Combustão e Propulsão (LCP) is subordinated to INPE. It was founded in 1968 and relocated from São José dos Campos to Cachoeira Paulista in 1976. LCP is engaged in research activities and developments in combustion, propulsion of satellites, auxiliary propulsion and catalysis with applications in combustion and propulsion. The Brazilian government considers LCP as an institution of the core of excellence in the propulsion of satellites, coordinated by its researchers and composed of national institutes and foreigners such as the Moscow Aviation Institute, the University Pierre et Marie Curie, the Faculty of Chemical Engineering in Lorraine, the Universidad Politecnica of Madrid and the Center National d'Etudes Spatiales (CNES). The LCP includes a specialized library, a research building, a chemical laboratory, a mechanical workshop, the BTSA (Satellite Thrusters Test Facility with Altitude Simulation), which is the only one in South America, and a test bank for research under atmospheric conditions (BTCA).

LIT – Integration and Test Laboratory
Address: Av. dos Astronautas, 1758, Caixa Postal 515, CEP 12227-010 São José dos Campos, SP, Brazil
Tel: +55(12)39456270
Fax: +55(12)39411884
URL: http://www.lit.inpe.br
E-mail:

Location: 23°12′S/45°51′W (São José dos Campos). The Laboratório de Integração e Testes (LIT) at INPE was designed and built (in 1988) to meet the needs of the Brazilian space program. LIT is considered to be one of the best laboratories in the world in this category, being the only one in South America. Created in 1987, the laboratory performs tests, integration and mounting of different types of satellite, subsystems and space components. It represents one of the most sophisticated and powerful facilities in the qualification of industrial products that demand a high reliability. The facility has two main areas in which integration and test activities are carried out. The larger is a class 100,000 clean area of 1,600 m^2. The second area is 450 m^2 and operates under class 10,000 conditions. The types of tests performed at LIT include thermal vacuum, climatic, vibration/shock, electromagnetic interference and electromagnetic compatibility tests, alignment measurements, and center of gravity and moment of inertia tests. There are also checkout equipment rooms, test control rooms, and additional laboratories. The total area of the facility is some 10,000 m^2. This includes a filtered compressed air system, closed circuit water system, a grounding system with less than 1 Ohm resistance and an uninterruptable power supply.

MECB – Complete Brazilian Space Mission
The Missão Espacial Completa Brasileira (MECB), created in 1981, was Brazil's ambitious US$1bn long-term space project. The MECB intends to develop the whole cycle of space technology, aimed at the development and operation of small satellites with applications in environmental data collection and remote sensing, a launch vehicle to bring them into space indigenously and the necessary ground infrastructure (satellite operation, data reception and use).

SINDAE – National System for the Development of Space Activities
By decree 1953 of 10th July 1996, the Sistema Nacional de Desenvolvimento das Atividades Spaciais (SINDAE) was established to organize the execution of activities aimed at the space development of Brazil. As the representative of the private sector, the Associação das Indústrias Aeroespaciais do Brasil, the Brazilian Aerospace Industries Association (AIAB), an association of companies working in aerospace engineering and already founded in 1993, became part of SINDAE.

UCA – Col Abner Propellant Plant
Address: São José dos Campos, SP, Brazil
Tel: +55( )
Fax: +55( )
URL: http://www.iae.cta.br
E-mail:

The Usina de Propelentes Coronel Abner (UCA) produces solid fuels (propellants) for rockets. UCA is installed in a reserved area of the CTA, São José dos Campos. It conducts the preparation of rocket engines, produces raw materials for propellants, and macerates and loads the rocket engines. The plant has a two test benches on which rocket motors can be completely tested. Currently, scientists of UCA are, in cooperation with Russia, also researching and developing liquid propellants.

Industries

Brazilian industries have always been part of the development and production of space-related commodities. Following is an, albeit incomplete, list of industries that are essential to the Brazilian space policy and its execution. This list has been composed from internet data.

Aeroeletrônica Indústria de Componentes Aviônicos S.A. (AEL)
Address: Av Sertório, 4400, Jd Floresta, CEP 91040-620 Porto Alegre, RS, Brazil
Tel: +55(51)21011200
Fax: +55(51)33612773

URL: http://www.aeroeletronica.com.br
E-mail: aeroel@aeroeletronica.com.br

AEL is a subsidiary of Elbit Systems Ltd. It produced products for the MECB (Missão Espacial Completa Brasileira) and CBERS (China–Brazil Earth Resources Satellites) programs.

AKROS Engenharia Indústria e Comércio Ltda
Address: Rua Euclides Miragaia 433, CEP 12245-902 São José dos Campos, SP, Brazil
Tel: +55(12)3411312
Fax: +55(12)3223699
URL:
E-mail:

Andrade Gutierrez Quimica Ltda (AGQ)
Address: Rua Dos Pampas, 484 Sala 63 Prado, CEP 30410-580 Belo Horizonte, Mg, Brazil
Tel: +55(31)2906777
Fax: +55(31)2906775
URL:
E-mail:

AGQ, based at Jacarei, São Paulo, founded in 1985, manufactures ammonium perchlorate solid rocket propellants for the Sonda sounding rockets and VLS orbital launchers.

Avibras Indústria Aeroespacial SA
Address: Rodovia dos Tamoios, Km 14, PO Box 278, CEP 12300-000 Jacarei, SP, Brazil
Tel: +55(12)39556000
Fax: +55(12)39516277
URL: http://www.avibras.com.br
E-mail: gspd@avibras.com.br

AVIBRAS has developed 10-m-diameter antennae and related equipment for satellite communications since 1976. In the early 1980s, AVIBRAS was chosen by the General Staff of the Brazilian Armed Forces to develop and supply communications antennae for the Brazilian Military Satellite Communications System.
The Jacarei Division at Avibras manufactures rockets for scientific applications.

Bernardini SA
Address: Rua Hipólito Soares 79, 04201-090 Ipiranga, São Paulo, SP, Brazil
Tel: +55(11)2748033
Fax: +55(11)2748567

URL:
E-mail:

Brasil Telecom
Address: SIA/SUL, ASP, Lote D, Bloco A 1° andar, CEP 71215-000 Brasilia, DF, Brazil
Tel: +55(61)34151414
Fax: +55(61)34158418
URL: http://wwwbrasiltelecom.com.br
E-mail:

Brasil Telecom is one of the companies in Brazil that emerged from the break-up of Telebrás.

Cenic Engenharia Indústria e Comércio Ltda
Address: AV Marginal B 1648, Chacaras Reundas, CEP 12238-390 São José dos Campos, SP, Brazil
Tel: +55(12)3314222
Fax: +55(12)3333101
URL: http://www.cenic.biz
E-mail:

Elebras Sistemas De Defesa e Controles Ltda (ESDC)
Address: Rua Bogaert 326, CEP 04298-020 São Paulo, SP, Brazil
Tel: +55(11)9691664
Fax: +55(11)9691966
URL:
E-mail:

ESDC designs and manufactures space equipment for satellite communications and specializes in high-tech systems and products primarily for air, space and military industries.

Equatorial Sistemas Ltda
Address: Av. Shishima Hifumi 2911, Pq. Tecnológico, UNIVAP Urbanova, CEP 12244-000 São José dos Campos, SP, BrazilTel: +55(12)39499060
Fax: URL: http://www.equatorialsistemas.com.br
E-mail:

Equatorial Sistemas was created 1996 in São José dos Campos to meet the requirements of the Brazilian space program. The company designs, manufactures, tests and qualifies aerospace components. Equatorial Sistemas established itself as prime contractor for the WFI (Wide Field Instrument) imager for the China–Brazil Earth Resources Satellites program and the humidity sounder for NASA's AQUA satellite. Equatorial Sistemas is also a supplier of satellite propellants and develops mono- and bipropellant propulsion systems.

Esca Engenharia de Sistemas de Controle e Automação SA (ESCA)
Address: Alameda Araguaia 1142, Alphaville, CEP 06455-940 Barueri, São Paulo, Sp, Brazil
Tel: +55(11)7254100
Fax: +55(11)7252545
URL: http://www
E-mail:

ESCA was created as a private company by the Brazilian government in 1972.

Mectron – Engenharia Indústria e Comércio Ltda (MEIC)
Address: Av. Brigadeiro Faria Lima 1399, Parque Flamboyant, CEP 12227-000 São José dos Campos, SP, Brazil
Tel: +55(12)39253500
Fax: +55(12)39253535
URL: http://www.mectron.com.br
E-mail:

MEIC is an aerospace industry that, amongst others, produces satellite subsystems and onboard equipment such as power supplies and TT&C subsystems.

Promon Eletronica Ltda
Address: Av. Pres. Juscelino Kubitscheck 1830, São Paulo, São Paulo, Brazil
Tel: +55(11)8274411
Fax: +55(11)8270472
URL: http://www
E-mail:

ANNEXE 4: SPACE INSTITUTES IN NORTH KOREA

Based on Daniel A. Pinkston, *The North Korean Ballistic Missile Program*, Strategic Studies Institute, Carlisle, Pennsylvania, February 2008. This detailed work refers to a great variety of sources about DPRK activities and achievements in advanced weaponry and military technologies. The organizational system to educate students in science and engineering looks like the military industrial system of the Soviet Union.

Kim Il-sun University at Pyongyang has a school for natural sciences, divided into eight departments: atomic energy, automation (computer science), biology, chemistry, geography, geology, mathematics and physics.

Kim Chaek University of Technology, also located at Pyongyang, has specialized in applied technologies for industry. It has 19 departments, including computer science, electrical engineering, electronics, materials and nuclear engineering. It consists of 19 research institutes in such fields as computers, electric components, materials, metals, numerically controlled devices, robotics, semi-conductors. About 1,500 students graduate from this university annually.

Kanggye Defense College in Pyongyang appears to be the source of high-tech specialists in rocket technology.

The State Academy of Sciences, through some 10 research institutes, is responsible for national research and development efforts in six strategic areas: aerospace engineering, biotechnology, electronics, information technology, material science and thermal engineering.

The Second Natural Sciences Academy, established as the National Defense Science Academy, is in charge of all applied military research. It is directly subordinate to the Central Military Committee of the KWP (Korean Workers Party). It conducts all the scientific and technological efforts in weapons systems for the DPRK. The Hamhung Branch of the Academy, with the Institutes of National Defense Sciences and of Chemical Materials, conducts missile research and development.

Sanum-Dong Design/Research Center, located in Kangso-kun, about 20 km west of downtown Pyongyang, is probably part of the Factory no. 125 or "Pyongyang Pig Factory", which produces Nodong and Hwasong missiles.

Mangyongdae Electric Machinery Factory, not so far away from Factory no. 125, is described as the major missile production plant in the DPRK.

The Second Economic Committee, subordinate to the KWP Central Committee Munitions Industry Department and to the National Defense Commission, manages the military economy based upon weapons production in North Korea. It operates up to 130 munitions factories and some 60 facilities for the production of parts and components for the rockets and their payloads.

ANNEXE 5: UNITED NATIONS TREATIES RELATED TO OUTER SPACE

The Treaty on Principles Governing the Activities of States in the Exploration and Use of Outer Space, including the Moon and Other Celestial Bodies (the "Outer Space Treaty", adopted by the General Assembly in its Resolution 2222 (XXI))
Opened for signature: 27th January 1967
Entry into force: 10th October 1967
98 ratifications and 27 signatures (as of 1st January 2008)
Iran: Signatory
(http://www.unoosa.org/oosa/en/SpaceLaw/outerspt.html)

The Agreement on the Rescue of Astronauts, the Return of Astronauts and the Return of Objects Launched into Outer Space (the "Rescue Agreement", adopted by the General Assembly in its Resolution 2345 (XXII))
Opened for signature: 22nd April 1968
Entry into force: 3rd December 1968
90 ratifications, 24 signatures, and one acceptance of rights and obligations (as of 1st January 2008)
Iran: Ratified
(http://www.unoosa.org/oosa/en/SpaceLaw/rescue.html)

The Convention on International Liability for Damage Caused by Space Objects (the "Liability Convention", adopted by the General Assembly in its Resolution 2777 (XXVI))
Opened for signature: 29th March 1972
Entry into force: 1st September 1972
86 ratifications, 24 signatures, and three acceptances of rights and obligations (as of 1st January 2008)
Iran: Ratified
(http://www.unoosa.org/oosa/en/SpaceLaw/liability.html)

The Convention on Registration of Objects Launched into Outer Space
(the "Registration Convention", adopted by the General Assembly in its Resolution 3235 (XXIX))
Opened for signature: 14th January 1975
Entry into force: 15th September 1976
51 ratifications, four signatures, and two acceptances of rights and obligations (as of 1st January 2008)
Iran: Signatory
(http://www.unoosa.org/oosa/en/SORegister/regist.html)

The Agreement Governing the Activities of States on the Moon and Other Celestial Bodies (the "Moon Agreement", adopted by the General Assembly in its Resolution 34/68)
Opened for signature: 18th December 1979
Entry into force: 11th July 1984
13 ratifications and four signatures (as of 1st January 2008)

Iran: Non-party
(http://www.unoosa.org/oosa/en/SpaceLaw/moon.html)

United Nations declarations and legal principles related to outer space

The Declaration of Legal Principles Governing the Activities of States in the Exploration and Uses of Outer Space
General Assembly Resolution 1962 (XVIII) of 13th December 1963
Iran: Non-party

The Principles Governing the Use by States of Artificial Earth Satellites for International Direct Television Broadcasting
Resolution 37/92 of 10th December 1982
Iran: Non-party

The Principles Relating to Remote Sensing of the Earth from Outer Space
Resolution 41/65 of 3rd December 1986
Iran: Non-party

The Principles Relevant to the Use of Nuclear Power Sources in Outer Space
Resolution 47/68 of 14th December 1992
Iran: Non-party

The Declaration on International Cooperation in the Exploration and Use of Outer Space for the Benefit and in the Interest of All States, Taking into Particular Account the Needs of Developing Countries
Resolution 51/122 of 13th December 1996
Iran: Non-party

Principles and international agreements related to outer space

Convention relating to the Distribution of Program-Carrying Signals transmitted by Satellite (BRS) 1974
Opened for signature: 21st May 1974 in Brussels
Entry into force: 25th August 1979
Depositary: Secretary-General of the UN
Iran: Non-party

Agreement on the Establishment of the INTERSPUTNIK International System and Organization of Space Communications (INTR) 1971
Opened for signature: 15th November 1971 in Moscow
Entry into force: 12th July 1972
Depositary: Russian Federation
Iran: Non-party

Convention for the Establishment of a European Space Agency (ESA) 1975
Opened for signature: 30th May 1975 in Paris
Entry into force: 30th October 1980
Depositary: France
Iran: Non-party

Agreement of the Arab Corporation for Space Communications (ARABSAT) 1976
Opened for signature: 14th April 1976 (14 Rabi' II 1396 H) in Cairo
Entry into force: 16th July 1976
Depositary: League of Arab States
Iran: Non-party

Agreement on Cooperation in the Exploration and Use of Outer Space for Peaceful Purposes (INTERCOSMOS) 1976
Opened for signature: 13th July 1976 in Moscow
Entry into force: 25th March 1977
Depositary: Russian Federation
Iran: Non-party

Convention Establishing the European Telecommunications Satellite Organization (EUTELSAT) 1982
Opened for signature: 15th July 1982 in Paris
Entry into force: 1st September 1985
Depositary: France
Iran: Non-party

Convention for the Establishment of a European Organization for the Exploitation of Meteorological Satellites (EUMETSAT) 1983
Opened for signature: 24th May 1983 in Geneva
Entry into force: 19th June 1986
Depositary: Switzerland
Iran: Non-party

Treaty Banning Nuclear Weapon Tests in the Atmosphere, in Outer Space and under Water (NTB) 1963
Opened for signature: 5th August 1963 in Moscow
Entry into force: 10th October 1963
Depositaries: USSR, UK and USA
Iran: Ratified

Agreement Relating to the International Telecommunications Satellite Organization (ITSO) 1971
Opened for signature: 20th August 1971 in Washington, DC
Entry into force: 12th February 1973
Depositary: USA
Iran: Ratified

Convention on the International Mobile Satellite Organization (IMSO) 1976
Opened for signature: 3rd September 1976 in London
Entry into force: 16th July 1979
Depositary: Secretary-General of the International Maritime Organization
Iran: Ratified

International Telecommunication Constitution and Convention (ITU) 1992
Opened for signature: 22nd December 1992 in Geneva
Date of entry into force: 1st July 1994
Depositary: Secretary-General of the International Telecommunication Union and Convention
Iran: Ratified

Convention on Registration of Objects Launched into Outer Space
Opened for signature: 14th January 1975 in New York
Date of entry into force: 15th September 1976
Depositary: Secretary-General of the UN
Iran: Signatory
According to the Online Index of Objects Launched into Outer Space, Iran did not register Omid with the UN. The launch of Sina-1 was reported by the Russian Federation; however, it is not registered with the UN either.

Bibliography

Reports, theses and major articles

Bandeira, I.; Bogossian, O.; Corrêa, F.: Centenary Mission – First Brazilian Microgravity Experiments at ISS. *Microgravity – Science and Technology Journal*, **19**(5–6), 42–48, 2007.

Barry, Dan; Konnici Wa, Kibo. *Air & Space*, May, 2008.

Bermudez, Joseph S., Jr: *A History of Ballistic Missile Development in the DPRK*. Center for Nonproliferation Studies, Monterey Institute of International Studies, Monterey, California, 1999.

Bhardwaj, Ashwani: *Great Scientists of the World*. New Delhi, Goodwill Publishing, 2004.

Câmara, Gilberto: *Annual Report to INPÉs Scientific and Technological Advisory Council for 2008*. INPE, São José dos Campos, May 2009.

Dolinsky, Mauro M.: *IAE – Presença Brasileira no Espaço/IAE – The Brazilian Presence in Space*. 21 August 1989 (in Portuguese).

Gerl, Florian Albin: *Outer Space – The Emerging Market*. Diploma Thesis, GRIN Dokument V1166794, Munich/Ravensburg, 2008.

India, Government of: *Indo – Soviet Joined Manned Spaceflight, April 1984*. New Delhi, 1984.

India's Prolific Space Program. Feature in *Aviation Week & Space Technology*, 22nd November 2004.

Israel at Sixty: From Modest Beginning to a Vibrant State 1948–2008. Boston Hannah Chicago, 2008. Available on website www.israels60th.net/, Chapter 7.6. Out of This World: Israel Space Program, pp. 99–102.

Israel Missile Update – 2005, *The Risk Report*, **11**(6), November–December 2005.

ISRO: *Sarabhai on Space – A Selection of Writings and Speeches*. Bangalore, 1979.

ISRO: *Survey of the Japanese Space Program, with an Emphasis on Kappa and Lambda Rockets*. NASA, translation services, TTF 303, 1963.

Itokawa, Hideo: *Development of Sounding Rockets in Japan*. NASA, translation services, TTF 87, 1963.

Joshi, Padmanabh K.: Dr Vikram Sarabhai – A Study in Innovative Leadership and Institution Building. Ph.D. thesis, Gujarat University, Ahmedabad, 1985.

L'industrie Spatiale en Corée du Sud, Etude (68 pages) de la Mission économique de Séoul pour Spheris, Paris, juin 2009.

Matogawa, Yasunori: Minoru Oda and his Pioneering Role in Space Science in Japan. Paper presented to the 59th conference of the International Astronautical Federation, Glasgow, 29th September–2nd October 2008.

Matogawa, Yasunori: Shusui – Japanese Rocket Fighter in World War II. Paper presented to the 50th conference of the International Astronautical Federation, Amsterdam, 4–8 October 1999.

Mecham, Michael: Indian Space – Success on a Shoestring (series of three articles). *Aviation Week & Space Technology*, 12 August 1996.

Moraes Jr, Paulo: *An Overview of the Brazilian Launch Vehicle Program Cruzeiro do Sul*. CTA Space Directorate, IAF, 2 October 2006, São José dos Campos, Brazil.

National Papers – Second United Nations Conference on the Exploration and Peaceful Uses of Outer Space. Vienna, August 1982.

National Program of Space Activities 2005–2014. PNAE/Brazilian Space Agency. Brasília: Ministry of Science and Technology, Brazil 2005.

Padmanabhan, Anil: *Kalpana Chawla, a Life*. Puffin books, New Delhi, 2003.

Pal, Yash (ed.): Space for development. COSPAR Advances in Space Exploration Series, Vol. 6, Pergamon Press, 1979.

Rajan, Mohan Sundara: *India in Orbit*. Government of India, New Delhi, 1997.

Reichl, Eugen: *Das Raketentypenbuch*. Motorbuch Verlag, Stuttgart, 2007.

Reichl, Eugen; Schiessl, Stefan: *Space2006/Space 2007/Space2008/Space2009*, annual German chronicle of the astronautics in the world, VFR e.V., Munich, published every year.

Sarabhai, Vikram; Chitnis, E.V.; Rao, B.S.; Kale, P.P.; Karnik, K.S.: INSAT – A National Satellite for Television and Development. Presented by Vikram Sarabhai at the national conference on electronics, Bombay, 24–28 March 1970.

Sharma, Jagannath: *Space Research in India*. Physical Research Laboratory, Ahmedabad, 1968.

Unispace III: *National Papers*. Vienna, June 1999.

UNOOSA (United Nations Office for Outer space Affairs)/COPUOS (Committee on the Peaceful Uses of Outer Space, national space activities.

Watanabe, Hirotaka: Japan–US Space Relations during the 1960s – Dependence or Autonomy? Paper presented to the International Astronautical Conference, 2003.

Yiftah, Shapir: *Iran's Effort to Counter Space: Strategic Assessment*. The Institute for National Security Studies 8, no. 3, November 2005.

Yoshikawa, Makoto; Yano, Hajime; Abe, Sasanao; Kawaguchi, Junichiro: Japan's Missions to Primitive Bodies in the Solar System – From Hayabusa to Marco Polo. Paper presented to the 59th Conference of the International Astronautical Federation, Glasgow, 29th September–2nd October 2008.

Zorn, E.L.: *Expanding the Horizon: Israel's Quest for Satellite Intelligence*. CSI (Center for the Study of Intelligence) of the CIA, 2001.

Books

Behnam, Farjam: *Iran Almanac and the Book of Facts 2007–2008*. Iranalmanac.com, ISBN 978-964-2865-16-1.
Bond, Peter: *Jane's Space Recognition Guide*. Harper Collins Publishers, London, 2008.
Burleson, Daphne: *Space Programs Outside the United States*. McFarland & Company Inc. Publishers, Jefferson, North Carolina, 2005.
Gatland, Kenneth: *Missiles and Rockets*. London, Blandford, 1975.
Gatland, Kenneth (ed.): *The Illustrated Encyclopedia of Space Technology*. Salamander, 1981.
Gatland, Kenneth (ed.): *The Illustrated Encyclopedia of Space Technology*, 2nd edn. Salamander, 1989.
Johnson, Nicholas L.; Rodvold, David M.: *European & Asia in Space 1993–1994*. Kaman Sciences Corporation/USAF Phillips Laboratory, 1995.
Joshi, Padmanabh K.: *Vikram Sarabhai – The Man and the Vision*. Mapin Publishers, Ahmedabad, 1992.
Oliveira, Fabiola de: *Caminhos Para O Espaço/Pathways to Space*. Instituto Nacional de Pesquisas Espaciais (INPE) – São José dos Campos, Brasil, 1991.
Oliveira, Fabiola de: *O Brasil Chega Ao Espaço/Brazil Reaches The Space*. Instituto Nacional de Pesquisas Espaciais (INPE) – São José dos Campos, Brasil, 1996.
Sarabhai, Mrinalini: *The Voice of the Heart*. Harper Collins, New Delhi, 2004.
Turnill, Reginald: *The Observer's Book of Unmanned Spaceflight*. Frederick Warne, London, 1974.
Winter, Othon C.; Almeida Prado, Antonio F.B. de: *A Conquista do Espaco*. Brazilian Space Agency (AEB), São Paulo, Brasil, 2007.

Periodicals and regular reports

Air & Cosmos
Aviation Week & Space Technology
Electronics today, notably Homage to Dr Sarabhai, February 1972
Espace
European Space Directory, annually published by ESD Partners, Paris
Flight International
Institute of Space & Astronautical Science (ISAS): *Annual reports*.
International Astronautical Congress: *Proceedings*
ISRO: *Space India*. Quarterly, Bangalore, ISRO, 1988
Jane's Spaceflight Directory
Jane's Space Directory 2006–2007. Bill Sweetman (ed.) and regularly updated, Coulsdon (Surrey), 2006
Jonathan's Space Report
NASDA: *NASDA report*. No 1. Monthly, Tokyo, NASDA, 1988
Physics News (notably Vol. 3, §1, March 1972)

Space, Department of: *Annual Reports*. Annual, from 1988. ISRO, Bangalore
Spaceflight
World-Wide Space Activities, report of Committee on Science & Technology, US House of Representatives, published in September 1977

Television programs

Discovery channel Science: *Man-Made Marvels of Asia*, 2006

Websites and internet

Armed Forces of the Islamic Republic of Iran, Wikipedia, 2008

Global security reports at http://www.globalsecurity.org/space/world/

Norbert Brügge on Space Launch Vehicles website, http://www.b1443.de/Space-rockets_1/index.htm

Index